Mathematics for Physicists

The Manchester Physics Series

General Editors
J.R. FORSHAW, H.F. GLEESON, F.K. LOEBINGER

School of Physics and Astronomy,
University of Manchester

Properties of Matter	B.H. Flowers and E. Mendoza
Statistical Physics *Second Edition*	F. Mandl
Electromagnetism *Second Edition*	I.S. Grant and W.R. Phillips
Statistics	R.J. Barlow
Solid State Physics *Second Edition*	J.R. Hook and H.E. Hall
Quantum Mechanics	F. Mandl
Computing for Scientists	R.J. Barlow and A.R. Barnett
The Physics of Stars *Second Edition*	A.C. Phillips
Nuclear Physics	J.S. Lilley
Introduction to Quantum Mechanics	A.C. Phillips
Particle Physics *Third Edition*	B.R. Martin and G. Shaw
Dynamics and Relativity	J.R. Forshaw and A.G. Smith
Vibrations and Waves	G.C. King
Mathematics for Physicists	B.R. Martin and G. Shaw

Mathematics for Physicists

B.R. MARTIN
Department of Physics and Astronomy
University College London

G. SHAW
Department of Physics and Astronomy
Manchester University

WILEY

Registered office
John Wiley & Sons Ltd, The Atrium, Southern Gate, Chichester, West Sussex, PO19 8SQ, United Kingdom

For details of our global editorial offices, for customer services and for information about how to apply for permission to reuse the copyright material in this book please see our website at www.wiley.com.

Library of Congress Cataloging-in-Publication Data applied for.

ISBN hardback 9780470660232
ISBN paperback 9780470660225

Set in 11/13pt Computer Modern by Aptara Inc., New Delhi, India.

MIX
Paper from responsible sources
FSC® C013604
www.fsc.org

2 2015

Contents

Editors' preface to the Manchester Physics Series

The Manchester Physics Series is a set of textbooks at first degree level. It grew out of the experience at the University of Manchester, widely shared elsewhere, that many textbooks contain much more material than can be accommodated in a typical undergraduate course; and that this material is only rarely so arranged as to allow the definition of a short self-contained course. The plan for this series was to produce short books so that lecturers would find them attractive for undergraduate courses, and so that students would not be frightened off by their encyclopaedic size or price. To achieve this, we have been very selective in the choice of topics, with the emphasis on the basic physics together with some instructive, stimulating and useful applications.

Although these books were conceived as a series, each of them is self-contained and can be used independently of the others. Several of them are suitable for wider use in other sciences. Each Author's Preface gives details about the level, prerequisites, etc., of that volume.

The Manchester Physics Series has been very successful since its inception over 40 years ago, with total sales of more than a quarter of a million copies. We are extremely grateful to the many students and colleagues, at Manchester and elsewhere, for helpful criticisms and stimulating comments. Our particular thanks go to the authors for all the work they have done, for the many new ideas they have contributed, and for discussing patiently, and often accepting, the suggestions of the editors.

Finally, we would like to thank our publisher, John Wiley & Sons, Ltd., for their enthusiastic and continued commitment to the Manchester Physics Series.

J. R. Forshaw
H. F. Gleeson
F. K. Loebinger
August 2014

Authors' preface

Our aim in writing this book is to produce a relatively short volume that covers all the essential mathematics needed for a typical first degree in physics, from a starting point that is compatible with modern school mathematics syllabuses. Thus, it differs from most books, which include many advanced topics, such as tensor analysis, group theory, etc., that are not required in a typical physics degree course, except as specialised options. These books are frequently well over a thousand pages long and contain much more material than most undergraduate students need. In addition, they are often not well interfaced with school mathematics and start at a level that is no longer appropriate. Mathematics teaching at schools has changed over the years and students now enter university with a wide variety of mathematical backgrounds.

The early chapters of the book deliberately overlap with senior school mathematics, to a degree that will depend on the background of the individual reader, who may quickly skip over those topics with which he or she is already familiar. The rest of the book covers the mathematics that is usually compulsory for all students in their first two years of a typical university physics degree, plus a little more. Although written primarily for the needs of physics students, it would also be appropriate for students in other physical sciences, such as astronomy, chemistry, earth science, etc.

We do not try to cover all the more advanced, optional courses taken by some physics students, since these are already well treated in more advanced texts, which, with some degree of overlap, take up where our book leaves off. The exception is statistics. Although this is required by undergraduate physics students, we have not included it because it is usually taught as a separate topic, using one of the excellent specialised texts already available.

The book has been read in its entirety by one of the editors of the Manchester Physics Series, Jeff Forshaw of Manchester University, and we are grateful to him for many helpful suggestions that have improved the presentation.

B.R. Martin
G. Shaw
April 2015

Notes and website information

'Starred' material

Some sections of the book are marked with a star. These contain more specialised or advanced material that is not required elsewhere in the book and may be omitted at a first reading.

Website

Any misprints or other necessary corrections brought to our attention will be listed on www.wiley.com/go/martin/mathsforphysicists. We would also be grateful for any other comments about the book.

Examples, problems and solutions

Worked examples are given in all chapters. They are an integral part of the text and are designed to illustrate applications of material discussed in the preceding section. There is also a set of problems at the end of each chapter. Some equations which are particularly useful in problem solving are highlighted in the text for ease of access and brief 'one-line' answers to most problems are given at the end of the book, so that readers may quickly check whether their own answer is correct. Readers may access the full solutions to all the odd-numbered problems at www.wiley.com/go/martin/mathsforphysicists. Full solutions to all problems are available to instructors at the same website, which also contains electronic versions of the figures.

1

Real numbers, variables and functions

In this chapter we introduce some simple ideas about real numbers, i.e. the ordinary numbers used in arithmetic and measurements, real variables and algebraic functions of a single variable. This discussion will be extended in Chapter 2 by considering some important examples in more detail: polynomials, trigonometric functions, exponentials, logarithms and hyperbolic functions. Much of the material in these first two chapters will probably already be familiar to many readers and so is covered briefly, but even if this is the case, it is useful revision and sets the scene for later chapters.

1.1 Real numbers

This section starts from the basic rules of arithmetic and introduces a number of essential techniques for manipulating real numerical quantities. We also briefly consider number systems other than the decimal system.

1.1.1 Rules of arithmetic: rational and irrational numbers

The first contact with mathematics is usually via counting, using the positive integers $1, 2, 3, 4, \ldots$ (also called *natural numbers*). Later, fractional numbers such as $\frac{1}{2}, \frac{3}{5}$, etc. and negative numbers -1, -3, $-\frac{1}{3}$, $-\frac{7}{9}$, etc. are introduced, together with the rules for combining positive and negative numbers and the basic laws of arithmetic. As we will build on these laws later in this chapter, it is worth reminding oneself of what they are by stating them in a somewhat formal way as follows.

Mathematics for Physicists, First Edition. B.R. Martin and G. Shaw.

Companion website: www.wiley.com/go/martin/mathsforphysicists

(i) *Commutativity*: The result of subtracting or dividing two integers is dependent on the order in which the operations are performed, but addition and multiplication are independent of the order. For example,

$$3 + 6 = 6 + 3 \quad \text{and} \quad 3 \times 6 = 6 \times 3, \tag{1.1a}$$

but

$$5 - 3 \neq 3 - 5 \quad \text{and} \quad 5 \div 3 \neq 3 \div 5, \tag{1.1b}$$

where $\neq$ means *not equal to.*

(ii) *Associativity*: The result of subtracting or dividing three or more integers is dependent on the way the integers are associated, but is independent of the association for addition and multiplication. Examples are

$$(2 + 3) + 4 = 2 + (3 + 4) \quad \text{and} \quad 2 \times (3 \times 4) = (2 \times 3) \times 4, \tag{1.2a}$$

but

$$6 - (3 - 2) \neq (6 - 3) - 2 \quad \text{and} \quad 12 \div (6 \div 2) \neq (12 \div 6) \div 2. \tag{1.2b}$$

(iii) *Distributivity*: Multiplication is distributed over addition and subtraction from both left and right, whereas division is only distributed over addition and subtraction from the right. For example, for multiplication:

$$2 \times (4 + 3) = (2 \times 4) + (2 \times 3) \tag{1.3a}$$

and

$$(3 - 2) \times 4 = (3 \times 4) - (2 \times 4), \tag{1.3b}$$

but for division, from the right we have

$$(60 + 15) \div 3 = (60 \div 3) + (15 \div 3), \tag{1.3c}$$

whereas division from the left gives

$$60 \div (12 + 3) = 60 \div 15 \neq (60 \div 12) + (60 \div 3). \tag{1.3d}$$

Positive and negative integers and fractions can all be expressed in the general form n/m, where n and m are integers (with $m \neq 0$ because division by zero is undefined). A number of this form is called a *rational number.* The latter is said to be *proper* if its numerator is less than its denominator, otherwise it is said to be *improper.* The operations of addition, subtraction, multiplication and division, when applied to rational numbers, always result in another rational number. In the case of fractions, multiplication is applied to the numerators and denominators separately; for division, the fraction is inverted and then multiplied. Examples are

$$\frac{3}{4}\times\frac{5}{7}=\frac{3\times 5}{4\times 7}=\frac{15}{28}\quad\text{and}\quad\frac{3}{4}\div\frac{5}{7}=\frac{3}{4}\times\frac{7}{5}=\frac{21}{20}. \tag{1.4a}$$

For addition (and subtraction) all the terms must be taken over a common denominator. An example is:

$$\frac{3}{4}-\frac{5}{7}+\frac{1}{3}=\frac{(3\times 3\times 7)-(5\times 3\times 4)+(1\times 4\times 7)}{3\times 4\times 7}=\frac{31}{84}. \tag{1.4b}$$

Not all numbers can be written in the form n/m. The exceptions are called *irrational numbers*. Examples are the square root of 2, that is, $\sqrt{2}=1.414\ldots$, and the ratio of the circumference of a circle to its diameter, that is, $\pi=3.1415926\ldots$, where the dots indicate a non-recurring sequence of numbers. Irrational numbers, when expressed in decimal form, always lead to such non-recurrence sequences, but even rational numbers when expressed in this form may not always terminate, for example $\frac{2}{11}=0.1818\ldots$. The proof that a given number is irrational can be very difficult, but is given for one particularly simple case in Section 1.2.2.

In practice, an irrational number may be represented by a rational number to any accuracy one wishes. Thus π is often represented as $\frac{22}{7}=3.143$ in rough calculations, or as $\frac{355}{113}=3.141593$ in more accurate work. Rational and irrational numbers together make up the class of so-called *real numbers* that are themselves part of a larger class of numbers called *complex numbers* that we will meet in Chapter 6. It is worth remarking that infinity, denoted by the symbol ∞, is not itself a real number. It is used to indicate that a quantity may become arbitrarily large.

In the examples above of irrational numbers, the sequence of numbers after the decimal point is endless and so, in practice, one has to decide where to terminate the string. This is called *rounding*. There are two methods of doing this: quote either the *number of significant figures* or the *number of decimal places*. Consider the number 1234.567.... To two decimal places this is 1234.57; the last figure has been rounded up to 7 because the next number in the string after 6 is 7, which is greater than 5. Likewise, we would round down if the next number in the string were less than 5. If the next number in the string were 5, then the 5 and the next number following it are rounded up or down to the nearest even multiple of 10, and the zero dropped. For example, 1234.565 to two decimal places would be 1234.56, whereas 1234.575 would be rounded to 1234.58. If we were to quote 1234.567 to five significant figures, it would be 1234.6 and to three significant figures it would be 1230.

1.1.2 Factors, powers and rationalisation

Integer numbers may often be represented as the product of a number of smaller integers. This is an example of a process called *factorisation*, that is, decomposition into a product of smaller terms, or *factors.* For example, 24 is equal to $2 \times 2 \times 2 \times 3$. In this example, the integers in the product cannot themselves be factorised further. Such integers are called *prime numbers.* (By convention, unity is not considered a prime number.) By considering all products of the prime numbers in the factorisation, we arrive at the result that the *factors* of 24 are 1, 2, 3, 4, 6, 8, 12 and 24, that is, these are all the numbers that divide exactly into 24. If we have several numbers, the *highest common factor* (HCF) is the largest factor that can divide exactly into all the numbers. The *lowest common multiple* (LCM) is the smallest number into which all the given numbers will divide exactly. Thus the HCF of 24 and 36 is 12 and the LCM of all three numbers is 72.

In the example of the factorisation of the number 24 above, the factor 2 occurs three times. It is common to encounter situations where a number is multiplied by itself several times. A convenient notation for this is to introduce the idea of a *power* (or *index*) n, such that, for example, $5^n \equiv 5 \times 5 \times 5 \ldots n$ times. To emphasise that this relation *defines* the index n, the usual two-line equality sign has been replaced by a three-line equality sign ($\equiv$). So, using powers, we could also write 24 in the compact prime-number factorised form $24 = 3 \times 2^3$. Any real number p to power zero is *by definition* equal to unity, that is, $p^0 \equiv 1$ for any p.

By writing out in full, it is easy to see that multiplying the *same* integers each raised to a power is equivalent to adding the powers. Thus

$$\begin{aligned} 5^n \times 5^m &= (5 \times 5 \times 5 \ldots n \text{ times}) \times (5 \times 5 \times 5 \ldots m \text{ times}) \\ &= 5 \times 5 \times 5 \ldots (n+m) \text{ times} \\ &= 5^{(n+m)} \end{aligned} \tag{1.5a}$$

and analogously for division,

$$5^n/5^m = 5^{(n-m)}. \tag{1.5b}$$

A power can also be a fraction or rational number, since, for example, the combination rule (1.5a) implies $5^{1/2} \times 5^{1/2} = 5^1$, so that $5^{1/2} = \sqrt{5}$. Similarly the expression $3^{1/3}3^0 3^{4/3}/27^{1/3}$, for example, can be simplified to give

$$3^{1/3}3^0 3^{4/3}/27^{1/3} = 3^{(1/3+0+4/3-1)} = 3^{2/3}. \tag{1.5c}$$

An example of the use of factors is to express numbers in so-called *scientific notation* (also called *normal form*). In this representation, any real number is written as the product of a number between -10 and $+10$ (excluding the numbers ± 10 themselves), with as many decimal places as required, and a power of 10. The number 1245.678 to four significant figures in scientific notation is therefore 1.246×10^3.

It is conventional to write arithmetical forms in a compact form and to remove as far as possible fractional powers from the denominator of a fraction, a process called *rationalisation.* For example, consider the form

$$\frac{1}{2\sqrt{5}+1} - \frac{1}{2\sqrt{5}+2}. \tag{1.6a}$$

By taking the terms over a common denominator and then multiplying numerator and denominator by $(11 - 3\sqrt{5})$, we have

$$\begin{aligned}\frac{1}{2\sqrt{5}+1} - \frac{1}{2\sqrt{5}+2} &= \frac{(2\sqrt{5}+2)-(2\sqrt{5}+1)}{(2\sqrt{5}+1)(2\sqrt{5}+2)} = \frac{1}{2(11+3\sqrt{5})} \\ &= \frac{(11-3\sqrt{5})}{2(11+3\sqrt{5})(11-3\sqrt{5})} = \frac{11-3\sqrt{5}}{152},\end{aligned} \tag{1.6b}$$

which is the rationalised form of (1.6a).

Example 1.1

Simplify the following forms:

(a) $(7\frac{1}{9})^{3/2}$, (b) $5^{1/3}25^{-1/2}/5^{-1/6}5^{-2/3}$, (c) $3^{1/2}27^{-2/3}9^{-1/2}$.

Solution

(a) $(7\frac{1}{9})^{3/2} = (64/9)^{3/2} = (8/3)^3 = 512/27$,

(b) $5^{1/3}25^{-1/2}/5^{-1/6}5^{-2/3} = 5^{(1/3-1+1/6+2/3)} = 5^{1/6}$,

(c) $3^{1/2}27^{-2/3}9^{-1/2} = 3^{1/2}3^{-2}3^{-1} = 3^{-5/2} = 3^{2/5}$.

Example 1.2

Rationalise the numerical expressions:

(a) $\dfrac{1}{\sqrt{3}-1} + \dfrac{1}{\sqrt{3}+1}$, (b) $\dfrac{1}{\sqrt{2}+1}$, (c) $\dfrac{(2-\sqrt{5})}{(3+2\sqrt{5})}$.

Solution

(a) Taking both terms over a common denominator gives

$$\frac{1}{\sqrt{3}-1} + \frac{1}{\sqrt{3}+1} = \frac{\sqrt{3}+1+\sqrt{3}-1}{(\sqrt{3}-1)(\sqrt{3}+1)} = \sqrt{3}.$$

(b) Multiplying numerator and denominator by $(\sqrt{2}-1)$ gives

$$\frac{1}{\sqrt{2}+1}=\frac{\sqrt{2}-1}{(\sqrt{2}+1)(\sqrt{2}-1)}=\sqrt{2}-1.$$

(c) Multiplying numerator and denominator by $(3-2\sqrt{5})$ gives

$$\frac{(2-\sqrt{5})}{(3+2\sqrt{5})}=\frac{(2-\sqrt{5})(3-2\sqrt{5})}{(3+2\sqrt{5})(3-2\sqrt{5})}=\frac{7\sqrt{5}-16}{11}.$$

*1.1.3 Number systems[1]

All the numbers in the previous sections are expressed in the decimal system, where the 'basis' set of integers is $0, 1, 2, \ldots, 9$. Real numbers in this system are said to be to '*base 10*'. In the number 234, for example, the integers 2, 3 and 4 indicate how many powers of 10 are present, reading from the right, i.e.

$$234=(2\times 10^2)+(3\times 10^1)+(4\times 10^0). \tag{1.7}$$

Any other base could equally well be used and in some circumstances other number systems are more appropriate. The most widely used number system other than base 10 is the *binary system*, based on the two integers 0 and 1, that is, *base 2*, so we will only discuss this case. Its importance stems from its use in computers, because the simplest electrical switch has just two states, 'open' and 'closed'. To distinguish numbers in this system we will write them with a subscript 2.

As an example, consider the number 123. In the binary system this is 1111011_2. To check: in the decimal system,

$$1111011_2=2^6+2^5+2^4+2^3+2^1+2^0=123. \tag{1.8}$$

Fractions are accommodated by using negative values for the indices. Thus the number 6.25 in the binary system is 110.01_2. To check: in the decimal system,

$$110.01_2=2^2+2^1+2^{-2}=4+2+0.25=6.25. \tag{1.9}$$

To convert a number in one basis to another is straightforward, if rather tedious. Consider, for example, the conversion of the number 51.78 to the binary system. We start with the integer 51 and find the largest value of an integer n such that 2^n is less than or equal to

[1] The reader is reminded that the results of starred sections are not needed later, except in other starred sections, and therefore they may prefer to omit them on a first reading.

51 and then note the remainder $R = 51 - 2^n$. This is then repeated by again finding the largest number n such that 2^n is less than or equal to R, and continued in this way until the remainder is zero. We thus obtain:

$$51 = 2^5 + 19 = 2^5 + 2^4 + 3 = 2^5 + 2^4 + 2^1 + 1 = 2^5 + 2^4 + 2^1 + 2^0,$$

so that in the binary system

$$51 = 110011_2. \tag{1.10a}$$

Similarly, we can convert the numbers after the decimal point using negative powers. This gives

$$0.78 = 2^{-1} + 0.28 = 2^{-1} + 2^{-2} + 0.03 \approx 2^{-1} + 2^{-2} + 2^{-5},$$

so again in the binary system,

$$0.78 \approx 0.11001_2 \tag{1.11a}$$

and finally,

$$51.78 = 110011.11001_2, \tag{1.11b}$$

in the binary system, which represents the decimal number to an accuracy of two decimal places.

All the normal arithmetic operations of addition, subtraction, multiplication and division can be carried out in any number system. For example, in the binary system, we have the basic result $1_2 + 1_2 = 10_2$. So adding the numbers 101_2 and 1101_2 gives $101_2 + 1101_2 = 10010_2$. To check, we can again use the decimal system. Thus,

$$101_2 = 2^2 + 2^0 = 5, \text{ and } 1101_2 = 2^3 + 2^2 + 2^0 = 13, \tag{1.12a}$$

with

$$10010_2 = 2^4 + 2^1 = 18. \tag{1.12b}$$

As an example of multiplication, consider the numbers 5 and 7. In the binary system these are 101_2 and 111_2, respectively, and multiplying them together gives, using $1_2 + 1_2 = 10_2$,

$$\begin{array}{r} 101 \\ 111 \\ \hline 10100 \\ 1010 \\ 101 \\ \hline 100011 \end{array}$$

Once again, we can check the result using the decimal system:

$$100011_2 = 2^5 + 2^1 + 2^0 = 35. \tag{1.13}$$

As an example of division, consider the numbers 51 and 3. In the binary system these are 110011_2 and 11_2, respectively, and dividing them we have

$$\begin{array}{r} 10001 \\ 11\overline{)110011} \\ \underline{110000} \\ 11 \\ \underline{11} \\ 00 \end{array}$$

So the quotient is 10001_2, which in the decimal system is $2^4 + 2^0 = 17$, as required.

Example 1.3

Write the decimal number 100 in base 3 and base 4.

Solution

(a) The decimal number 100 written as powers of 3 is $100 = 3^4 + (2 \times 3^2) + 3^0$, so to base 3 it is 10201_3.

(b) The decimal number 100 written as powers of 4 is $100 = 4^3 + (2 \times 4^2) + 4^1$, so to base 4 it is 1210_4.

Example 1.4

Consider the base 3 numbers $p = 201_3$ and $q = 112_3$. Find (a) $p + q$, (b) $p - q$, (c) $p \times q$ and (d) p/q to two decimal places and check your results in the decimal system.

Solution

In base 3,

(a) $p + q = 201_3 + 112_3 = 1020_3$, which as a decimal number is $3^3 + (2 \times 3) = 33$,

(b) $p - q = 201_3 - 112_3 = 12_3$, which as a decimal number is $3^1 + (2 \times 3^0) = 5$,

(c) $p \times q = 201_3 \times 112_3 = 100212_3$, which as a decimal number is $3^5 + (2 \times 3^2) + 3^1 + (2 \times 3^0) = 266$,

(d) $p/q = 201_3/112_3 = 1.1002_3 \ldots$, which as a decimal number is $3^0 + 3^{-1} + (2 \times 3^{-4}) \cdots = 1.357 \ldots = 1.36$ to two decimal places.

To check these, we have $p = 201_3$ as a decimal number is $(2 \times 3^2) + 3^0 = 19$ and $q = 112_3$ is $3^2 + 3^1 + (2 \times 3^0) = 14$. Thus in the decimal system, (a) $p + q = 19 + 14 = 33$, (b) $p - q = 19 - 14 = 5$, (c) $p \times q = 19 \times 14 = 266$, and (d) $p/q = 19/14 = 1.36$ to two decimal places, as required.

1.2 Real variables

The work in Section 1.1 can be generalised by representing real numbers as symbols, x, y, etc. Thus we are entering the field of *algebra*. This section starts by generalising the methods of Section 1.1 for real numbers to algebraic quantities and also discusses the general idea of algebraic expressions and the important result known as the binomial theorem.

1.2.1 Rules of elementary algebra

Algebra enables us to consider general expressions like, for example, $(x + y)^2$, where x and y can be any real number. When manipulating real numbers as symbols, the fundamental rules of algebra apply. These are analogous to the basic rules of arithmetic given in Section 1.1 and can be summarised as follows.[2]

(i) *Commutativity:* Addition and multiplication are commutative operations, i.e.

$$x + y = y + x \quad \text{commutative law of addition} \tag{1.14a}$$

and

$$xy = yx \quad \text{commutative law of multiplication.} \tag{1.14b}$$

In contrast, subtraction and division are only commutative operations under special circumstances. Thus,

$$x - y \neq y - x \quad \text{unless } x = y$$

and

$$x \div y \neq y \div x \quad \text{unless } x = y \textit{ and } \text{neither equals zero.}$$

(ii) *Associativity:* Addition and multiplication are associative operations, i.e.,

$$x + (y + z) = (x + y) + z \quad \text{associative law of addition} \tag{1.15a}$$

[2] Here and in what follows the explicit multiplication signs between terms are usually omitted if there is no loss of clarity, so that xy is equivalent to $x \times y$ and so on.

and

$$x(yz) = (xy)z \quad \text{associative law of multiplication.} \tag{1.15b}$$

Subtraction and division are not associative operations except in very special circumstances. Thus,

$$x - (y - z) \neq (x - y) - z \quad \text{unless } z = 0$$

and

$$x \div (y \div z) \neq (x \div y) \div z \quad \text{unless } z = 1 \textit{ and } y \neq 0,$$

as is easily verified by choosing any particular values for x, y and z.

(iii) *Distributivity*: The basic rule is

$$x(y + z) = xy + xz \quad \text{distributive law.} \tag{1.16a}$$

Together with the commutative law of multiplication, this implies

$$(x + y)z = xz + yz, \tag{1.16b}$$

since

$$(x + y)z = z(x + y) = zx + zy = xz + yz.$$

In addition, by noting that $(y - z) = (y + (-z))$ etc., one sees that these results imply that multiplication is distributed over addition *and* subtraction from both the left and the right, i.e.

$$x(y \pm z) = xy \pm xz \text{ and } (x \pm y)z = xz \pm yz. \tag{1.16c}$$

Finally, since $(x + y)/z = (x + y)z^{-1}$, equation (1.16b) implies that division is distributed over addition and subtraction from the right, i.e.

$$(x \pm y) \div z = (x \div z) \pm (y \div z), \tag{1.16d}$$

but not from the left, i.e.

$$x \div (y + z) \neq (x \div y) + (x \div z).$$

(iv) *The law of indices:* This is

$$x^n x^m = x^{(n+m)} \quad \text{law of indices,} \tag{1.17}$$

with $x^n/x^m = x^{n-m}$, and where, by definition, $x^0 \equiv 1$.

The nine laws (1.14)–(1.17) are the fundamental laws of elementary algebra. To illustrate their use, consider the proof of the familiar result

$$(x + y)^2 = x^2 + 2xy + y^2.$$

We have,

$$
\begin{aligned}
(x+y)^2 &= (x+y)(x+y) && \text{by the index law (1.17)}\\
&= (x+y)x+(x+y)y && \text{by the distributive law (1.16a)}\\
&= x(x+y)+y(x+y) && \text{by the commutative law (1.14b)}\\
&= x^2+xy+yx+y^2 && \text{by the distributive law (1.16b)}\\
&= x^2+2xy+y^2. && \text{by the commutative law (1.14b)}
\end{aligned}
$$

It should be emphasised that although the above rules are obeyed by the real variables of elementary algebra, in later chapters we will encounter other mathematical quantities, such as vectors and matrices, that do not necessarily obey all these rules.

*1.2.2 Proof of the irrationality of $\sqrt{2}$

Now we have introduced algebraic symbols and the idea of powers, we can return to the discussion of Section 1.1.1 and prove that $\sqrt{2}$ is an irrational number. The proof uses a general method called *reductio ad absurdum*, or *proof by contradiction*; that is, we assume the opposite, and prove it leads to a contradiction. This is a commonly used method of proof in mathematics. Suppose $\sqrt{2}$ is rational. It then follows that

$$\sqrt{2} = p/q, \tag{1.18}$$

where p and q are integers, and we may, without loss of generality, assume that they are the smallest integers for which this is possible, that is, they have no common factors. Then from (1.18), we have

$$p^2 = 2q^2, \tag{1.19}$$

so that p^2 is even. Furthermore, since the square of an odd number is odd and the square of an even number is even, p itself must be even; and since p and q have no common factors, q must be odd, since otherwise both would be divisible by 2. On the other hand, since p is even, we can write $p = 2r$, where r is an integer. Substituting this in (1.19) now gives $q^2 = 2r^2$, so that q is even, in contradiction to our previous result. Hence the assumption that $\sqrt{2}$ is rational must be false, and $\sqrt{2}$ can only be an irrational number.

1.2.3 Formulas, identities and equations

The use of symbols enables general *algebraic expressions* to be constructed. An example is a *formula*, which is an algebraic expression relating two or more quantities. Thus the volume of a rectangular solid, given by volume = length × breadth × height, may be written $V = lbh$. Given numerical values for l, b and h, we can calculate a value for the volume V. Formulas may be manipulated to more convenient forms providing certain rules are respected. These include (1) taking terms from one side to the other reverses their sign; and (2) division (multiplication) on one side becomes multiplication (division) on the other. For example, if $S = ab + c$, then $S - c = ab$, $a = (S - c)/b$ etc.

As with numerical forms, it is usual to rationalise algebraic expressions where possible. Thus,

$$\begin{aligned}\frac{x}{\sqrt{x}+\sqrt{y}}+\frac{y}{\sqrt{x}-\sqrt{y}}&=\frac{x(\sqrt{x}-\sqrt{y})+y(\sqrt{x}+\sqrt{y})}{(\sqrt{x}-\sqrt{y})(\sqrt{x}+\sqrt{y})}\\&=\frac{\sqrt{x}(x+y)-\sqrt{y}(x-y)}{x-y}.\end{aligned}\tag{1.20}$$

Sometimes factorisation may be used to simplify the results. For example,

$$\begin{aligned}\frac{1}{x^2-3x+2}-\frac{1}{x^2+x-2}&=\frac{(x^2+x-2)-(x^2-3x+2)}{(x-1)^2(x-2)(x+2)}\\&=\frac{4}{(x-1)(x^2-4)},\end{aligned}\tag{1.21}$$

where we have used the results

$$x^2 - 3x + 2 = (x-1)(x-2) \tag{1.22a}$$

and

$$x^2 + x - 2 = (x-1)(x+2). \tag{1.22b}$$

Equations (1.21)–(1.22) are examples of *identities*, because they are true for *all* values of x, and the three-line equality symbol (mentioned earlier) is also sometimes used to emphasise this, although in this book we will reserve its use for definitions.

In contrast, the expression on the left-hand side of (1.22a) can also be written $f(x) = x^2 - 3x + 2$ and setting $f(x)$ equal to a specific value gives an *equation* that will only have *solutions* (or *roots*) for specific values of x. In the case of (1.22a), setting $f(x) = 0$ yields the two solutions $x = 1$ and $x = 2$.

Example 1.5

Simplify: (a) $\dfrac{2x-y}{x-y}-\dfrac{2x-y}{x+y}$,

(b) $\dfrac{1}{x^3+2x^2+x+2}-\dfrac{1}{2x^2+x-6}$.

Solution

(a) Taking both terms over a common denominator, gives

$$\frac{2x+y}{x-y} - \frac{2x-y}{x+y}$$
$$= \frac{(2x^2+xy+2xy+y^2)-(2x^2-2xy-xy+y^2)}{(x-y)(x+y)}$$
$$= \frac{6xy}{(x^2-y^2)}.$$

(b) Taking both terms over a common denominator,

$$\frac{1}{x^3+2x^2+x+2} - \frac{1}{2x^2+x-6}$$
$$= \frac{(2x^2+x-6)-(x^3+2x^2+x+2)}{(x^2+1)(x+2)^2(2x-3)}$$
$$= \frac{-(x^3+8)}{(x^2+1)(x+2)^2(2x-3)}.$$

Example 1.6

Which of the following are equations and which are identities?

(a) $\dfrac{2x}{x^2-1} = \dfrac{1}{x-1} + \dfrac{1}{x+1}$, (b) $\dfrac{2}{x^2-1} = \dfrac{1}{x-1} + \dfrac{1}{x+1}$,

(c) $x^2 - 2x - 15 = (x+3)(x-5)$?

Solution

(a) Right-hand side $= \dfrac{1}{x-1} + \dfrac{1}{x+1} = \dfrac{2x}{x^2-1}$ which equals the left-hand side for all x, so it is an identity.

(b) Right-hand side $= \dfrac{1}{x-1} + \dfrac{1}{x+1} = \dfrac{2x}{x^2-1}$ which does not equal the left-hand side for all x, so it is an equation.

(c) Right-hand side $= (x+3)(x-5) = x^2 - 2x - 15$ which equals the left-hand side for all x, so it is an identity.

1.2.4 The binomial theorem

An important class of algebraic expressions consists of the *binomials* $(x+y)^n$, where the integer $n \geq 0$. These can be built up starting from $(x+y)^0 = 1$ by successively multiplying by $(x+y)$ to give

$$\begin{array}{lc} (x+y)^0 & 1 \\ (x+y)^1 & x+y \\ (x+y)^2 & x^2+2xy+y^2 \\ (x+y)^3 & x^3+3x^2y+3xy^2+y^3 \\ (x+y)^4 & x^4+4x^3y+6x^2y^2+4xy^3+y^4 \end{array}$$

and so on. The coefficients of the terms in this sequence form the *Pascal triangle* are

$$\begin{array}{lc} n=0 & 1 \\ n=1 & 1\ \ 1 \\ n=2 & 1\ \ 2\ \ 1 \\ n=3 & 1\ \ 3\ \ 3\ \ 1 \\ n=4 & 1\ \ 4\ \ 6\ \ 4\ \ 1 \\ \vdots & \vdots \end{array}$$

in which the elements of each row sum to 2^n, and those in the $(n+1)$th row are given by the sum of the neighbouring elements in the nth row. Thus, the fourth element for $n = 4$ is given by the sum of the neighbouring elements 3 and 1 in the row with $n = 3$.

These results are generalised to arbitrary n by the *binomial theorem*. In this theorem, the binomial expansion is written in the form

$$\begin{aligned} (x+y)^n &= \binom{n}{0} x^n + \binom{n}{1} x^{n-1}y + \binom{n}{2} x^{n-2}y^2 + \ldots + \binom{n}{n} y^n \\ &= \sum_{k=0}^{n} \binom{n}{k} x^{n-k}y^k, \qquad (n \geq 0) \end{aligned} \tag{1.23}$$

where the summation symbol $\sum\limits_{k=0}^{n}$ means that a sum is to be taken over all terms labelled by $k = 0, 1, 2, \ldots, n$. Here k is called a *dummy index* because the sum, that is, the left-hand side of (1.23), does not depend on the index k. The *binomial coefficients* are defined by

$$\binom{n}{k} \equiv \frac{n!}{(n-k)!k!}, \tag{1.24a}$$

where $n!$ indicates the *factorial*

$$n! \equiv n(n-1)(n-2)\cdots 1, \tag{1.25}$$

with $0! \equiv 1$ by definition, so that

$$(n+1)! = (n+1)n!, \quad n \geq 0. \tag{1.26}$$

An alternative notation that is used is

$$^nC_k \equiv \binom{n}{k}. \qquad (1.24b)$$

The binomial coefficients (1.24), which occur frequently in, for example, probability theory and statistical physics, have a number of important properties which include

$$\binom{n}{0} = \binom{n}{n} = 1, \qquad (1.27a)$$

$$\binom{n}{1} = \binom{n}{n-1} = n, \qquad (1.27b)$$

$$\binom{n}{k} = \binom{n}{n-k}, \qquad (1.27c)$$

and

$$\binom{n+1}{k} = \binom{n}{k} + \binom{n}{k-1}. \qquad (1.27d)$$

The first three of these follow trivially from the definition (1.24). The fourth, called *Pascal's rule,* is just the relation between the elements of the nth and $(n+1)$th rows of the Pascal triangle mentioned above. To prove (1.27d), we note that

$$\begin{aligned}\binom{n}{k} + \binom{n}{k-1} &= \frac{n!}{(n-k)!\,k!} + \frac{n!}{(n-k+1)!\,(k-1)!} \\ &= \frac{n!}{(n+1-k)!\,k!}[(n+1-k)+k] \\ &= \frac{(n+1)!}{(n+1-k)!\,k!} = \binom{n+1}{k},\end{aligned}$$

as required.

It remains to prove the binomial theorem (1.23). This is done by another general method, that of *induction*: one proves that if a proposition is true for a value n, it is true for a value $n+1$. Then provided it is true for $n=1$, its validity for all positive integers n is established. We therefore assume that (1.23) is valid for a value $n=m$. Multiplying by $(x+y)$ then gives

$$\begin{aligned}(x+y)^{m+1} &= (x+y)\sum_{k=0}^{m}\binom{m}{k}x^{m-k}y^k \\ &= \sum_{k=0}^{m}\binom{m}{k}\left(x^{m+1-k}y^k + x^{m-k}y^{k+1}\right) \\ &= \sum_{k=0}^{m}\binom{m}{k}x^{m+1-k}y^k + \sum_{j=1}^{m+1}\binom{m}{j-1}x^{m+1-j}y^j, \qquad (1.28)\end{aligned}$$

where we have substituted $j = k + 1$ in the second term. The value of this term is unchanged by relabeling the dummy index $j \to k$, so that (1.28) becomes

$$(x+y)^{m+1} = \binom{m}{0} x^{n+1} + \sum_{k=1}^{m} \left[\binom{m}{k} + \binom{m}{k-1} \right] x^{m+1-k} y^k + \binom{m}{m} y^{m+1}, \tag{1.29}$$

where we have separated off the first and last terms in the first and second summations in (1.28), respectively. Since (1.27a) holds for arbitrary n, we may replace m by $(m+1)$ in the first and last terms in (1.29); and substituting (1.27d) in the middle term then gives

$$\begin{aligned}(x+y)^{m+1} &= \binom{m+1}{0} x^{m+1} + \sum_{k=1}^{m} \binom{m+1}{k} x^{m+1-k} y^k + \binom{m+1}{m+1} y^{m+1} \\ &= \sum_{k=0}^{m+1} \binom{m+1}{k} x^{m+1-k} y^k.\end{aligned}$$

This is just the binomial theorem (1.23) for index $n = m + 1$, so that if the theorem holds for index $n = m$, it holds for $n = m + 1$. Since it is trivially true for $n = 1$, this implies it holds for all positive integers n, as required.

Example 1.7

Find the values of a and n that ensure the expansions of the expressions $(1 + 2x + x^2)^4$ and $(1 + ax)^n$ agree up to and including terms in x^2.

Solution

The two binomial expansions are:

$$(1 + 2x + x^2)^4 = 1 + 4(2x + x^2) + 6(2x + x^2)^2 + \cdots,$$

and

$$(1 + ax)^n = 1 + nax + \frac{n(n-1)}{2} a^2 x^2 + \cdots.$$

Equating coefficients of x gives $8 = na$, and equating coefficients of x^2 gives $56 = na^2(n-1)$. So, $56n = n^2a^2(n-1) = 64(n-1)$, giving $n = 8$ and hence $a = 1$.

Example 1.8
Find the coefficient of the term that is independent of x in the binomial expansion of $(x^2 + 1/x)^6$.

Solution
The expansion is

$$\left(x^2 + \frac{1}{x}\right)^6 = \sum_{k=0}^{6} \binom{6}{k} x^{2(6-k)} \frac{1}{x^k}.$$

The term that is independent of x has $2(6-k) - k = 0$, implying $k = 4$, and hence its coefficient is

$$\binom{6}{4} = \frac{6!}{4!2!} = 15.$$

1.2.5 Absolute values and inequalities

We are often interested in the numerical values of real numbers and variables without regard to their signs. This is called the *modulus* (or *absolute value*), with the notation $|x|$ or mod (x). We will also be using *inequalities*, with the symbols $>$ meaning 'greater than' and $<$ meaning 'less than'. Thus $3 < 4 < 7$ is the statement that 3 *is less than* 4 *which in turn is less than* 7. A related statement is $7 > 4 > 3$, that is, 7 *is greater than* 4, *which in turn is greater than* 3.

Using algebraic quantities, the definition of the modulus is

$$|x| \equiv \begin{cases} x & x > 0 \\ -x & x < 0 \end{cases} \tag{1.30}$$

Therefore,

$$|x| < a \quad \Rightarrow \quad -a < x < a, \tag{1.31}$$

where a is a real number and the symbol $\Rightarrow$ means 'implies'. Generalising further to include the possibility that $|x| = a$, that is $|x| \leq a$, we have $-a \leq x \leq a$, where we have used the obvious notation $\leq$ to mean 'less than or equal to'. In general, if $a \leq x \leq b$, where a and b are real numbers, then we say that x lies in a *closed interval* (or *range*) of length $(b - a)$. Likewise, if $a < x < b$, the interval is said to be *open*. Using the definition of the modulus, gives

$$|x - a| < b \quad \Rightarrow \quad -b < x - a < b. \tag{1.32}$$

The manipulation of inequalities differs from the manipulation of equalities, so we will discuss it in some detail. Terms may be taken from one side of an inequality to the other if their sign is changed. Also, adding a constant (positive or negative) to the terms of an inequality, or multiplying it by a positive constant, does not alter its validity. Thus, by adding a to each part of the inequality (1.32), we have $a - b < x < a + b$. However, multiplying or dividing by a negative number will reverse the sense of the inequality. For example, multiplying both sides of the inequality $x < 6$ by -1 does ***not*** imply $-x < -6$, which obviously contradicts the original inequality, but rather $-x > -6$, that is, the sense of the inequality is reversed. For this reason, particular care should be taken when simplifying an inequality involving algebraic quantities, such as

$$\frac{3}{3x-1} > \frac{1}{x+1}. \tag{1.33}$$

Cross-multiplying is not permitted, because the denominators may be negative. Rather, the inequality should be simplified by taking the terms over a common denominator. For (1.33),

$$\frac{3}{3x-1} - \frac{1}{x+1} > 0, \tag{1.34a}$$

so that

$$\frac{3(x+1)-(3x-1)}{(3x-1)(x+1)} = \frac{4}{(3x-1)(x+1)} > 0, \tag{1.34b}$$

which implies that the inequality is true only for $x > \frac{1}{3}$ or $x < -1$.

To illustrate these results, consider the inequality

$$\left|\frac{2}{x} + 3\right| < 5 \Rightarrow -5 < \frac{2}{x} + 3 < 5, \text{ that is, } -4 < \frac{1}{x} < 1. \tag{1.35}$$

There are two possible cases:

$$x > 0 \Rightarrow -4x < 1 < x, \tag{1.36a}$$

i.e. $x > 1$, and

$$x < 0 \Rightarrow -4x > 1 > x, \tag{1.36b}$$

where we now have to reverse the direction of the inequalities, i.e. $x < -\frac{1}{4}$. Another example is

$$2x^2 + 5x > 12 \Rightarrow 2x^2 + 5x - 12 = (2x-3)(x+4) > 0. \tag{1.37}$$

Thus either both brackets are positive, or both are negative. In the first case $x > \frac{3}{2}$ and in the second case $x < -4$.

Care must also be taken when manipulating pairs of inequalities. Thus for addition, while

$$x > y \text{ and } u > v \quad \Rightarrow \quad (x+u) > (y+v), \tag{1.38a}$$

on adding the two inequalities, we cannot deduce by subtraction that $(x-u) > (y-v)$. Likewise, if x, y, u, v, are positive quantities, then

$$x > y \text{ and } u > v \quad \Rightarrow \quad xu > yv, \tag{1.38b}$$

but this conclusion does not follow if any of x, y, u, v are negative numbers. For division, $x > y$ and $u > v$ do not imply $x/u > y/v$, even for positive numbers. The validity of these statements can be verified by some simple numerical examples. Thus if we take $x = 3$, $y = 2$ and $u = 5, v = 1$, then $x + u = 8 > y + v = 3$, but $x - u = -2 \not> y - v = 1$, where the symbol $\not>$ means 'not greater than'. The other statements can also be confirmed by using specific numbers.

Example 1.9

A number p is 37 when expressed to two significant figures. Deduce the closed interval allowed for p.

Solution

Using the rounding rules, $p_{\min} = 37.4\dot{9}$, where the dot over the 9 means that the figure 9 recurs repeatedly, and $p_{\max} = 36.5$. The allowed closed interval is thus $37.4\dot{9} \leq p \leq 36.5$.

Example 1.10

Find the range of real values of x that satisfies the following inequalities:

$$\text{(a)}\ -1 < \frac{4x+5}{2x-3} < 1, \quad \text{(b)}\ \frac{3}{3x-1} < \frac{5}{5x-1}.$$

Solution

$$\text{(a)}\ -1 < \frac{4x+5}{2x-3} < 1 \quad \Rightarrow \quad -1 < 2 + \frac{11}{2x-3} < 1$$

$$\Rightarrow \quad -3 < \frac{11}{2x-3} < -1$$

and hence, since the middle term is negative,

$$-1/3 > (2x-3)/11 > -1.$$

Then multiplying throughout by 11 and adding 3 to each term in the inequality, gives the solution $-1/3 > x > -4$.

$$\text{(b)}\quad \frac{3}{3x-1} - \frac{5}{5x-1} < 0 \quad \Rightarrow \quad \frac{2}{(3x-1)(5x-1)} < 0.$$

Thus, one bracket has to be positive and the other negative. In fact this is only possible if the first bracket is negative and the second one positive, for which $1/5 < x < 1/3$.

1.3 Functions, graphs and co-ordinates

In this section, we introduce the fundamental idea of functions and illustrate some of their properties by the use of graphs. We then briefly discuss co-ordinates and their use in describing geometrical forms.

1.3.1 Functions

Suppose two variables, x and y, are related in such a way that there is a single value of y corresponding to each value of x that lies within a given range $a < x < b$. Then the *dependent variable* y is said to be a *single-valued function* of the *independent variable* x, whose value may be varied at will within the allowed range. This is written

$$y = f(x),$$

where $f(x)$ specifies the particular function. In many cases, $f(x)$ is an *explicit function*, for example,

$$f(x) = x^3 - 3x^2 - 6x + 8. \tag{1.39}$$

Functions can also be defined *implicitly*, for example as the solutions of a given equation. A simple example would be to define $y = f(x)$ as the solution of the equation $y^2 = x - 1$ for $x > 1$. In this case, there are two solutions, $y = +\sqrt{x-1}$ and $y = -\sqrt{x-1}$, and such cases are referred to as *multi-valued functions*.[3] Alternatively, one can impose a subsidiary condition, for example $y > 0$, to ensure that the solution is unique, in accord with our original definition.

It is often useful to represent functions by *graphs*, which summarise, and give considerable insight into their properties. Figure 1.1 shows a graph of the function (1.39) in the range $-2.5 < x < 4.5$. The graph shows that the function has one maximum and one minimum in this range and that the solutions of the equation $f(x) = 0$ are $x = -2, 1$ and 4.

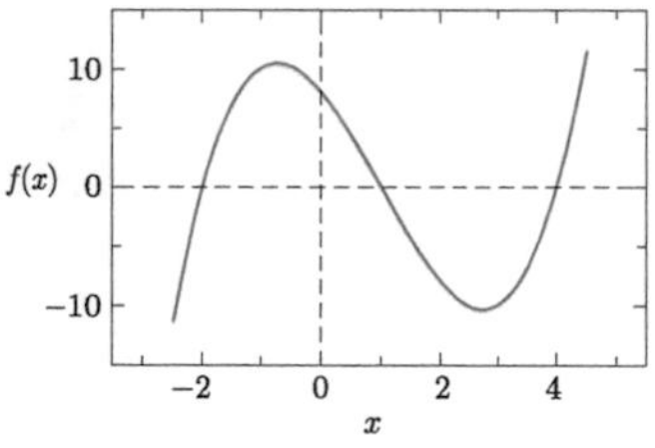

Figure 1.1 Graph of the function $f(x) = x^3 - 3x^2 - 6x + 8$.

Functions, whether of algebraic form or not, may be characterised by a variety of general properties and below we list some of these for use in later chapters.

If $f(-x) = f(x)$ for all values of x, the function is said to be *even* (or *symmetric*), whereas if $f(-x) = -f(x)$ for all values of x, the function is said to be *odd* (or *antisymmetric*). The simple examples

$$f(x) = 3x^2 - 15 \quad \text{(even)}, \qquad f(x) = x^3 + 4x \quad \text{(odd)}$$

---

[3] Mathematicians include the condition of being single-valued as part of the definition of a function, so a multi-valued function would be a misnomer. Nevertheless, in physical science it is still normal, and useful, to use the latter term. In general, we will use the word 'function' to refer to both types.

are shown in Figure 1.2. Although most functions have no specific symmetry, any function can always be written as the sum of even and odd functions. To see this we can write

$$f(x) = f_S(x) + f_A(x), \tag{1.40a}$$

where

$$f_S(x) \equiv \tfrac{1}{2}[f(x) + f(-x)] \quad \text{and} \quad f_A(x) \equiv \tfrac{1}{2}[f(x) - f(-x)] \tag{1.40b}$$

are symmetric and anti-symmetric functions by construction. As an example, consider the function

$$f(x) = (3x^3 - 2x^2 + 5)/(x - 1), \tag{1.41a}$$

from which we have

$$f(-x) = (3x^3 + 2x^2 - 5)/(x + 1) \tag{1.41b}$$

and hence from (1.40),

$$f_S(x) = \frac{3x^4 - 2x^2 + 5}{x^2 - 1} \quad \text{and} \quad f_A(x) = \frac{x^3 + 5x}{x^2 - 1}. \tag{1.41c}$$

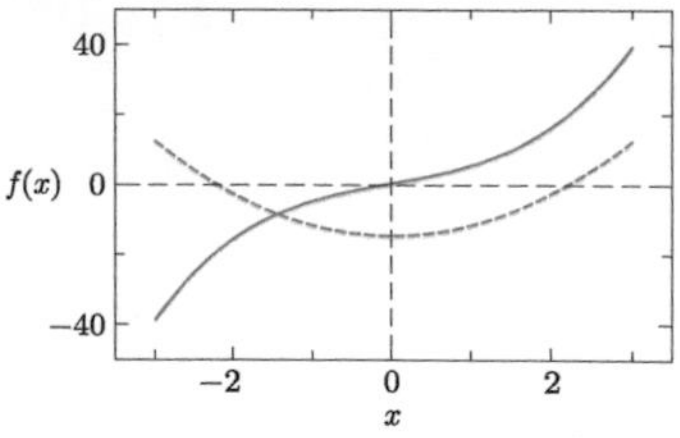

Figure 1.2 Graphs of the functions $f(x) = 3x^2 - 15$ (dashed line) and $f(x) = x^3 + 4x$ (solid line).

The usefulness of this decomposition is that exploitation of the symmetry of a function can often lead to simplifications in calculations. We will see examples of this in later chapters.

The function $f(x)$ is a prescription for calculating f given the value of x. We often need to know the prescription for the inverse process, that is, to find what value (or values) of x corresponds to a given value of f. This is called the *inverse function* of f and is written $f^{-1}(x)$. The notation is not perfect, because there is a danger of confusion with $1/f(x)$. It is important to remember that they are *not* the same. The inverse function corresponding to $y = f(x)$ is found by transposing the equation so that x is given as a function of y and then replacing y by x, and x by the inverse function $y^{-1}(x)$. Thus if $y(x) = x^3 + 3$, then $x = (y - 3)^{1/3}$ and hence the inverse function is $y^{-1}(x) = (x - 3)^{1/3}$. The inverse function may be multivalued. Thus, if $f(x) = x^2$, then 'inverting' gives the function $g(x) = \pm\sqrt{x}$, with two values.

Most functions we will discuss are *continuous*. We will define this term more precisely in Chapter 3, but roughly speaking it means that the values of y vary smoothly without sudden 'jumps' when the value of x is slowly varied. Functions that do not have this property are said to be *discontinuous* and we will see that frequently met functions are often of this type. A common situation is when a function is of the form $1/f(x)$, where $f(x)$ is zero at some point x_0 and changes sign as x passes through the point; for example $f(x) = (x - x_0)$. In this case the function will pass from $+\infty$ to $-\infty$ as x passes through the value x_0.

Finally, the argument of a function can itself be another function, in which case we speak of a 'function of a function'. Thus, if

$$q(x) = x^2 \quad \text{and} \quad p(x) = 3x + 2, \tag{1.42}$$

then p as a function of q is given by

$$p[q(x)] = 3q + 2 = 3x^2 + 2 \tag{1.43a}$$

and likewise q as a function of p is

$$q[p(x)] = p^2 = 9x^2 + 12x + 4. \tag{1.43b}$$

Example 1.11

Transpose the following functions to give x as an explicit function of y:

$$\text{(a) } y = \frac{3x-2}{x+4}, \quad \text{(b) } y = \sqrt{x^3+6}, \quad \text{(c) } y = \left(\frac{x^2-1}{3x^2-2}\right)^{1/3}.$$

Solution

(a) Cross multiplying gives $xy + 4y = 3x - 2$ and collecting terms in x on one side yields

$$x(y) = \frac{2(2y+1)}{3-y}.$$

(b) Squaring both sides gives $y^2 = x^3 + 6$, i.e. $x^3 = y^2 - 6$, and hence taking the cube root of both sides, $x(y) = (y^2 - 6)^{1/3}$.

(c) Cubing both sides and cross multiplying gives $y^3(3x^2 - 2) = (x^2 - 1)$. Then collecting terms in x on one side, we have $x^2(3y^2 - 1) = 2y^3 - 1$, and finally taking square roots,

$$x(y) = \left(\frac{2y^3-1}{3y^3-1}\right)^{1/2}.$$

Example 1.12

Write the function $f(x) = 2x/(x+1)$ as a sum of functions $f_S(x)$ and $f_A(x)$ having even and odd symmetry, respectively.

Solution

If $f(x) = f_S(x) + f_A(x)$, with $f_S(x)$ and $f_A(x)$ even and odd functions, respectively, then using

$$f(x) = \frac{2x}{x+1} \quad \text{and} \quad f(-x) = \frac{2x}{x-1},$$

in (1.40b) gives

$$f_S(x) = \frac{2x^2}{x^2 - 1} \quad \text{and} \quad f_A(x) = \frac{-2x}{x^2 - 1}.$$

Example 1.13

Find the inverse functions for: (a) $y(x) = 2x^2 - 3$, (b) $y(x) = (x-2)(x-4), x \geq 4$.

Solution

(a) $y(x) = 2x^2 - 3$, so $x = \pm[(y+3)/2]^{1/2}$ and the inverse function is the multi-valued form

$$y^{-1}(x) = \pm\sqrt{\frac{x+3}{2}}.$$

(b) $y(x) = (x-2)(x-4)$, so $x^2 - 6x + (8 - y) = 0$ and for $x \geq 4$, $x = 3 + \sqrt{(1+y)}$, $y \geq 0$. Hence, the inverse function is

$$y^{-1}(x) = 3 + \sqrt{(1+x)}, \quad x \geq 0.$$

1.3.2 Cartesian co-ordinates

Algebra and geometry are united by the use of co-ordinates, which enable geometrical forms to be described by algebraic equations. Here we illustrate this by considering Cartesian co-ordinates, mainly in two dimensions, leaving other co-ordinate systems to later chapters.

In two-dimensional Cartesian co-ordinates the position of a point P in a plane is specified relative to a chosen pair of horizontal and vertical axes, called the x- and y-axes respectively. The corresponding co-ordinates are written $P = (x, y)$, where x and y are the projections of the point onto the x and y-axes respectively, as shown for two points $A(x_1, y_1)$ and $B(x_2, y_2)$ in Figure 1.3. The axes themselves intersect at the *origin*, that is, the point $(x, y) = (0, 0)$.

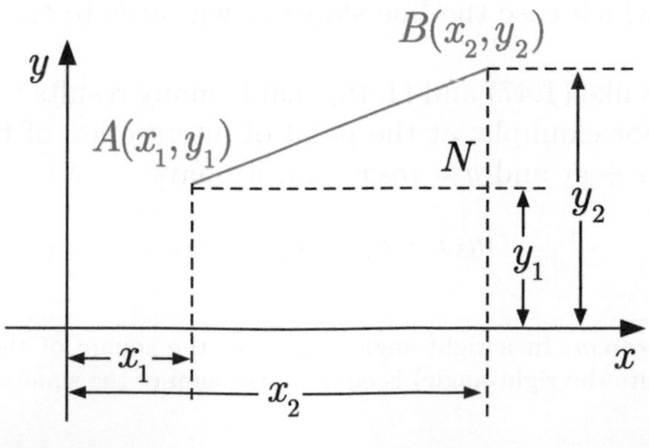

Figure 1.3 Cartesian co-ordinate system for the points $A(x_1, y_1)$ and $B(x_2, y_2)$.

Using Cartesian co-ordinates we can deduce a number of useful results. Thus the distance between any two points $A(x_1, y_1)$ and $B(x_2, y_2)$ is given by

$$AB = \sqrt{(x_2 - x_1)^2 + (y_2 - y_1)^2}, \tag{1.44}$$

which follows from using the Pythagoras Theorem[4] for the triangle ABN in Figure 1.3. Likewise, the *gradient*, or *slope*, of the straight line AB joining A and B is given by

$$\text{gradient} = \frac{\text{increase in } y \text{ co-ordinate}}{\text{increase in } x \text{ co-ordinate}} = \frac{y_2 - y_1}{x_2 - x_1}, \tag{1.45}$$

and the co-ordinates of the midpoint of AB are

$$\left[\tfrac{1}{2}(x_1 + x_2), \tfrac{1}{2}(y_1 + y_2)\right]. \tag{1.46}$$

Any line in the xy-plane implies an equation relating the x and y co-ordinates of any point which lies upon it. Consider for example a circle centre $C(a, b)$ and radius r, as shown in Figure 1.4. If $P(x, y)$ is any point on the circumference, then by using the Pythagoras Theorem in the triangle PCN, we have

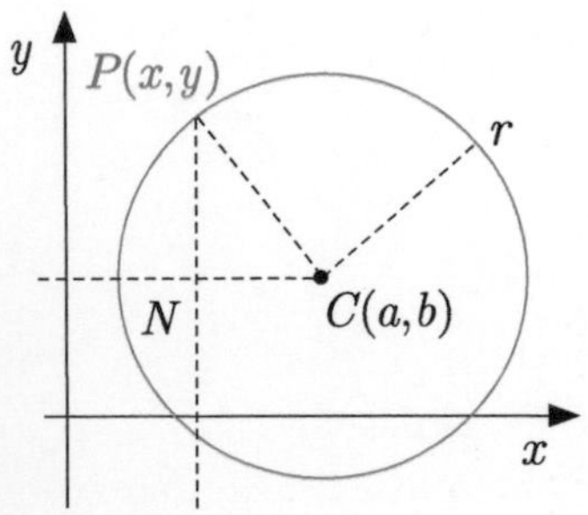

Figure 1.4 Construction to deduce the equation of a circle in Cartesian co-ordinates.

$$(x - a)^2 + (y - b)^2 = r^2, \tag{1.47}$$

which is therefore the equation of the circle in Cartesian co-ordinates.

An even simpler geometrical figure is a straight line. In this case, the co-ordinates $P(x, y)$ of any point lying on a straight line satisfy a linear equation of the form

$$y = mx + c, \tag{1.48}$$

where m and c are constants. In Figure 1.5(a) the resulting lines are shown for $m = 1$ and different values of c, and in Figure 1.5(b) for $c = 2$ but with m varying. It can be seen from Figure 1.5(a) that c is the y co-ordinate of the point where the line cuts the vertical (i.e. y) axis (this is called the *intercept*) and m is the gradient. In Figure 1.5 the gradients are all positive, but m can also take negative values (or zero) in which case the line slopes downwards to the right (or is horizontal).

Equations like (1.47) and (1.48) enable many results to be derived very easily. For example, at the point of intersection of two straight lines $y = m_1 x + c_1$ and $y = m_2 x + c_2$, we have

$$m_1 x + c_1 = m_2 x + c_2,$$

[4] *Pythagoras theorem:* In a right-angled triangle, the square of the hypotenuse (the side opposite the right-angle) is equal to the sum of the squares of the other two sides.

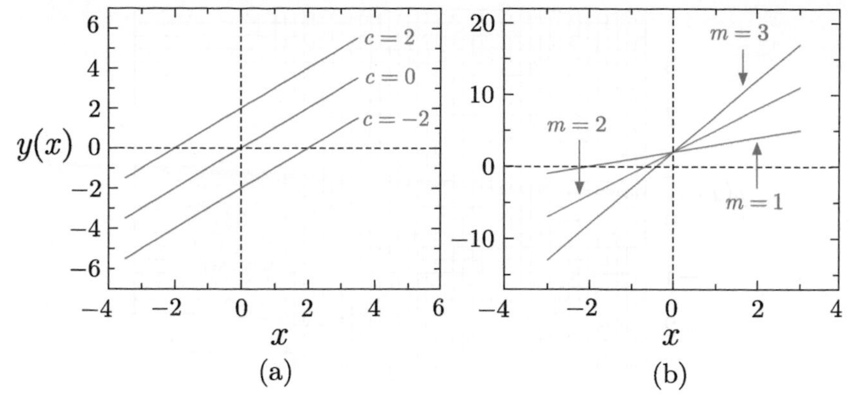

Figure 1.5 The linear function $y = mx + c$ with parameters: (a) $m = 1$ and $c = -2, 0, 2$ and (b) $m = 1, 2, 3$ and $c = 2$.

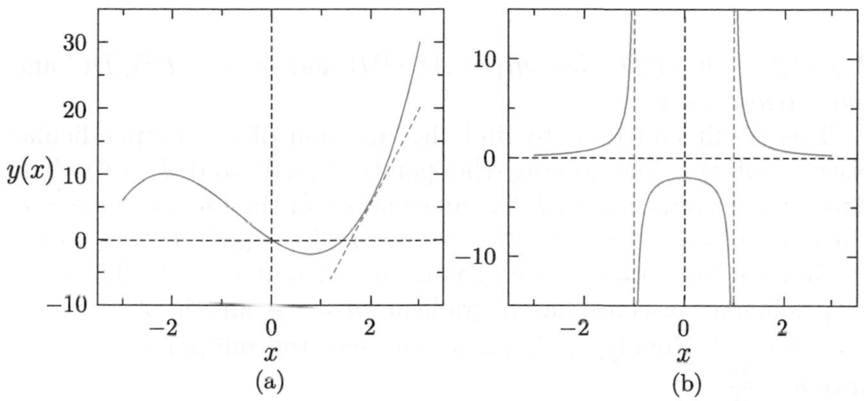

Figure 1.6 The functions: (a) $y = x^3 + 2x^2 - 5x$; and (b) $y = 2/(x^2 - 1)$. The blue dashed lines show the tangents at the points (2, 6) and ± 1, for curves (a) and (b), respectively.

so that $x = (c_1 - c_2)/(m_2 - m_1)$ and the value of y follows from the equation of either straight line. In particular we see that for *parallel* lines, that is, lines which have the same slopes $m_1 = m_2$, but different intercepts, $c_1 \neq c_2$, there is no solution, thus proving that 'parallel lines never meet'.

In general, for any curve $y = f(x)$, we define the *tangent* at a point as the straight line that just touches the curve at the point, so that the gradient of a curve at any point is equal to the gradient of the tangent at that point. This is illustrated in Figure 1.6(a), which shows the cubic polynomial $x^3 + 2x^2 - 5x$, together with the tangent drawn at the point (2,6). Finding the gradient by graphical methods will only give an estimate, because the accuracy depends on how well one can draw the tangent. We will see in later chapters that there are better methods for finding gradients. Figure 1.6(b) shows the function $2/(x^2 - 1)$, together with tangents drawn at the points $x = \pm 1$. Notice that this function is discontinuous at $x = \pm 1$, and the gradients at these points are infinite.

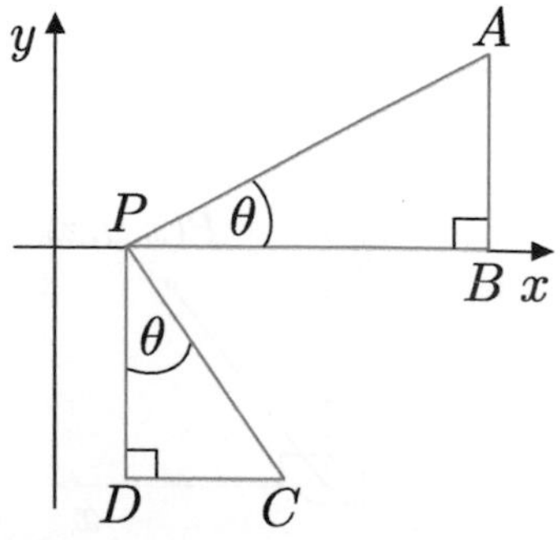

Figure 1.7 Construction to show that the product of the gradients of perpendicular lines is −1.

Other results can be found by geometrical methods. One example is to prove that the product of the gradients of two perpendicular lines is -1. Let the gradients of the two perpendicular lines PA and PC in Figure 1.7 be m_1 and m_2, respectively. Since the two lines are perpendicular, the triangles PAB and PCD are similar, with

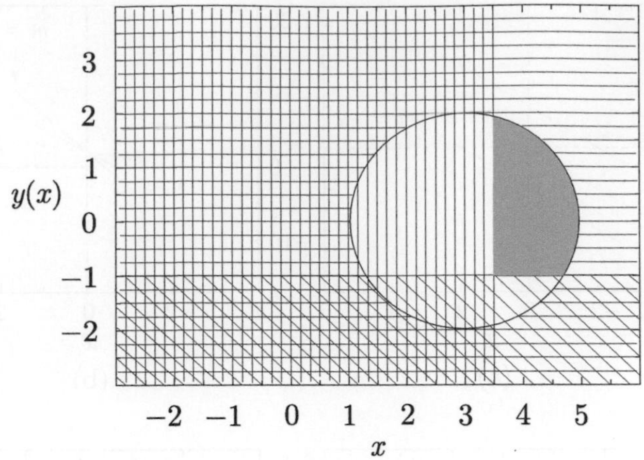

Figure 1.8 Use of inequalities to define regions of the xy-plane.

$AB/PB = DC/PD$. Now $m_1 = AB/PB$ and $m_2 = -PD/DC$ and thus $m_1m_2 = -1$.

This result can used to find the equation of the perpendicular bisector of the line joining the points $E(3,0)$ and $F(5,6)$. The straight line connecting EF has an equation of the form $y = mx + c$. Since it passes through the points $E(3,0)$ and $F(5,6)$, then $6 = 5m + c$ and $0 = 3m + c$, giving $m = 3$ and $c = -9$. Thus the perpendicular bisector has a gradient $m = -\frac{1}{3}$ and is of the form $y = -\frac{1}{3}x + d$. Finally, as it passes through the midpoint $(4,3)$, we have $d = \frac{13}{3}$.

Inequalities in x and y define regions of the xy-plane, that can be combined to find areas of allowed values. For example, Figure 1.8 shows the xy-plane with a number of shaded areas. These indicate the areas satisfying the set of inequalities

$$x \leq 3.5 \text{ (vertical hatching),}$$

$$y \leq -1.0 \text{ (diagonal hatching)}$$

and

$$y^2 \geq 4 - (x-3)^2 \text{ (horizontal hatching).}$$

where the last equation restricts $P(x,y)$ to points outside the circle $(x-3)^2 + y^2 = 4$. The coloured region thus represents the area occupied by all points that simultaneously satisfy the three inequalities $x > 3.5$, $y > -1.0$ and $y^2 < 4 - (x-3)^2$.

All the above discussion has been in the context of two dimensions, but it can easily be generalised to three dimensions. In this case we construct three axes x, y and z, with the property that if the thumb and first two fingers of the right hand are arranged so that they are mutually perpendicular, then the first and second fingers point along the positive x- and y-axes, respectively, and the thumb points along the positive z-axis. This is called a *right-handed Cartesian co-ordinate system* and is shown in Figure 1.9. Alternatively, one can say that the rotations $x \to y$, $y \to z$ and $z \to x$ are all in the

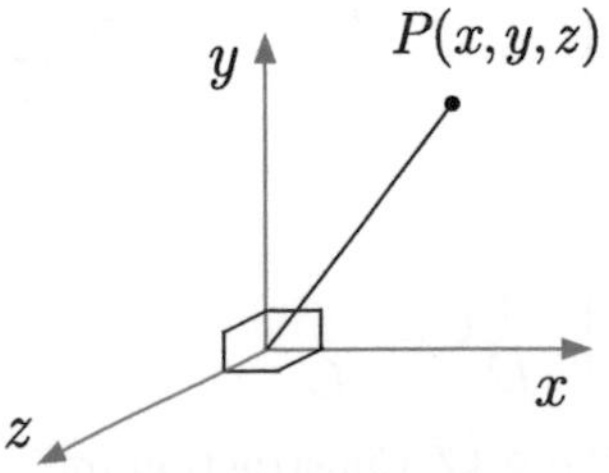

Figure 1.9 A right-handed three-dimensional Cartesian co-ordinate system.

sense of a right-handed screw. A point P in three-dimensional space is then described by co-ordinates (x, y, z), as shown in Figure 1.9.

As examples of the generalisation, the distance between any two points in two dimension (1.44) becomes the distance between any two points $A(x_1, y_1, z_1)$ and $B(x_2, y_2, z_1)$ in three dimensions, and is given by

$$AB = \sqrt{(x_2 - x_1)^2 + (y_2 - y_1)^2 + (z_2 - z_1)^2}. \tag{1.49}$$

Similarly, the equation

$$(x - a)^2 + (y - b)^2 + (z - c)^2 = r^2 \tag{1.50}$$

is the generalisation of the equation of a circle (1.47) and describes a sphere with centre at the point $(x, y, z) = (a, b, c)$. Finally, if the equation of a straight line in two dimensions (1.48) is generalised to

$$ax + by + cz = d, \tag{1.51}$$

where a, b, c and d are constants, it describes a plane in three dimensions. To describe a straight line in three dimensions requires two equations, for example,

$$y = ax + b \quad \text{and} \quad z = cx + d, \tag{1.52}$$

which determine both the y and z co-ordinates for a given value of x.

Example 1.14

Sketch the region bounded by the inequalities $y - x \leq 4$, $2y + x \leq 6$, $4y + x \geq -4$ and $x \leq 3$.

Solution

The required area is the coloured one in Figure 1.10 below.

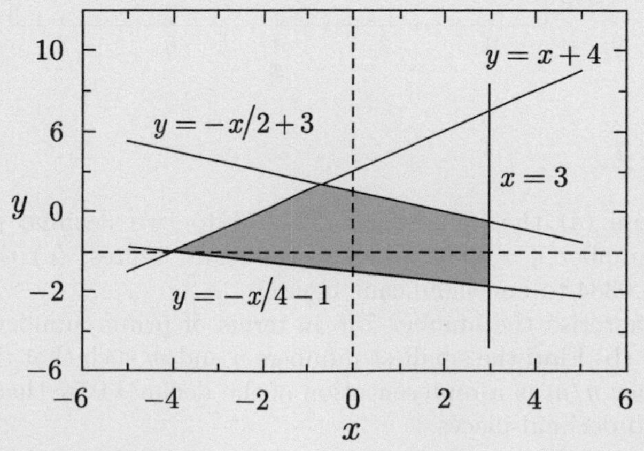

Figure 1.10

Example 1.15

Find the distance from the point $(x, y) = (2, 9)$ to the nearest point on the line $y = 2x - 1$.

Solution

If the nearest point on the line $y = 2x - 1$ is (a, b), then the line through (a, b) and $(2, 9)$ will be perpendicular to the line $y = 2x - 1$ and hence have a gradient $-1/2$, because the product of the gradients of two perpendicular lines is -1. Since the perpendicular line goes through $(2, 9)$, its equation is $y = -x/2 + 10$, and to find (a, b), we solve simultaneously this equation and $y = 2x - 1$ to give $(a, b) = (22/5, 39/5)$. Finally, the shortest distance d is given by

$$d = \sqrt{(2-a)^2 + (9-b)^2} = 6/\sqrt{5}.$$

Example 1.16

Draw the graph of the function:

$$y(x) = \begin{cases} x^3 & 0 \le x < 2 \\ x + 6 & 2 \le x < 5 \\ -\frac{1}{2}(x^2 - 47) & 5 \le x \le 8 \end{cases}$$

Solution

The graph is shown in Figure 1.11 below.

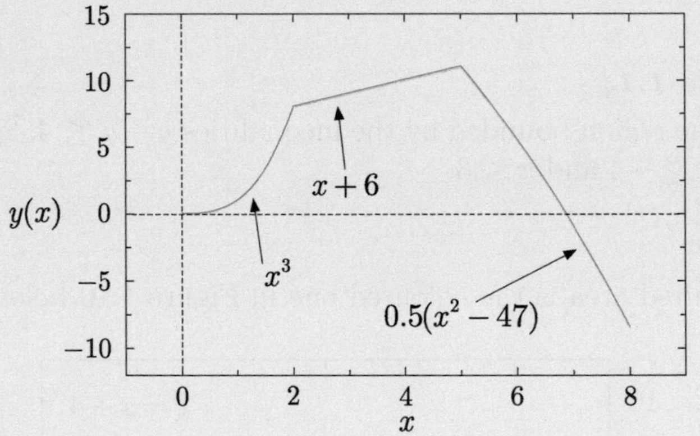

Figure 1.11

Problems 1

1.1 Express (a) the number $p = 124.675$ to two decimal places; (b) the number $q = 395.2$ to two significant figures; (c) the number $r = 0.0384$ to one significant figure.

1.2 (a) Factorise the number 756 in terms of prime numbers in index form. (b) Find the smallest numbers n and m such that the rational number n/m is a representation of the decimal 0.57 that is correct to two decimal places.

1.3 Express in scientific notation: (a) the number $p = 1245.78$ to four significant figures; (b) the product of 1.31×10^5 and -1.2×10^{-2} to two decimal places; (c) $5 \times 10^{-4} + 4 \times 10^{-3}$.

***1.4** Use the method of *reductio ad absurdum* (proof by contradiction) to show that if m is any integer, then $\sqrt{m}$ cannot be a rational number unless $\sqrt{m}$ is itself an integer.

1.5 Simplify the following:

(a) $12^{3/4}9^{-1/3}6^{1/2}$, (b) $3^{1/2}5^2/(25^{-1/5}3^{2/3})$,

(c) $49^{3/2}7^{-1/2}/(14^{1/2}7^{-1/3})$.

1.6 Simplify and/or rationalise as appropriate the following forms:

$$\text{(a)}\ \frac{3-5^{1/2}}{1+125^{1/2}},\quad \text{(b)}\ \frac{1}{2-\sqrt{3}+\sqrt{5}}.$$

***1.7** Convert: (a) the decimal number 10.31 to base 2, (b) the binary number 1101.01_2 to base 10.

***1.8** If in base 4, $p = 201_4$ and $q = 130_4$, find (a) $p \times q$ and (b) $p - q$. Check your answers in the decimal system.

1.9 Simplify and/or rationalise as appropriate the following forms:

$$\text{(a)}\ \frac{\sqrt{x}\sqrt{x^3}}{x^5},\quad \text{(b)}\ \frac{1+3\sqrt{x}}{3+\sqrt{x}},\quad \text{(c)}\ \frac{\sqrt{3a}(3a)^{-5/3}a^{3/2}}{(a/3)^2}.$$

1.10 Which of the following are identities and which are equations?

$$\text{(a)}\ \frac{1}{x-2}+\frac{1}{x+3}=\frac{2x+1}{x^2+x-6},\quad \text{(b)}\ \frac{1}{x+2}+\frac{1}{x+3}=\frac{3x}{x^2-6}.$$

1.11 Find the range of values of x that satisfy the following inequalities:

(a) $|2x-3| < 5$, (b) $|x^2-10| < 6$, (c) $x/(x+2) < 2$.

1.12 Use the binomial expansion to find the value of $(1.996)^5$ correct to five significant figures

1.13 In an expansion of $(3a-b)^{14}$ in ascending powers of b, what is the term containing a^3?

1.14 Show that

$$\sum_{k=0}^{n}\binom{n}{k} = \binom{n}{0}+\binom{n}{1}+\cdots\binom{n}{k}+\cdots\binom{n}{n} = 2^n$$

and find the value of $\sum\limits_{k=0}^{n} 2^n \binom{n}{k}$.

1.15 Use induction to show that $(5^n + 7)$ is divisible by 4 for all integer $n \geq 0$.

1.16 Prove that

$$\sum_{k=1}^{n} k^2 = \frac{n(n+1)(n+2)}{6}.$$

1.17 Transpose the following functions to give x as an explicit function of y.

$$\text{(a) } y = \left(\frac{x^2-6}{x^2-2}\right)^{1/2}, \text{ (b) } y = \frac{x^2-3x+2}{x^2-4}, \text{ (c) } y^2 = \left(\frac{x^2-4}{x+2}\right)^{1/3}.$$

1.18 Write the function $f(x) = (4-x)/(x-3)$ as the sum of functions $f_S(x)$ and $f_A(x)$ having even and odd symmetry, respectively.

1.19 Find the inverse of the functions:

$$\text{(a) } f(x) = (x-2)/(x+2),\ (x \neq -2), \quad \text{(b) } f(x) = \tfrac{1}{3}(x+4)^{1/3}.$$

1.20 Find the equations of the straight lines that satisfy the following conditions:
(a) passes through the points $(x, y) = (1, 2)$ and $(4, -1)$,
(b) passes through the point $(x, y) = (2, -1)$ with slope $\frac{2}{3}$,
(c) passes through the point $(1, 4)$ and is parallel to the line $2y + 5x = 7$,
(d) passes through the point (1,4) and is perpendicular to the line $2y + 5x = 7$.

1.21 Find the shortest distance from the point $(2, 2)$ to the line $y = 3x + 6$.

1.22 Sketch the triangle bounded by the lines $5y = 3x + 2$, $y = -x + 2$ and $y = 3x - 14$ and calculate its area.

1.23 Find the equations of the circles centred at $(x, y) = (1, 3)$ that (a) have radius 2 and (b) pass through the point $(-2, -5)$.

1.24 Show that the equation

$$x^2 + y^2 - ax - by + c^2 = 0$$

describes a circle if $a^2 + b^2 - 4ac > 0$. What is the radius of the circle described by the equation

$$x^2 + y^2 - 2x - 3y + 1 = 0$$

and what are the co-ordinates of its centre?

1.25 On a graph show the area corresponding to

$$(2.5 - 0.5x) \geq y(x) \geq |x^2 - 5x + 6|.$$

2

Some basic functions and equations

The discussion of functions of one real variable is continued in this chapter by considering three classes of functions that play an important role in physical sciences. They are: simple algebraic functions; trigonometric functions; logarithms and exponentials.

2.1 Algebraic functions

Here we discuss polynomials and the more complicated functions that can be defined in terms of them.

2.1.1 Polynomials

The *polynomial* function has the general form

$$P_n(x) = \sum_{i=0}^{n} a_i x^i = a_0 + a_1 x + a_2 x^2 + \cdots + a_n x^n, \qquad (2.1)$$

where $a_i (i = 0,\ 1,\ 2, \ldots,\ n)$ are constants, n is a non-negative integer (i.e. including zero), and the symbol Σ means that a sum is to be taken of all terms labelled by the indices $0,\ 1,\ 2, \ldots,\ n$. The value of n defines the *order* (or *degree*) of the polynomial. The expression $x^3 - 3x^2 - 6x + 8$ plotted in Figure 1.1 is therefore a polynomial of order 3.

The *roots* of polynomials are defined as the solutions of the equation

$$P_n(x) = 0, \qquad (2.2)$$

Mathematics for Physicists, First Edition. B.R. Martin and G. Shaw.

Companion website: www.wiley.com/go/martin/mathsforphysicists

and correspond to the points where a graph of $P_n(x)$ crosses the x-axis. For first-order polynomials, (2.2) is a linear equation of the form

$$ax + b = 0, \tag{2.3}$$

where a and b are constants. This has one root, which is trivially given by $x = -b/a$.

For second-order polynomials, (2.2) leads to a quadratic equation of the form

$$ax^2 + bx + c = 0, \tag{2.4}$$

where a, b and c are numerical constants. When the coefficients are simple, for example integers, one can sometimes spot that the quadratic form factorises, i.e.

$$ax^2 + bx + c = a(x - \alpha)(x - \beta), \tag{2.5}$$

where α and β are real numbers. The solutions are then clearly $x_1 = \alpha$ and $x_2 = \beta$. For example, $2x^2 - 7x + 3 = (2x - 1)\ (x - 3)$ and so the roots of this polynomial are $x = \frac{1}{2}$, and $x = 3$. In some circumstances a solution can be 'lost' if care is not taken. For example, the equation $x^2 - 5x = 0$ factorises to $x(x - 5) = 0$, with solutions $x = 0$ and $x = 5$. However, had we divided both sides of the original equation by x, we would have only found the solution $x = 5$. So when manipulating the original equation, particularly when using division to simplify it, one should always check that no solutions are thereby being omitted.

If there is no obvious factorisation, the general solution of the quadratic is obtained by a process known as 'completing the square', as follows. By taking the constant c onto the right-hand side of the equation and then adding $b^2/4a$ to both sides, the left-hand side may be written as a perfect square. We have,

$$ax^2 + bx + \frac{b^2}{4a} = a\left(x + \frac{b}{2a}\right)^2 = \frac{b^2 - 4ac}{4a} \tag{2.6a}$$

and hence

$$x = \frac{-b \pm \sqrt{b^2 - 4ac}}{2a}. \tag{2.6b}$$

We thus see that for $b^2 > 4ac$ there are two solutions, which coincide when $b^2 = 4ac$. If we denote these two solutions by α and β, then from (2.6b) one easily confirms that the polynomial can be written in the factorised form (2.5) and that

$$\alpha + \beta = -\frac{b}{a} \quad \text{and} \quad \alpha\beta = \frac{c}{a}. \tag{2.7}$$

These results are sometimes useful because a quadratic equation may be written

$$x^2 - (\text{sum of roots})\ x + (\text{product of roots}) = 0, \qquad (2.8)$$

so that, for example, the polynomial with roots 2.1 and 3.2 is $x^2 - 5.30x + 6.72$. If $b^2 - 4ac < 0$, the argument of the square root in (2.6b) is negative, and (2.4) has no solutions for real x. For example, the quadratic equation $x^2 - 3x + k = 0$ has two roots if $9 - 4k \geq 0$, i.e. $k \leq 9/4$, but no real roots if $9 - 4k < 0$.

Exact solutions for cubic and quartic equations exist, so that the roots of third-order and fourth-order polynomials may be determined exactly. However, the solutions are algebraically very complicated and we will not pursue them further. Except in special cases, the roots of higher-order polynomials are found by approximate methods. However, one can establish some important general results. To obtain these we initially consider the result obtained by dividing a polynomial $P_n(x)$ of order n by a factor $(x - a)$ using long division, until only a constant remainder is left. For example, on dividing $x^4 - 2x^3 + 3x^2 - 4x + 5$ by $(x - 1)$ one obtains

$$\begin{array}{r|l}
 & x^3 - x^2 + 2x - 2 \\
\hline
(x-1) & x^4 - 2x^3 + 3x^2 - 4x + 5 \\
 & x^4 - x^3 \\
\hline
 & \quad -x^3 + 3x^2 - 4x + 5 \\
 & \quad -x^3 + x^2 \\
\hline
 & \qquad 2x^2 - 4x + 5 \\
 & \qquad 2x^2 - 2x \\
\hline
 & \qquad\quad -2x + 5 \\
 & \qquad\quad -2x + 2 \\
\hline
 & \qquad\qquad 3
\end{array}$$

so that

$$x^4 - 2x^3 + 3x^2 - 4x + 5 = (x - 1)(x^3 - x^2 + 2x - 2) + 3.$$

More generally, dividing any polynomial of order n by $(x - a)$ leads to an expression of the form

$$P_n(x) = (x - a)Q(x) + R, \qquad (2.9)$$

where the quotient $Q(x)$ is a polynomial of order $(n - 1)$ and the remainder $R = P_n(a)$. This result is called the *remainder theorem* and implies that if $a = \alpha$ is a root of $P_n(x)$, then $R = 0$ and (2.9) reduces to the partially factorised form

$$P_n(x) = (x - \alpha)P_{n-1}(x), \qquad (2.10a)$$

where $P_{n-1}(x)$ is a polynomial of order $n-1$. This is called the *factor theorem*. Furthermore, repeating the process for all m roots $\alpha_1, \alpha_2, \ldots, \alpha_m$ gives

$$P_n(x) = a(x-\alpha_1)(x-\alpha_2)\ldots\ (x-\alpha_m)P_{n-m}(x), \quad (2.10b)$$

and since the highest power on the left is x^n, there are at most n real roots. Thus a polynomial $P_n(x)$ of order n has at most n real roots.[1] Beyond this, one can only say that the number of roots is odd or even, corresponding to whether the order of the polynomial is odd or even, respectively, provided that if two or more factors in (2.10b) are equal, we still count them separately. This is most easily seen by considering a graph of the polynomial, in which the roots correspond to the values at which the curve intercepts the x-axis, as illustrated in Figure 1.1. The results then follow by considering the asymptotic behaviour of the polynomial as $x \to \pm\infty$, which is dominated by the term $a_n x^n$ in (2.1). Hence if n is even, the polynomial has the same sign in the limits $x \to \pm\infty$, and since it is continuous, it must either not cross the x-axis at all, corresponding to no roots, or cross it an even number of times, corresponding to an even number of roots. A similar argument shows that a polynomial whose order is odd must have at least one root and there can only be an odd number of roots.

We now return to the problem of finding the roots. As noted above, the general solution for third and fourth order polynomials is very complicated, and for higher orders no general exact solution is known. However, for simple cases it may still be possible to find exact solutions of higher-order polynomials by spotting factors and using the factor theorem. For example, consider the fourth-order polynomial

$$f(x) = 2x^4 - x^3 - 8x^2 + x + 6. \quad (2.11)$$

By inspection, $f(1) = 0$, so $(x-1)$ is a factor. To find the quotient $Q(x)$ we need to carry out a long division, which yields $(2x^3 + x^2 - 7x - 6)$, so that

$$f(x) = 2x^4 - x^3 - 8x^2 + x + 6 = (x-1)(2x^3 + x^2 - 7x - 6). \quad (2.12)$$

We now repeat the process by finding factors (if they exist) of the cubic. The final result is

$$f(x) = 2x^4 - x^3 - 8x^2 + x + 6 = (x-1)(x+1)(x-2)(2x+3),$$

so the solutions are $x = 1, -1,\ 2$ and $-\frac{3}{2}$. (2.13)

[1] For readers already familiar with complex numbers, which are discussed in Chapter 6, we stress that we are restricting ourselves here to real variables x and real roots α_i. If complex roots are allowed, then a polynomial of order n always has precisely n roots. This result is called *the fundamental theorem of algebra.*

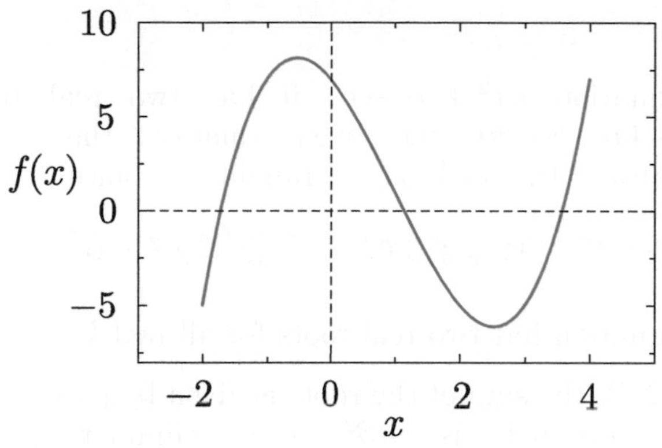

Figure 2.1 Plot of the function $y = f(x) = x^3 - 3x^2 - 4x + 7$.

Not all polynomials factorise, and in practice equations involving higher order polynomials are solved by approximate methods, either graphical or numerical. In the former, the function is plotted and the points where $f(x)$ crosses the x-axis are found. In Figure 2.1 the function $x^3 - 3x^2 - 4x + 7$ is plotted and it is seen that it crosses the x-axis at the values $x \approx -1.7$, 1.1 and 3.6, which are therefore the approximate solutions of the equation $x^3 - 3x^2 - 4x + 7 = 0$. It is worth reiterating that a polynomial of order n does not necessarily have n real roots, and so a graph of the function will not necessarily cut the x-axis at n points.

Although only approximate, the graphical solutions are still useful, as numerical methods for finding accurate roots often rely on knowing approximate solutions as starting values. One simple technique is the so-called *bisection method*, which can be applied to any continuous function. In this method one starts by finding two values of x, say x_1 and x_2, that straddle the position of a zero. Thus $f(x_1)$ and $f(x_2)$ will have opposite signs and so $f(x_1)f(x_2) < 0$. Now let $x_m = \frac{1}{2}(x_1 + x_2)$ and calculate $f(x_m)$. If $f(x_1)f(x_m) < 0$, then the root lies between x_1 and x_m. In this case, the mid-point of x_1 and x_m is found and the calculation repeated. If, however, $f(x_1)f(x_m) > 0$, then the root lies between x_2 and x_m, and in this case, the mid-point of x_2 and x_m is found and the calculation repeated. This iterative method can be rapidly implemented on a computer, and when applied to any of the roots, the range of values of x that produces a value of $f(x)$ as close to zero as desired may be found.

Example 2.1

(a) Prove that the equation $4kx^2 + 8x - (k - 4) = 0$ has two real roots for all real k.

(b) If the quadratic equation $ax^2 + bx + c = 0$ has two real roots β and $n\beta$, show that $(n + 1)^2ac = nb^2$.

Solution

(a) The equation $ax^2 + bx + c = 0$ has two real roots provided $b^2 > 4ac$. So for the given equation, the condition is $64 > -4(4k)(k-4)$, which on rearranging becomes

$$k^2 - 4k + 4 > 0, \quad \text{or} \quad (k-2)^2 > 0.$$

Thus the equation has two real roots for all real k.

(b) From (2.7), the sum of the roots is given by $(\beta + n\beta) = -b/a$ and their product is $n\beta^2 = c/a$, eliminating β gives $(n+1)^2 ac = nb^2$.

Example 2.2

Find the points of intersection of the curve $y = x^2 + 1$ (a parabola) and the straight line $y = 2x + 3$.

Solution

At the points of intersection, $x^2 + 1 = 2x + 3$, i.e. $x^2 - 2x - 2 = 0$, with solutions

$$x = \frac{2 \pm \sqrt{4+8}}{2} = 1 \pm \sqrt{3},$$

giving

$$y = 2x + 3 = 5 \pm 2\sqrt{3}.$$

So the points of intersection are

$$(x, y) = [(1+\sqrt{3}), 5 + 2\sqrt{3})] \text{ and } [(1-\sqrt{3}), 5 - 2\sqrt{3})].$$

Example 2.3

A third-order polynomial $P_3(x) = 2x^3 - 4x^2 + 3x - 1$ has a root $x = 1$. Use the factor theorem to find its quotient $P_2(x)$ and show that $x = 1$ is the only real root of $P_3(x)$.

Solution

From the factor theorem we can write

$$\begin{aligned} P_3(x) = 2x^3 - 4x^2 + 3x - 1 &= (x - \alpha)P_2(x) \\ &= (x-1)(ax^2 + bx + c), \end{aligned}$$

where a, b and c are constants. Multiplying the expressions on the right-hand side and equating the coefficients of powers of x on both sides gives $a = 2,\ b = -2,\ c = 1$, and since $b^2 < 4ac$ there are no other real roots.

Example 2.4

From Figure 2.1 we see that the polynomial $x^3 - 3x^2 - 4x + 7$ has a root near $x = 1.1$. Use the bisection method to find the value of the root accurate to four significant figures.

Solution

For example, if we start with $x_1 = 1$ and $x_2 = 1.2$, we find $f(x_1) = 1$ and $f(x_2) = -0.392$. Since $f(x_1)f(x_2) < 0$, we set $x_3 = \frac{1}{2}(x_1 + x_2)$ and calculate $f(x_3) = 0.301$. Now $f(x_1)f(x_3) > 0$, so we set $x_4 = \frac{1}{2}(x_2 + x_3)$ and calculate $f(x_4) = -0.046625$. Repeating this technique we can construct the table below (which of course can be constructed very rapidly on a computer).

n	x_n		$f(x_n)$
1	$x_1 =$	1.000000000	1.00000000
2	$x_2 =$	1.200000000	−0.39200000
3	$x_3 = \frac{1}{2}(x_1 + x_2) =$	1.100000000	0.30100000
4	$x_4 = \frac{1}{2}(x_2 + x_3) =$	1.150000000	−0.04662500
5	$x_5 = \frac{1}{2}(x_3 + x_4) =$	1.125000000	0.12695313
6	$x_6 = \frac{1}{2}(x_4 + x_5) =$	1.137500000	0.04009961
7	$x_7 = \frac{1}{2}(x_4 + x_6) =$	1.143750000	−0.00327954
8	$x_8 = \frac{1}{2}(x_6 + x_7) =$	1.140625000	0.01840591
9	$x_9 = \frac{1}{2}(x_7 + x_8) =$	1.142187500	0.00756215
10	$x_{10} = \frac{1}{2}(x_7 + x_9) =$	1.142968750	0.00214104
11	$x_{11} = \frac{1}{2}(x_7 + x_{10}) =$	1.143359375	−0.00056932

The root correct to four significant figures is thus 1.143.

2.1.2 Rational functions and partial fractions

Given two polynomials $P(x)$ and $Q(x)$, we can form the *rational function* $f(x)$, defined by $f(x) \equiv P(x)/Q(x)$. These are generalisations of numerical fractions and, by analogy with those, the rational expression P/Q is said to be *proper* if the order of the numerator is less than the order of the denominator. Otherwise it is called an *improper* fractional expression. Examples are:

$$\frac{3x^2 - 4}{x^5 + x - 3} \text{ (proper)}; \qquad \frac{5x^7 - x^3 - x + 1}{x^2 + 7} \text{ (improper)}. \qquad (2.14)$$

In contrast to polynomials, rational functions are not in general continuous, but can have discontinuities corresponding to the roots of the denominator function $Q(x)$, that is, where $Q(x) = 0$, and so are undefined at those points. For example, the rational function

$$\frac{2}{(x^2-1)} = \frac{2}{(x-1)(x+1)} \tag{2.15}$$

has discontinuities at $x = \pm 1$, where the denominator vanishes, as shown in Figure 1.6b.

Rational expressions where the denominator is itself the product of polynomials may often usefully be decomposed into a sum of simpler terms called *partial fractions.* Assume for the moment that the initial expression is a proper fraction. There are several possible forms this can take and we will look at each in turn, before illustrating them with specific examples.

(i) The first form is

$$\frac{P(x)}{(x-a)(x-b)\cdots(x-n)}, \tag{2.16}$$

where $a,\ b, \ldots,\ n$ are constants, and because the fraction is proper, $P(x)$ is a polynomial of lower order than the product of factors in the denominator. In this case, we may write the identity

$$\frac{P(x)}{(x-a)(x-b)\cdots(x-n)} = \frac{A}{(x-a)} + \frac{B}{(x-b)} + \cdots + \frac{N}{(x-n)}, \tag{2.17}$$

where A, B, ... N are constants. By putting the terms on the right-hand side over a common denominator, (2.17) may be written

$$P(x) = A[(x-b)(x-c)\cdots(x-n)] + B[(x-a)(x-c)\cdots (x-n)] + \cdots + N[(x-a)(x-b)\cdots(x-n+1)]. \tag{2.18}$$

Because this is an identity, it is true for *all* values of x. Thus we can choose any values of x to evaluate it. So, in particular, if we choose $x = a,\ x = b, \ldots$ in turn, in each case all the terms on the right-hand side are zero except one and we can solve for the coefficients A, B, etc.

(ii) A second common occurrence is where one of the factors in the denominator is a quadratic of the form $\alpha x^2 + \beta x + \gamma$ that cannot be factored further, since $\beta^2 < 4\alpha\gamma$. In this case, we may write

$$\frac{P(x)}{(x-a)(x-b)\cdots(x-n)(\alpha x^2+\beta x+\gamma)} = \frac{A}{(x-a)} + \frac{B}{(x-b)} + \cdots + \frac{N}{(x-n)} + \frac{A'x+B'}{\alpha x^2+\beta x+\lambda} \tag{2.19}$$

and again, by using the fact that this is an identity, we can take both sides over a common denominator and equate the coefficients of the same powers of x on both sides. Then by choosing suitable values for x, values of the constants $A,\ B, \ldots,\ N$, and A' and B' can be found.

(iii) The third type is when there are repeated factors $(ax+b)^n$ in the denominator. These will give rise to partial fractions of the form

$$\frac{P(x)}{(ax+b)^n} = \frac{A}{ax+b} + \frac{B}{(ax+b)^2} + \cdots + \frac{N}{(ax+b)^n} \tag{2.20}$$

that again have to be added to any other terms of the type (2.19) and the sum treated in an analogous way to the previous cases, that is, take both sides over a common denominator and using the fact that the expression is an identity, equate coefficients of the same powers of x on both sides, and then choose suitable values of x to determine the coefficients.

In all the above cases the original fractional function was proper. If the fraction is improper, then an initial long division must be made to write it as the sum of a polynomial and a proper fraction. The latter is then decomposed into partial fractions as above.

Example 2.5

Write the following expressions as partial fractions:

(a) $\dfrac{3x-2}{(x-3)(x+4)}$ (b) $\dfrac{11x^2-5x+14}{(x-2)(3x^2+4)}$,

(c) $\dfrac{15x^2-x-9}{(x-2)(3x+1)^2}$ (d) $\dfrac{x^3-6x-17}{(x+1)(x-3)}$.

Solution

(a) This is of type (i) above, and from (2.17) takes the form

$$\frac{3x-2}{(x-3)(x+4)} = \frac{A}{(x-3)} + \frac{B}{(x+4)}.$$

Then taking both sides over a common denominator gives

$$3x-2 = A(x+4) + B(x-3).$$

Because this is an identity, it is true for *all* values of x. Thus we can choose any values of x to evaluate it. Setting $x=3$ and $x=-4$ in succession, so that one term is zero in each case, we find $B=2A=2$ and hence

$$\frac{3x-2}{(x-3)(x+4)} = \frac{1}{(x-3)} + \frac{2}{(x+4)}.$$

(b) This is of type (ii) and from (2.19) can be written

$$\frac{11x^2 - 5x + 14}{(x-2)(3x^2+4)} = \frac{A}{(x-2)} + \frac{Bx + C}{(3x^2+4)}.$$

Again, taking both sides over a common denominator gives

$$\begin{aligned} 11x^2 - 5x + 14 &= A(3x^2+4) + (x-2)(Bx+C) \\ &= (3A+B)x^2 + (C-2B)x + (4A-2C). \end{aligned}$$

Using the fact that this is an identity, we can equate the coefficients of the same powers of x on both sides, which gives

$$3A + B = 11, \; C - 2B = -5, \; 4A - 2C = 14.$$

Solving for A, B and C gives $A = 3, \; B = 2, \; C = -1$ and hence

$$\frac{11x^2 - 5x + 14}{(x-2)(3x^2+4)} = \frac{3}{(x-2)} + \frac{2x - 1}{(3x^2+4)}.$$

(c) This is of type (iii) and in accordance with (2.20) can be written as

$$\frac{15x^2 - x - 9}{(x-2)(3x+1)^2} = \frac{A}{(x-2)} + \frac{B}{(3x+1)} + \frac{C}{(3x+1)^2}.$$

Taking both sides over a common denominator gives

$$15x^2 - x - 9 = A(3x+1)^2 + B(x-2)(3x+1) + C(x-2),$$

and hence by equating coefficients of powers of x as before (or setting $x = 2$ and then $x = -\frac{1}{3}$ so that some terms are zero), $A = 1$, $B = 2$ and $C = 3$, and finally

$$\frac{15x^2 - x - 9}{(x-2)(3x+1)^2} = \frac{1}{(x-2)} + \frac{2}{(3x+1)} + \frac{3}{(3x+1)^2}.$$

(d) Because the expression is improper, we first have to perform a long division, which gives

$$\frac{x^3 - 6x - 17}{(x+1)(x-3)} = x + 2 + \frac{\text{remainder}}{(x+1)(x-3)}.$$

The remainder does not have to be found, because by construction the fraction on the right-hand side is a proper fraction. Thus, using the result (2.17) above, we may write

$$\frac{x^3 - 6x - 17}{(x+1)(x-3)} = x + 2 + \frac{A}{(x+1)} + \frac{B}{(x-3)}$$

and proceeding as in (a) gives $A = 3$ and $B = -2$ and finally

$$\frac{x^3 - 6x - 17}{(x+1)(x-3)} = x + 2 + \frac{3}{(x+1)} - \frac{2}{(x-3)}.$$

2.1.3 Algebraic and transcendental functions

Polynomials and rational functions are the simplest examples of a broader class of functions, called *algebraic functions.* An algebraic function is any function y that can be defined by an equation of the form

$$P^{(0)}(x)y^n + P^{(1)}(x)y^{n-1} + \cdots + P^{(n-1)}(x)y + P^{(n)}(x) = 0,$$

where $P^{(i)}(x)$ $(i = 0, 1, \ldots, n)$ are given polynomials of any order. This definition is implicit, and for any x the function can be evaluated by first evaluating the polynomials $P^{(i)}(x)$, and then finding the roots of the resulting polynomial in y.

For $n = 1$, one easily sees that the above definition reduces to a rational function, or a polynomial in the case of $P^{(0)} = 1$. More generally, it implies that any algebraic function can be defined in terms of a finite number of the basic operations of algebra (i.e. addition, subtraction, multiplication and division).

In contrast, functions that are not of the above form cannot be defined by a finite sequence of basic algebraic operations. Such functions are called *transcendental functions* and are somewhat analogous to irrational numbers, which cannot be evaluated from integers by a finite sequence of the operations of arithmetic. The functions to be discussed in the next two subsections – trigonometric functions, logarithms and exponential functions – are all examples of transcendental functions.

2.2 Trigonometric functions

The trigonometric functions, sine, cosine and others (also called *circular functions*) have many applications. In particular, because of their periodic behaviour, they play a central role in the mathematical description of the phenomena of waves and oscillations that permeate the whole of physical science. Here we discuss their basic properties and some of their important applications in geometry.

2.2.1 Angles and polar co-ordinates

Trigonometry is the study of angles, and before turning to the trigonometric functions themselves, it will be useful to consider

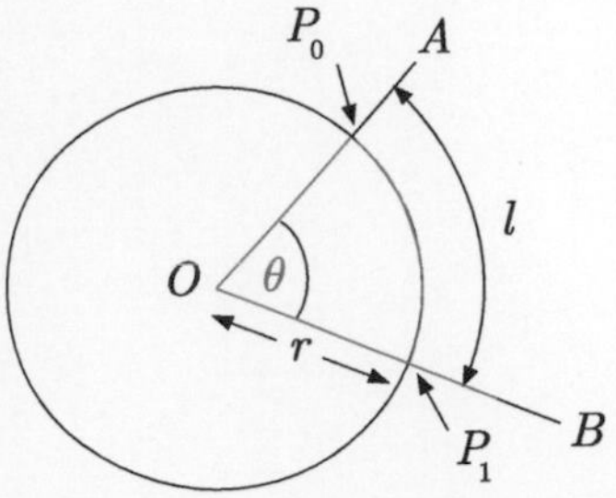

Figure 2.2 Angle, arc and sector.

angles and their use as co-ordinates. In doing so, we will make reference to Figure 2.2, which shows the angle of intersection θ between two lines OA, OB, together with a circle of radius r whose centre lies at the point of intersection.

One unit of angle is the *degree*, which is defined to be a $\frac{1}{360}$ part of a complete rotation and is denoted $1°$. Thus a *right-angle* corresponds to $90°$. In scientific work it is more usual to work in terms of the *radian*, which is defined as the angle when the length l of the arc P_0P_1 is equal to the radius r. Since the circumference of a circle of radius r is $2\pi r$ and corresponds to an angle of 2π radians, it follows that the length of an arc of a circle of radius r that subtends an angle θ at the centre of the circle is

$$\text{arc length} = l = r\theta. \tag{2.21}$$

It also follows that 2π radians = 360 degrees, so that a right angle is $\pi/2$ radians and 1 radian $\approx 57.3°$. In addition, the area of the corresponding sector shown in Figure 2.2 is

$$\text{area of sector} = A = \pi r^2(\theta/2\pi) = \tfrac{1}{2}r^2\theta. \tag{2.22}$$

We stress that, like many other equations in this book, (2.21) and (2.22) are only valid if the angles are expressed in radians. Unless stated otherwise it will be assumed from now on that *all angles are expressed in radians.*

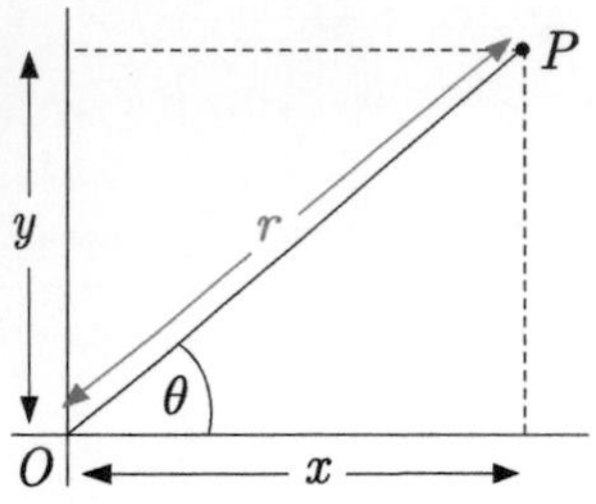

Figure 2.3 Plane polar co-ordinates (r, θ).

Angles can also be used as co-ordinates, provided we adopt a convention to specify their sign. This is illustrated in Figure 2.3, where the position of the point P can be specified by the Cartesian co-ordinates (x, y) used in Section 1.3.1, or the *plane polar co-ordinates* (r, θ). Here, $r > 0$ is the distance of P from the origin O, with

$$r^2 = x^2 + y^2, \tag{2.23}$$

by Pythagoras' theorem, and θ is the angle between the line OP and the x-axis *measured in a counter-clockwise sense.* Thus in Figure 2.4a the point P corresponds to $\theta = -\pi/4$, since OP is at an angle $\pi/4$ to the x-axis when measured in a *clockwise* direction. However, the polar angle is not unique, and P also corresponds to $\theta = 7\pi/4$, since OP is at an angle $7\pi/4$ to the x-axis when measured in the counter-clockwise direction, as shown in Figure 2.4b. In general, the points (r, θ) and $(r, \theta + 2n\pi)$ correspond to the same point in the plane for any integer n. This is illustrated for the case $n = 1$ in Figures 2.4c and 2.4d.

The ambiguity in the value of the polar angle corresponding to a given point can be removed by restricting the range of θ to $0 < \theta < 2\pi$. However, this is not always convenient. Consider, for example, a particle moving in a circular orbit of constant radius r with constant speed v, as shown in Figure 2.5.

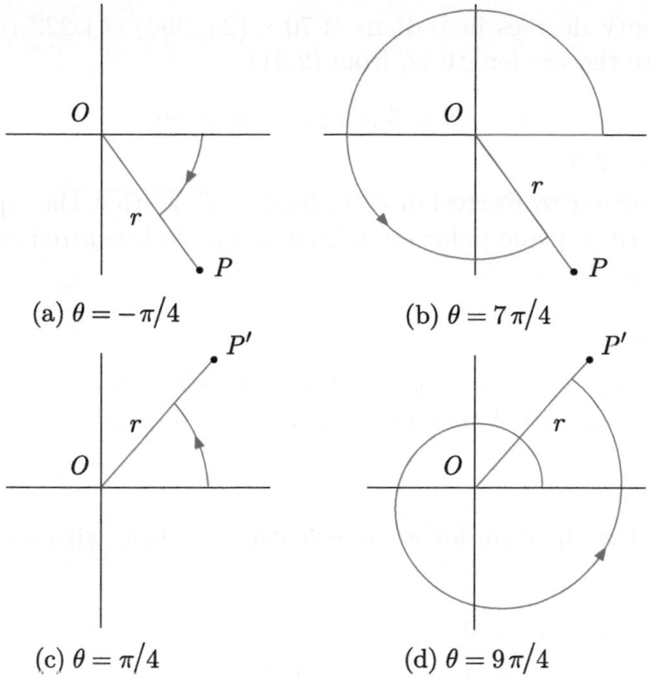

Figure 2.4 Polar angles θ, for constant r; diagrams (a) and (b) represent the same point P, and diagrams (c) and (d) represent the same point P'.

Assuming that $\theta = 0$ at time $t = 0$, the motion is described in polar co-ordinates by the simple equations

$$r = \text{constant}, \quad \theta = vt/r, \tag{2.24}$$

where we have deduced the equation for θ from (2.21) together with the fact that the particle traverses a length of arc $l = vt$ in time t. The angle increases indefinitely as t increases and $\theta = 2n\pi + \phi$, with $0 < \phi < 2\pi$, corresponds to the particle arriving at the point (r, ϕ) after n complete revolutions since $t = 0$.

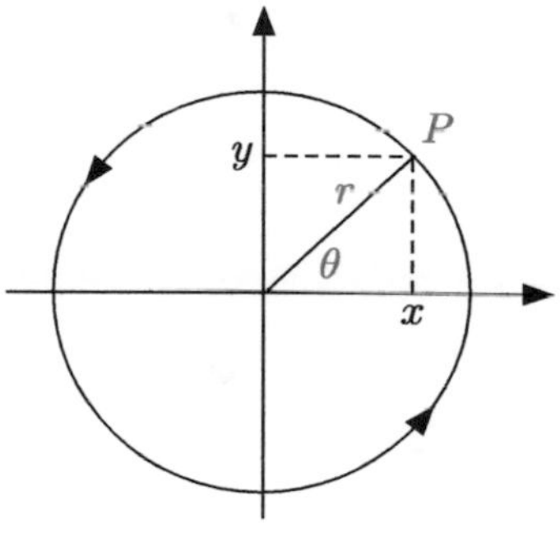

Figure 2.5 A particle/point moving counter-clockwise in a circular trajectory with constant radius r.

Example 2.6

(a) Find the angles 45°, 60°, 90°, 150° and 180° in radians, expressing your answers in multiples of π.

(b) A circle has a radius of 5 cm. What is the length of arc subtended by an angle of 70°? Give your answer to three significant figures.

Solution

(a) Since $360° = 2\pi$ radians, α degrees $= 2\pi\alpha/360$ radians, giving $\pi/4$, $\pi/3$, $\pi/2$, $5\pi/6$ and π for the angles 45°, 60°, 90°, 150° and 180°, respectively.

(b) Seventy degrees in radians is $70 \times (2\pi/360) = 1.222$ radians. Therefore the arc length is, from (2.21),

$$l = r\,\theta = 5 \times 1.222 = 6.11 \text{ cm}.$$

Example 2.7

By converting to Cartesian co-ordinates, show that the equation $r = 2b\sin\theta$ in plane polar co-ordinates is a circle centred on $(0, b)$ with radius b.

Solution

Substituting $r = \sqrt{x^2 + y^2}$ and $\sin\theta = y/r$ gives $r^2 = 2by$ or $x^2 + y^2 - 2by = 0$. This may be rearranged to give

$$x^2 + (y - b)^2 = b^2,$$

which is the equation for a circle centred on $(0, b)$ with radius b.

2.2.2 Sine and cosine

For angles less than $\pi/2$, the *sine* and *cosine* functions (written 'sine' and 'cosine') are defined in terms of the sides of a right-angled triangle by

$$\text{sine} \equiv \text{length of opposite side/length of hypotenuse}$$

$$\text{cosine} \equiv \text{length of adjacent side/length of hypotenuse},$$

and applying this to the triangle in Figure 2.3 we obtain

$$\sin\theta = y/r, \qquad \cos\theta = x/r, \tag{2.25}$$

where x and y are the projections of OP onto the x-axis and y-axis, respectively, and r is the length of OP. However, if we consider a point P rotating in a counter-clockwise direction about the origin at $(0, 0)$, as shown in Figure 2.5, then (2.25) allows us to extend the definitions of sine and cosine to all angles provided the signs of x and y are taken into account. For example, in the fourth quadrant $(3\pi/2 < \theta < 2\pi)$, we see from Figure 2.4(a) that $x > 0$ while $y < 0$, so that $\cos\theta > 0$ and $\sin\theta < 0$. More generally, as θ increases in Figure 2.5, one sees that x and y oscillate between r and $-r$, and hence $\cos\theta$ and $\sin\theta$ oscillate between -1 and $+1$, with a period of 2π corresponding to a single revolution. In other words,

$$-1 < \sin\theta,\ \cos\theta < 1 \tag{2.26}$$

and they are *periodic* with a period of 2π, that is, the form of the function repeats at intervals of 2π, so that

$$\sin(\theta + 2\pi) = \sin\theta, \quad \cos(\theta + 2\pi) = \cos\theta. \tag{2.27}$$

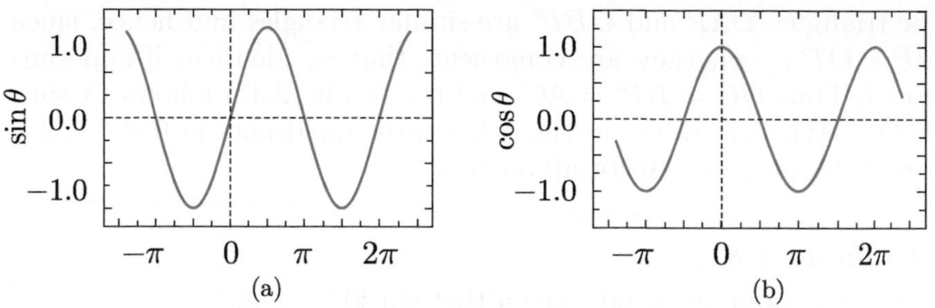

Figure 2.6 The circular functions $\sin\theta$ and $\cos\theta$ as functions of θ.

In addition, together with (2.23), the definitions (2.25) imply the important relation

$$\sin^2\theta + \cos^2\theta = 1 \qquad (2.28)$$

for all values of θ.

The graphical forms of the sine and cosine functions follow from the definitions (2.25), together with the construction in Figure 2.5. They are shown in Figures 2.6a and 2.6b, respectively, and have a number of other general features which, in view of the enormous importance of these functions in physical science, are worth emphasising.

(i) If we replace θ by $-\theta$ in Figure 2.5, then $y \to -y$, but x remains unchanged. Hence from (2.25), $\sin\theta$ and $\cos\theta$ are odd and even functions, respectively:

$$\sin(-\theta) = -\sin\theta \quad \text{and} \quad \cos(-\theta) = \cos\theta. \qquad (2.29)$$

(ii) One sees from (2.28) and Figure 2.5 that y, and hence $\sin\theta$, vanishes when $\theta = 0$ and π. Similarly $y = r$ and $-r$, and hence $\sin\theta = 1$ and -1 at $\theta = \pi/2$ and $3\pi/2$, respectively. Together with (2.27) this implies the results

$$\sin(n\pi) = 0 \quad \text{and} \quad \sin\left[\left(\frac{2n+1}{2}\right)\pi\right] = (-1)^n. \qquad (2.30a)$$

for any integer n. The corresponding results for the cosine function are

$$\cos(n\pi) = (-1)^n \quad \text{and} \quad \cos\left[\left(\frac{2n+1}{2}\right)\pi\right] = 0. \qquad (2.30b)$$

(iii) By inspection, it can be seen that the forms of the sine and cosine curves in Figure 2.6 are the same but displaced by a distance $\pi/2$ along the θ-axis. So we deduce that

$$\sin\theta = \cos(\theta - \pi/2). \qquad (2.31)$$

For the first quadrant, $0 < \theta < \pi/2$, this result follows from the construction of Figure 2.7, where P and P' correspond to polar angles θ and $(\theta - \pi/2)$. From this diagram, one easily sees that

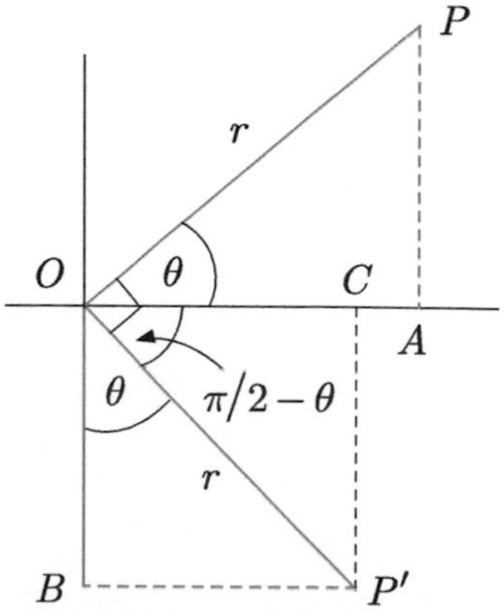

Figure 2.7 The construction used to establish the result (2.31).

the triangles OAP and OBP' are similar triangles and hence, since $OP = OP'(= r)$, they are congruent, that is, identical if superimposed. Thus $OC = BP' = AC$, and the result (2.31) follows. A similar construction works in the other three quadrants in $0 < \theta < 2\pi$, establishing the result for all angles.

Example 2.8

Find the value of $\sin 60°$, given that $\sin 30° = 1/2$.

Solution

If the angles are measure in degrees, equation (2.31) becomes $\sin\theta = \cos(\theta - 90°)$, so that $\cos(-60°) = \sin 30° = 1/2$. Hence $\cos 60° = 1/2$, since $\cos\theta$ is an even function, by (2.29). The value $\sin 60° = \sqrt{3}\big/2$ then follows from (2.28) since both sine and cosine are positive in the first quadrant, $0° < \theta < 90°$.

2.2.3 More trigonometric functions

Sine and cosine are not the only important circular functions, but the others can be defined in terms of them. In particular, we define the *tangent* and *cotangent*, written as 'tan' and 'cot', respectively, as

$$\tan\theta \equiv \frac{\sin\theta}{\cos\theta} \quad \text{and} \quad \cot\theta \equiv \frac{\cos\theta}{\sin\theta} = \frac{1}{\tan\theta} \tag{2.32a}$$

and the *secant* and *cosecant*, written as 'sec' and 'cosec', by

$$\sec\theta \equiv \frac{1}{\cos\theta} \quad \text{and} \quad \operatorname{cosec}\theta \equiv \frac{1}{\sin\theta}, \tag{2.32b}$$

which, together with (2.28), lead to the relations

$$1 + \tan^2\theta = \sec^2\theta, \quad \text{and} \quad 1 + \cot^2\theta = \operatorname{cosec}^2\theta. \tag{2.33}$$

The behaviours of these functions follow from the behaviour of sine and cosine shown in Figures 2.6a and 2.6b. The functions $\tan\theta$ and $\cot\theta$ are plotted in Figures 2.8a and 2.8b, respectively.

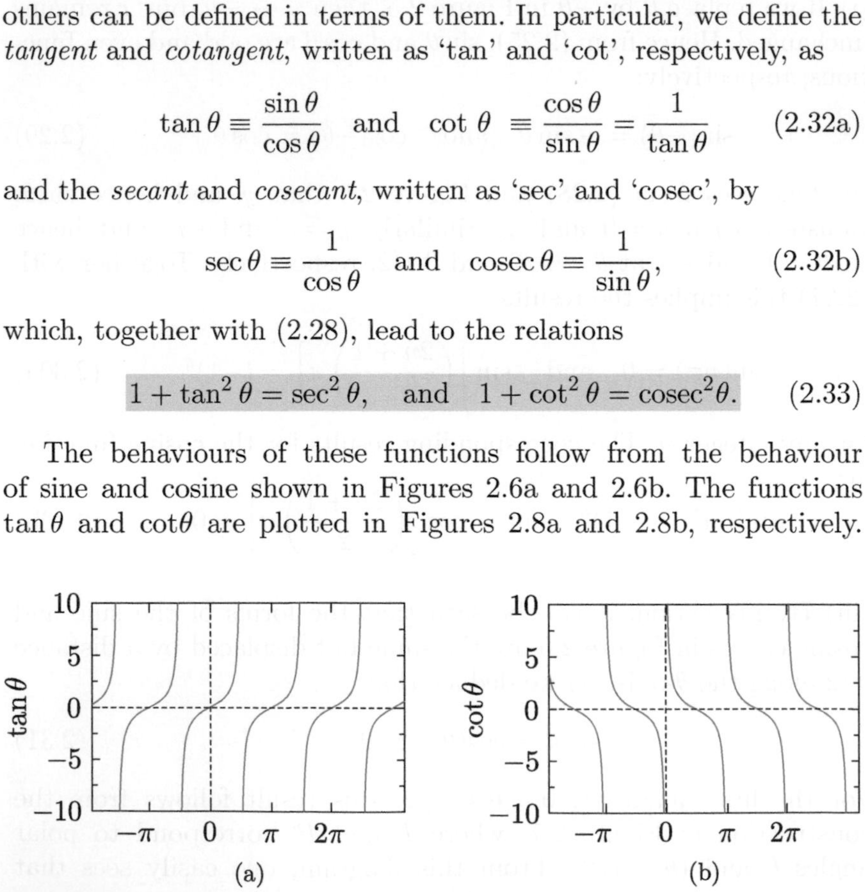

Figure 2.8 The circular functions (a) $\tan\theta$ and (b) $\cot\theta$ as functions of θ.

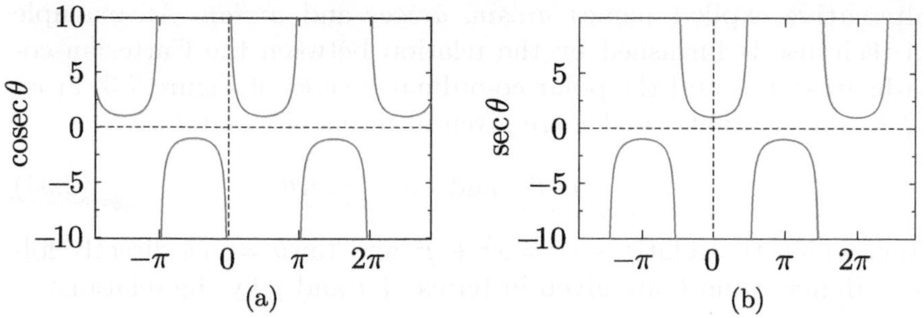

Figure 2.9 The circular functions (a) $\mathrm{cosec}\,\theta$ and (b) $\sec\theta$ as functions of θ.

Like the sine and cosine functions, they are periodic, but with a period of π rather than 2π. However, unlike those functions, $\tan\theta$ is unbounded and is discontinuous at the points $\theta = (2n+1)\pi/2$, for $n = 0, \pm 1, \pm 2, \ldots$, where $\cos\theta$ vanishes. Similarly, $\cot\theta$ is discontinuous at the points where $\sin\theta$ vanishes. The remaining circular functions may also be deduced from (2.28) and are shown in Figure 2.9.

In Section 1.3.1 we defined inverse functions. In the case of the circular functions this must be done with care, because it is clear from Figures 2.6, 2.8 and 2.9 that there are an infinite number of angles for a given value of sine, cosine or tangent. To obtain a single-valued function, we would therefore have to formally restrict the angular range of θ. The corresponding inverse circular functions for sine, cosine and tangent are shown for convenient choices in Figures 2.10a–2.10c, respectively. Using the notation of Section 1.3, it would be natural to refer to these as $\sin^{-1}$, $\cos^{-1}$ and $\tan^{-1}$, but to avoid ambiguity with $1/\sin$, etc. it is probably better to always use their

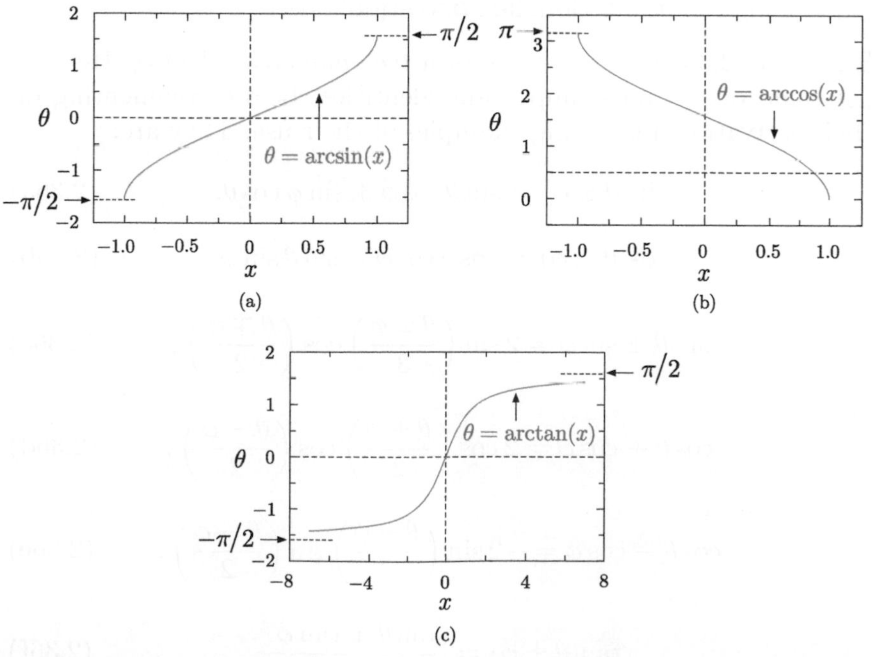

Figure 2.10 Inverse circular functions: (a) $\arcsin x$, (b) $\arccos x$ and (c) $\arctan x$ as functions of x.

alternative explicit names *arcsin*, *arccos* and *arctan*. An example of their use is furnished by the relation between the Cartesian co-ordinates (x,y) and the polar co-ordinates (r,θ) of Figure 2.3. From (2.25) we see that x and y are given in terms of r and θ by

$$x = r\cos\theta \quad \text{and} \quad y = r\sin\theta, \tag{2.34}$$

from which the relations $r^2 = x^2 + y^2$ and $\tan\theta = x/y$ directly follow. Hence r and θ are given in terms of x and y by the relations

$$r = +\sqrt{(x^2+y^2)} \quad \text{and} \quad \theta = \arctan(y/x), \tag{2.35}$$

respectively.

Example 2.9

Prove that $\sec\theta = -\mathrm{cosec}\ (\theta - \pi/2)$.

Solution

From (2.32b),

$$\sec\theta = \frac{1}{\cos\theta} = \frac{1}{\sin(\theta+\pi/2)},$$

which, using (2.30) and (2.32), is

$$\sec\theta = \frac{1}{\sin[(\theta-\pi/2)+\pi]} = \frac{1}{-\sin(\theta-\pi/2)} = -\mathrm{cosec}\ (\theta-\pi/2).$$

2.2.4 Trigonometric identities and equations

Equation (2.33) is an example of a *trigonometric identity*. Here we list some of the most important identities, before commenting on their derivation and giving examples of their use. They are:

$$\sin(\theta \pm \phi) = \sin\theta\cos\phi \pm \sin\phi\cos\theta, \tag{2.36a}$$

$$\cos(\theta \pm \phi) = \cos\theta\cos\phi \mp \sin\theta\sin\phi, \tag{2.36b}$$

$$\sin\theta \pm \sin\phi = 2\sin\left(\frac{\theta\pm\phi}{2}\right)\cos\left(\frac{\theta\mp\phi}{2}\right), \tag{2.36c}$$

$$\cos\theta + \cos\phi = 2\cos\left(\frac{\theta+\phi}{2}\right)\cos\left(\frac{\theta-\phi}{2}\right), \tag{2.36d}$$

$$\cos\theta - \cos\phi = -2\sin\left(\frac{\theta+\phi}{2}\right)\sin\left(\frac{\theta-\phi}{2}\right), \tag{2.36e}$$

$$\tan(\theta \pm \phi) = \frac{\tan\theta \pm \tan\phi}{1 \mp \tan\theta\tan\phi}. \tag{2.36f}$$

Specific useful cases that follow directly from (2.36a) (2.36b) and (2.36f) are the 'double-angle' formulas obtained by setting $\phi = \theta$:

$$\sin 2\theta = 2\sin\theta\cos\theta, \quad \cos 2\theta = 1 - 2\sin^2\theta = 2\cos^2\theta - 1 \quad (2.37a)$$

and

$$\tan 2\theta = \frac{2\tan\theta}{1-\tan^2\theta}. \quad (2.37b)$$

The analogous 'half-angle' formulas are

$$\sin\theta = 2\sin(\theta/2)\cos(\theta/2), \quad \cos\theta = 1 - 2\sin^2(\theta/2) = 2\cos^2(\theta/2) - 1 \quad (2.37c)$$

and

$$\tan\theta = \frac{2\tan(\theta/2)}{1-\tan^2(\theta/2)}. \quad (2.37d)$$

These identities can be proved by simple geometrical methods. To illustrate this we will prove (2.36a) by referring to Figure 2.11. From triangle ABC, we have

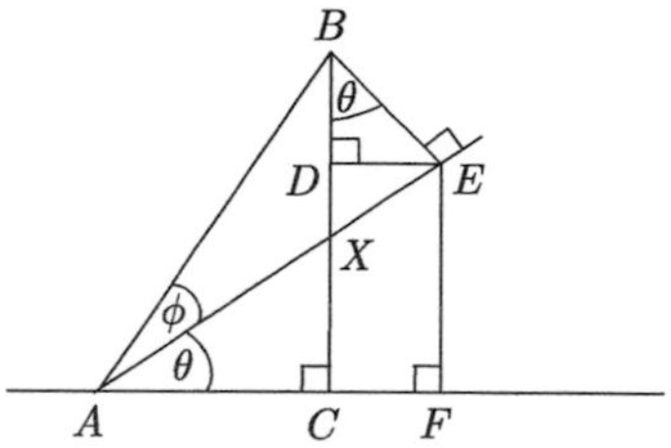

Figure 2.11 Construction to prove the identity (2.36a).

$$\sin(\theta+\phi) = \frac{BC}{AB} = \frac{DC+BD}{AB} = \frac{EF+BD}{AB} = \frac{EF}{AB} + \frac{BD}{AB}, \quad (2.38a)$$

since $DC = EF$. But from the triangles BDE and AEF,

$$BD = BE\cos\theta \quad \text{and} \quad EF = AE\sin\theta, \quad (2.38b)$$

so that

$$\sin(\theta+\phi) = (AE/AB)\sin\theta + (BE/AB)\cos\theta. \quad (2.38c)$$

Also, from triangle ABE,

$$\sin\phi = BE/AB \quad \text{and} \quad \cos\phi = AE/AB \quad (2.38d)$$

so finally, using these relations in (2.38c), we have

$$\sin(\theta+\phi) = \sin\theta\cos\phi + \cos\theta\sin\phi. \quad (2.38e)$$

This derivation establishes (2.38e) for acute angles only, since this is what we have assumed in Figure 2.11. However, the proof can be extended to all angles. The result (2.36a) with a minus sign follows by letting $\phi \to -\phi$ and using the odd and even properties of the sine and cosine functions (2.29).

The rest of the formulas (2.36) follow from (2.30a) using our previous results. For example, to derive (2.36b) we write

$$\begin{aligned}\cos(\theta\pm\phi) &= \sin[(\theta\pm\pi/2)+\phi] \\ &= \sin(\theta\pm\pi/2)\cos\phi \pm \cos(\theta\pm\pi/2)\sin\phi \\ &= \cos\theta\cos\phi \pm \sin(\theta+\pi)\sin\phi,\end{aligned}$$

using (2.30b) and (2.36a). Equation (2.36b) then follows, since

$$\sin(\theta + \pi) = \sin\theta\cos\pi + \cos\theta\sin\pi = -\sin\theta,$$

using (2.36a) and (2.29). The other identities follow in a similar way, and can be used to derive many more results, and solve trigonometric equations, as the following examples illustrate.

Example 2.10

Prove the relation $\sin 3\theta = 3\sin\theta - 4\sin^3\theta$.

Solution

Using (2.36a) we obtain

$$\sin 3\theta = \sin(2\theta + \theta) = \sin 2\theta\cos\theta + \cos 2\theta\sin\theta.$$

Then using the double-angle formula (2.37a), we have

$$\begin{aligned}\sin 3\theta &= 2\sin\theta\cos^2\theta + (1 - 2\sin^2\theta)\sin\theta\\ &= 2\sin\theta(1 - \sin^2\theta) + (1 - 2\sin^2\theta)\sin\theta\\ &= 3\sin\theta - 4\sin^3\theta.\end{aligned}$$

Example 2.11

Prove the identity

$$\tan(x + y) - \tan(y) = \frac{\sin x}{\cos y\cos(x + y)}.$$

Solution

Using (2.36f) on the left-hand side we obtain

$$\frac{\tan x + \tan y}{1 - \tan x\tan y} - \tan y = \frac{\tan x(1 + \tan^2 y)}{1 - \tan x\tan y}.$$

Then, using the identities

$$\tan x = \sin x/\cos x \quad \text{and} \quad 1 + \tan^2 x = \sec^2 x = 1/\cos^2 x$$

gives

$$\frac{\sin x}{\cos y(\cos x\cos y - \sin x\sin y)} = \frac{\sin x}{\cos y\cos(x + y)},$$

where we have used (2.36b) in the final step.

Example 2.12

Solve the equation $\sin\theta - \sin 2\theta + \sin 3\theta = 0$ for $0 < \theta < 2\pi$.

Solution

Combining the first and third terms using (2.36c), and using (2.37a) gives

$$2\sin 2\theta\cos\theta - 2\sin\theta\cos\theta = 0.$$

So the first solution is

$$\cos\theta = 0 \Rightarrow \theta = \pi/2,\ 3\pi/2.$$

If $\cos\theta \neq 0$, then we may divide by $\cos\theta$ and use (2.37a) to give $\sin\theta(2\cos\theta - 1) = 0$. The two possibilities are

$$\sin\theta = 0 \Rightarrow \theta = \pi \quad \text{or} \quad \cos\theta = \tfrac{1}{2} \Rightarrow \theta = \pi/3, 5\pi/3.$$

So $\theta = \pi/3, \pi/2, \pi, 3\pi/2$ and $5\pi/3$.

Example 2.13

Derive the identity

$$\arctan x + \arctan\left(\frac{1-x}{1+x}\right) = \frac{\pi}{4} + n\pi,$$

where n is an integer.

Solution

If we set

$$\arctan x = \theta \Rightarrow \tan\theta = x \quad \text{and}$$
$$\arctan\left(\frac{1-x}{1+x}\right) = \phi \Rightarrow \tan\phi = \left(\frac{1-x}{1+x}\right).$$

Then using (2.36f) in $\tan(\theta+\phi)$, we have

$$\tan(\theta+\phi) = \left(\frac{\tan\theta + \tan\phi}{1 - \tan\theta\tan\phi}\right) = \frac{x + (1-x)/(1+x)}{1 - x(1-x)/(1+x)} = 1.$$

Hence $\theta + \phi = \arctan 1 = \pi/4$ and

$$\arctan x + \arctan\left(\frac{1-x}{1+x}\right) = \frac{\pi}{4} + n\pi.$$

2.2.5 Sine and cosine rules

The trigonometric functions enable the discussion of co-ordinate geometry to be extended in a number of ways. One is to solve triangles, that is, to determine completely the lengths of their sides and the magnitude of all three angles. If two angles and one side, or two sides and a non-included angle are given, this can be done using the *sine rule*,

$$\frac{a}{\sin A} = \frac{b}{\sin B} = \frac{c}{\sin C}, \tag{2.39}$$

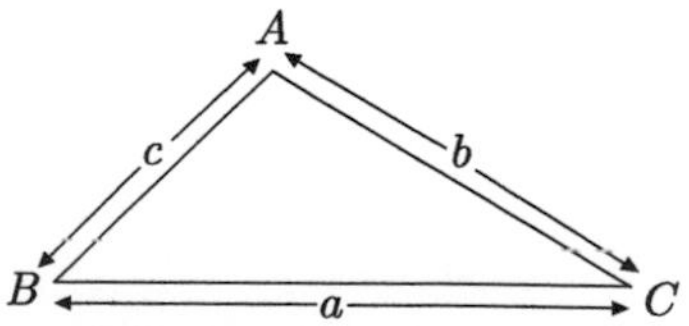

Figure 2.12 Definition of the angles and lengths of the sides for the sine and cosine rules. The angles are labelled by the same letters A, B, C as the vertices, while the opposing sides are labelled by the corresponding lower case letters a, b, c, respectively.

where the definitions of the angles A, B and C and lengths of the sides a, b and c are specified in Figure 2.12.

Alternatively, if the three sides, or two sides and the included angle are known, we can use the *cosine rule*,

$$\cos A = \frac{b^2 + c^2 - a^2}{2bc}, \tag{2.40a}$$

together with its permutations

$$\cos B = \frac{c^2 + a^2 - b^2}{2ca} \quad \text{and} \quad \cos C = \frac{a^2 + b^2 - c^2}{2ab}. \tag{2.40b}$$

In what follows we shall prove these rules then illustrate their use by examples.

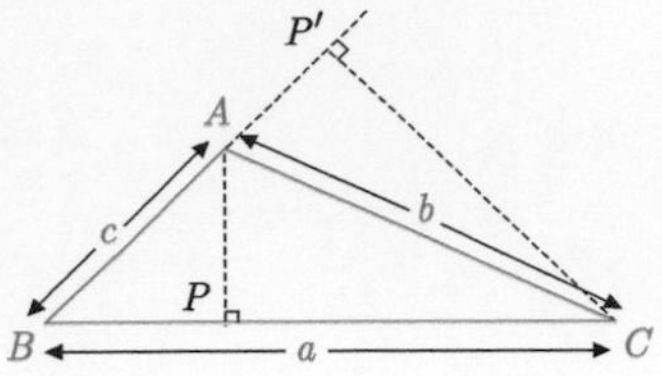

Figure 2.13 Construction to prove the sine and cosine rules.

The sine and cosine rules can both be derived from the construction of Figure 2.13. To obtain the sine rule, we infer the length AP by using the definition of sine in the triangles PAB and PAC to give

$$c \sin B = AP = b \sin C, \tag{2.41a}$$

from which the second equality in (2.39) immediately follows. Applying the same argument to the triangles BCP' and ACP' gives

$$a \sin B = CP' = b \sin(\pi - A) = b \sin A, \tag{2.41b}$$

where we have used (2.36a) and (2.30a) to show that $\sin(\pi - A) = \sin A$. The first equality in (2.39) then follows directly, completing the derivation of the sine rule.

To prove the cosine rule (2.40a), we apply Pythagoras' theorem to the triangle $P'BC$ to obtain

$$\begin{aligned} a^2 &= (P'C)^2 + (P'B)^2 \\ &= [b \sin(\pi - A)]^2 + [c + b \cos(\pi - A)]^2 \\ &= b^2 + c^2 + 2bc \ \cos(\pi - A), \end{aligned}$$

where we have multiplied out the brackets and used (2.31). Since $\cos(\pi - A) = -\cos A$, from (2.36b) and (2.30b), this gives

$$a^2 = b^2 + c^2 - 2bc \cos A$$

and the cosine rule (2.40a). In a similar way, applying Pythagoras' theorem to the triangle PAC gives

$$\begin{aligned} b^2 = (AP)^2 + (CP)^2 &= [c \sin B]^2 + [a - c \cos(B)]^2 \\ &= c^2 + a^2 - 2ac \cos(B), \end{aligned}$$

thus establishing the first part of (2.40b). The second equation in (2.40b) follows by the same argument applied to the triangle PAB.

Finally, before giving examples of the application of the sine and cosine rules, we note that in proving them we have assumed that one of the angles A is obtuse.[2] The corresponding proofs for the case of three acute angles are very similar and are left as exercises for the reader.

[2] It is because of this that we have treated (2.40a) and (2.40b) separately, rather than assuming that (2.40a) implies (2.40b).

Example 2.14

Solve the triangle shown in Figure 2.12, for the two cases: (a) $b = 17$ cm, $A = 0.61$ rad and $B = 1.33$ rad; and (b) $a = 16$ cm, $b = 7$ cm and $C = 1.20$ rad.

Solution

(a) Since $A + B + C = \pi(\text{i.e } 180^\circ), C = 1.20$ rad. Then, using the sine rule (2.40)

$$a/\sin A = b/\sin B, \text{that is, } a = b\sin A/\sin B = 10.03 \text{ cm.}$$

Likewise,

$$c = b\sin C/\sin B = 16.32 \text{ cm.}$$

(b) Using the cosine rule,

$$c^2 = a^2 + b^2 - 2ab\cos C,$$

we obtain $c = 11.56$ cm. Likewise, from (2.40a),

$$\cos A = (b^2 + c^2 - a^2)/2bc = -0.45333,$$

which gives $A = 2.04$ rad. Finally, since the angles of a triangle sum to π,

$$B = \pi - 2.04 - 0.70 = 0.40 \text{ rad.}$$

Example 2.15

Prove that the length l of the chord subtended by an angle θ on a circle of radius r is given by $l = 2r\sin(\theta/2)$.

Solution

Using the cosine rule in Figure 2.14 gives

$$l^2 = r^2 + r^2 - 2r^2\cos\theta = 2r^2(1 - \cos\theta) = 4r^2\sin^2(\theta/2),$$

where we have used the double-angle formula (2.36). Therefore, $l = 2r\sin(\theta/2)$, as required.

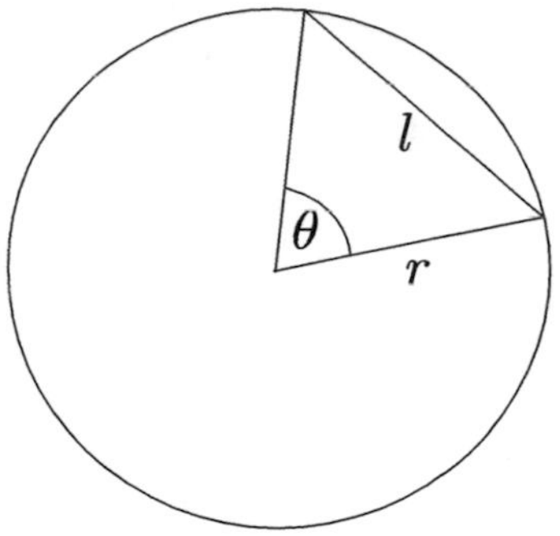

Figure 2.14

2.3 Logarithms and exponentials

In this section we first define *logarithms* with respect to an arbitrary base and obtain the 'laws of logarithms'. We then introduce the irrational number e as a favoured base, in order to discuss *natural logarithms* and *exponentials*, and the *hyperbolic functions* defined in terms of them.

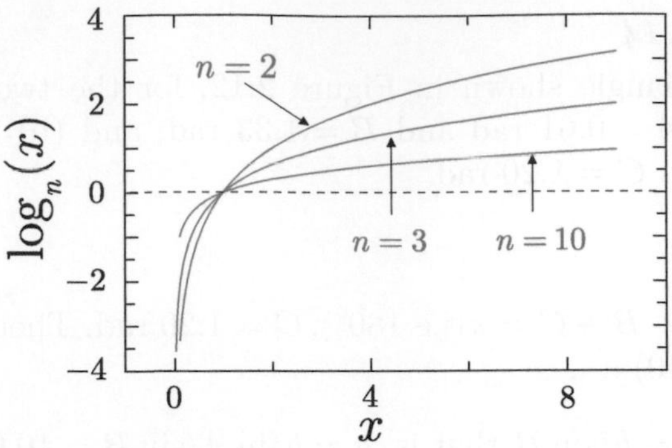

Figure 2.15 Plots of $\log_n(x)$ for $n = 2,\ 3$ and 10.

2.3.1 The laws of logarithms

In Section 1.1.2 we met expressions of the form a^b, where the index, or power, b was a rational number. For $a > 0$, this definition can be extended to irrational numbers, since, for example, 2^π can be evaluated to arbitrary precision by exploiting the fact that π itself, like any irrational number, can be approximated to arbitrary accuracy by a rational number.[3] Since we wish to include irrational numbers in our discussion, we restrict ourselves to the case $a > 0$, when the expression $c = a^b$ is called an *exponential* expression with a the *base* and b the *index*. Conversely, b is called the *logarithm of c to base a* and is written $b = \log_a c$. To summarise,

$$c = a^b \Leftrightarrow b = \log_a c, \tag{2.42}$$

where the symbol $\Leftrightarrow$ means 'implies and is implied by', that is, the expressions on either side of the symbol are equivalent. Graphs of $\log_a x$ for various integer values of a are shown in Figure 2.15.

Logarithms obey a number of laws that are easily derived from the basic result (2.42). If we set

$$\log_a A = A', \quad \log_a B = B', \quad \log_a(AB) = C, \tag{2.43a}$$

then

$$A = a^{A'}, \quad B = a^{B'}, \quad AB = a^C \tag{2.43b}$$

and hence, from the result on indices (1.5), $A' + B' = C$, i.e.

$$\log_a A + \log_a B = \log_a(AB) \tag{2.44a}$$

[3]For $a < 0$ this procedure is ambiguous. For example, $(-1)^\pi = -1$ if we use the approximation $\pi = 22/7$, but $(-1)^\pi = +1$ if we use the identical approximation $\pi = 220/70$. This is because in the former case, $(-1)^\pi$ would become $(-1)^{22}(-1)^{-7} = (-1)^{15} = -1$, since 15 is an odd number, whereas using the approximation $\pi = 220/70$ would lead to an even exponent and hence the result $+1$. We will not discuss this possibility further.

and in general

$$\log_a A + \log_a B + \log_a C + \cdots = \log_a (ABC \ldots) \qquad (2.44b)$$

By setting $A = B = C \ldots$ in (2.44b), it follows that

$$\log_a (A^n) = n \log_a A, \qquad (2.44c)$$

a result that holds also for fractional and negative values of n. In a similar way to the proof of (2.44b), we can show that

$$\log_a A - \log_a B = \log_a \left(\frac{A}{B} \right). \qquad (2.44d)$$

Finally, setting $A = B$ in (2.44d) gives $\log_a 1 = 0$ for any base a. The results (2.44) are referred to as the *laws of logarithms*.

These relations may be used to simplify expressions and solve equations involving logarithms, as we shall illustrate by examples below. They may also be used to derive the general formula for changing a logarithm from base a to base b. Thus, if $\log_a c = x$, then $c = a^x$. So $\log_b c = x \log_b a$ and

$$\log_a c = \frac{\log_b c}{\log_b a}, \qquad (2.45)$$

which, for the special case $b = c$, implies

$$\log_b a = 1/\log_a b. \qquad (2.46)$$

Because the decimal system is so widespread, logarithms to base 10 are called *common logarithms* and the base is usually omitted. For example, $\log 7 = 0.845$. In the binary system, it would be equally appropriate to use base 2, when

$$\log_2 c = \frac{\log c}{\log 2} = 3.32 \log c$$

by (2.46). However, it is usual instead to choose the irrational number $e = 2.71828\ldots$ as the base, for reasons to be explained in the next section.

Example 2.16

Simplify the expression

$$\log \left(\frac{5x}{3} \right) + \frac{1}{3} \log(27x) - \log \left(\frac{1}{x} \right),$$

and write it as a single logarithm.

Solution

Using (2.44), this is

$$\begin{aligned}&\log(5x/3) + \tfrac{1}{3}\log(27x) - \log(1/x)\\ &\quad = \log 5 + \log x - \log 3 + \tfrac{1}{3}\log x + \log 3 - \log 1 + \log x\\ &\quad = \tfrac{7}{3}\log x + \log 5 = \log(5x^{7/3}).\end{aligned}$$

Example 2.17

Solve the equation

$$\log(5-t) + \log(5+t) = 1.3.$$

Solution

Using (2.44), this is

$$\log[(5-t)(5+t)] = \log(25-t^2) = 1.3,$$

which, from (2.42) implies

$$25 - t^2 = 10^{1.3},$$

and hence

$$t = \pm\sqrt{25 - 10^{1.3}} = \pm 2.2466\ldots$$

2.3.2 Exponential function

We next consider the *exponential function* a^x, where again $a > 0$, but now the exponent is a real variable x. The resulting function is plotted for the values $a = 1/2$, 1, 3/2 and 2 in Figure 2.16. As can be seen, a^x increases rapidly for large positive x if $a > 1$, but decreases for all $a < 1$. In addition, $a^x = 1$ for $x = 0$ for all x, and the behaviours for positive and negative x are related by

$$a^{-x} = \frac{1}{a^x} = \left(\frac{1}{a}\right)^x, \tag{2.47}$$

so that, for example, the curves 2^{-x} and $(1/2)^x$ are identical.

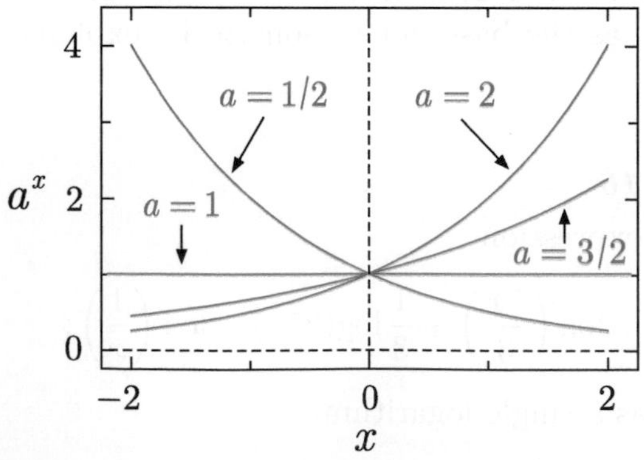

Figure 2.16 The exponential function a^x for $a = 1/2$, 1, 3/2 and 2.

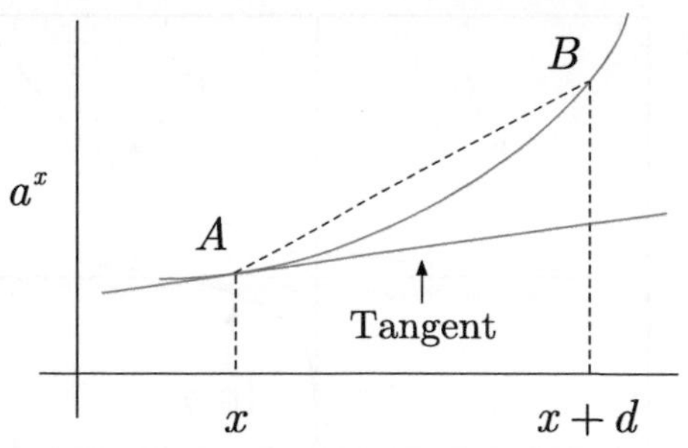

Figure 2.17 Construction to show that the exponential function is proportional to its gradient.

Perhaps the most important property of the exponential function is that it is proportional to its own gradient. To see this, consider the line AB joining the function a^x at x and $x+d$, as shown in Figure 2.17. As can be seen, the gradient of this line becomes a better and better approximation to the gradient at x itself as d gets increasingly smaller. Hence, since

$$\text{slope}(AB) = \frac{a^{x+d} - a^x}{d} = a^x \left(\frac{a^d - 1}{d} \right), \tag{2.48}$$

by (1.45) and (1.17), we immediately obtain the desired result

$$\text{slope } (a^x) = k a^x, \tag{2.49}$$

where the constant of proportionality

$$k = \lim_{d \to 0} \left(\frac{a^d - 1}{d} \right) \tag{2.50}$$

and the notation means 'take the limit of the term in the brackets as d approaches zero'.[4]

At this point, we note that the constant k depends on the base a, and we define the *Euler number* e such that $k = 1$ for $a = e$. To find this number, for any given d, we choose a value

$$a = (1+d)^{1/d}, \tag{2.51}$$

so that (2.48) gives

$$\text{slope } (AB) = a^x \left(\frac{a^d - 1}{d} \right) = a^x$$

for any given d. Since as d approaches zero, the slope (AB) approaches the slope of the curve, this implies that

$$\text{slope}(e^x) = e^x \tag{2.52}$$

[4]The concept of a limiting value will be discussed in more detail in Chapter 3.

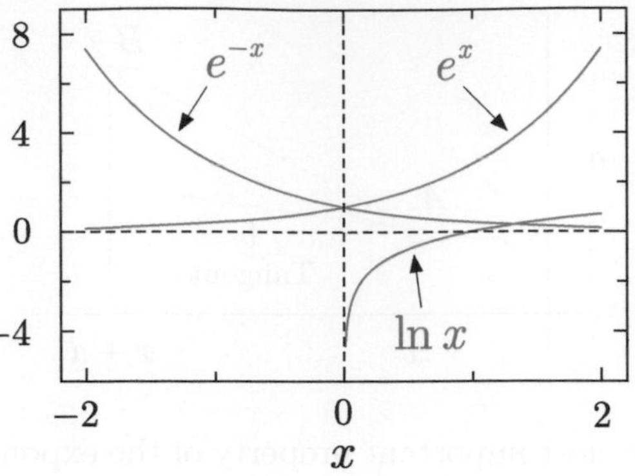

Figure 2.18 The functions e^x, e^{-x} and $\ln x$.

as required, if

$$e \equiv \lim_{d \to 0} (1 + d)^{1/d}. \tag{2.53}$$

The number e can now be estimated with increasing accuracy by choosing smaller and smaller d, the values $d = 0.1$, 0.01, 0.001, 0.0001, ... giving 2.594, 2.705, 2.717, 2.718, ... A better method for evaluating Euler's number will be given in Section 5.3.4, and to 6 significant figures

$$e = 2.71828\ldots \tag{2.54}$$

The corresponding behaviour of e^x and of the closely related function $e^{-x} = (e^x)^{-1}$ is shown in Figure 2.18.

Because of the property (2.52), the number e is almost always chosen as a base in physical science work and the corresponding function $\exp(x)$, defined by

$$\exp(x) \equiv e^x, \tag{2.55}$$

is called the *natural exponential function*, or more usually, but imprecisely, just the *exponential function*. The corresponding inverse function

$$\ln x \equiv \log_e x, \tag{2.56}$$

is referred to as the *natural logarithmic function*, or simply the *natural logarithm*. Since it is the inverse of e^x, its behaviour can be inferred from the plot of e^x, and is shown in Figure 2.18. Finally, from (2.45), natural and common logarithms are related by

$$\ln x = \frac{\log x}{\log e} = (2.303\ldots) \log x.$$

Example 2.18

The decay of a radioactive substance is governed by the exponential decay law

$$\frac{N(t)}{N_0} = \exp\left(-\frac{t}{\tau}\right),$$

where N_0 is the initial number of atoms at time $t = 0$, $N(t)$ the number remaining after time t, and the constant τ is the lifetime. The time after which half the initial sample has decayed is called the half-life, denoted $\tau_{1/2}$. Show that $\tau_{1/2} = \tau \ln 2$ and hence that

$$\frac{N(t)}{N_0} = \left(\frac{1}{2}\right)^{\tau/\tau_{1/2}}.$$

Carbon-14, used in dating organic artefacts, has a half-life of 5730 years. How much time would have to elapse before 99% of a specimen of carbon-14 has decayed?

Solution

At $t = \tau_{1/2}$, $N(t) = N_0/2$, so that

$$\frac{1}{2} = \exp\left(-\frac{\tau_{1/2}}{\tau}\right) \Rightarrow -\frac{\tau_{1/2}}{\tau} = \ln\left(\frac{1}{2}\right) = -\ln 2.$$

Hence $\tau_{1/2} = \tau \ln 2$, as required. To derive the second result, take the logarithm of the exponential decay law, giving

$$\ln\left(\frac{N(t)}{N_0}\right) = -\frac{t}{\tau}.$$

Then using (2.45) to change the base from e to $\frac{1}{2}$ gives

$$\ln_{1/2}\left(\frac{N}{N_0}\right) = \frac{\ln(N/N_0)}{\ln(1/2)} = \frac{t}{\tau/\ln 2} = \frac{t}{\tau_{1/2}},$$

and therefore

$$\frac{N(t)}{N_0} = \left(\frac{1}{2}\right)^{\tau/\tau_{1/2}}.$$

Taking logarithms of this equation gives the time elapsed as

$$t = \tau_{1/2}\frac{\ln(N/N_0)}{\ln(1/2)} = 5730 \times \frac{\ln(0.01)}{\ln(0.5)} = 38069 \text{ yrs.}$$

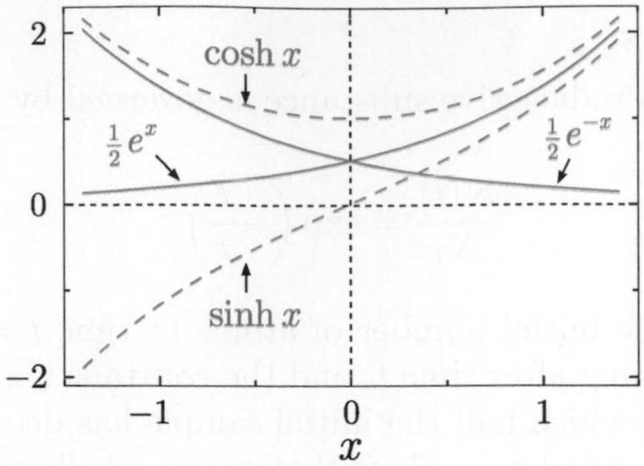

Figure 2.19 The functions $\sinh x$ and $\cosh x$ and their relation to the exponential functions. Note that $\cosh x$ and $\sinh x$ become equal at large positive values of x.

2.3.3 Hyperbolic functions

Given the natural logarithms, we can define *hyperbolic functions* as follows.

$$\sinh x \equiv \frac{e^x - e^{-x}}{2}, \quad \cosh x \equiv \frac{e^x + e^{-x}}{2},$$

$$\tanh x \equiv \frac{\sinh x}{\cosh x} = \frac{e^x - e^{-x}}{e^x + e^{-x}}. \tag{2.57}$$

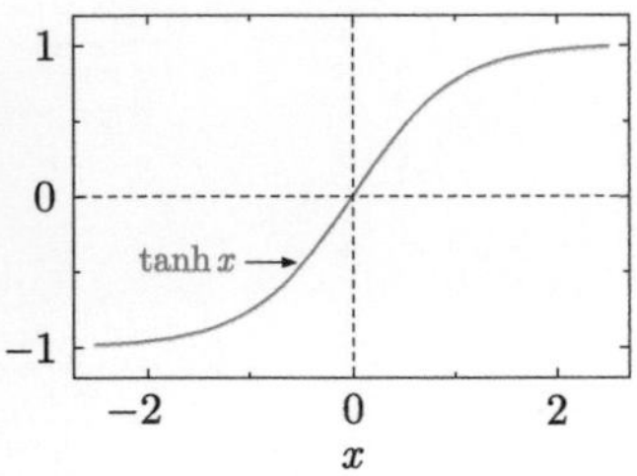

Figure 2.20 The function $\tanh x$.

These are called the *hyperbolic sine, hyperbolic cosine* and *hyperbolic tangent*, respectively, and are shown in Figures 2.19 and 2.20. Their inverses are defined as

$$\text{sech } x \equiv \frac{1}{\cosh x}, \quad \text{cosech } x \equiv \frac{1}{\sinh x}, \quad \coth x \equiv \frac{1}{\tanh x}.$$

We will show in Section 6.4.1 that the hyperbolic functions are related to the circular functions, hence the origin of their names. The word 'hyperbolic' appears because they are also related to the equation for a hyperbola. This follows from the first of the identities

$$\cosh^2 x - \sinh^2 x = 1, \tag{2.58a}$$

and

$$\text{sech}^2 x = 1 - \tanh^2 x, \tag{2.58b}$$

which may be checked using the definitions (2.57). Equations (2.57) and (2.58a) imply that if $x = \cosh\theta$ and $y = \sinh\theta$, then the point (x, y) lies on the branch of the rectangular hyperbola $x^2 - y^2 = 1$ for which $x + y > 0$. (Hyperbolas are discussed in Section 2.4.)

By analogy with the circular functions, we have

$$\coth x = 1/\tanh x, \text{sech } x = 1/\cosh x \text{ and cosech } x = 1/\sinh x \tag{2.59}$$

and we can also form the inverse hyperbolic functions by 'inverting' (2.59). For example, let

$$y = \sinh x = \frac{e^x - e^{-x}}{2} = \frac{(e^x)^2 - 1}{2e^x}. \tag{2.60}$$

This can be written as a quadratic in the variable $z = e^x$, treating y as if it were a constant, leading to the solution $z = y + \sqrt{y^2 + 1}$. Hence

$$x = \ln(y + \sqrt{y^2 + 1}).$$

and

$$\text{arcsinh}\, x = \sinh^{-1} x = \ln(x + \sqrt{x^2 + 1}). \tag{2.61}$$

In a similar way we find that

$$\text{arccosh}\, x = \cosh^{-1} x = \ln(x \pm \sqrt{x^2 - 1}), \tag{2.62}$$

where unlike in the case of $\sinh^{-1} x$, both signs of the square root lead to positive values for e^x. However, because

$$\frac{1}{x + \sqrt{x^2 - 1}} = x - \sqrt{x^2 - 1}, \tag{2.63a}$$

the result for $\cosh^{-1} x$ may also be written

$$\cosh^{-1} x = \pm \ln\left(x + \sqrt{x^2 - 1}\right), \tag{2.63b}$$

which shows explicitly that the two values are equal in magnitude but with opposite signs. Finally,

$$\text{arctanh}\, x = \tanh^{-1} x = \frac{1}{2} \ln\left(\frac{1 + x}{1 - x}\right). \tag{2.64}$$

Just as for the inverse trigonometric functions, both the 'arc' and '-1' notations used in (2.61b), (2.62) and (2.64) are in common use.

The hyperbolic functions satisfy a number of identities. By analogy with (2.36), they are

$$\sinh(x \pm y) = \sinh x \cosh y \pm \cosh x \sinh y, \tag{2.65a}$$

$$\cosh(x \pm y) = \cosh x \cosh y \pm \sinh x \sinh y, \tag{2.65b}$$

$$\sinh x \pm \sinh y = 2 \sinh\left(\frac{x \pm y}{2}\right) \cosh\left(\frac{x \mp y}{2}\right), \tag{2.65c}$$

$$\cosh x + \cosh y = 2 \cosh\left(\frac{x + y}{2}\right) \cosh\left(\frac{x - y}{2}\right), \tag{2.65d}$$

$$\cosh x - \cosh y = 2\sinh\left(\frac{x+y}{2}\right)\sinh\left(\frac{x-y}{2}\right), \tag{2.65e}$$

$$\tanh(x \pm y) = \frac{\tanh x \pm \tanh y}{1 \pm \tanh x \tanh y}. \tag{2.65f}$$

Specific useful cases that follow directly from (2.65) by setting $x = y$ are the double-argument identities,

$$\sinh 2x = 2\sinh x \cosh x, \tag{2.66a}$$

$$\cosh 2x = 2\cosh^2 x - 1, \tag{2.66b}$$

and

$$\tanh(2x) = \frac{2\tanh x}{1 + \tanh^2 x}. \tag{2.66c}$$

Example 2.19

Prove the identities (2.58a) and (2.66a).

Solution

To prove (2.58a), we use the definitions of $\sinh x$ and $\cosh x$ to give

$$\begin{aligned}\cosh^2 x - \sinh^2 x &= (\cosh x + \sinh x)(\cosh x - \sinh x) \\ &= \frac{(2e^x)}{2}\frac{(2e^{-x})}{2} = 1.\end{aligned}$$

Similarly, to prove (2.66a), we use the definitions of $\sinh x$ and $\cosh x$ to give

$$\begin{aligned}2\sinh x \cosh x &= 2\left(\frac{e^x - e^{-x}}{2}\right)\left(\frac{e^x + e^{-x}}{2}\right) \\ &= \frac{e^{2x} - e^{-2x}}{2} = \sinh 2x.\end{aligned}$$

Example 2.20

Solve the equation

$$\cosh^2 x + \sinh x = 7$$

and express the answers in terms of logarithms.

Solution

Using (2.58a) in the equation gives

$$\sinh^2 x + \sinh x - 6 = 0,$$

which factorises in the form

$$(\sinh x - 2)(\sinh x + 3) = 0,$$

with solutions $x = \text{arcsinh } (2)$ and $x = \text{arcsinh } (-3)$. But from (2.61b),

$$\text{arcsinh } x = \ln\left(x + \sqrt{x^2+1}\right),$$

so the two solutions are

$$x = \ln\left(2+\sqrt{5}\right) \quad \text{and} \quad x = \ln\left(-3+\sqrt{10}\right).$$

2.4 Conic sections

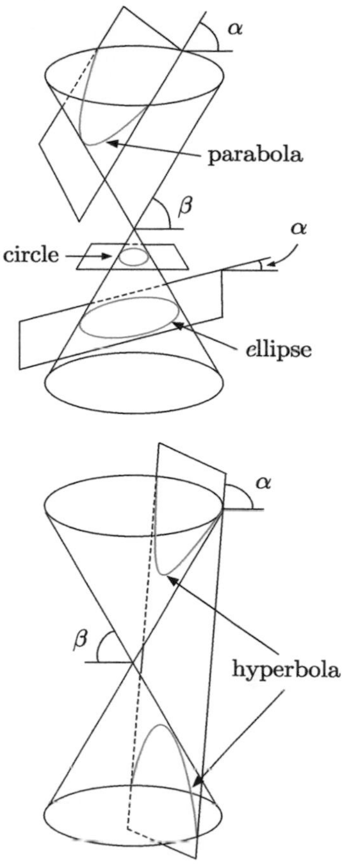

Figure 2.21 Geometrical interpretation of conic sections.

Another class of functions that is commonly met in physics are the *conic sections*. Their name derives from the fact that they are formed by the intersection of a plane with a double circular cone, that is, a pair of symmetric cones that are constructed by rotating a straight line through one revolution about an axis through the vertex of the cones and the centre of the base of the cones. This is illustrated in Figure 2.21 where β is the angle of rotation relative to the bases, which are taken to be horizontal when viewed in profile. Also shown in this figure are the intersections of a plane oriented at different angles α relative to the horizontal. The resulting curves are of four possible types. If $\alpha < \beta$, the plane intersects only one cone and the resulting closed curve is called an *ellipse*. In the limiting case where $\alpha = 0$, that is, the plane is horizontal, the closed curve is a *circle*. If $\alpha = \beta$, that is, the plane is parallel to the edge of the cone, it again only intersects one cone, but the resulting curve is now open. It is called a *parabola*. Finally, if $\alpha > \beta$, the plane intersects both cones and results in two non-intersecting branches of an open curve called a *hyperbola*.

All conic sections can be shown to have the property that there exists in the plane of the curve a point F called the *focus*, and a straight line d, called the *directrix*, such that if P is any point on the curve, the ratio of the distance from P to F to that of the perpendicular distance from P to a point N on the directrix is a fixed number e called the *eccentricity*. In this section, we shall take this as a definition of a conic section, rather than the geometrical constructions of Figure 2.21, and use it to derive the functions that describe them.

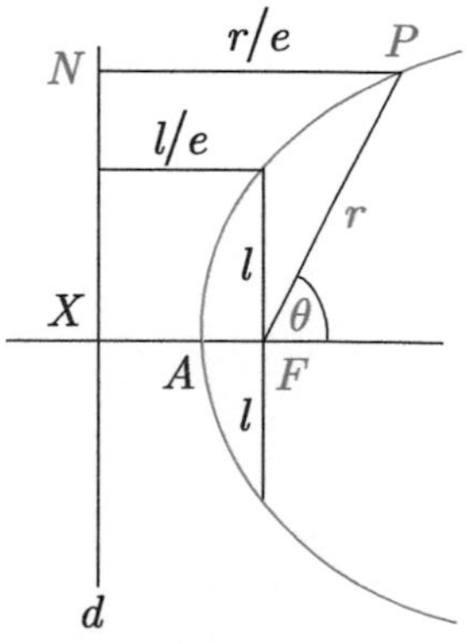

Figure 2.22 Derivation of the polar equation for a conic section.

Consider a point P lying on a conic section, where F is the focus and we assume that P and F are on the same side of the directrix d, as shown in Figure 2.22. We then introduce polar co-ordinates $P(r, \theta)$ where the focus F is taken to be at the origin and $\theta = 0$ corresponds to the direction XA. From the general property of a conic section

$$FP/PN = e, \tag{2.67}$$

so the curve is specified by e and the distance XF, or equivalently the length $L = 2l$ of the chord parallel to d through F. From Figure 2.22, $XF = l/e$ and so the length NP is given by

$$NP = l/e + r\cos\theta, \text{ if } P \text{ and } F \text{ are on the same side of } d. \tag{2.68a}$$

But $NP = r/e$, and so

$$l/r = 1 - e\cos\theta, \text{ if } P \text{ and } F \text{ are on the same side of } d \tag{2.68b}$$

If, on the other hand, we consider the case where P and F are on opposite sides of d, a similar argument leads to

$$NP = -l/e - r\cos\theta, \text{ if } P \text{ and } F \text{ are on the opposite sides of } d \tag{2.69a}$$

and

$$l/r = -1 - e\cos\theta, \text{ if } P \text{ and } F \text{ are on the opposite sides of } d. \tag{2.69b}$$

The above equations define the functions $r(\theta)$ that describe conic sections, where, since $l,\ r > 0$, the second result (2.69b) applies only when $e\cos\theta < -1$, that is, when $e > 1$ and $\cos\theta < -1/e$. However, in both cases multiplying by r and rearranging gives

$$r^2 = (l + er\cos\theta)^2 \tag{2.70}$$

which applies for any e and θ. Equivalently, if we consider Cartesian co-ordinates with the origin at the focus F and the positive x-axis in the direction of the line XF, the corresponding equation is found by substituting

$$r^2 = x^2 + y^2 \quad \text{and} \quad r\cos = x.$$

into (2.70) to give

$$x^2(1 - e^2) + y^2 - 2lex = l^2. \tag{2.71}$$

The properties of the different types of conic sections are now obtained by choosing different values of the eccentricity, starting with $e = 1$.

(i) *Parabola.* For $e = 1$, (2.71) becomes

$$y^2 = 2l(x + l/2),$$

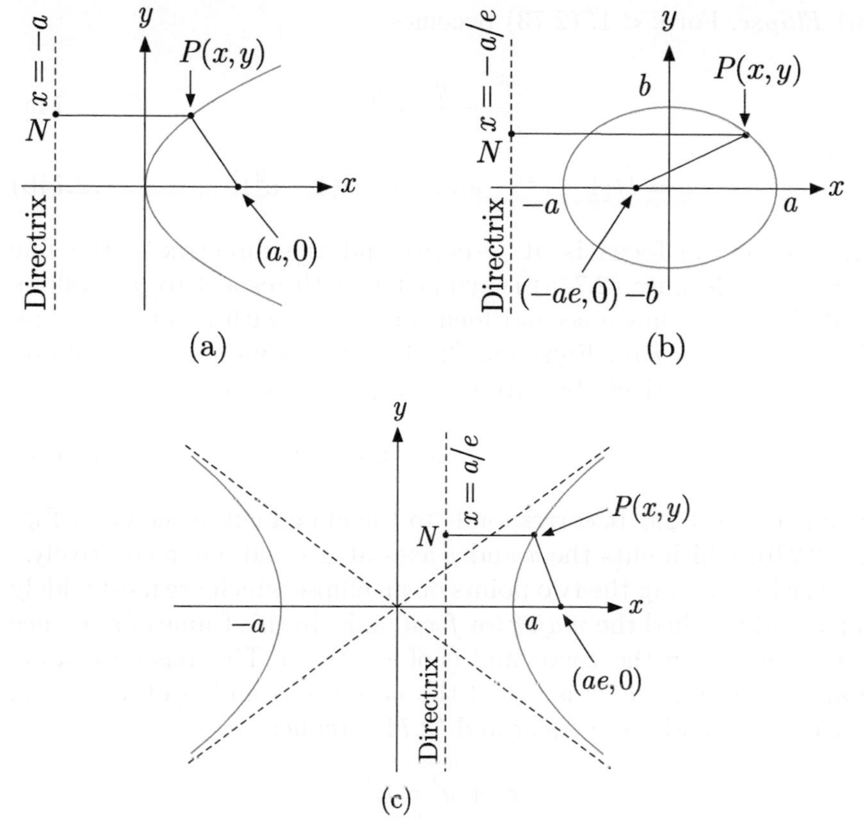

Figure 2.23 The standard forms for: (a) the parabola ($e = 1$); (b) the ellipse ($e < 1$); and (c) the hyperbola ($e > 1$). Only one focus and directrix is shown for the ellipse and hyperbola.

which is the implicit function for a parabola. A simpler form is obtained by writing $a = l/2$ and shifting the origin of the co-ordinate system to $(-a, 0)$, so that in the new variables

$$y^2 = 4ax. \tag{2.72}$$

This corresponds to the unbounded curve in Figure 2.23a, where, in this frame of reference, the focus is $F = (a, 0)$ and the directrix d is the line $x = -a$, as shown.

The second possibility is that $e \neq 1$. In this case, (2.71) may be written

$$\left(x - \frac{le}{1-e^2}\right)^2 + \frac{y^2}{1-e^2} = \frac{l^2}{(1-e^2)^2},$$

which becomes

$$\frac{x^2}{l^2/(1-e^2)^2} + \frac{y^2}{l^2/(1-e^2)} = 1 \tag{2.73}$$

on shifting the origin to $(le/(1-e^2), 0)$, that is, the centre of the conic, while the focus and directrix are now at $-el/(1-e^2)$ and $-1/[e(1-e^2)]$, respectively. There are now two possibilities $e < 1$ and $e > 1$ and we consider each in turn.

(ii) *Ellipse.* For $e < 1$, (2.73) becomes

$$\frac{x^2}{a^2} + \frac{y^2}{b^2} = 1, \tag{2.74a}$$

with

$$a = l/(1 - e^2) \quad \text{and} \quad b = a(1 - e^2)^{1/2}, \tag{2.74b}$$

and where the focus is at $(-ae, 0)$ and the directrix is the line $x = -a/e$. Because (2.74a) is symmetric with respect to y it follows that the ellipse has a second focus at $(ae, 0)$, with a corresponding directrix at $x = a/e$. Equation (2.74) is the equation of an ellipse, and may alternatively be expressed in parametric form

$$x = a\cos\phi \quad \text{and} \quad y = b\sin\phi, \tag{2.74b}$$

where $0 \leq \phi \leq 2\pi$. It corresponds to the closed curve shown in Figure 2.23b, which cuts the x and y axes at $\pm a$ and $\pm b$, respectively.

The line joining the two points on an ellipse which are most widely separated is called the *major* (or *focal*) axis. In this frame of reference it coincides with the x-axis and is of length $2a$. The axis perpendicular to the major axis is called the *minor* axis and is of length $2b$. For $e = 0$, a and b are equal and (2.74a) reduces to

$$x^2 + y^2 = a^2,$$

which is the equation of a circle centred at the origin. Hence a circle can be regarded as an ellipse with zero eccentricity. This allows us to infer the formula

$$A = \pi ab \tag{2.75}$$

for the area of an ellipse, since the area must be proportional to both a and b and reduce to the area of a circle when $a = b$.

(iii) *Hyperbola.* For $e > 1$, (2.73) becomes

$$\frac{x^2}{a^2} - \frac{y^2}{b^2} = 1, \tag{2.76a}$$

where a and b are now defined by

$$a = l/(e^2 - 1) \quad \text{and} \quad b = a(e^2 - 1)^{1/2}. \tag{2.76b}$$

This is the equation of a hyperbola. The corresponding curves are shown in Figure 2.23c. It is clear that the hyperbola has two distinct branches because for $y = 0$ there are two solutions for x, but for $x = 0$ there are no real solutions for y. In this reference frame, the focus F and the directrix are at $(ae, 0)$ and $x = a/e$, respectively, and as for the ellipse, the symmetry of (2.76a) implies that the hyperbola has a second focus at $(-ae, 0)$, with a corresponding directrix at $x = -a/e$. Equation (2.76a) can be written in the parametric form

$$x = a\cosh u \quad \text{and} \quad y = b\sinh u, \tag{2.76c}$$

where $-\infty < u < \infty$.

Equations (2.72), (2.74a) and (2.76a) are called the standard forms for the parabola, ellipse and hyperbola. They apply in co-ordinates systems chosen so that the directrix is parallel to the y axis and the focus is at $(-a, 0)$ for the parabola, $(-ae, 0)$ for the ellipse and $(ae, 0)$ for the hyperbola. In an arbitrary Cartesian co-ordinate system, the three conic sections are described by second order equations of the form

$$Ax^2 + 2Hxy + By^2 + 2Fy + 2Gx + C = 0, \tag{2.77}$$

where A, B, C, F, G, and H are constants. The following conditions, which we state without proof, determine which conic section this function represents:

(i) $H = 0,\ A = B \neq 0$ is a circle; (ii) $H^2 = AB$ is a parabola;

(iii) $H^2 < AB$ is an ellipse; (iv) $H^2 > AB$ is a hyperbola.

For example, the equation

$$3x^2 + 2xy + 3y^2 - 8x + 2y + 4 = 0$$

represents an ellipse because $H^2 < AB$, but the non-zero terms in x and y indicate that the centre of the ellipse is not at the origin, and those in xy indicate that the major and minor axes do not coincide with co-ordinate axes, that is, the ellipse has been rotated.

Example 2.21

Show that the curve with parametric equations

$$x = 1 + 2\cos\theta, \quad y = 2 - 3\sin\theta$$

is an ellipse. Find the co-ordinates of its centre, the lengths of its two axes, its eccentricity and the positions of the focus and directrix.

Solution

Using $\sin^2\theta + \cos^2\theta = 1$, we have

$$\left(\frac{x-1}{2}\right)^2 + \left(\frac{y-2}{3}\right)^2 = 1, \quad (1)$$

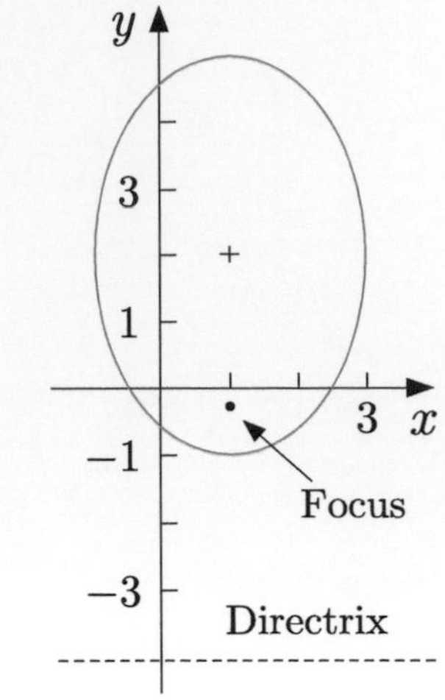

Figure 2.24

which is the Cartesian equation of an ellipse with centre (1,2) and axes of length 4 and 6, where the major axis in the y-direction, rather than the x-direction. To see this more explicitly, introduce new variables $x' = y - 2$ and $y' = x - 1$. In terms of these variables, (1) is of the form (2.74) with $a = 3$ and $b = 2$ and the eccentricity is

$$e = \sqrt{1 - b^2/a^2} = \sqrt{5}/3.$$

Hence the focus and directrix are given by

$$(x', y') = (-ae, 0) = (-\sqrt{5}, 0) \quad \text{and } x' = -a/e = -9/\sqrt{5},$$

respectively, or in terms of the original variables,

$$(x, y) = (1, 2 - \sqrt{5}) \quad \text{and } y = 2 - 9/\sqrt{5}.$$

The ellipse is shown in Figure 2.24.

Problems 2

2.1 If α and β are the roots of the equation $x^2 - 2x - 3 = 0$, find the equation whose roots are $1/\alpha$ and $1/\beta$.

2.2 If x is real and $p = 5(x^2 + 2)/(3x - 1)$, show that $9p^2 \geq 20(p + 10)$.

2.3 Find the gradients of the tangents to the circle $x^2 + y^2 = r^2$ that intersect the y-axis at the point $(0, c)$, where c is greater than r.

2.4 Two circles of radius 2 are centred at $(x, y) = (0, 0)$ and $(1, -1)$, respectively. What are the co-ordinates of their points of intersection? The line joining these points is a chord of both circles. What angle does it subtend at their centres?

2.5 (a) Write $x^3 + x^2 - x - 4$ in the form $(x - 1)Q(x) + R(x)$, where the quotient $Q(x)$ is a polynomial of order 2.
(b) Show that the quartic $f(x) = 3x^4 - x^3 + 4x^2 + 5x + 15$ can be written in the form $(x^2 - 2x + 3)\, Q(x)$, where the quotient $Q(x)$ is a polynomial of order two, and hence show that $f(x)$ has no real roots.

2.6 Determine the integer roots of $x^4 - 2x^3 - 2x^2 + 5x - 2$ and find its other two roots.

2.7 The function $f(x) = x^3 - 2x^2 + 4x - 5$ has a real root in the range $1.5 < x < 1.6$. Find the value of the root correct to three decimal places.

2.8 Express in partial fractions:

(a) $\dfrac{2(x^2 - 9x + 11)}{(x - 2)(x - 3)(x + 4)}$, (b) $\dfrac{7x^2 + 6x - 13}{(2x + 1)(x^2 + 2x - 4)}$,

(c) $\dfrac{2(3x^2 + 4x + 2)}{(x - 1)(2x + 1)^2}$.

2.9 Express in partial fractions:

(a) $\dfrac{x^3 - 2x^2 + 10}{(x - 1)(x + 2)}$, (b) $\dfrac{3x^2 - 5x - 4}{(x + 2)(3x^2 + x - 1)}$, (c) $\dfrac{3x^2 - x + 2}{(x - 1)(x - 3)^3}$.

2.10 Prove the identities:
(a) $\cos 4\theta \equiv 8\cos^4\theta - 8\cos^2\theta + 1$,
(b) $\dfrac{\sin(n\theta) + \sin[(n+2)\theta] + \sin[(n+4)\theta]}{\cos(n\theta) + \cos[(n+2)\theta] + \cos[(n+4)\theta]} \equiv \tan[(n+2)\theta]$.
(c) $\left(\dfrac{\sin 5\theta}{\sin\theta}\right)^2 - \left(\dfrac{\cos 5\theta}{\cos\theta}\right)^2 = 8\cos 2\theta(4\cos^2 2\theta - 1)$,

2.11 Solve the following equations for angles in the range $0 < \theta < 2\pi$.
(a) $2\cos\theta\cos 2\theta + \sin 2\theta = 2(3\cos^3\theta - \cos\theta)$,
(b) $\sin\theta - \sin 2\theta + \sin 3\theta = 0$.

2.12 Find the general solution of the equation $\sin k\theta = \sin\theta$.

2.13 Show that the straight line $x\sin\theta + y\cos\theta = p$ is a tangent to the hyperbola $(x/a)^2 - (y/b)^2 = 1$ if $(a\sin\theta)^2 - (b\cos\theta)^2 = p^2$ and find the co-ordinates of the point of contact.

2.14 Prove the identity

$$\frac{1 + \sin\theta + \cos\theta}{1 + \sin\theta - \cos\theta} \equiv \frac{1 + \cos\theta}{\sin\theta}.$$

2.15 The triangle ABC has lengths $a = BC = 5$ cm, $b = AC = 4$ cm and the angle B is 0.5 rad or 28.65 degrees. Find the length $c = AB$ and the angles A and C.

2.16 The co-ordinates (x, y) of a triangle ABC are $A = (1, 3)$, $B = (5, 6)$ and $C = (7, 2)$. Find the angles at the vertices.

2.17 Use the method of induction to show that $\sin[(2n+1)\theta]$ and $\cos[(2n+1)\theta]/\cos\theta$ can be expressed as polynomials in $\sin\theta$ for all $n \geq 0$.

2.18 Simplify the expressions:
(a) $\log(xy) + 3\log(x/y) + 2\log(y/x)$, (b) $6\log x^{1/3} + 2\log(1/x)$

2.19 Solve the equations:
(a) $\log(x+3) + \log(x-3) = 3$ to five significant figures,
(b) $\ln(\log x) = -3$ to four significant figures.

2.20 Solve the equations:
(a) $\ln x + \log x = 5$, (b) $3\ln\left(\dfrac{2}{x}\right) - \frac{1}{3}\ln x = -2$.

2.21 (a) Verify that $\text{sech}\ x < \text{cosech}\ x < \coth x$, for $x > 0$.
(b) Prove the identity $\cosh 3x = 4\cosh^3 x - 3\cosh x$.

2.22 Solve for real values of x the equations:
(a) $3\cosh 2x + 2\sinh 2x = 3$, and (b) $\text{arctanh}\ x = \ln 5$.
(c) For what values of c does the equation $\cosh(\ln x) = \sinh[\ln(x/2)] + c$ have real roots?

2.23 Solve the equation $\cosh 4x + 4\cosh 2x - 125 = 0$.

2.24 A straight line passes through the focus of the parabola $y^2 = 4ax$ and cuts the parabola at the points $P_i(at_i^2,\ 2at_i)$, $i = 1,\ 2$. Find the relationship between t_1 and t_2.

2.25 Find the equation of the tangent and the normal to the parabola $y^2 = 4x$ at the point $(x, y) = (1, 2)$.

3

Differential calculus

The introduction of the infinitesimal calculus, independently by Newton and Leibnitz in the late seventeenth century, was one of the most important events not only in the history of mathematics but also of physics, where it has been an indispensable tool ever since.

In this chapter and the one that follows, we introduce the formalism in the context of functions of a single variable. We start by considering *differentiation*, the calculation of the instantaneous rate of change of a function as its argument changes. So, for example, given a function $x(t)$, which specifies the position of a particle moving in one dimension as a function of time t, the operation of differentiation yields a function $v(t)$ representing the velocity. The inverse operation, called *integration*, will be discussed in Chapter 4 and enables the position $x(t)$ to be deduced from $v(t)$ and the value of x at some time, for example $t = 0$. These two operations – differentiation and integration – play a crucial role in understanding not only mechanics, but the whole of physical science. Both rest on ideas of limits and continuity, to which we now turn.

3.1 Limits and continuity

In previous chapters, we have used the ideas of limits and continuity in simple cases where their meaning is obvious. In this section we shall define them more precisely, before showing how they lead naturally to the idea of differentiation in Section 3.2

3.1.1 Limits

If a function $f(x)$ approaches arbitrarily close to a fixed value α as x approaches arbitrarily close to a constant a, then α is said to be

Mathematics for Physicists, First Edition. B.R. Martin and G. Shaw.

Companion website: www.wiley.com/go/martin/mathsforphysicists

the *limit* of $f(x)$ as x approaches a, with the notation

$$\lim_{x \to a} f(x) = \alpha, \tag{3.1a}$$

or equivalently

$$f(x) \to \alpha \quad \text{as}\, x \to a. \tag{3.1b}$$

More precisely, (3.1) means that for any $\varepsilon > 0$, however small, we can always find a number $\delta > 0$, depending on ε, such that

$$|f(x) - \alpha| < \varepsilon \quad \text{for any}\, |x - a| < \delta. \tag{3.2}$$

For example, the obvious result

$$\lim_{x \to 1} \left[f(x) = x^2 - 2x + 3 \right] = 2$$

is formally verified by noting that $f(x) = (x - 1)^2 + 2$, so that

$$|f(x) - 2| = (x - 1)^2$$

and thus for any ε, however small, $|f(x) - 2| < \varepsilon$, provided that $|x - 1| < \delta = +\sqrt{\varepsilon}$.

In this example, the limit of $f(x)$ as $x \to 1$ is equal to the value at $x = 1$, i.e.

$$\lim_{x \to 1} f(x) = f(1).$$

However, the existence of the limit (3.1) does not in general imply that $f(a) = \alpha$, and indeed $f(a)$ may not even exist. For example, consider the function

$$f(x) = \frac{x^2 - 9}{x - 3}. \tag{3.3a}$$

Taking the limit as $x \to 3$ gives

$$\lim_{x \to 3} \left[f(x) = \frac{x^2 - 9}{x - 3} \right] = 6 \tag{3.3b}$$

because

$$\frac{x^2 - 9}{x - 3} = \frac{(x - 3)(x + 3)}{(x + 3)} = x + 3$$

However, direct evaluation of (3.3a) at $x = 3$, gives $f(3) = 0/0$ and is undefined.

A number of important results follow directly from the definition of a limit. With the notations

$$\lim_{x \to a} f(x) = \alpha \quad \text{and} \quad \lim_{x \to a} g(x) = \beta, \tag{3.4}$$

and taking c as a constant, these are

(i) if $f(x) = c$, then $\lim_{x\to a} f(x) = c$;

(ii) $\lim_{x\to a} [c f(x)] = c \lim_{x\to a} f(x) = c\,\alpha$;

(iii) $\lim_{x\to a} [f(x) \pm g(x)] = \lim_{x\to a} f(x) \pm \lim_{x\to a} g(x) = \alpha \pm \beta$;

(iv) $\lim_{x\to a} [f(x)g(x)] = \lim_{x\to a} f(x) \lim_{x\to a} g(x) = \alpha\beta$;

(v) $\lim_{x\to a} \left[\dfrac{f(x)}{g(x)}\right] = \dfrac{\lim_{x\to a} f(x)}{\lim_{x\to a} g(x)} = \dfrac{\alpha}{\beta}, (\beta \neq 0)$;

and, if n is an integer,

(vi) $\lim_{x\to a} \left[f^{1/n}(x)\right] = \left[\lim_{x\to a} f(x)\right]^{1/n} = \alpha^{1/n}$, if $\alpha^{1/n}$ is defined.

The proof of these results is straightforward. As an example, we will prove (iv) as follows. From the definition (3.2) and (3.4),

$$|f(x) - \alpha| < \varepsilon \text{ for } |x - a| < \delta_1 \text{ and } |g(x) - \beta| < \varepsilon \text{for } |x - a| < \delta_2.$$

Let $\delta_1 < \delta_2$. Then

$$\begin{aligned} |f(x)g(x) - \alpha\beta| &= |[f(x) - \alpha][g(x) - \beta] + \alpha[g(x) - \beta] + \beta[f(x) - \alpha]| \\ &\leq |f(x) - \alpha||g(x) - \beta| + |\alpha||g(x) - \beta| \\ &+ |\beta||f(x) - \alpha| \leq \varepsilon^2 + \alpha\varepsilon + \beta\varepsilon. \end{aligned} \tag{3.5}$$

If ε is chosen to be the positive root of

$$\varepsilon^2 + |\alpha|\varepsilon + |\beta|\varepsilon = \eta,$$

where η is any small quantity, then (3.5) may be written

$$|f(x)g(x) - \alpha\beta| < \eta, \quad \text{whenever } |x - a| < \delta_1,$$

which concludes the proof.

The definition of a limit can be extended to the case where x increases indefinitely, either positively or negatively. For example,

$$\lim_{x\to+\infty} f(x) = \alpha$$

means that, for any $\varepsilon > 0$, however small, a number $l > 0$ can be found such that $|f(x) - \alpha| < \varepsilon$ for any $x > l$. If $f(x)$ increases indefinitely, positively or negatively, as $x \to a$, we will use the notation

$$\lim_{x\to a} f(x) = \pm\infty,$$

with the appropriate sign.

The following examples illustrate these results.

Example 3.1

Evaluate the following limits:

(a) $\lim_{x\to 1}\left[\frac{3x^2+2x+1}{x^2-2x+3}\right]$, (b) $\lim_{x\to -\infty}\left[\frac{2x^5+10x-3}{3x^3+1}\right]$

(c) $\lim_{x\to -2}[(2x^2+2x+4)^{1/3}]$.

Solution

(a) From (v) above this is

$$\frac{\lim_{x\to 1}(3x^2+2x+1)}{\lim_{x\to 1}(x^2-2x+3)} = \frac{6}{2} = 3.$$

(b) As $x \to \infty$ only the highest powers of x need be retained, so

$$\lim_{x\to -\infty}\left[\frac{2x^5 10x-3}{3x^3+1}\right] = \lim_{x\to -\infty}\left[\frac{2x^2}{3}\right] = +\infty.$$

(c) From (vi),

$$\lim_{x\to -2}\left[(2x^2+2x+4)^{1/3}\right] = \left[\lim_{x\to -2}(2x^2+2x+4)\right]^{1/3} = 8^{1/3} = 2.$$

Example 3.2

Show that

$$\lim_{x\to\infty}\left[x^n e^{-x}\right] = 0 \tag{3.6a}$$

for any finite n, and hence that

$$\lim_{x\to 0}(x\ln x) = 0. \tag{3.6b}$$

Solution

(a) Taking logarithms, we have

$$\lim_{x\to\infty}\left[\ln(x^n e^{-x})\right] = \lim_{x\to\infty}(n\ln x - x) = -\infty,$$

using (iii) above. Hence

$$\lim_{x\to\infty}\left[x^n e^{-x}\right] = e^{-\infty} = 0.$$

(b) Substituting $x = e^{-z}$, so that $x \to 0$ as $z \to \infty$, we obtain

$$\lim_{x\to 0}(x\ln x) = \lim_{z\to\infty}(-z\,e^{-z}) = 0$$

by (3.6a).

3.1.2 Continuity

So far, we have not specified the path taken as $x \to a$. There are two possibilities. Firstly, x could tend to a via values less than a. This is referred to as *approaching a from the left* (or *below*) and the limit is denoted $\lim\limits_{x\to a^-}$. Alternatively, if x approaches a via values greater than a, then x is said to *approach a from the right* (or *above*) and the limit is written $\lim\limits_{x\to a^+}$. For the limit (3.1) to exist, these two limits must be identical, since the defining condition (3.2) is independent of the sign of $(x-a)$. However, in practice, the two limits are not always the same. As an example, consider the function $f(x) = x/|x|$. At $x = 0$, $f(x)$ is undefined and in addition

$$\lim_{x\to 0+} f(x) = \lim_{x\to 0^+}\left(\frac{x}{x}\right) = 1, \tag{3.7a}$$

but

$$\lim_{x\to 0^-} f(x) = \lim_{x\to 0^-}\left(\frac{-x}{x}\right) = -1. \tag{3.7b}$$

In general, a function $f(x)$ is said to be *continuous* at the point $x = x_0$ if the following conditions are satisfied:

$$\text{(i) } f(x_0) \text{ is defined; and (ii) } \lim_{x\to x_0^+} f(x) = \lim_{x\to x_0^-} f(x) = f(x_0). \tag{3.8}$$

If a function $f(x)$ is defined in the interval (a, x_0) to the left of x_0, and in the interval (x_0, b) to the right of x_0, then $f(x)$ is said to be *discontinuous* at x_0 if either of the above conditions fails at $x = x_0$. Thus the function $f(x) = x/|x|$ above is said to be discontinuous at the point $x = 0$, as shown in Figure 3.1. Another example is the function

$$f(x) = (x^2 - 9)\big/(x-3).$$

In this case, $f(3) = 0/0$ and is undefined, and $f(x)$ is discontinuous at $x = 3$. However, we saw in (3.4b) that in the limit as $x \to 3$, $f(x) \to 6$, so that in this case we could define a function $g(x)$ that is identical to $f(x)$ except at $x = 3$, where we define $g(x)$ to be 6. Then the function $g(x)$ would be continuous. This type of discontinuity, which can be removed by redefining the value of the function at the point of discontinuity, is said to be *removable*. In the case of the function $f(x) = x/|x|$ at $x = 0$, the function is discontinuous because the limits from above and below are not equal (cf. Eqs. 3.7). This is called a *jump discontinuity* and is not removable. Another type of jump discontinuity is illustrated by the plot of $\tan\theta$, shown in Figure 2.8(a). At $\theta = \pi/2$, for example, $\tan\theta$ is ill defined, since

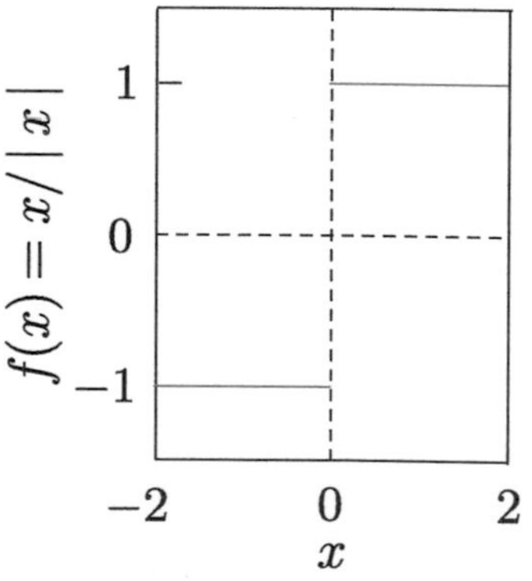

Figure 3.1 The function $f(x) = x/|x|$ in the vicinity of $x = 0$.

$$\tan\pi/2 = \frac{\sin\pi/2}{\cos\pi/2} = \frac{1}{0},$$

while $\tan\theta \to \infty$ as $\theta \to (\pi/2)^-$ and to $-\infty$ as $\theta \to (\pi/2)^+$. This type of discontinuity, associated with divergent behaviour, is called an *infinite discontinuity.*

It follows from the properties of limits discussed in Section 3.1.1, that the sum, product, difference or quotient of two functions that are both continuous at a point are themselves continuous, provided in the case of a quotient that the denominator does not vanish at the point.

Example 3.3

What is the limiting behaviour of the function

$$f(x) = \frac{2x^2 - 1}{x^2 + x - 2}$$

as $x \to \pm\infty$? Find any discontinuities in $f(x)$ and the limiting behaviours as they are approached from the left and from the right.

Solution

When $x \to \pm\infty$, only the leading powers of x need be retained, so that

$$\lim_{x\to\pm\infty} = \frac{2x^2}{x^2} = 2$$

in both cases. The numerator and denominator are both polynomials and hence continuous functions. Discontinuities can only arise at points where the denominator vanishes, when $f(x)$ is ill defined. Since

$$x^2 + x - 2 = (x - 1)(x + 2),$$

zeros occur at $x = 1$ and $x = -2$. To investigate the behaviour near $x = 1$, we write $x = 1 - \delta$, so that the limits $\delta \to 0^+$ and $\delta \to 0^-$ correspond to the limits $x \to 1^+$ and $x \to 1^-$, respectively. This gives

$$f(x) = \frac{1 + 4\delta + 2\delta^2}{3\delta + \delta^2} \to \frac{1}{3\delta} \quad \text{as } \delta \to 0.$$

Hence,

$$\lim_{x\to 1^-} f(x) = -\infty, \qquad \lim_{x\to 1^+} f(x) = +\infty$$

and we have an infinite discontinuity. In a similar way we obtain

$$\lim_{x\to -2^-} f(x) = +\infty, \qquad \lim_{x\to -2^+} f(x) = -\infty,$$

so that again we have an infinite discontinuity.

3.2 Differentiation

The aim in this section is, given a function $f(x)$, to find a function that gives the gradient of $f(x)$ at a given value of the independent variable x. This is achieved by a limiting procedure. Consider the change in the function in going from x to $x+\delta x$, where δx is a small quantity, positive or negative, in a region where the function is continuous. Then the average rate of change of $f(x)$ in the range x to $x+\delta x$ is clearly

$$\frac{f(x+\delta x)-f(x)}{\delta x} \tag{3.9}$$

and the instantaneous rate of change at x, denoted by $\mathrm{d}f/\mathrm{d}x$, is given by

$$\frac{\mathrm{d}f}{\mathrm{d}x}=\lim_{\delta x\to 0}\left[\frac{f(x+\delta x)-f(x)}{\delta x}\right], \tag{3.10}$$

provided the limit exists. In this case, the function is said to be *differentiable* and $\mathrm{d}f/\mathrm{d}x$ is called the *derivative* of $f(x)$ with respect to x. In calculating it, we say that we have *differentiated* $f(x)$ with respect to x. This is illustrated in Figure 3.2, from which we see that the term in square brackets in (3.10) is just the gradient of the straight line AB, which approaches the gradient of the tangent at x as $\delta x \to 0$. In other words, the derivative at x is the gradient of the curve at x.

There are several equivalent notations used to denote a derivative. Each is convenient for different circumstances. If, as in Chapter 1, we introduce the dependent variable $y=f(x)$, these are

$$y'=\frac{\mathrm{d}y}{\mathrm{d}x}=f'(x)=\frac{\mathrm{d}}{\mathrm{d}x}f(x)=\frac{\mathrm{d}f}{\mathrm{d}x}. \tag{3.11a}$$

It is also useful to define

$$\delta f(x)=f(x+\delta x)-f(x) \tag{3.11b}$$

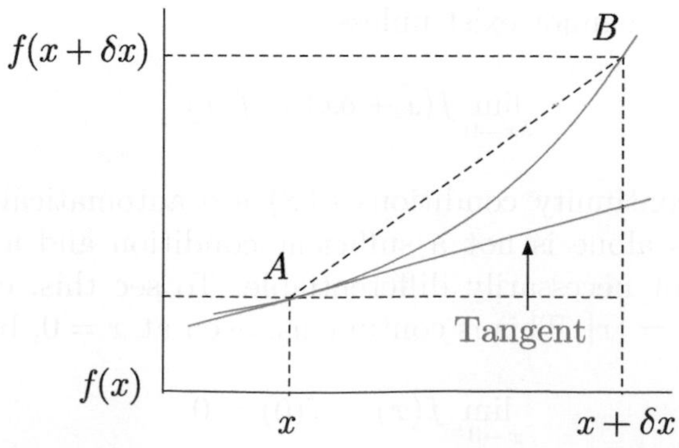

Figure 3.2 Geometrical interpretation of the derivative $\mathrm{d}f/\mathrm{d}x$.

for a change in $f(x)$ corresponding to a change from x to $(x + \delta x)$ in the independent variable, so that (3.10) becomes

$$\frac{df}{dx} = \lim_{\delta x \to 0} \frac{\delta f}{\delta x}. \tag{3.12}$$

Example 3.4

Differentiate from first principles with respect to x the following functions: (a) $3x^2 + 2$ and (b) $1/x$.

Solution

(a) We have $f(x) = 3x^2 + 2$ and hence

$$f(x + \delta x) = 3[x^2 + 2x\delta x + (\delta x)^2] + 2.$$

So from (3.11) and (3.12),

$$\frac{df}{dx} = \lim_{\delta x \to 0} \left[\frac{6x\delta x + 3(\delta x)^2}{\delta x} \right] = 6x.$$

(b) We have $f(x) = 1/x$, so that

$$f(x + \delta x) = \frac{1}{x + \delta x}$$

and

$$\delta f = f(x + \delta x) - f(x) = -\frac{\delta x}{x(x + \delta x)}.$$

So

$$\frac{df}{dx} = \lim_{\delta x \to 0} \left[-\frac{1}{x(x + \delta x)} \right] = -\frac{1}{x^2}.$$

3.2.1 Differentiability

Differentiability is closely related to continuity. A necessary condition for a function to be differentiable is that it must be continuous, since the limit (3.10) cannot exist unless

$$\lim_{\delta x \to 0} f(x + \delta x) = f(x),$$

so that the continuity conditions (3.8) are automatically satisfied. However, this alone is not a sufficient condition and a continuous function is not necessarily differentiable. To see this, consider the function $f(x) = |x|$. This is continuous, even at $x = 0$, because

$$\lim_{x \to 0^{\pm}} f(x) = f(0) = 0.$$

However, one easily verifies that the quantity in brackets in (3.10) is equal to $+1$ for all $x > 0$, but -1 for all $x < 0$. Hence $f(x)$ is not differentiable at $x = 0$, since

$$\lim_{x\to 0^+} \frac{\delta f}{\delta x} \neq \lim_{x\to 0^-} \frac{\delta f}{\delta x}.$$

But it is differentiable at $x \neq 0$.

Example 3.5

Prove that at the points $x = 2$ and $x = 5$ the function

$$y(x) = \begin{cases} x^3 & 0 \le x < 2 \\ x + 6 & 2 \le x < 5 \\ -\frac{1}{2}(x^2 - 47) & 5 \le x \le 8 \end{cases}$$

is (a) continuous but (b) not differentiable, and comment on these results, given the plot of the function given in Figure (1.10).

Solution

(a) Consider firstly the point $x = 2$. For $x \to 2^-$, i.e. from below, we can set $x \to 2 - \delta x$ ($\delta x > 0$) and consider the limit as $\delta x \to 0$. Then

$$\lim_{x\to 2^-} y(x) = \lim_{\delta x\to 0} (2 - \delta x)^3 = 8.$$

Similarly for $x \to 2^+$, that is, the limit from above, we can set $x \to 2 + \delta x$ ($\delta x > 0$) and again consider the limit as $\delta x \to 0$. This gives

$$\lim_{x\to 2^+} y(x) = \lim_{\delta x\to 0} [(2 + \delta x) + 6] = 8,$$

so the two limits are equal. In addition, at $x = 2$, $y(2) = 8$, so from the continuity conditions (3.8), $y(x)$ is continuous at the point $x = 2$. Proceeding in the same way for the point $x = 5$, we have

$$\lim_{x\to 5^-} y(x) = \lim_{\delta x\to 0} [(5 - \delta x) + 6] = 11,$$

and

$$\lim_{x\to 5^+} y(x) = \lim_{\delta x\to 0} \{-\tfrac{1}{2}[(5 + \delta x)^2 - 47]\} = 11,$$

so again the two limits are equal and since at $x = 5$, $y(5) = 11$ the function $y(x)$ is also continuous at the point $x = 5$.

(b) For a function $y(x)$ to be differentiable, the limit of $\delta y/\delta x$ as x tends to the limit point both from above and below must be equal. Using the previous notations, for $x \to 2^-$

$$\lim_{x\to 2^-} \frac{\delta y}{\delta x} = \lim_{\delta x\to 0} \left[\frac{(2 - \delta x)^3 - 2^3}{\delta x}\right] = 12$$

and for $x \to 2^+$,

$$\lim_{x\to 2^+}\frac{\delta y}{\delta x}=\lim_{\delta x\to 0}\left[\frac{\{(2+\delta x)+6\}-8}{\delta x}\right]=1.$$

So the two limits are not equal and therefore $y(x)$ is not differentiable at $x=2$. In a similar way, for $x \to 5^-$

$$\lim_{x\to 5^-}\frac{\delta y}{\delta x}=\lim_{\delta x\to 0}\left[\frac{\{(5-\delta x)+6\}-11}{\delta x}\right]=1$$

and for $x \to 5^+$,

$$\lim_{x\to 5+}\frac{\delta y}{\delta x}=\lim_{\delta x\to 0}\left[\frac{-\frac{1}{2}\{(5+\delta x)^2-47\}-11}{\delta x}\right]=-5.$$

Again the two limits are not equal and therefore $y(x)$ is not differentiable at $x=5$.

These conclusion can also be deduced without formal proof from Figure 1.10, which shows no discontinuity at either point, but a clear break in slope at both points.

3.2.2 Some standard derivatives

Although (3.10) is the fundamental definition, it is not necessary to use it directly in most cases. Rather, one uses it to deduce the derivatives for a number of important standard functions. These are then used, together with general properties that follow from (3.10), to deduce the result for other cases, as we shall see.

Here we shall consider some of these standard derivatives starting with simple powers $f(x)=x^n$, where n is any integer. From the binomial theorem (1.23) we have

$$f(x+\delta x)=(x+\delta x)^n=x^n+nx^{n-1}\delta x+O[(\delta x)^2],$$

where $O[(\delta x)^2]$ means terms that are at most of order $(\delta x)^2$, that is, are proportional to $(\delta x)^2$, and so can be neglected compared to terms that are linear in δx as $\delta x \to 0$. Hence,

$$\frac{\mathrm{d}f}{\mathrm{d}x}=\lim_{\delta x\to 0}\frac{\delta f(x)}{\delta x}=\lim_{\delta x\to 0}\left\{\frac{nx^{n-1}\delta x+O[(\delta x)^2]}{\delta x}\right\}=nx^{n-1}.$$

In other words

$$\frac{\mathrm{d}}{\mathrm{d}x}(x^n)=nx^{n-1},\qquad n=0,\ 1,\ 2,\ \ldots \tag{3.13}$$

This result also holds when n is not an integer, as we shall show in Section 3.3.5.

Next we consider the more difficult case of $f(x) = \sin x$. Using (2.36c), we have

$$\sin(x + \delta x) - \sin x = 2\sin\left(\frac{\delta x}{2}\right)\cos\left(\frac{2x + \delta x}{2}\right),$$

and since $\cos\left[(2x + \delta x)/2\right] \to \cos x$ as $\delta x \to 0$, we have

$$\frac{\mathrm{d}f}{\mathrm{d}x} = \cos x \lim_{\delta x \to 0}\left[\frac{\sin(\delta x/2)}{\delta x/2}\right]. \tag{3.14}$$

It remains to find the limit of the term in the brackets. This is done by the construction of Figure 3.3, from which we see that the ratio of the length of the line AB to the length of the arc AC is given by

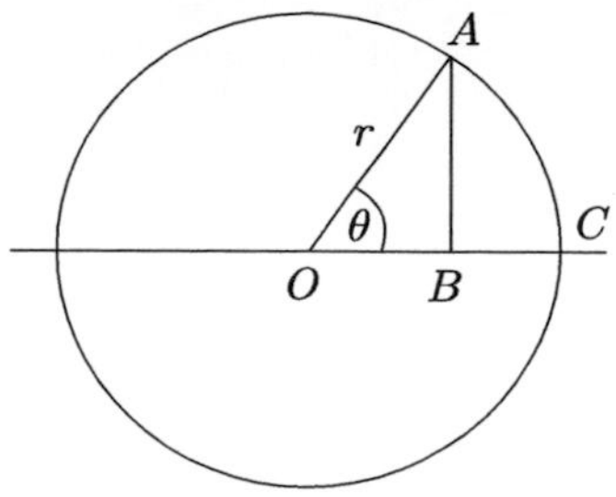

Figure 3.3 Construction to find the limit of $\sin\theta/\theta$ as $\theta \to 0$.

$$\frac{AB}{AC} = \frac{r\sin\theta}{r\theta} = \frac{\sin\theta}{\theta},$$

where the angle is as usual measured in radians. We also see that, as $\theta \to 0$, the lengths of AB and AC tend to equality, giving the important result

$$\lim_{\theta \to 0}\left(\frac{\sin\theta}{\theta}\right) = 1. \tag{3.15}$$

Applying this to (3.14), with $\theta = \delta x/2$, gives the desired result

$$\frac{\mathrm{d}}{\mathrm{d}x}(\sin x) = \cos x. \tag{3.16a}$$

A similar argument, left to the reader, leads to the result

$$\frac{d}{dx}(\cos x) = -\sin x. \tag{3.16b}$$

Finally, in Section 2.3.2 we showed, using essentially the argument formulated more generally at the beginning of this section, that the slope or gradient of e^x was e^x (cf. Eq. 2.49). In the present notation, this result is written

$$\frac{\mathrm{d}}{\mathrm{d}x}e^x = e^x. \tag{3.17}$$

Example 3.6

If $f(x) = 2^x$, show that $\mathrm{d}f/\mathrm{d}x = kf(x)$ and evaluate the constant k to two decimal places.

Solution

From (3.10),

$$\delta f = 2^{x+\delta x} - 2^x = 2^x(2^{\delta x} - 1),$$

so that

$$\frac{\mathrm{d}f}{\mathrm{d}x} = kf(x),$$

where

$$k = \lim_{\delta x \to 0} \left[\frac{2^{\delta x} - 1}{\delta x} \right].$$

For $\delta x = 0.1, 0.01, 0.001, 0.0001, \ldots$ the value of the square bracket to three decimal places is $0.718, 0.696, 0.693, 0.693, \ldots$, so that $k = 0.69$ to two decimal places.

3.3 General methods

Methods for differentiating other functions may be derived using general properties of derivatives that follow from their definition, together with standard results like (3.13)–(3.16).

Suppose we have a function of the general form

$$f(x) = a_1 f_1(x) + a_2 f_2(x) + \cdots + a_N f_N(x), \tag{3.18}$$

where $a_1,\ a_2 \cdots a_N$ are constants and $f_1(x),\ f_2(x), \ldots, f_N(x)$ are differentiable functions. Then from (3.11) we easily see that

$$\delta f(x) = a_1 \delta f_1(x) + a_2 \delta f_2(x) + \cdots + a_N \delta f_N(x),$$

so that

$$\frac{\mathrm{d}f}{\mathrm{d}x} = a_1 \frac{\mathrm{d}f_1(x)}{\mathrm{d}x} + a_2 \frac{\mathrm{d}f_2(x)}{\mathrm{d}x} + \cdots + a_N \frac{\mathrm{d}f_N(x)}{\mathrm{d}x} \tag{3.19}$$

by (3.12). Hence, for example, if

$$f(x) = 3\sin x + 4\cos x$$

then

$$\frac{\mathrm{d}f}{\mathrm{d}x} = 3\frac{\mathrm{d}\sin x}{\mathrm{d}x} + 4\frac{\mathrm{d}\cos x}{\mathrm{d}x} = 3\cos x - 4\sin x$$

by (3.19), together with the standard derivatives (3.16) and (3.17) for $\sin x$ and $\cos x$, respectively. Similarly, for an arbitrary polynomial of order N, i.e.

$$f(x) = \sum_{n=0}^{N} a_n x^n = a_0 + a_1 x + \cdots + a_N x^N,$$

we have

$$\frac{\mathrm{d}f(x)}{\mathrm{d}x} = \sum_{n=0}^{N} n a_n x^{n-1} = a_1 + 2a_2 x + \cdots + N a_N x^{N-1}$$

by (3.19) and (3.13).

In what follows, we shall introduce a series of general results analogous to (3.19) and illustrate their use in finding the derivatives of specific functions by examples.

3.3.1 Product rule

Consider a function of the form

$$f(x) = u(x)v(x),$$

where $u(x)$ and $u(x)$ are differentiable functions. Then

$$\begin{aligned}\delta f(x) &= u(x+\delta x)v(x+\delta x) - u(x)v(x)\\ &= [u(x)+\delta u(x)]\,[v(x)+\delta v(x)] - u(x)v(x)\\ &= u(x)\delta v(x) + v(x)\delta u(x) + \delta u(x)\delta v(x).\end{aligned}$$

Since the last term is second order in small quantities,[1] it can be neglected in taking the limit (3.11), which then gives the *product rule*

$$\frac{\mathrm{d}}{\mathrm{d}x}(uv) = u\frac{\mathrm{d}v}{\mathrm{d}x} + v\frac{\mathrm{d}u}{\mathrm{d}x}. \tag{3.20}$$

Example 3.7

Differentiate the function (a) $f(x) = 3x^2 \sin x$.

Solution

By (3.19) and (3.20)

$$\begin{aligned}\frac{\mathrm{d}}{\mathrm{d}x}f(x) &= 3\frac{\mathrm{d}}{\mathrm{d}x}(x^2\sin x) = 3\left(\frac{\mathrm{d}x^2}{\mathrm{d}x}\right)\sin x + 3x^2\left(\frac{\mathrm{d}\sin x}{\mathrm{d}x}\right)\\ &= 6x\sin x + 3x^2\cos x,\end{aligned}$$

where we have used the standard derivatives (3.13) and (3.16a).

3.3.2 Quotient rule

We next consider a quotient

$$f(x) = \frac{u(x)}{v(x)},$$

[1] As we saw in Section 3.2.1, the condition that $u(x)$ and $v(x)$ are differentiable requires that both δu and δv vanish in the limit $\delta x \to 0$.

where $u(x)$ and $v(x)$ are again arbitrary differentiable functions and $v(x) \neq 0$. Then

$$\delta f(x) = \frac{u(x) + \delta u(x)}{v(x) + \delta v(x)} - \frac{u(x)}{v(x)} = \frac{v(x)\delta u(x) - u(x)\delta v(x)}{v(x)[v(x) + \delta v(x)]},$$

so that (3.12) gives the *quotient rule*

$$\frac{\mathrm{d}}{\mathrm{d}x}\left(\frac{u}{v}\right) = \frac{v\mathrm{d}u/\mathrm{d}x - u\mathrm{d}v/\mathrm{d}x}{v^2}, \tag{3.21}$$

where in the denominator we have used the fact that $\delta v(x) \to 0$ as $\delta x \to 0$ for any differentiable function $v(x)$. Setting $u = 1$ leads immediately to the *reciprocal rule*:

$$\frac{\mathrm{d}}{\mathrm{d}x}\left(\frac{1}{v}\right) = -\frac{1}{v^2}\frac{\mathrm{d}v}{\mathrm{d}x}, \tag{3.22}$$

since clearly $\mathrm{d}u/\mathrm{d}x = 0$ for any constant u.

Example 3.8

Differentiate $\tan x$.

Solution

Since $\tan x = \sin x/\cos x$, the quotient rule, together with (3.16a) and (3.16b), gives

$$\frac{\mathrm{d}}{\mathrm{d}x}\tan x = \frac{\mathrm{d}}{\mathrm{d}x}\left(\frac{\sin x}{\cos x}\right) = \frac{\cos^2 x + \sin^2 x}{\cos^2 x} = \sec^2 x\,, \tag{3.23}$$

using (2.28) and (2.32b).

Example 3.9

Differentiate $\sec x$.

Solution

Since $\sec x = 1/\cos x$, from (3.22) we have

$$\frac{\mathrm{d}}{\mathrm{d}x}\sec x = -\frac{1}{\cos^2 x}\frac{\mathrm{d}}{\mathrm{d}x}\cos x = \frac{\sin x}{\cos^2 x}.$$

3.3.3 Reciprocal relation

The derivatives of functions and their inverses are closely related. Consider a function $y = f(x)$ and its inverse $x = f^{-1}(y)$, where we assume that both f and f^{-1} are differentiable functions in what follows.[2] When $x \to x + \delta x$, the function $y \to y + \delta y$ and

$$\frac{dy}{dx} = \lim_{\delta x \to 0} \frac{\delta y}{\delta x},$$

while similarly

$$\frac{dx}{dy} = \lim_{\delta y \to 0} \frac{\delta x}{\delta y}.$$

Since $\delta x \to 0$ corresponds to $\delta y \to 0$ and vice versa, the trivial relation

$$\frac{\delta y}{\delta x} = \left(\frac{\delta x}{\delta y} \right)^{-1}$$

leads immediately to the *reciprocal relation*

$$\frac{dy}{dx} = \frac{1}{dx/dy} \tag{3.24}$$

for differentiable functions. Since $y = f(x)$ and $x = f^{-1}(y)$, this can alternatively be written

$$\frac{df(x)}{dx} = \frac{1}{df^{-1}(y)/dy}, \tag{3.25}$$

so that df/dx is easily obtained if the derivative of the inverse $1/f$ is known.

We will illustrate this for the important case of $y = f(x) = \ln x$. Then,

$$x = f^{-1}(y) = e^y,$$

so that

$$\frac{dx}{dy} = e^y$$

by (3.17) and

$$\frac{dy}{dx} = \frac{1}{e^y} = \frac{1}{x},$$

giving the important result

$$\frac{d}{dx} \ln x = \frac{1}{x}. \tag{3.26}$$

[2] Note that f^{-1} is the inverse function as defined in Section 1.3.1 and not $1/f$.

Example 3.10
Differentiate $y = \arcsin x$.

Solution
Inverting the relation, we have $x = \sin y$, so that $\mathrm{d}x/\mathrm{d}y = \cos y$, by (3.16). Using (2.28), we obtain

$$\frac{\mathrm{d}y}{\mathrm{d}x} = \frac{1}{\cos y} = \frac{1}{\sqrt{1-\sin^2 y}} = \frac{1}{\sqrt{1-x^2}},$$

and so

$$\frac{\mathrm{d}}{\mathrm{d}x}\arcsin x = \frac{1}{\sqrt{1-x^2}}. \tag{3.27}$$

3.3.4 Chain rule

We next consider a function y that is itself a function of a second function $z(x)$, that is, $y[z(x)]$, or more explicitly

$$y = f(z), \qquad z = g(x), \tag{3.28}$$

where f and g are continuous, differentiable functions of x. For such functions, when $x \to x + \delta x$ there are corresponding changes $z \to z + \delta z$, $y \to y + \delta y$, such that δy, $\delta z \to 0$ when $\delta x \to 0$. Hence

$$\lim_{\delta x\to 0}\frac{\delta y}{\delta x} = \lim_{\delta x\to 0}\left[\frac{\delta y}{\delta z}\frac{\delta z}{\delta x}\right] = \lim_{\delta z\to 0}\frac{\delta y}{\delta z}\lim_{\delta x\to 0}\frac{\delta z}{\delta x},$$

i.e.

$$\frac{\mathrm{d}y}{\mathrm{d}x} = \frac{\mathrm{d}y}{\mathrm{d}z}\frac{\mathrm{d}z}{\mathrm{d}x}. \tag{3.29}$$

Equation (3.29) is called *the chain rule.* When used together with judiciously chosen substitutions, it is a key tool in evaluating derivatives, as we shall immediately illustrate.

Example 3.11
Differentiate the function

$$y = \frac{3}{(x^3+2x+1)^{1/2}}.$$

Solution
This can be written

$$y = 3z^{-1/2}, \quad z = (x^3+2x+1),$$

so that

$$\frac{\mathrm{d}y}{\mathrm{d}z} = -\frac{3}{2}z^{-3/2} = -\frac{3}{2}\frac{1}{(x^3+2x+1)^{3/2}}$$

and

$$\frac{dz}{dx} = 3x^2 + 2.$$

Hence the chain rule (3.29) gives

$$\frac{dy}{dx} = -\frac{3}{2}\frac{(3x^2+2)}{(x^3+2x+1)^{3/2}}.$$

Example 3.12

Differentiate the function $y = \sin(3x^2 - 2)$.

Solution

This can be written

$$y = \sin z, \qquad z = 3x^2 - 2,$$

so that

$$\frac{dy}{dz} = \cos z = \cos(3x^2 - 2)$$

and

$$\frac{dz}{dx} = 6x.$$

Hence by the chain rule

$$\frac{dy}{dx} = 6x\cos(3x^2 - 2).$$

3.3.5 More standard derivatives

In this section we obtain some more standard derivatives, this time involving logarithms and exponentials. We start by considering functions of the form $y = \ln f(x)$, which using (3.28) may be written

$$y = \ln z\,, \qquad z = f(x).$$

Hence the chain rule (3.29) gives

$$\frac{dy}{dx} = \frac{1}{z}\frac{df(x)}{dx} = \frac{1}{f(x)}\frac{df(x)}{dx},$$

i.e.

$$\frac{d}{dx}\ln f(x) = \frac{1}{f(x)}\frac{df(x)}{dx}. \tag{3.30}$$

Equation (3.30) is called a *logarithmic derivative.* If, for example, we choose $f(x) = 3x^2$, (3.30) gives

$$\frac{d}{dx}\ln(3x^2) = \frac{1}{3x^2}6x = \frac{2}{x}.$$

Another class of functions is $\exp[f(x)]$, when (3.29) gives

$$\frac{\mathrm{d}}{\mathrm{d}x}e^{f(x)} = e^{f(x)}\frac{\mathrm{d}f(x)}{\mathrm{d}x}. \tag{3.31}$$

For the simple case $f(x) = -x$, this gives

$$\frac{\mathrm{d}}{\mathrm{d}x}e^{-x} = -e^{-x},$$

which, together with the corresponding result (3.18) for e^x, enables the hyperbolic functions to be differentiated. In this way, starting from the definitions (2.57), and using (3.19), one obtains the standard results:

$$\frac{\mathrm{d}}{\mathrm{d}x}\sinh x = \cosh x \tag{3.32a}$$

and

$$\frac{\mathrm{d}}{\mathrm{d}x}\cosh x = \sinh x. \tag{3.32b}$$

The corresponding result for tanh x,

$$\frac{\mathrm{d}}{\mathrm{d}x}\tanh x = \mathrm{sech}^2 x, \tag{3.32c}$$

follows from $\tanh x = \sinh x/\cosh x$ using the quotient rule (3.22).

Another important result that follows from (3.31) is

$$\frac{\mathrm{d}x^\alpha}{\mathrm{d}x} = \alpha x^{\alpha-1} \tag{3.33}$$

for any real number α. Previously we obtained this result for integer $\alpha = n$. To establish it in general, we note that

$$y = x^\alpha = (e^{\ln x})^\alpha = e^{\alpha \ln x},$$

which is of the form $e^{f(x)}$ with $f(x) = \alpha \ln x$. Relation (3.31) then gives

$$\frac{\mathrm{d}y}{\mathrm{d}x} = e^{\alpha \ln x}\frac{\alpha}{x} = \alpha x^{\alpha-1}.$$

The result (3.33) is the last of a set of 'standard derivatives' that we have derived in this and previous sections and which are extremely useful in calculating the derivatives of other functions, using the product, quotient and chain rules, and the reciprocal relation (3.24). They are listed in Table 3.1 for later convenience.

Table 3.1 Some standard derivatives

y	x^α	e^x	$\ln x$	$\sin x$	$\cos x$	$\tan x$	$\sinh x$	$\cosh x$	$\tanh x$
$\mathrm{d}y/\mathrm{d}x$	$\alpha x^{\alpha-1}$	e^x	$1/x$	$\cos x$	$-\sin x$	$\sec^2 x$	$\cosh x$	$\sinh x$	$\mathrm{sech}^2 x$

3.3.6 Implicit functions

So far we have discussed the techniques available to differentiate explicit functions. Here we briefly extend the discussion to include functions defined implicitly as the solution of an equation, or by parametric forms.

In the latter cases, both x and y are defined in terms of a third variable, a parameter t, say. That is, by equations of the form

$$x = f(t), \qquad y = g(t), \tag{3.34}$$

where we assume $f(t)$ and $g(t)$ are themselves continuous differentiable functions. For example, x and y could specify the positions of a point in a plane as a function of the time t. Equations (3.34) imply a functional relationship between x and y that can be written as the explicit function $y = g[f^{-1}(x)]$ if the function f has an inverse. However, to find the derivative of y with respect to x, it is easier to note that if a small change δt leads to changes δx, δy in x and y, then the trivial relation

$$\frac{\delta y}{\delta x} = \frac{\delta y}{\delta t} \bigg/ \frac{\delta x}{\delta t},$$

implies

$$\frac{\mathrm{d}y}{\mathrm{d}x} = \frac{\mathrm{d}y}{\mathrm{d}t} \bigg/ \frac{\mathrm{d}x}{\mathrm{d}t}, \tag{3.35}$$

since $\delta t \to 0$ implies $\delta x,\ \delta y \to 0$ for continuous functions f, g.

Alternatively, a function might be defined implicitly as a solution of an equation of the form

$$f(x, y) = c, \tag{3.36}$$

where c is a constant. The derivative of y with respect to x can then be deduced from the equation

$$\frac{\mathrm{d}f(x, y)}{\mathrm{d}x} = 0, \tag{3.37}$$

which follows directly from (3.36).

Example 3.13

Find the gradient of the tangent to the circle $x^2 + y^2 = 25$ at $x = 3,\ y = 4$.

Solution

Differentiating this with respect to x gives

$$2x + 2y\frac{\mathrm{d}y}{\mathrm{d}x} = 0,$$

where we have used the chain rule to differentiate y^2 with respect to x. Hence

$$\frac{\mathrm{d}y}{\mathrm{d}x} = -\frac{x}{y} = -\frac{3}{4}.$$

Example 3.14

Find $\mathrm{d}y/\mathrm{d}x$, given that $x = t + 1/t$, $y = 3t^{1/2} + t^{3/2}$.

Solution

Differentiating with respect to t gives

$$\frac{\mathrm{d}x}{\mathrm{d}t} = 1 - \frac{1}{t^2} \quad \text{and} \quad \frac{\mathrm{d}y}{\mathrm{d}t} = \frac{3}{2}\left(t^{1/2} + t^{-1/2}\right).$$

Then

$$\frac{\mathrm{d}y}{\mathrm{d}x} = \frac{\mathrm{d}y}{\mathrm{d}t}\frac{1}{\mathrm{d}x/\mathrm{d}t} = \frac{3\,(t^{1/2} + t^{-1/2})}{2(1 - t^{-2})} = \frac{3\,t^{3/2}}{2(t-1)}.$$

3.4 Higher derivatives and stationary points

We have seen above how to differentiate a function $y = f(x)$ to yield its derivative

$$f'(x) = \frac{\mathrm{d}y}{\mathrm{d}x} = \frac{\mathrm{d}f}{\mathrm{d}x}.$$

This derivative itself is often a differentiable function, in which case it may also be differentiated to give a *second derivative*,

$$\frac{\mathrm{d}^2y}{\mathrm{d}x^2} = \frac{\mathrm{d}}{\mathrm{d}x}\left(\frac{\mathrm{d}y}{\mathrm{d}x}\right), \tag{3.38a}$$

which, like the first derivative (cf. Eq. 3.9) can be written in the alternative forms

$$f''(x) = \frac{\mathrm{d}^2f}{\mathrm{d}x^2} = \frac{\mathrm{d}^2y}{\mathrm{d}x^2}. \tag{3.38b}$$

The first derivative $\mathrm{d}y/\mathrm{d}x$ specifies the gradient or instantaneous rate of change of the function $y(x)$ at any given x. Similarly, the second derivative (3.38) gives the instantaneous rate of change of the gradient itself. So, for example, if

$$y = x^2, \quad \mathrm{d}y/\mathrm{d}x = 2x \quad \text{and} \quad \mathrm{d}^2y\big/\mathrm{d}x^2 = 2,$$

implying that the slope of x^2 itself increases as x increases at a constant rate 2, independent of x.

If the second derivative is differentiable, one can similarly define a third derivative

$$\frac{d^3y}{dx^3} = \frac{d}{dx}\left(\frac{d^2y}{dx^2}\right), \tag{3.39}$$

or, more generally, an *n*th *derivative*

$$\frac{d^ny}{dx^n} = \frac{d}{dx}\left(\frac{d^{n-1}y}{dx^{n-1}}\right), \tag{3.40a}$$

provided that all the lower derivatives exist and are differentiable. Using 'primes' as superscripts, as in (3.38b), is impractical for the general case, and an alternative notation is

$$y^{(n)}(x) = \frac{d^ny}{dx^n}. \tag{3.40b}$$

Such higher derivatives, with $n \geq 3$, can be important in applications, as we shall see in Chapter 5. Here we shall give one worked example, which we will require later, and then describe an important application that depends on the first and second derivatives only.

Example 3.15

Find expressions for the nth derivatives of $\sin x$ and $\cos x$.

Solution

For $\sin x$, we have

n	1	2	3	4
$d^n \sin x/dx^n$	$\cos x$	$-\sin x$	$-\cos x$	$\sin x$

after which the pattern repeats. So,

$$\frac{d^{2n}}{dx^{2n}} \sin x = (-1)^n \sin x, \tag{3.41a}$$

and

$$\frac{d^{2n+1}}{dx^{2n+1}} \sin x = (-1)^n \cos x. \tag{3.41b}$$

Since $d(\sin x)/dx = \cos x$, we have

$$\frac{d^n \cos x}{dx^n} = \frac{d^{n+1} \sin x}{dx^{n+1}},$$

so that

$$\frac{d^{2n}}{dx^{2n}} \cos x = (-1)^n \cos x, \tag{3.42a}$$

and

$$\frac{d^{2n+1}}{dx^{2n+1}} \cos x = (-1)^{n+1} \sin x. \tag{3.42b}$$

3.4.1 Stationary points

In examining the form of a given function $y = f(x)$, it is often useful to consider not only its roots defined by the requirement $y = 0$, but also the points x_0 defined by the condition

$$f'(x_0) = \left.\frac{dy}{dx}\right|_{x=x_0} = 0.$$

These are called *stationary points*, because the instantaneous rate of change of $f(x)$ with respect to x vanishes at $x = x_0$, and the tangent to the curve is horizontal, as shown in Figure 3.4. The figure shows four types of stationary point, corresponding to different behaviours of the gradient $f'(x)$ immediately below and immediately above the stationary point $x = x_0$.

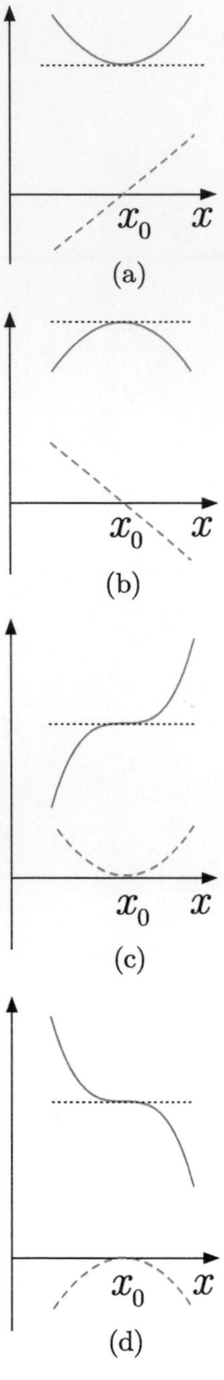

Figure 3.4 The behaviour of a function (solid line) and its derivative (dashed line) in the vicinity of a stationary point $x = x_0$, together with the gradient at x_0 (dotted line), for (a) a minimum, (b) a maximum and (c, d) points of inflection.

(i) *Local minima*[3]
In this case, the gradient $f'(x)$ is negative immediately below and positive immediately above the stationary point $x = x_0$, as shown in Figure 3.4(a). Because $f''(x)$ is the instantaneous rate of change of $f'(x)$, and $f'(x_0) = 0$, this implies

$$f''(x_0) = \left.\frac{d^2y}{dx^2}\right|_{x=x_0} \geq 0,$$

since otherwise $f'(x)$ would be negative immediately above $x = x_0$, in contradiction to our assumption. In other words, the existence of a local minimum at $x = x_0$ implies

$$\frac{dy}{dx} = 0 \quad \text{and} \quad \frac{d^2y}{dx^2} \geq 0 \quad \text{at the minimum.} \tag{3.43a}$$

(ii) *Local maxima*
In this case, the gradient of the function is positive immediately above and negative immediately below the stationary point, as shown

[3] A *local* minimum at x_0 means that the function takes its smallest value in a range $x_0 - \delta < x < x_0 + \delta$, where δ is finite, as opposed to the *global* minimum, which is the smallest value for any value of x.

in Figure 3.4(b). An argument similar to that given above for local maxima leads to

$$\frac{\mathrm{d}y}{\mathrm{d}x} = 0 \quad \text{and} \quad \frac{\mathrm{d}^2 y}{\mathrm{d}x^2} \leq 0 \quad \text{at the maximum.} \tag{3.43b}$$

(iii) *Stationary points of inflection*[4]
These correspond to the case where $f'(x)$ has the same sign on both sides of the stationary point, and can be positive, as shown in Figure 3.4(c), or negative, as shown in Figure 3.4(d). Consider the first case, in which $f'(x)$ is positive both immediately below and above $x = x_0$. Since $f'(x) = 0$ at $x = x_0$, it follows that x_0 is a stationary point (a minimum) of $f'(x)$, implying that its derivative $f''(x_0) = 0$. A similar argument applies to Figure 3.4(d), corresponding to the case where $f'(x)$ is negative on both sides of the stationary point, leading again to the result $f''(x_0) = 0$. Hence for a stationary point of inflection[5]

$$\frac{\mathrm{d}y}{\mathrm{d}x} = 0 \quad \text{and} \quad \frac{\mathrm{d}^2 y}{\mathrm{d}x^2} = 0 \quad \text{at the point of inflection.} \tag{3.43c}$$

The three cases (i), (ii) and (iii), exhaust all possibilities for the signs of $f'(x)$ in the immediate vicinity of the stationary point. To summarise, from (3.43) the conditions

$$\frac{\mathrm{d}y}{\mathrm{d}x} = 0 \quad \text{and} \quad \frac{\mathrm{d}^2 y}{\mathrm{d}x^2} > 0. \tag{3.44a}$$

at $x = x_0$ unambiguously identifies the stationary point as a minimum and

$$\frac{\mathrm{d}y}{\mathrm{d}x} = 0 \quad \text{and} \quad \frac{\mathrm{d}^2 y}{\mathrm{d}x^2} < 0 \tag{3.44b}$$

unambiguously identifies the stationary point as a maximum. On the other hand, the combination

$$\frac{\mathrm{d}y}{\mathrm{d}x} = 0 \quad \text{and} \quad \frac{\mathrm{d}^2 y}{\mathrm{d}x^2} = 0 \tag{3.44c}$$

---

[4] A stationary point of inflection is also called a *saddle point*, although this term is usually used for functions of two or more variables. The name is derived from the shape of the corresponding surface near the point in the case of two variables.
[5] The qualification 'stationary', although often omitted, is necessary, because there are functions for which $\mathrm{d}^2y/\mathrm{d}x^2 = 0$, but $\mathrm{d}y/\mathrm{d}x \neq 0$. These are called *general points of inflection.* They resemble the examples shown in Figure 3.5c and Figure 3.5d, but where the tangent at x_0 is not parallel to the x-axis.

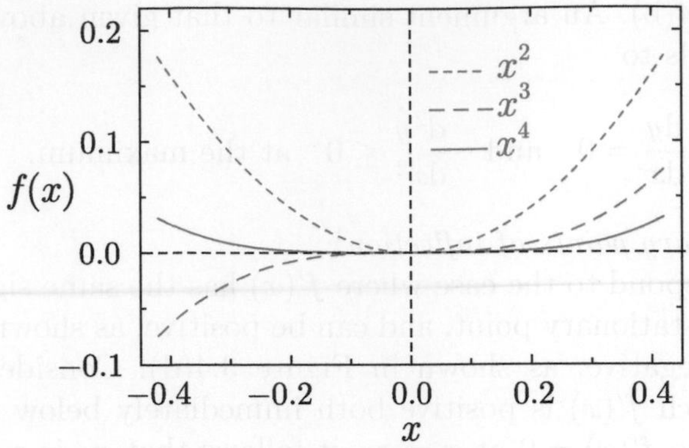

Figure 3.5 The functions $f(x) = x^2$, x^3 and x^4.

can correspond to a maximum, a minimum or a point of inflection and we must examine the behaviour of the derivative $f'(x)$ on both sides of the stationary point to distinguish them.[6]

To illustrate this, consider the simple cases $f(x) = x^2$, x^3 and x^4, which all have a stationary point at $x = 0$. In the first case, we have

$$f(x) = x^2, \quad f'(x) = 2x, \quad f''(x) = 2,$$

which satisfies (3.44a) at $x = 0$, and so is a minimum. In the second case,

$$f(x) = x^3, \quad f'(x) = 3x^2, \quad f''(x) = 6x,$$

so that (3.44c) is satisfied at $x = 0$. However, $f'(x) > 0$ both immediately below and above $x = 0$, so that we have a point of inflection. In the final case we have

$$f(x) = x^4, \quad f'(x) = 4x^3, \quad f''(x) = 12x^2,$$

so that (3.44c) is again satisfied at $x = 0$. However, in this case $f'(x) < 0$ for $x < 0$ and $f'(x) > 0$ for $x > 0$, so that $x = 0$ is a minimum. These three functions are plotted in Figure 3.5, where their behaviours at the stationary point are clearly seen.

Example 3.16

Find the values of x and y at the stationary points of the function

$$y = x^3 - 3x^2 - 4x + 7 \tag{3.45}$$

and identify them as maxima, minima or points of inflection.

[6] An alternative method will be discussed in Section 7.5.

Solution

To find the stationary points we solve the equation

$$\frac{dy}{dx} = 3x^2 - 6x - 4 = 0 \tag{3.46}$$

and then evaluate

$$\frac{d^2y}{dx^2} = 6x - 6 \tag{3.47}$$

to characterise them. Equation (3.46) is a quadratic of the form $ax^2 + bx + c = 0$, with $b^2 > 4ac$, so there are two solutions

$$x = \frac{-b \pm \sqrt{b^2 - 4ac}}{2a} = -0.53 \text{ and } 2.53.$$

From (3.47), $d^2y/dx^2 > 0$ at $x = 2.53$ and $d^2y/dx^2 < 0$ at $x = -0.53$, so that $x = 2.53$ and $x = -0.53$ are minimum and maximum points, respectively. The function (3.45) is plotted in Figure 2.1, where the stationary points can be clearly seen.

3.5 Curve sketching

Curve sketching is a very useful way of understanding and summarising the main features of a given function $y = f(x)$. When doing so, it is important to pay attention to

(i) the limiting behaviour of the function as $x \to \pm\infty$,

(ii) any roots, where $y = 0$,

(iii) any stationary points, where $f'(x) = 0$,

as well as any other general features, for example if the function is symmetric or antisymmetric, or if there are any discontinuities.

In the rest of this section, we shall illustrate the above points by a series of examples. In so doing, we shall assume that the main features of the plots of $\sin x$, $\cos x$, e^x and $\ln x$, given in Figures 2.6 and 2.8, may be used without citation. These functions permeate the whole of physical science and their characteristic forms are well worth memorising.

Example 3.17

Sketch the function $y = (1 + x)/(1 - x)$.

Solution

(a) As $x \to \pm\infty$, $y \to -1$, so that the function approaches the line $y = -1$ both as $x \to \infty$ and $x \to -\infty$.

(b) There is a singular point at $x = 1$, where $y \to 2/(1-x)$, so that

$$\lim_{x \to 1^-} y = +\infty \qquad \lim_{x \to 1+} y = -\infty.$$

(c) Using the quotient rule (3.21), we see that

$$\frac{\mathrm{d}y}{\mathrm{d}x} = \frac{2}{1-x^2} \neq 0$$

for any x, so that there are no turning points.

These features are sufficient to determine the general shape of the function. This is illustrated in Figure 3.6, where they are supplemented by the calculated values at $x = 0$, 2 and 3, respectively.

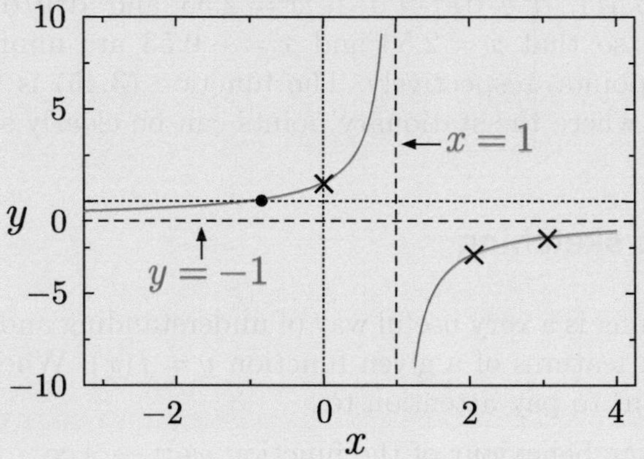

Figure 3.6 The function $y = (1+x)/(1-x)$, showing the asymptotes $x = 1$ and $y = -1$ (dashed lines), the root $x = -1(\bullet)$ and three sample values at $x = 0,\ 2,\ 3\,(\times)$. There are no stationary points.

Example 3.18

In Section 2.1.1 we saw that a cubic polynomial has either three real roots or one real root. Show, using a sketch, that the polynomial

$$f(x) = x^3 - \tfrac{3}{2}x^2 - 6x + 3 \tag{3.48}$$

has three real roots and estimate their values for possible use as the starting points for a more precise evaluation using, for example, the bisection method described at the end of Section 2.1.

Solution

In this case, there are no singular points and the roots are non-trivial. However,

(a) $f(x) \to x^3$ for large x, so that $f(x) \to \pm\infty$ as $x \to \pm\infty$, respectively.

(b) $f'(x) = 3x^2 - 3x - 6 = 3(x+1)(x-2)$ and $f''(x) = 6x - 3$, so there is a maximum $(f' = 0,\ f'' < 0)$ at $x = -1$, with $f(x) = 13/2$ from (3.46); and a minimum $(f' = 0,\ f'' > 0)$ at

$x = 2$, where $f(x) = -7$. Results (a) and (b) imply that there must be three roots, that is, one in each of the regions $x < -1$, $-1 < x < 2$ and $x > 2$. They also determine the general behaviour of the function as shown in Figure 3.7, where we have included the sample points $x = -2,\ 0,\ 1, 3$ at which $y = 1,\ 3,\ -3.5,\ -1.5$, respectively. From this we see that the approximate values of the roots are -2.1, 0.4 and 3.1.

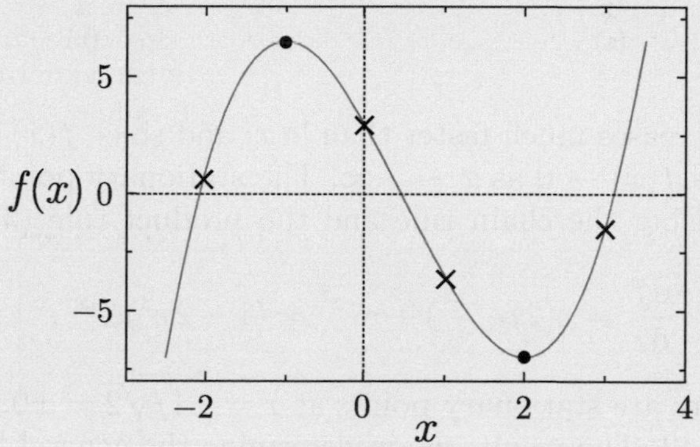

Figure 3.7 The function (3.48) showing the roots, stationary points (•) at $x = -1$, 2 and four sample values at $x = -2,\ 0,\ 1,\ 3$, (×).

Example 3.19

Sketch the functions (a) $\exp(-x^2)$ and (b) $x\exp(-x^2)$.

Solution

(a) The function $\exp(-x^2)$ is symmetric, $f(x) = f(-x)$, with $f(0) = 1$ and $f(1) = e^{-1} = 0.37$ to two significant figures. Since the exponential function is finite and positive for all finite x, there are no roots or divergences. Further, using the chain rule (3.29),

$$\frac{\mathrm{d}f}{\mathrm{d}x} = -2x\exp(-x^2).$$

Thus there is a single stationary point at $x = 0$ and

$$\frac{\mathrm{d}^2 f}{\mathrm{d}x^2} = -2e^{+x^2} + 4x^2 e^{+x^2} < 0 \quad \text{at} x = 0,$$

so that it is a maximum. These results imply the symmetrical bell-shaped curve of Figure 3.8(a), to which we have added the additional sample points $f = 0.78$ and 0.02 at $1/2$ and 2, respectively. Note that the function falls off rapidly beyond $|x| = 1$.

(b) The function $x\exp(-x^2)$ is antisymmetric, $f(x) = -f(-x)$, with $f(0) = 0$ and $f(1) = e^{-1} = 0.37$ to two significant figures. Furthermore, as $x \to \infty$

$$x\,e^{-x^2} = e^{\ln x}e^{-x^2} = e^{-x^2+\ln x} \to 0,$$

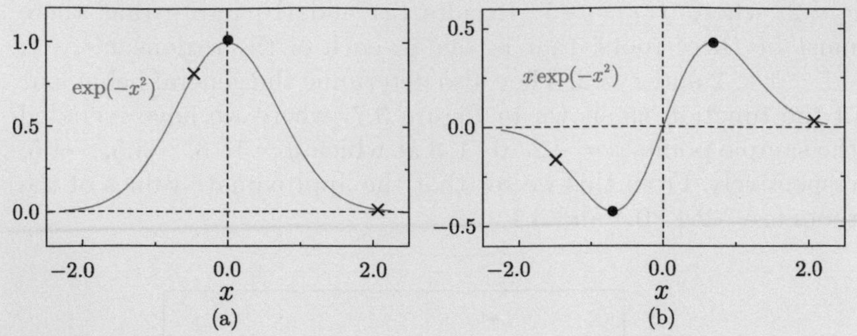

Figure 3.8 The functions $\exp(-x^2)$ and $x\exp(-x^2)$, showing the roots, stationary points $(\cdot)$ and sample points $(\times)$ used, together with the asymptotic behaviour to define the shapes of the curves.

since x^2 increases much faster than $\ln x$; and since $f(x) = f(-x)$, this implies $f(x) \to 0$ as $x \to -\infty$. The stationary points may be found by using the chain rule and the product rule (3.20). One obtains

$$\frac{\mathrm{d}f}{\mathrm{d}x} = x(2xe^{-x^2}) + e^{-x^2} = (1 - 2x^2)e^{-x^2},$$

so that there are stationary points at $x = \pm 1/\sqrt{2} = \pm 0.71$, where $f(x) = \pm 0.43$. The results again determine the general behaviour of the curve, as shown in Figure 3.8(b), where we have also included the sample points $f(-1.5) = -0.16$, $f(2) = 0.037$. We note that the curve still falls off very rapidly beyond $|x| = 1$, despite the extra factor of x.

Problems 3

3.1 Find the limits of

$$\left(\frac{2x^3 - 3x + 1}{x^3 + x^2 - 1}\right)^2$$

as (a) $x \to 0$, (b) $x \to 1$ (c) $x \to \infty$.

3.2 Find the following limits:

$$\text{(a)} \lim_{x\to 0}\left[\frac{(x+5)^2 - 25}{x}\right] \quad \text{(b)} \lim_{x\to 1}\left[\frac{1+\cos(\pi x)}{\tan^2(\pi x)}\right] \quad \text{(c)} \lim_{x\to 0}\left(\frac{\arcsin x}{x}\right)$$

3.3 If $f(x) = x^2$, prove from first principles that $\lim_{x\to 2} f(x) = 4$. (Hint: it is sufficient to prove Eq. 3.1 assuming that $\delta = |x - 2| < 1$, and you may use the general relation $|a + b| \le |a| + |b|$ for any a, b.)

3.4 Find the locations x_0 of any discontinuities in the following functions and classify them as removable or non-removable. In the former case, specify the redefined value $f(x_0)$ required to remove the discontinuity.

$$\text{(a) } f(x) = \begin{cases} x^2 - 1 & x < 0 \\ (x-1)^2 & x \ge 0 \end{cases}, \quad \text{(b) } \frac{x^3 - 3x^2 + 3x - 2}{x^2 + 3x - 10}$$

3.5 Identify the locations x_0 of any discontinuities in the following functions and classify them as removable or non-removable. In the case of removable discontinuities, find the redefined value $f(x_0)$ required to remove the discontinuity.

$$\text{(a)}\ \frac{x^2}{|x|}, \quad \text{(b)}\ \frac{x+3}{4-\sqrt{x^2+7}}.$$

3.6 Consider the function

$$f(x) = \begin{cases} \cos x & x < 0 \\ A + Bx^n & x \geq 0 \end{cases}$$

where A, B are constants and the integer $n \geq 1$. For what values of A, B and n is the function (a) continuous, (b) differentiable, both at $x = 0$?

3.7 Use the limiting procedure of Eqn. (3.10) to differentiate:
(a) $2x^3 + 4x + 3$ (b) x^{-2} and (c) $f(x) = 5\cos(3x)$.

3.8 Differentiate:
(a) $x^3 e^x$ (b) $\sinh(x)/x$ (c) $\arccos x$
(d) $\operatorname{arcsinh} x$ (e) e^{2x^3} (f) $3\ln(1+x^2)$

3.9 Differentiate:
(a) $\sin(\ln x)$ (b) $x\cos x/(1-x^2)$ (c) $\arctan(1+x^2)$ (d) $\ln(\ln x)$

3.10 Differentiate:
(a) x^x (b) $x^{\cos x}$ (c) $\ln\left[\sin(1/x^2)\right]$ (d) $\ln(\sec x)$

3.11 Differentiate (a) $y = a^x$ and (b) $y = \log_a x$. For $y = a^x$, compare your result with that obtained from the limiting procedure of Example 3.6 for $a = 2$.

3.12 Find $\mathrm{d}y/\mathrm{d}x$ when $x = t(1+t^2)$ and $y = 2/(1+t^2)$. Express your answer in terms of x and y.

3.13 Neglecting air resistance, the path of a projectile moving under gravity is given by

$$x = ut, \qquad y = wt - \tfrac{1}{2}gt^2,$$

where x is the horizontal distance travelled, y is the height, t is the time, and u and w are constants. Calculate the angle of flight and the rate of loss of kinetic energy at time t.

3.14 Find the points where the tangent to the curve

$$3x^2 - 6y^2 + 3xy - 8x + 8 = 0$$

at $(x, y) = (1, 1)$ intercept the x and y axes, respectively.

3.15 Find the value of $\mathrm{d}y/\mathrm{d}x$ at the point $(x, y) = (2, -1)$, where

$$2x^2 + y^3 - xy^2 - y + 10 = 0$$

and the equation of the normal to the curve at this point.

3.16 How many derivatives of the function

$$f(x) = \begin{cases} x\sin x & x < 0 \\ x^2 & 0 \leq x \end{cases}$$

exist at $x = 0$?

3.17 Find general formulas for $f^{(n)}(x)$ where
(a) $f(x) = \sinh 2x$, (b) $f(x) = \ln x$.

3.18 Show that

$$\frac{\mathrm{d}^3x}{\mathrm{d}y^3} = \frac{3}{(\mathrm{d}y/\mathrm{d}x)^5}\left(\frac{\mathrm{d}^2y}{\mathrm{d}x^2}\right)^2 - \frac{1}{(\mathrm{d}y/\mathrm{d}x)^4}\frac{\mathrm{d}^3y}{\mathrm{d}x^3}.$$

3.19 (a) Prove the Leibnitz formula:

$$\frac{\mathrm{d}^n\left[f(x)g(x)\right]}{\mathrm{d}x^n} = \sum_{r=0}^{n}\binom{n}{r}\frac{\mathrm{d}^{n-r}f(x)}{\mathrm{d}x^{n-r}}\frac{\mathrm{d}^r g(x)}{\mathrm{d}x^r},$$

where $\binom{n}{r}$ are the binomial coefficients.

(b) Hence evaluate the fourth derivative of $x^2 \ln x$.

3.20 (a) Find the stationary points of the function $f(x) = x^2 \exp(-x^2)$ and identify them as maxima or minima. (b) Sketch the resulting curve.

3.21 Locate the maxima and minima of $f(x) = e^{-x}\sin x$.

3.22 Make a sketch of the function $y(x) = x^3/(x^2 - x - 2)$ showing clearly its main features.

3.23 Find the stationary points of the function

$$f(x) = \frac{(2x^2 - 5x - 25)}{(x^2 + x - 2)}$$

and state their nature. Locate any points of discontinuity and evaluate the limits of $f(x)$ as $x \to \pm\infty$. Sketch the form of $f(x)$.

3.24 Use a graphical method to find approximate values for the real solutions of the equation $x^3 - 4x^2 + x + 4 = 0$.

3.25 Sketch the function $f(x) = \cos(10\pi x)\exp(-x^2)$.

3.26 The normal at a point $P(x_1, y_1)$ on an ellipse of eccentricity e and centre at the origin, as shown in Figure 3.9, meets the major axis at a point$A$. If F is a focus, show that $AF = ePF$.

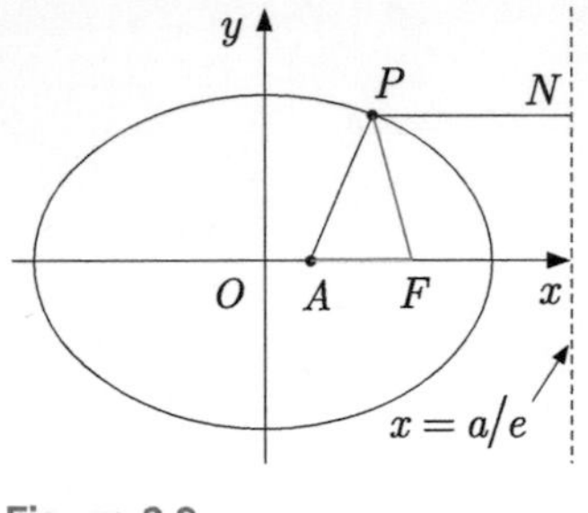

Figure 3.9

3.27 The tangent at a point P on the parabola $y^2 = 4ax$ meets the directrix at a point Q. The straight line through Q parallel to the axis of the parabola meets the normal at P at a point R. Show that the locus of R is a parabola and find its vertex.

4

Integral calculus

We now turn from differentiation to the other crucial ingredient of the infinitesimal calculus, namely *integration*. This may be approached in two ways: either as the inverse process of differentiation; or as the means of calculating the area under a given curve, using an argument that serves as a template for many other important applications. However, before discussing these, we must develop the above two approaches, and the relations between them.

4.1 Indefinite integrals

Given a function $f(x)$, the *indefinite integral* $F(x)$ is defined as the most general solution of the equation

$$\frac{\mathrm{d}F(x)}{\mathrm{d}x} = f(x). \tag{4.1}$$

It is not unique. Suppose we have a particular solution $F_0(x)$, with

$$\frac{\mathrm{d}F_0(x)}{\mathrm{d}x} = f(x). \tag{4.2}$$

Then the most general solution can be written

$$F(x) = F_0(x) + G(x), \tag{4.3}$$

where $G(x)$ is any function such that (4.1) is satisfied. On substituting (4.3) into (4.1) and using (4.2), one obtains $\mathrm{d}G(x)/\mathrm{d}x = 0$, and so $G(x) = c$, where c is a constant. Hence the indefinite integral is given by

$$F(x) = F_0(x) + c, \tag{4.4}$$

Mathematics for Physicists, First Edition. B.R. Martin and G. Shaw.

Companion website: www.wiley.com/go/martin/mathsforphysicists

where $F_0(x)$ is any particular solution of (4.1) and c is an arbitrary constant. An alternative way of writing the indefinite integral is

$$\int f(x)\,\mathrm{d}x \tag{4.5}$$

for reasons that will become clearer in the next section. Thus (4.4) is often written in the form

$$\int f(x)\,\mathrm{d}x = F_0(x) + c. \tag{4.6}$$

The process of determining the indefinite integral of a given function $f(x)$ is called *integration*, $f(x)$ is called the *integrand*[1], and c the *integration constant*. In simple cases, it can be done using the standard table of derivatives, Table 3.1, together with the basic properties

$$\int a\,f(x)\,\mathrm{d}x = a\int f(x)\,\mathrm{d}x, \tag{4.7a}$$

or, more generally,

$$\int [af(x) + bg(x)]\,\mathrm{d}x = a\int f(x)\,\mathrm{d}x + b\int g(x)\,\mathrm{d}x, \tag{4.7b}$$

which follow directly from the definition (4.1), where a, b are arbitrary constants and $f(x)$ and $g(x)$ are arbitrary functions. For example, from Table 3.1 we have

$$\frac{\mathrm{d}}{\mathrm{d}x}x^{n+1} = (n+1)x^n,$$

and so deduce that

$$\int (n+1)x^n\,\mathrm{d}x = x^{n+1} + c,$$

and hence, using (4.7a),

$$\int x^n\,\mathrm{d}x = \frac{x^{n+1}}{n+1} + c' \qquad (n \neq -1),$$

where $c' = c/(n+1)$. For the case $n = -1$, we can use the result

$$\frac{\mathrm{d}}{\mathrm{d}x}\ln x = \frac{1}{x} \qquad (x > 0),$$

[1] The order of the terms in the integrals (4.5) and (4.6) is irrelevant, provided they are both after the integral sign, and some authors place the differential element $\mathrm{d}x$ before the integrand $f(x)$, i.e.

$$\int \mathrm{d}x\,f(x) = \int f(x)\mathrm{d}x \neq f(x)\int \mathrm{d}x.$$

Table 4.1 Standard indefinite integrals

Integrand	Integral
$x^n (n \neq -1)$	$x^{n-1}/(n+1) + c$
e^x	$e^x + c$
$1/x$	$\ln\lvert x\rvert + c$
$\sin x$	$-\cos x + c$
$\cos x$	$\sin x + c$
$\sec^2 x$	$\tan x + c$
$\cosh x$	$\sinh x + c$
$\sinh x$	$\cosh x + c$
$\text{sech}^2\, x$	$\tanh x + c$

from Table 3.1, so that

$$\int \frac{\mathrm{d}x}{x} = \ln x + c \qquad (x > 0). \tag{4.8}$$

In this way, one builds up the table of standard indefinite integrals shown in Table 4.1, from which other integrals may be deduced using (4.7). More complicated integrals will be considered later. In all cases, there is an undetermined integration constant c, whose value can only be determined given additional information, as illustrated in Example 4.2 below.

Example 4.1

Evaluate the indefinite integrals:

$$(a) \int \tan^2\theta\,\mathrm{d}\theta\,, \quad (b) \int [5x^3 + 2e^x]\,\mathrm{d}x\,, \quad (c) \int \frac{(1+x)^3}{x^4}\,\mathrm{d}x\,.$$

Solution

(a) Setting $\tan^2\theta = \sec^2\theta - 1$, and using Table 4.1 gives

$$I = \int \tan^2\theta\,\mathrm{d}\theta = \int (\sec^2\theta - 1)\,\mathrm{d}\theta = \tan\theta - \theta + c\,,$$

where c is a constant.

(b) Using (4.7b),

$$I = \int [5x^3 + 2e^x]\,\mathrm{d}x = 5\int x^3\mathrm{d}x + 2\int e^x\mathrm{d}x\,,$$

and then from Table 4.1

$$I = 5x^4/4 + c' + 2e^x + c'' = 5x^4/4 + 2e^x + c\,,$$

where $c = c' + c''$ is a constant.

(c)

$$I=\int\frac{(1+x)^3}{x^4}\,\mathrm{d}x=\int\frac{(1+3x+3x^2+x^3)}{x^4}\,\mathrm{d}x\,,$$

which using (4.7b) is

$$I=\int x^{-4}\,\mathrm{d}x+3\int x^{-3}\,\mathrm{d}x+3\int x^{-2}\,\mathrm{d}x+\int x^{-1}\,\mathrm{d}x,$$

and finally, using Table 4.1,

$$I=-\frac{1}{3x^3}-\frac{3}{2x^2}-\frac{3}{x}+\ln x+c\,,$$

where c is a constant.

Example 4.2

A car, initially at rest at $x=0$, $t=0$, moves with acceleration

$$\frac{\mathrm{d}v}{\mathrm{d}t}=\alpha\,(t_0^2-t^2)\,,\qquad\qquad t\le t_0$$

where v is its velocity, until it reaches its maximum velocity at $t=t_0$. If $\alpha=0.5\,\mathrm{ms}^{-4}$ and $t_0=5\,\mathrm{s}$, what is the maximum velocity and how far does the car travel before it reaches it?

Solution

The velocity is given by

$$v=\int\alpha(t_0^2-t^2)\,\mathrm{d}t=\alpha\,t_0^2t-\alpha\,t^3\big/3+c=\alpha\,t_0^2t-\alpha\,t^3\big/3\,,$$

where the integration constant $c=0$ because $v=0$ at $t=0$. Similarly, since

$$v=\frac{\mathrm{d}x}{\mathrm{d}t}=\alpha\,t_0^2t-\frac{\alpha\,t^3}{3},$$

integrating gives

$$x=\alpha\,t_0^2t^2\big/2-\alpha\,t^4\big/12+c=\alpha\,t_0^2t^2\big/2-\alpha\,t^4\big/12$$

and $c=0$ because $x=0$ at $t=0$. Then, using $\alpha=0.5\,\mathrm{ms}^{-4}$ and $t=t_0=5\,\mathrm{s}$ gives

$$v_{\mathrm{max}}=v(t=5\mathrm{s})=41.7\,\mathrm{ms}^{-1}\quad\text{and}\quad x(t=5\mathrm{s})=130\,\mathrm{m}.$$

4.2 Definite integrals

In this section we introduce the 'definite integral', defined in terms of areas, and relate it to the indefinite integral of the previous section.

4.2.1 Integrals and areas

Consider a function $f(x)$ that is continuous in the interval $a \leq x \leq b$. Then the *definite integral*, written as,

$$\int_a^b f(x)\,\mathrm{d}x \tag{4.9}$$

is defined to be the area between the curve $y = f(x)$ and the x-axis between the points $x = a$ and $x = b$, where the areas above and below the axis are defined to be positive and negative, respectively. The quantities a and b are called the *lower and upper limits of integration*, respectively. Thus in Figure 4.1,

$$\int_a^b f(x)\,\mathrm{d}x = A - B + C,$$

where A, B, C, are the magnitudes of the areas shown. Two simple results that follow directly from this definition are

$$\int_a^a f(x)\,\mathrm{d}x = 0, \tag{4.10a}$$

and for $a < b < c$,

$$\int_a^c f(x)\mathrm{d}x = \int_a^b f(x)\mathrm{d}x + \int_b^c f(x)\mathrm{d}x\,. \tag{4.10b}$$

The definition of the definite integral can be extended to the case where $a > b$ in a way consistent with (4.10a) and (4.10b) by defining

$$\int_a^b f(x)\mathrm{d}x \equiv -\int_b^a f(x)\mathrm{d}x\,. \tag{4.10c}$$

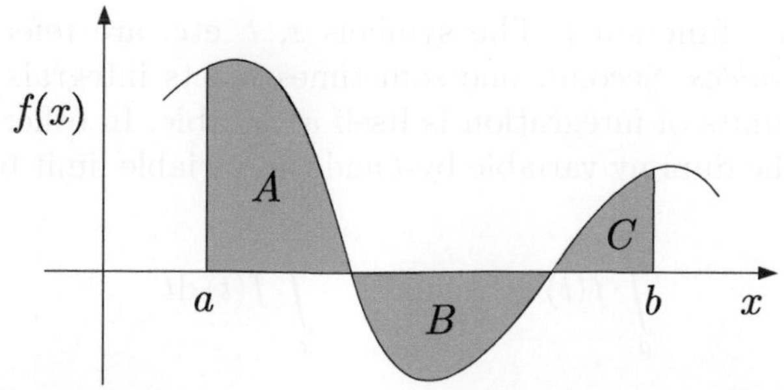

Figure 4.1 Integral of the function $f(x)$ between the limits a and b.

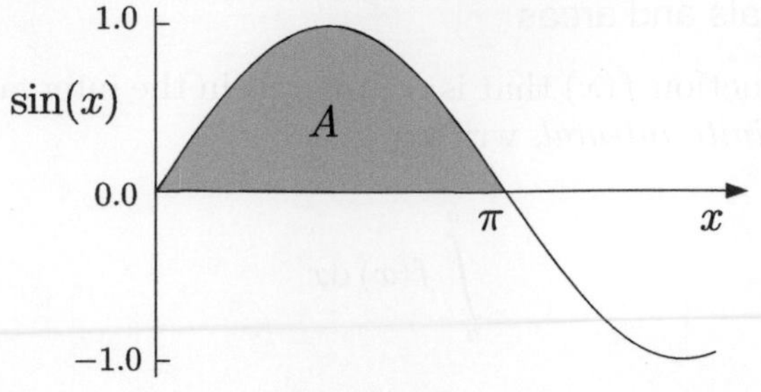

Figure 4.2 Integral of the function $\sin x$ between the limits 0 and π.

However, the key result, which will be derived in the next section, is

$$\int_a^b f(x)\mathrm{d}x = F(b) - F(a) = F_0(b) - F_0(a), \qquad (4.11)$$

where $F(a)$ is the indefinite integral (4.6) and $F_0(x)$ is any solution of (4.2), since the constant c in (4.6) cancels on taking the difference $F_0(b) - F_0(a)$. For example, the area under the curve $y = \sin x$ between $x = 0$ and $x = \pi$, shown in Figure 4.2, is given by

$$A = \int_0^\pi \sin x \, \mathrm{d}x,$$

and since, from Table 4.1,

$$\int \sin x \, \mathrm{d}x = -\cos x + c,$$

Equation (4.11) gives

$$A = (-\cos\pi) - (-\cos 0) = 2.$$

Before proceeding to derive (4.11), it is worth mentioning here three further points. First, integrals like (4.9) are not functions of x, and the variable in the integrand is therefore not significant, i.e.

$$F_{ab} \equiv \int_a^b f(x)\mathrm{d}x = \int_a^b f(t)\,\mathrm{d}t\,,$$

for any given function f. The symbols x, t, etc. are referred to as *dummy variables*. Second, one sometimes meets integrals in which one of the limits of integration is itself a variable. In other words, if we denote the dummy variable by t and the variable limit by x, then

$$\int_a^x f(t)\,\mathrm{d}t \quad \text{and} \quad \int_x^b f(t)\,\mathrm{d}t$$

are both functions of x, and on differentiating with respect x using (4.11) and (4.1) one obtains

$$\frac{\mathrm{d}}{\mathrm{d}x}\int_a^x f(t)\,\mathrm{d}t = f(x) \quad \text{and} \quad \frac{\mathrm{d}}{\mathrm{d}x}\int_x^b f(t)\,\mathrm{d}t = -f(x). \tag{4.12}$$

Finally, when evaluating definite integrals it is useful to introduce the notation

$$[f(x)]_a^b \equiv f(b) - f(a). \tag{4.13}$$

For example,

$$\left[\frac{\sin x}{x} + \tan x\right]_a^b = \frac{\sin b}{b} + \tan b - \frac{\sin a}{a} - \tan a.$$

Example 4.3

Find the area between the curves:

(a) $y = \sqrt{x}$ and $y = x^2$ in the range $0 < x < 1$,

(b) $y = (1+x)^3/x^4$ and $y = \sinh x$ in the range 1.1 to 1.6.

Solution

The two curves are shown in Figure 4.3, and the coloured areas are those required.

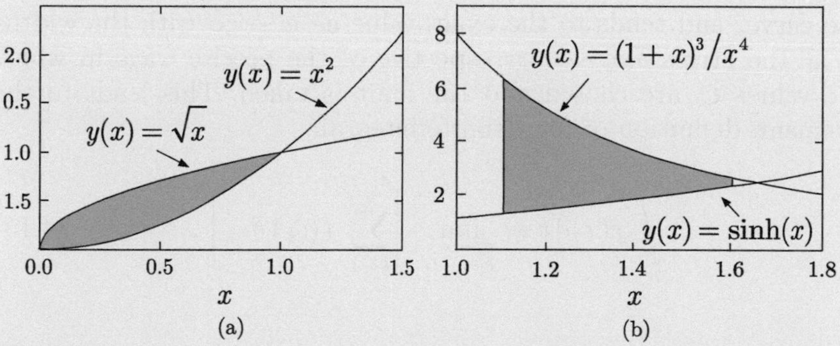

Figure 4.3

(a)

$$A = \int_0^1 \sqrt{x}\,\mathrm{d}x - \int_0^1 x^2\,\mathrm{d}x = \left[\frac{2}{3}x^{3/2}\right]_0^1 - \left[\frac{x^3}{3}\right]_0^1 = \frac{1}{3}.$$

(b) Using the results of Example 4.1(c) and Table 4.1, we have,

$$A = \left[-\frac{1}{3x^3} - \frac{3}{2x^2} - \frac{3}{x} + \ln x \right]_{1.1}^{1.6} - [\cosh]_{1.1}^{1.6}$$

$$= [(-2.0723) - (-4.1221)] - [(2.5775) - (1.6685)]$$

$$= 1.1408\,.$$

4.2.2 Riemann integration

It remains to derive (4.11), relating the indefinite integral (4.6), defined as the inverse of differentiation, to the definite integral (4.9), defined as the area under the curve $f(x)$ in the given range $a \leq x \leq b$. To do this, we need to express the area under the curve explicitly in terms of the function $f(x)$ itself. This is achieved by the construction of Figure 4.4, in which the range $a \leq x \leq b$ has been divided into n strips of width

$$\delta x_k = x_k - x_{k-1} \qquad k = 1, 2, \ldots, n$$

where $x_0 = a$ and $x_n = b$. We then consider the quantity

$$\sum_{k=1}^{n} f(\zeta_k)\, \delta x_k, \tag{4.14}$$

where ζ_k is any point within the kth strip, i.e. $x_{k-1} \leq \zeta_k \leq x_k$. For continuous functions, (4.14) is an approximation to the area under the curve, and tends to the exact value as $n \to \infty$ with the widths of all the strips $\delta x_k \to 0$, irrespective of the precise ways in which the values ζ_k are chosen and the limit is taken. This leads to the Riemann definition of the definite integral:

$$\int_b^b f(x)\mathrm{d}x \equiv \lim_{\substack{n\to\infty \\ \delta x_k \to 0}} \left[\sum_{k=1}^{n} f(\zeta_k)\, \delta x_k \right], \tag{4.15}$$

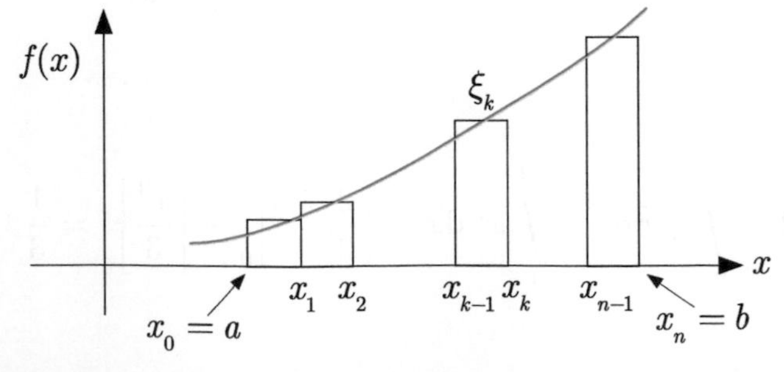

Figure 4.4 Construction to derive Eq. (4.11).

where $f(x)$ is, for the moment, assumed to be continuous in the interval $a \leq x \leq b$. It leads directly to the important general results (4.10a) and (4.10b), which we obtained in the previous section from the intuitive understanding of the area under the curve. In addition, it enables the proof of the crucial result (4.11). However, to do this, we must first prove another important result, called the *first mean value theorem for integration.*

Suppose f_m and f_M are the minimum and maximum values of $f(x)$ in the interval $a \leq x \leq b$. Then clearly

$$(b-a)\, f_m \leq \int_a^b f(x)\,\mathrm{d}x \leq (b-a)\, f_M,$$

and since $f(x)$ varies continuously between f_m and f_M, there must be at least one value $x = \zeta$ in this range for which

$$\int_a^b f(x)\mathrm{d}x = (b-a)\, f(\zeta)\,. \qquad a \leq \zeta \leq b \tag{4.16}$$

This result is called the *first mean value theorem for integration*, and $f(\zeta)$ is the mean value of $f(x)$ in the interval $a \leq x \leq b$.

We can now derive the key result (4.11) by defining

$$\tilde{F}(a,b) \equiv \int_a^b f(x)\mathrm{d}x$$

and considering

$$\frac{\mathrm{d}\tilde{F}(a,b)}{\mathrm{d}b} = \lim_{\delta b \to 0} \left[\frac{\tilde{F}(a, b+\delta b) - \tilde{F}(a,b)}{\delta b}\right], \tag{4.17}$$

where a is kept fixed. By (4.10b) and the mean value theorem (4.16),

$$\tilde{F}(a, b+\delta b) - \tilde{F}(a,b) = \int_b^{b+\delta b} f(x)\mathrm{d}x = \delta b\, f(\zeta),$$

with $b \leq \zeta \leq b + \delta b$. Hence $f(\zeta) \to f(b)$ as $\delta b \to 0$, and (4.17) gives

$$\frac{\mathrm{d}\tilde{F}(a,b)}{\mathrm{d}b} = f(b).$$

Integrating with respect to b using (4.4) then gives

$$\tilde{F}(a,b) = F_0(b) + c_a, \tag{4.18}$$

where F_0 is any indefinite integral of f and where the integration constant c_a is independent of b. A similar argument, treating a as a variable with b fixed, leads to

$$\frac{\mathrm{d}\tilde{F}(a,b)}{\mathrm{d}a} = -f(a)$$

and hence

$$\tilde{F}(a,b) = -F_0(a) + c_b, \tag{4.19}$$

where c_b is independent of a. Equations (4.18) and (4.19) are only compatible if

$$\tilde{F}(a,b) \equiv \int_a^b f(x)\mathrm{d}x = F_0(b) - F_0(a) + c,$$

where c is independent of a and b, and using (4.10a) one sees that $c = 0$. Hence, using (4.4), we obtain

$$\int_a^b f(x)\mathrm{d}x = F(b) - F(a).$$

This is the desired relation between the definite integral (4.9) and the indefinite integral (4.6).

Example 4.4

Use the first mean value theorem for integration to find the upper (U) and lower (L) bounds of the integral

$$I = \int_0^\pi (9 - \cos^2 x)^{-1/2}\mathrm{d}x.$$

Solution

By the mean value theorem (4.16), there exists a value of x, say ζ, in the range $0 \le \zeta \le \pi$ such that,

$$I = \int_0^\pi \frac{1}{(9-\cos^2 x)^{1/2}}\mathrm{d}x = \frac{\pi}{(9-\cos^2\zeta)^{1/2}}, \quad 0 \le \zeta \le \pi. \tag{4.20}$$

The maximum and minimum values of the integrand occur for x equal to π or 0, and $x = \pi/2$. Setting ζ equal to these values in turn gives $U = \pi\big/2\sqrt{2}$ and $L = \pi/3$. Hence the integral must lie between these two values. (The value of I is in fact 1.08, which does indeed lie between $L = 1.05$ and $U = 1.11$.)

4.3 Change of variables and substitutions

In this and the following two sections we will discuss techniques for integrating given functions, beginning with methods based on a change of variables. We start by summarising the approach in general and then consider specific applications. In all cases, the aim is to relate the given integral to the standard integrals of Table 4.1. The latter are well worth remembering and will be assumed in what follows. More complicated integrals may often be evaluated using the methods in Sections 4.3 and 4.4, together with the results of Table 4.1, but if this fails, there exist several useful reference books that include tables of integrals and there are also online resources[2].

4.3.1 Change of variables

Suppose that

$$F(x) = \int f(x)\mathrm{d}x$$

is the indefinite integral of a given function $f(x)$, where $x \equiv x[z]$ is itself a function of another variable z. Then, using the chain rule (3.29), we have

$$\frac{\mathrm{d}F}{\mathrm{d}z} = \frac{\mathrm{d}F}{\mathrm{d}x}\frac{\mathrm{d}x}{\mathrm{d}z} = f(x)\frac{\mathrm{d}x}{\mathrm{d}z}$$

and so from (4.1),

$$F(x[z]) = \int f(x[z])\,\frac{\mathrm{d}x[z]}{\mathrm{d}z}\,\mathrm{d}z \tag{4.21a}$$

is the original integral expressed in terms of the new variable z. In addition, in the case of a definite integral, we must also modify the limits of integration, i.e.

$$\int_a^b f(x)\,\mathrm{d}x = \int_{z(a)}^{z(b)} f(x[z])\frac{\mathrm{d}x[z]}{\mathrm{d}z}\,\mathrm{d}z\,, \tag{4.21b}$$

where $z(a)$ and $z(b)$ are the values of z at $x = a$ and $x = b$, respectively.

---

[2]Examples are S.M. Selby (1967) Editor *Standard Mathematical Tables*, The Chemical Rubber Company, Cleveland, Ohio, and the definitive volume, I.S. Gradshteyn and I.M. Ryzhik (1980) *Tables of Integrals, Series, and Products*, Academic Press Inc., New York. Examples of online resources are *integrals.wolfram.com* and *www.wolframalpha.com*.

Changes of variable are often used in evaluating integrals. As an example, consider

$$F(x) = \int \cosh(3x+2)\mathrm{d}x.$$

Substituting $z = 3x + 2$ and using (4.21a) gives

$$F(x) = \frac{1}{3}\int \cosh z\,\mathrm{d}z = \frac{1}{3}\sinh z + c,$$

where c is an integration constant and we have used the standard integral for $\cosh z$. Substituting for z gives the final result

$$F(x) = \int \cosh(3x+2)\mathrm{d}x = \frac{1}{3}\sinh(3x+2) + c.$$

In general, integration is a more difficult process than differentiation. So after evaluating an integral, it is always good practice to differentiate the result to check that the original function is recovered. This is easily done in this case, i.e.

$$\frac{\mathrm{d}}{\mathrm{d}x}\left[\frac{1}{3}\sinh(3x+2)+c\right] = \cosh(3x+2),$$

but other cases can be more complicated.

Example 4.5

Evaluate (a) the indefinite integrals

$$\text{(i)}\ \int (2x-3)^5\mathrm{d}x \qquad \text{(ii)}\ \int x\exp(-x^2)\mathrm{d}x,$$

and (b) the integral of $\tan^4\theta$ between the limits $\theta = 0$ and $\theta = \pi/4$.

Solution

(a) (i) Substituting $z = 2x - 3$ and using (4.21) gives

$$\int (2x-3)^5\mathrm{d}x = \tfrac{1}{2}\int z^5\mathrm{d}z = \tfrac{1}{12}z^6 + c$$
$$= \tfrac{1}{12}(2x-3)^6 + c.$$

(ii) Substituting $z = -x^2$ and using (4.21a) gives

$$\int x\exp(-x^2)\mathrm{d}x = -\tfrac{1}{2}\int \mathrm{d}z\,e^z = -\tfrac{1}{2}e^z + c$$
$$= -\tfrac{1}{2}\exp(-x^2) + c.$$

(b) We have

$$I = \int_0^{\pi/4} \tan^4\theta \, \mathrm{d}\theta = \int_0^{\pi/4} \tan^2\theta \, (\sec^2\theta - 1) \, \mathrm{d}\theta$$

$$= \int_0^{\pi/4} \tan^2\theta \, \sec^2\theta \, \mathrm{d}\theta - \int_0^{\pi/4} \sec^2\theta \, \mathrm{d}\theta + \int_0^{\pi/4} \mathrm{d}\theta.$$

Evaluating the first integral using (4.21) with $z = \tan\theta$ and $\mathrm{d}z/\mathrm{d}\theta = \sec^2\theta$, and using Table 4.1 to evaluate the second integral, gives

$$\int_0^{\pi/4} \tan^2\theta \sec^2\theta \, \mathrm{d}\theta = \int_0^1 z^2 \mathrm{d}z = \left[\tfrac{1}{3}z^3\right]_0^1$$

and so

$$I = \left[\tfrac{1}{3}\tan^3\theta\right]_0^{\pi/4} - [\tan\theta - \theta]_0^{\pi/4}$$
$$= 1/3 - [1 - \pi/4] = (3\pi - 8)/12 = 0.1187\,.$$

4.3.2 Products of sines and cosines

Next we consider integrals of the form

$$\int \sin^m x \cos^n x \, \mathrm{d}x, \qquad (4.22)$$

where m and n are integers with $m,\ n \geq 0$. If *m is odd*, then independent of whether n is odd or even, the integral can be evaluated by substituting $z = \cos x$. Similarly, if *n is odd*, independent of the value of m, it can be evaluated by substituting $z = \sin x$. As an example, consider

$$F(x) = \int \sin^3 x \cos^4 x \, \mathrm{d}x.$$

If we set $z = \cos x$ so that $\sin^2 x = 1 - z^2$ and $\mathrm{d}z/\mathrm{d}x = -\sin x$, then (4.21a) gives

$$F(x[z]) = -\int (1 - z^2) z^4 \, \mathrm{d}z = \int z^6 \mathrm{d}z - \int z^4 \mathrm{d}z\,,$$

where we have used the relation (4.7b). Hence

$$\int \sin^3 x \cos^4 x \, \mathrm{d}x = \tfrac{1}{7}z^7 - \tfrac{1}{5}z^5 + c = \tfrac{1}{7}\cos^7 x - \tfrac{1}{5}\cos^5 x + c,$$

where, as usual, c is an integration constant.

The only remaining possibility is if m and n are both even in (4.22). In this case, neither of the substitutions $z = \sin x$ or $z = \cos x$ helps. However, in these cases the integrals can be evaluated by exploiting the relations

$$\cos^2 x = \tfrac{1}{2}(1 + \cos 2x) \text{ and } \sin^2 x = \tfrac{1}{2}(1 - \cos 2x), \qquad (4.23)$$

which follow from (2.37a). For example,

$$\int \sin^2 x \, \mathrm{d}x = \frac{1}{2}\int \mathrm{d}x - \frac{1}{2}\int \cos 2x \, \mathrm{d}x = \frac{x}{2} - \frac{\sin 2x}{4} + c,$$

where we have used the substitution $z = 2x$ to evaluate the second integral.

Finally, we note that integrals of the type

$$\int \sinh^m x \cosh^n x \, \mathrm{d}x$$

can be evaluated by similar methods, where the sines and cosines are replaced by their hyperbolic analogues, as we demonstrate in Example 4.6(a) below.

Example 4.6

Evaluate the indefinite integrals

$$\text{(a)} \int \cosh^3 x \, \mathrm{d}x \text{ and (b)} \int \sin^2 x \cos^2 x \, \mathrm{d}x.$$

Solution

(a) Setting $z = \sinh x$, $\cosh^2 x = 1 + \sinh^2 x = 1 + z^2$ and $\mathrm{d}z/\mathrm{d}x = \cosh x$, we obtain

$$\int \cosh^2 x \, \mathrm{d}x = \int (1 + z^2)\, \mathrm{d}z$$
$$= z + \tfrac{1}{3}z^3 + c = \sinh x + \tfrac{1}{3}\sinh^3 x + c.$$

(b) On using (4.23) repeatedly we obtain

$$\int \sin^2 x \cos^2 x \, \mathrm{d}x = \tfrac{1}{4}\int (1 - \cos^2 2x)\, \mathrm{d}x$$
$$= \tfrac{1}{8}\int (1 - \cos 4x)\, \mathrm{d}x = \tfrac{1}{8}x - \tfrac{1}{32}\sin 4x + c.$$

4.3.3 Logarithmic integration

Consider integrals of the form

$$F(x) = \int \frac{\phi'(x)}{\phi(x)}\,\mathrm{d}x,$$

where $\phi(x)$ is any differentiable function of x. The substitution $z = \phi(x)$ gives

$$F(x[z]) = \int \frac{\mathrm{d}z}{z} = \ln z + c = \ln \phi + c,$$

provided $z = \phi(x) > 0$; while substituting $z = -\phi(x)$ gives

$$F(x[z]) = \int \frac{\mathrm{d}z}{z} = \ln z + c = \ln(-\phi) + c,$$

providing $z = -\phi(x) > 0$. Combining these results gives the *logarithmic integral*

$$\int \frac{\phi'(x)}{\phi(x)}\mathrm{d}x = \ln|\phi(x)\,| + c. \tag{4.24}$$

For the particular case $\phi(x) = x$, this reduces to

$$\int \frac{1}{x}\,\mathrm{d}x = \ln|x\,| + c, \tag{4.25}$$

which generalises the standard integral (4.8) to all non-zero x.

Example 4.7

Integrate the functions

$$\text{(a)}\ \frac{3x}{x^2+9}, \quad \text{(b)}\ \cot x, \quad \text{(c)}\ \frac{3x(x+2)}{x^3+3x^2+1}.$$

Solution

(a) The integral is

$$I = \int \frac{3x}{x^2+9}\,\mathrm{d}x = \frac{3}{2}\frac{\phi'(x)}{\phi(x)}\,\mathrm{d}x,$$

where $\phi(x) = x^2 + 9$. Hence

$$I = \frac{3}{2}\ln(x^2+9) + c,$$

since $x^2 + 9 > 0$.

(b) The integral is

$$I = \int \cot x \mathrm{d}x = \int \frac{\cos x}{\sin x}\, \mathrm{d}x\,,$$

which is of the form (4.23) with $\phi(x) = \sin x$. Hence $I = \ln|\sin x\,| + c$.

(c) The integral is

$$I = \int \frac{3x(x+2)}{x^3+3x^2+1}\, \mathrm{d}x = \int \frac{3x^2+6x}{x^3+3x^2+1}\, \mathrm{d}x = \int \frac{u'}{u}\, \mathrm{d}x\,,$$

where $u = x^3 + 3x^2 + 1$. Thus

$$I = \ln|u| + c = \ln\left|(x^3+3x^2+1)\right| + c,$$

where c is a constant.

4.3.4 Partial fractions

Rational functions (i.e. ratios of polynomials) can often be decomposed into a sum of simpler functions called partial fractions. This was discussed in detail in Section 2.1.2. We shall not repeat that discussion here, but merely note its usefulness in evaluating integrals of rational functions. For example, consider the integral of the function

$$f(x) = \frac{5x+12}{x^2+5x+6}.$$

The denominator is $x^2 + 5x + 6 = (x+2)(x+3)$, so that

$$f(x) = \frac{5x+12}{(x+2)(x+3)} = \frac{A}{x+2} + \frac{B}{x+3}$$

by (2.17). Hence

$$5x + 12 = A(x+3) + B(x+2),$$

and setting $x = -2$ and then $x = -3$, gives $A = 2$ and $B = 3$, respectively. Hence the integral is

$$\int \frac{5x+12}{x^2+5x+6}\mathrm{d}x = 2\int \frac{\mathrm{d}x}{x+2} + 3\int \frac{\mathrm{d}x}{x+3}$$
$$= 2\ln|x+2\,| + 3\ln|x+3\,| + c.$$

Other examples are given in later sections.

4.3.5 More standard integrals

Consider an integral of the form

$$F(x) = \int \frac{1}{f(x)} \mathrm{d}x, \tag{4.26a}$$

where $f(x)$ is a given function. If we can find a substitution $x = g(z)$, such that $\mathrm{d}x/\mathrm{d}z = \alpha f(x)$, where α is a constant, then equation (4.21a) gives

$$F(x[z]) = \alpha \int \mathrm{d}z = \alpha z + c = \alpha\, g^{-1}(x) + c, \tag{4.26b}$$

where g^{-1} is the inverse function of g (not $1/g(x)$).

Several 'standard integrals' may be evaluated in this way. An example is

$$F(x) = \int \frac{1}{\sqrt{a^2 + x^2}}\, \mathrm{d}x. \tag{4.27a}$$

Substituting $x = a \sinh z$, we have

$$f(x) = \sqrt{a^2 + a^2 \sinh^2 z} = a \cosh z \text{ and } \mathrm{d}x/\mathrm{d}z = a \cosh z,$$

so that (4.26b) gives

$$F(x[z]) = z + c = \sinh^{-1}(x/a) + c, \tag{4.27b}$$

where we have used the notation $\sinh^{-1}$ to denote the inverse function of sinh, as an alternative to arcsinh.

Other standard integrals of a similar type, together with the substitutions required to derive then, are given in Table 4.2. Their use is illustrated in Example 4.8 below.

Table 4.2 More standard integrals and the substitutions used to derive them

Integrand	Substitution	Integral
$1/\sqrt{a^2 - x^2}$	$x = a \sin z$	$\sin^{-1}(x/a) + c$
$1/\sqrt{a^2 + x^2}$	$x = a \sinh z$	$\sinh^{-1}(x/a) + c$
$1/\sqrt{x^2 - a^2}$	$x = a \cosh z$	$\cosh^{-1}(x/a) + c,\ \lvert x \rvert > a$
$1/(a^2 + x^2)$	$x = a \tan z$	$(1/a) \tan^{-1}(x/a) + c$
$1/(x^2 - a^2)$	$x = a \coth z$	$(1/a) \coth^{-1}(x/a) + c,\ \lvert x \rvert < a$
$1/(a^2 - x^2)$	$x = a \tanh z$	$(1/a) \tanh^{-1}(x/a) + c,\ \lvert x \rvert < a$

Example 4.8

Evaluate the integrals

$$\text{(a)} \int \frac{dx}{\sqrt{x^2-4x+8}} \quad \text{and} \quad \text{(b)} \int \frac{2x+1}{(1+x^2)}\,dx\,.$$

Solution

(a) On completing the square,

$$x^2-4x+8=(x-2)^2+2^2,$$

and then setting $z=x-2$, using the standard integral (4.27b), gives

$$\int \frac{dx}{\sqrt{x^2-4x+8}}=\int \frac{dz}{\sqrt{z^2+2^2}}=\sinh^{-1}\left(\frac{x-2}{2}\right)+c\,.$$

(b) The integral is

$$\begin{aligned}\int \frac{2x+1}{(1+x^2)}\,dx &= \int \frac{2x}{(1+x^2)}\,dx+\int \frac{1}{(1+x^2)}\,dx \\ &= \ln(1+x^2)+\tanh^{-1}x+c,\end{aligned}$$

where we have used the logarithmic integration (4.25) and a standard integral from Table 4.2.

4.3.6 Tangent substitutions

It is sometime useful to convert integrals involving $\sin x$ and $\cos x$ into integrals over rational functions by the substitution

$$t=\tan(x/2), \qquad \frac{dt}{dx}=\frac{\sec^2(x/2)}{2}=\frac{1+t^2}{2}, \tag{4.28}$$

whence

$$\sin x = 2\sin(x/2)\cos(x/2) = \frac{2\tan(x/2)}{\sec^2(x/2)}=\frac{2t}{1+t^2}, \tag{4.29a}$$

$$\cos x = 1-2\sin^2(x/2) = 1-\frac{2\tan^2(x/2)}{\sec^2(x/2)}=\frac{1-t^2}{1+t^2}, \tag{4.29b}$$

and

$$\tan x = \frac{2t}{1-t^2}. \tag{4.29c}$$

Similarly, expressions involving $\sin^2 x$ and $\cos^2 x$ can sometimes be converted to rational functions by the substitutions

$$t=\tan x, \qquad \frac{dt}{dx}=\sec^2 x = 1+t^2, \tag{4.30}$$

whence

$$\cos^2 x = \frac{1}{1+t^2}, \tag{4.31a}$$

and

$$\sin^2 x = 1 - \cos^2 x = \frac{t^2}{1+t^2}. \tag{4.31b}$$

The resulting integrals can then be evaluated by the standard methods for rational functions.

Example 4.9

Evaluate the integrals

$$\text{(a)} \int \frac{1}{3+\cos x}\,\mathrm{d}x\,, \qquad \text{(b)} \int \frac{\tan x}{1+\sin^2 x}\,\mathrm{d}x.$$

Solution

(a) Making the substitutions (4.28) and (4.29) gives

$$\int \frac{1}{3+\cos x}\,\mathrm{d}x = \int \frac{1}{2+t^2}\,\mathrm{d}t$$
$$= \frac{1}{\sqrt{2}}\tan^{-1}\left(\frac{t}{\sqrt{2}}\right) + c = \frac{1}{\sqrt{2}}\tan^{-1}\left[\frac{\tan(x/2)}{\sqrt{2}}\right] + c.$$

(b) Making the substitutions (4.30) and (4.31) gives

$$\int \frac{\tan x}{1+\sin^2 x}\,\mathrm{d}x = \int \frac{t}{1+2t^2}\,\mathrm{d}t = \frac{1}{2}\int \frac{1}{1+2t^2}\,\mathrm{d}t^2$$
$$= \tfrac{1}{4}\ln\left|(1+2t^2)\right| + c = \tfrac{1}{4}\ln\left|(1+2\tan^2 x)\right| + c\,.$$

4.3.7 Symmetric and antisymmetric integrals

If an integrand has a definite symmetry, either even or odd, then this can be exploited to reduce the calculations involved in evaluating its integral. Thus, if $f_-(x) = -f_-(-x)$ is any odd function of x, for example $f_-(x) = \sin x$, a simple result that is often useful is

$$\int_{-a}^{a} f_-(x)\,\mathrm{d}x = 0, \tag{4.32a}$$

which follows directly from the definition of the integral as the area under the curve. Formally, it is obtained by making the substitution $z = -x$ in

$$\int_0^a f_-(x)\,\mathrm{d}x = -\int_0^{-a} f_-(-z)\,\mathrm{d}z = -\int_{-a}^{0} f_-(z)\,\mathrm{d}z,$$

where we have used $f_-(-z) = -f_-(z)$, together with (4.10c). Equation (4.32a) then follows from (4.10b). The corresponding result for even functions $f_+(x) = f_+(-x)$, obtained in the same way, is

$$\int_{-a}^{a} f_+(x)\,\mathrm{d}x = 2\int_{0}^{a} f_+(x)\,\mathrm{d}x. \tag{4.32b}$$

Example 4.10

Evaluate the integral

$$I = \int_{0}^{2} (x-1)^2 \sinh^3(x-1)\,\mathrm{d}x.$$

Solution

On substituting $z = x - 1$, one obtains

$$I = \int_{-1}^{1} z^2 \sinh^3 z\,\mathrm{d}z.$$

Since $\sinh(-z) = -\sinh(z)$, the integrand is an odd function and hence $I = 0$, by (4.32a), without the need for any explicit calculation.

4.4 Integration by parts

On integrating the product rule equation (3.20),

$$\frac{\mathrm{d}}{\mathrm{d}x}(uv) = u\frac{\mathrm{d}v}{\mathrm{d}x} + v\frac{\mathrm{d}u}{\mathrm{d}x},$$

one obtains the formula

$$\int u\frac{\mathrm{d}v}{\mathrm{d}x}\,\mathrm{d}x = uv - \int v\frac{\mathrm{d}u}{\mathrm{d}x}\,\mathrm{dx}, \tag{4.33a}$$

or its equivalent form for definite integrals,

$$\int_{a}^{b} u\frac{\mathrm{d}v}{\mathrm{d}x}\,\mathrm{d}x = [uv]_a^b - \int_{a}^{b} v\frac{\mathrm{d}u}{\mathrm{d}x}\,\mathrm{dx}, \tag{4.33b}$$

where we have used the notation (4.13).

Equations (4.33) are the basic formulas for *integration by parts*. They are often useful for integrals where the integrand can be written as the product of two terms, at least one of which can be easily integrated. For example, consider the integral

$$\int x \cos x \, \mathrm{d}x = \int x \frac{\mathrm{d} \sin x}{\mathrm{d}x} \, \mathrm{d}x \,.$$

On setting $u = x$ and $v = \sin x$, (4.33a) gives,

$$\int x \cos x \, \mathrm{d}x = x \sin x - \int \sin x \, \mathrm{d}x = x \sin x + \cos x + c \,.$$

Integration by parts is also sometimes useful to integrate functions that can be differentiated to give simpler functions. For example,

$$\int \ln x \, \mathrm{d}x = \int \ln x \frac{\mathrm{d}x}{\mathrm{d}x} \, \mathrm{d}x \,,$$

so that (4.33a) gives

$$\int \ln x \, \mathrm{d}x = x \ln x - \int \mathrm{d}x = x \ln x - x + c.$$

Finally, it can also be used to derive relations, called *reduction formulas* (also called *recurrence relations*), between families of integrals I_n whose members are characterised by an integer n. For example, consider the integrals

$$I_n \equiv \int_0^a x^n e^x \, \mathrm{d}x \,, \qquad n \geq 0.$$

Then

$$I_n = \int_0^a x^n \frac{\mathrm{d}e^x}{\mathrm{d}x} \, \mathrm{d}x = [x^n e^x]_0^a - n \int_0^a x^{n-1} e^x \mathrm{d}x,$$

i.e.

$$I_n = a^n e^a - n I_{n-1} \,.$$

This is a typical reduction formula, enabling I_n to be evaluated in terms of I_{n-1}, and hence by repeated application, I_n for any n to be found, starting from

$$I_0 = \int_0^a e^x \, \mathrm{d}x = e^a - 1 \,.$$

Although we have shown that reduction formulas can be obtained using integration by parts, they can sometimes be obtained in other ways, as illustrated in Example 4.12 below.

Example 4.11
Integrate the following functions by parts:

$$\text{(a) } e^x \sin x\,, \qquad \text{(b) } \sqrt{x^2 - 1}\,, \quad |x| > 1.$$

Solution

(a) The integral is

$$\begin{aligned} I = \int e^x \sin x \,\mathrm{d}x &= \int \sin x \frac{\mathrm{d}}{\mathrm{d}x} e^x \,\mathrm{d}x \\ &= e^x \sin x - \int e^x \cos x \,\mathrm{d}x. \end{aligned}$$

Integrating the last term by parts, using $u = \cos x$ and $v = e^x$, then gives

$$I = e^x \sin x - e^x \cos x - \int e^x \sin x \,\mathrm{d}x\,,$$

so that

$$\int e^x \sin x \,\mathrm{d}x = \tfrac{1}{2}\left(e^x \sin x - e^x \cos x\right) + c\,.$$

(b) The integral is

$$I = \int \sqrt{x^2 - 1}\,\mathrm{d}x = x\sqrt{x^2 - 1} - \int \frac{x^2}{\sqrt{x^2 - 1}}\,\mathrm{d}x,$$

where we have integrated by parts using $u = \sqrt{x^2 - 1}$ and $v = x$. Then writing

$$\frac{x^2}{\sqrt{x^2 - 1}} = \sqrt{x^2 - 1} + \frac{1}{\sqrt{x^2 - 1}}$$

in the integral gives

$$I = x\sqrt{x^2 - 1} - \int \sqrt{x^2 - 1}\,\mathrm{d}x - \int \frac{1}{\sqrt{x^2 - 1}}\,\mathrm{d}x,$$

so that

$$\int \sqrt{x^2 - 1}\,\mathrm{d}x = \frac{1}{2}\left[x\sqrt{x^2 - 1} - \cosh^{-1} x\right] + c,$$

where we have used a standard integral from Table 4.2.

Example 4.12
If

$$I_n = \int \frac{x^n}{1 + x^2}\,\mathrm{d}x,$$

show that

$$I_{n+2} = \frac{x^{n+1}}{n+1} - I_n$$

and find I_4.

Solution

This relation is found by forming I_{n+2} from the definition, i.e.

$$I_{n+2} = \int \frac{x^{n+2}}{1+x^2}\,\mathrm{d}x = \int \frac{x^n(1+x^2) - x^n}{1+x^2}\,\mathrm{d}x$$

$$= \int \left[x^n - \frac{x^n}{1+x^2}\right]\mathrm{d}x = \frac{x^{n+1}}{n+1} - I_n\,.$$

To find I_4 we start from

$$I_0 = \int \frac{\mathrm{d}x}{1+x^2} = \arctan x + c.$$

Then,

$$I_2 = x - I_0 = x - \arctan x - c$$

and finally

$$I_4 = x^3/3 - I_2 = x^3/3 - x + \arctan x + c\,.$$

4.5 Numerical integration

In practice, one often needs to evaluate the definite integral (4.9) for functions $f(x)$ where an explicit form for the corresponding indefinite integral (4.6) cannot be found. In these cases, one must resort to a numerical evaluation, usually with the aid of a computer.

There are many methods available for doing this. Here we shall consider only the two simplest, which are based on dividing the integral into strips, as shown in Figure 4.4, Section 4.4.2. We will assume that the widths of the strips are all equal, i.e. $x_n - x_{n+1} = h$ for all n. Then, if we approximate $f(x)$ between x_0 and x_1 by a straight line, the area of the first strip is approximately

$$\frac{h}{2}\left[f(x_0) + f(x_1)\right].$$

Repeating the procedure for the second strip gives its area as

$$\frac{h}{2}\left[f(x_1) + f(x_2)\right]$$

and so on. When the contributions of all the strips are added, we have in this approximation

$$\int_a^b f(x)\,\mathrm{d}x = \frac{h}{2}\left[f_0 + f_n + 2(f_1 + f_2 + \cdots + f_{n-1})\right] \tag{4.34}$$

where

$$f_k = f(x_k)\,, \qquad (k = 0\,, 1\,, 2,\, \ldots\,, n).$$

This approximation is called the *trapezium rule.* Its accuracy increases with n, becoming exact in the limit $n \to \infty$.

An alternative to the trapezium rule is obtained by choosing n to be even, and approximating $f(x)$ across each pair of strips by a quadratic form. Consider the first pair of strips as shown in Figure 4.4, spanning the interval $a \le x \le b$, where $x_0 = a$ and $x_2 = b$. Then approximating

$$f(x) = f(x_1) + \alpha(x - x_1) + \beta(x - x_1)^2,$$

and changing the variable to $y = x - x_1$ gives

$$\begin{aligned}\int_{x_0}^{x_2} f(x)\,\mathrm{d}x &= \int_{x_0}^{x_2} \left[f(x_1) + \alpha(x - x_1) + \beta(x - x_1)^2\right] \\ &= \int_{-h}^{h} \left[f(x_1) + \alpha y + \beta y^2\right] = 2h\left[f(x_1) + \beta h^3/3\right].\end{aligned} \tag{4.35}$$

In the same approximation, we have

$$f(x_2) = f(x_1) + \alpha h + \beta h^2$$

and

$$f(x_0) = f(x_1) - \alpha h + \beta h^2,$$

so that

$$2\beta h^2 = f(x_0) + f(x_2) - 2f(x_1). \tag{4.36}$$

Substituting (4.36) into (4.35) gives

$$\int_{x_0}^{x_2} f(x)\,\mathrm{d}x = \frac{h}{3}\left[f(x_0) + 4f(x_1) + f(x_2)\right],$$

as the contribution from the first two strips. Similarly, the second pair contributes

$$\int_{x_2}^{x_4} f(x)\,\mathrm{d}x = \frac{h}{3}\left[f(x_2) + 4f(x_3) + f(x_4)\right]$$

and so on, giving finally

$$\int_a^b f(x)\,\mathrm{d}x = \frac{h}{3}[f_0 + f_n + 4(f_1 + f_3 + \cdots + f_{n-1}) + 2(f_2 + f_4 + \cdots + f_{n-2})], \tag{4.37}$$

where, once again, $f_k \equiv f(x_k)$ and n is even.

Equation (4.37) is called *Simpson's rule* and is usually more precise than the trapezium rule for a given fixed n. However, like other methods we will not discuss, they are both easy to implement on a computer and tend to the exact result as $n \to \infty$. Hence using either method, one can simply keep increasing n until the resulting value of the integral is stable to the precision required.

Example 4.13

Use Simpson's rule with four intervals to integrate the function $(5x^3 + 2e^x)$ between the limits 0 and 1. Compare your result with the value obtained using the trapezium rule, and the exact value using the integral found in Example 4.1(b).

Solution

With $n = 4$, we find the following values of x_n and f_n:

$$\begin{aligned} x_0 &= 0.00 & f_0 &= 2.00000 \\ x_1 &= 0.25 & f_1 &= 2.64618 \\ x_2 &= 0.50 & f_2 &= 3.92244 \\ x_3 &= 0.75 & f_3 &= 6.34338 \\ x_4 &= 1.00 & f_4 &= 10.43656 \end{aligned}$$

Using these in (4.37) gives

$$\int_0^1 (5x^3 + 2e^x)\,\mathrm{d}x = \frac{0.25}{3}\,[f_0 + f_4 + 4(f_1 + f_3) + 2f_2] = 4.6866.$$

Using the same values of f_n in (4.34) gives the result from the trapezium rule as 4.7826. The exact value may be found using the result of Example 4.1(b) and is

$$\int_0^1 (5x^3 + 2e^x)\,\mathrm{d}x = \left[\frac{5x^4}{4} + 2e^x\right]_0^1 = 4.6866\,,$$

so Simpson's rule is accurate even for small n for this particular function. For other functions more intervals may be required.

4.6 Improper integrals

So far we have restricted the discussion of definite integrals to cases where the limits of the integration a, b are finite and the integrand $f(x)$ is continuous in the range $a \le x \le b$. Here we consider whether it is possible to define integrals when these conditions are not satisfied. Such integrals are sometimes called *improper integrals* and often occur in physical applications.

4.6.1 Infinite integrals

We first consider the case

$$I = \int_a^\infty f(x)\,\mathrm{d}x, \tag{4.38}$$

where $f(x)$ is continuous in the range $a \le x < \infty$. Then (4.38) can be defined by

$$I \equiv \lim_{b\to\infty} I(b) \equiv \lim_{b\to\infty} \int_a^b f(x)\,\mathrm{d}x, \tag{4.39}$$

provided the limit is well-defined and finite. If it is, then the integral (4.38) is said to be *convergent*. If the limit is not well-defined or is infinite, then the integral is said to be *divergent*. Similar considerations apply in an obvious way when the lower limit $a \to -\infty$.

It is easiest to determine whether an integral converges when the corresponding finite integral $I(b)$ can be evaluated explicitly. For example, consider the integral

$$I = \int_0^\infty \cos nx\,\mathrm{d}x. \tag{4.40}$$

The limit

$$\lim_{b\to\infty} \int_0^b \cos nx\,\mathrm{d}x = \lim_{b\to\infty} \left(\frac{\sin nb}{n}\right)$$

is ill-defined, so that (4.40) is divergent. On the other hand,

$$\int_0^b \mathrm{e}^{-\alpha x}\,\mathrm{d}x = \frac{1 - \mathrm{e}^{-\alpha b}}{\alpha},$$

so that

$$\int_0^\infty \mathrm{e}^{-\alpha x}\,\mathrm{d}x = \frac{1}{\alpha}, \qquad \alpha > 0 \tag{4.41}$$

is a convergent integral, but diverges if $\alpha \le 0$.

More generally, it should be clear from (4.39) that the convergence of (4.38) only depends on the behaviour of $f(x)$ as $x \to \infty$, that is, on its asymptotic behaviour. Some useful results then follow from the interpretation of the integral as the area under the curve $y = f(x)$ as $b \to \infty$, as shown in Figure 4.5. In particular, it is clear that (4.39) can only converge to a finite limit if $f(x) \to 0$ as $x \to \infty$. However, this is only a necessary, but not a sufficient condition. For example,

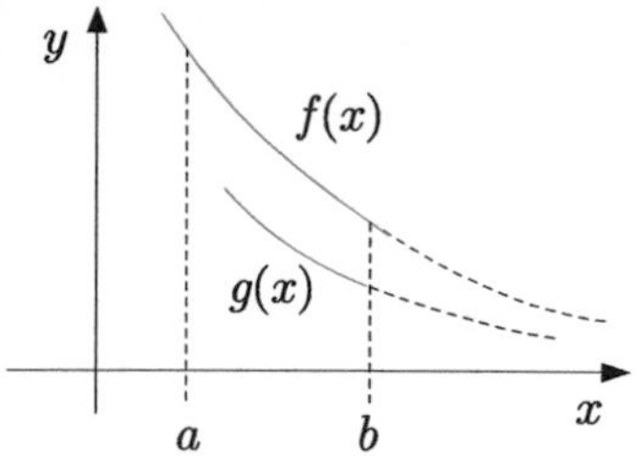

Figure 4.5 Infinite integrals over $f(x)$ and $g(x)$ (see text).

$$\int_1^\infty \frac{dx}{x} = \lim_{b\to\infty} \int_1^b \frac{dx}{x} = \lim_{b\to\infty} [\ln b] = \infty,$$

so that, even though $f(x) \to 0$, the integral does not converge.

Suppose we have a function $g(x)$ that is continuous in the range $a \leq x < \infty$, and suppose that f and g are such that the conditions

$$0 \leq g(x) \leq f(x) \tag{4.42}$$

are satisfied for large x. Then it is clear from Figure 4.5 that if (4.38) converges, then the integral

$$\int_a^\infty g(x)\, dx, \tag{4.43}$$

representing the area under the curve $y = g(x)$, also converges. Furthermore, from (4.7a) and (4.39), it follows that the convergence or otherwise of an integral is not affected by multiplying the integrand by a constant. Hence the condition (4.42) for the convergence of (4.43) can be relaxed to

$$0 \leq g(x) \leq k\, f(x), \tag{4.44}$$

where $k > 0$ is a positive constant. Thus, for example, from the convergence of (4.41) we can immediately infer that integrals like

$$\int_0^\infty \frac{e^{-\alpha x}}{1+x}\, dx \quad \text{and} \quad \int_0^\infty 2e^{-\alpha x} \sin^2 x\, dx$$

also converge if $\alpha > 0$.

A similar argument shows that if (4.38) diverges and $0 \leq f(x) \leq kg(x)$ at large x, with $k > 0$, then the corresponding integral (4.43) also diverges. However, the converse results do not apply; that is, if (4.38) converges and (4.44) is not satisfied, then it does not follow that (4.43) is divergent. On the other hand, if $g(x) \to c\, f(x)$ as $x \to \infty$, i.e. if

$$\lim_{x\to\infty} \frac{g(x)}{f(x)} = c,$$

where c is any finite non-zero constant, then the asymptotic behaviours of $f(x)$ and $g(x)$ are identical, apart from an irrelevant constant. Hence in this case, (4.38) and (4.43) are either both convergent or both divergent.

Example 4.14

(a) For what values of α does the integral

$$\int_1^\infty x^{-\alpha}\,\mathrm{d}x$$

converge?

(b) Show that the integral

$$\int_1^\infty x^n e^{-2x}\mathrm{d}x$$

converges for all finite integer n.

Solution

(a) We have shown above that the integral diverges for $\alpha = 1$. For $\alpha \neq 1$,

$$\int_1^\infty \frac{\mathrm{d}x}{x^\alpha} = \lim_{b\to\infty}\int_1^b \frac{\mathrm{d}x}{x^\alpha} = \lim_{b\to\infty}\left[\frac{x^{1-\alpha}}{1-\alpha}\right]_1^b = \lim_{b\to\infty}\left(\frac{b^{1-\alpha}-1}{1-\alpha}\right).$$

Hence,

$$\int_1^\infty \frac{\mathrm{d}x}{x^\alpha} = \frac{1}{\alpha-1}, \qquad \alpha > 1,$$

and so converges for $\alpha > 1$, but diverges for $\alpha \leq 1$.

(b) $\displaystyle\int_1^\infty x^n e^{-2x}\mathrm{d}x = \int_1^\infty \left(\frac{x^n}{e^x}\right) e^{-x}\,\mathrm{d}x\,.$

Hence, since the integral of e^{-x} converges, this integral converges if $0 < x^n e^{-x} < k$ for large x, where $k > 0$ is a constant. Taking logarithms gives

$$\ln\left(x^n e^{-x}\right) = n\,\ln\,x - x \to -\infty \text{ as } x \to \infty.$$

Hence $x^n e^{-x} \to 0$ as $x \to \infty$ and the condition is satisfied.

4.6.2 Singular integrals

In Section 3.1.2, we introduced the idea of an infinite discontinuity, that is, a point where the value of a given function tended to infinity. In this section we consider integrals in which the integrand is not continuous, but has an infinite discontinuity, also called a *singularity*, at some point x_s in the range of integration $a \leq x \leq b$. In this case we can write the integral in the form

$$\int_a^b f(x)\,\mathrm{d}x = \int_a^{x_s} f(x)\,\mathrm{d}x + \int_{x_s}^b f(x)\,\mathrm{d}x, \qquad (4.45)$$

provided the two integrals on the right-hand side are well-defined. Hence it is sufficient to investigate whether we can define integrals with a singular point at the end of the range of integration. This is done by a limiting process, as in the case of infinite integrals. For example, the second integral above is defined by

$$\int_{x_s}^b f(x)\mathrm{d}x \equiv \lim_{\varepsilon \to 0^+} \int_{x_s+\varepsilon}^b f(x)\,\mathrm{d}x, \qquad (4.46a)$$

assuming a well-defined finite limit exists. If it does, the integral (4.46a) is said to be *convergent* and is well-defined. On the other hand, if a well-defined finite limit does not exist, the integral is said to be *divergent* and is ill-defined. Similar considerations apply to the first integral in (4.45), where the limit analogous to (4.46a) is

$$\int_a^{x_s} f(x)\mathrm{d}x = \lim_{\eta \to 0^+} \int_a^{x_s-\eta} f(x)\,\mathrm{d}x. \qquad (4.46b)$$

The integral (4.45) is only defined if both the integrals on the right-hand side are convergent.

To illustrate these points, consider the family of integrals

$$\int_0^1 \frac{\mathrm{d}x}{x^\alpha}, \qquad \alpha > 0 \qquad (4.47)$$

where α is a real parameter, and the integrand diverges as $x \to x_s = 0$. For $\alpha = 1$ we have

$$\int_0^1 \frac{\mathrm{d}x}{x} = \lim_{\varepsilon \to 0^+} \int_\varepsilon^1 \frac{\mathrm{d}x}{x} = \lim_{\varepsilon \to 0^+} [\ln x]_\varepsilon^1 = -\infty, \qquad (4.48a)$$

so that the integral (4.47) is divergent for $\alpha = 1$. For $\alpha \neq 1$, we have

$$\int_0^1 \frac{dx}{x^\alpha} = \lim_{\varepsilon \to 0^+} \int_\varepsilon^1 \frac{dx}{x^\alpha} = \lim_{\varepsilon \to 0^+} \left[\frac{x^{1-\alpha}}{1-\alpha} \right]_\varepsilon^1 = \lim_{\varepsilon \to 0^+} \left(\frac{1-\varepsilon^{1-\alpha}}{1-\alpha} \right).$$

From this and (4.48a) we see that the integral (4.47) is divergent for $\alpha \geq 1$, whereas for $\alpha < 1$, it is convergent and given by

$$\int_0^1 \frac{dx}{x^\alpha} = \frac{1}{1-\alpha}, \qquad \alpha < 1. \tag{4.48b}$$

More generally, the divergence or otherwise of integrals like (4.46a) and (4.46b) depends solely on the behaviour of the integrand as the singularity is approached, that is, as $x \to x_s$. Hence it is not necessary to evaluate the entire integral explicitly to determine whether it converges. To illustrate this, consider the integral

$$\int_0^1 \frac{\sin x}{x^{3/2}} dx, \tag{4.49}$$

whose integrand is singular at $x = 0$, but is well-behaved as $x \to 1$. To determine its convergence, or otherwise, we therefore examine the behaviour of the integrand as $x \to 0$. To do this we write

$$\lim_{x \to 0} \frac{\sin x}{x^{3/2}} = \lim_{x \to 0} \left(\frac{\sin x}{x} \frac{1}{x^{1/2}} \right) = \lim_{x \to 0} \frac{1}{x^{1/2}},$$

using (3.15). Hence (4.50) has the same convergence properties at $x = 0$ as (4.49) with $\alpha = 1/2$, and, like (4.49), is convergent.

Finally, if both integrals on the right-hand side of (4.45) are divergent, it may be that the integral

$$\int_a^b f(x)dx = \lim_{\varepsilon \to 0^+} \left[\int_a^{x_s - \varepsilon} f(x)dx + \int_{x_s + \varepsilon}^b f(x)dx \right] \tag{4.50}$$

is well-defined. If this is the case, (4.50) is called the *principal value* of the integral and is written

$$P \int_a^b f(x)dx.$$

For example, the integral

$$\int_{-1}^{1} \frac{\mathrm{d}x}{x} = \lim_{\varepsilon \to 0^+} \int_{-1}^{-\varepsilon} \frac{\mathrm{d}x}{x} + \lim_{\eta \to 0^+} \int_{\eta}^{1} \frac{\mathrm{d}x}{x}$$

is ill-defined, since both integrals on the right-hand side are divergent. However, the corresponding principal value integral

$$P\int_{-1}^{1} \frac{\mathrm{d}x}{x} = \lim_{\varepsilon \to 0^+} \left[\int_{-1}^{-\varepsilon} \frac{\mathrm{d}x}{x} + \int_{\varepsilon}^{1} \frac{\mathrm{d}x}{x} \right] = 0$$

is well-defined, and in this case vanishes.

Example 4.15

Show that the integral

$$I = \int_{0}^{2} \frac{\mathrm{d}x}{x^2 - 1}$$

is divergent, but the corresponding principal value integral is well-defined, and find its value.

Solution

The integral is singular at $x = 1$, so by (4.45)

$$I = \int_{0}^{1} \frac{\mathrm{d}x}{x^2 - 1} + \int_{1}^{2} \frac{\mathrm{d}x}{x^2 - 1}$$

is well-defined if, and only if, both the integrals are convergent. Consider the second integral

$$\int_{1}^{2} \frac{\mathrm{d}x}{x^2 - 1} = \int_{1}^{2} \frac{\mathrm{d}x}{(x-1)(x+1)} = \int_{0}^{1} \frac{\mathrm{d}z}{z(z+2)},$$

where we have made the substitution $z = x - 1$. As $z \to 0$, $z(z+2) \to 2z$, so this integral has the same convergence properties as (4.47) for $\alpha = 1$. Hence by (4.48a) it is divergent and the integral I is not well-defined. However, using the indefinite integral

$$\int \frac{\mathrm{d}x}{x^2 - 1} = \frac{1}{2} \ln \left| \frac{x-1}{x+1} \right| + c,$$

where c is a constant, the principal value integral is

$$P\int_0^2 \frac{\mathrm{d}x}{x^2-1} = \lim_{\varepsilon\to 0^+}\left[\int_0^{1-\varepsilon} \frac{\mathrm{d}x}{x^2-1} + \int_{1+\varepsilon}^2 \frac{\mathrm{d}x}{x^2-1}\right] = -\frac{1}{2}\ln 3,$$

which is finite and well-defined.

4.7 Applications of integration

At the beginning of this chapter, we said that the calculation of the area under a curve, which we subsequently based on the Riemann definition (4.15), served as a template for many applications of integration. In this section we illustrate this with some important examples in physics and geometry.

4.7.1 Work done by a varying force

Consider a force $F(x)$, which acts in the x-direction and which varies continuously in the range $a \le x \le b$. To calculate the work done in moving from $x = a$ to $x = b$, we divide the interval into a large number n of small steps δx_k , $k = 1, 2, \ldots$, as shown in Figure 4.4 for $F(x) = f(x)$. If the strip widths δx_k are small, the variation of $F(x)$ across the strip can be neglected and the work done in moving from x_{k-1} to x_k is given by $\delta W_k = F(\zeta_k)\delta x_k$, where $x = \zeta_k$ is any point within the strip. Hence,

$$W = \sum_{\substack{n\to\infty \\ \delta x_k\to 0}} \delta W_k = \sum_{\substack{n\to\infty \\ \delta x_k\to 0}} F(\zeta_k)\delta x_k\,,$$

which on comparing to the Riemann definition (4.14), is just the integral

$$W = \int_a^b F(x)\,\mathrm{d}x\,. \tag{4.51}$$

In other words, the work done by the force $F(x)$ is the area under the curve $y = F(x)$ between $x = a$ and $x = b$.

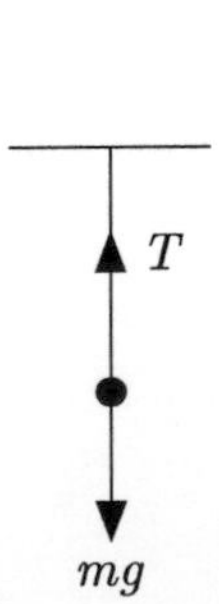

Figure 4.6

Example 4.16

The tension in an elastic string is given by $T = kx$, where k is a constant and x is the extension of the string beyond its natural length ℓ. If a ball of mass m hangs in equilibrium on the end of the string, as shown in Figure 4.6, what is the energy stored in the string?

Solution

If the string is extended by a length x_0, then the energy stored in the string is just the work needed to extend the string, i.e.

$$E = W = \int_0^{x_0} T\,\mathrm{d}x = \int_0^{x_0} kx\,\mathrm{d}x = \frac{1}{2}kx_0^2 .$$

In equilibrium, the forces balance so that $mg = kx_0$. Hence $x_0 = mg/k$ and the energy stored is

$$E = \frac{1}{2}kx_0^2 = \frac{m^2g^2}{2k} .$$

4.7.2 The length of a curve

The length of any curve $y = f(x)$ between any two points $x = a$ and $x = b > a$ may be found by reference to Figure 4.7. Let δl_k be the contribution to the length L arising from the small interval $\delta x_k = x_k - x_{k-1}$. Then from Figure 4.7, we see that in the limit $\delta x_k \to 0$, this is given by

$$\delta l_k = \sqrt{\delta x_k^2 + \delta x_y^2} = \delta x_k \left(1 + \frac{\delta x_y^2}{\delta x_k^2}\right)^{1/2} .$$

Hence, on summing over all segments δx_k and letting $\delta x_k \to 0$, we immediately obtain

$$L = \int_a^b \left[1 + \left(\frac{\mathrm{d}y}{\mathrm{d}x}\right)^2\right]^{1/2} \mathrm{d}x \qquad (4.52)$$

as the desired formula for the length of the curve. In many cases this integral will have to be evaluated numerically.

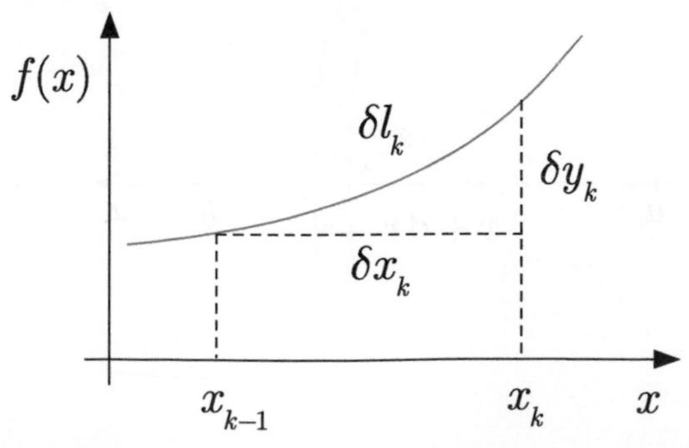

Figure 4.7 Construction used to obtain (4.52)

Example 4.17

Calculate the length of the curve $y = \cosh x$ between $x = 0$ and $x = 1$.

Solution

The integrand in (4.52) for L is

$$\sqrt{1 + \left(\frac{\mathrm{d}y}{\mathrm{d}x}\right)^2} = \sqrt{1 + \sinh^2 x} = \cosh x\,.$$

Therefore

$$L = \int_0^1 \cosh x \,\mathrm{d}x = [\sinh x]_0^1 = \frac{e - 1/e}{2} = 1.175\,.$$

*4.7.3 Surfaces and volumes of revolution

Suppose we form a three-dimensional shape by taking the curve $y = f(x)$, $z = 0$ in the range $a \le x \le b$ and rotating it about the x-axis, as shown in Figure 4.8. Then the area swept out by the curve is called the area of revolution and the volume enclosed is called the volume of revolution. Any three-dimensional shape with an axis of rotation symmetry, chosen to be the x-axis, can be constructed in this way. For example, $y = R$ produces a cylinder of radius R and length $(b - a)$, while

$$y = xR/h, \qquad 0 < x < h \tag{4.53}$$

produces a cone of length h with radius R at the base.

To calculate the surface and volume of revolution, we divide the interval $a \le x \le b$ into slices of infinitesimal width $\mathrm{d}x$. The slice

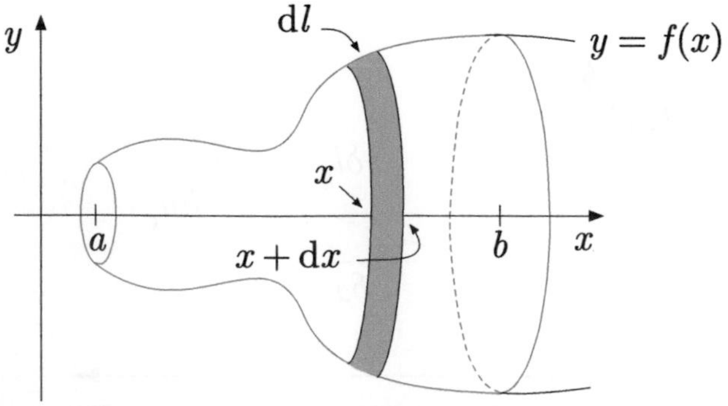

Figure 4.8 The surface of revolution generated by $y = f(x)$ for $a \le x \le b$ and the interval x to $(x + \mathrm{d}x)$.

between an arbitrary x and $x + \mathrm{d}x$ is shown in Figure 4.8. The area of its edge is

$$\mathrm{d}A = 2\pi y\,\mathrm{d}l = 2\pi y\left[1 + \left(\frac{\mathrm{d}y}{\mathrm{d}x}\right)^2\right]^{1/2}\mathrm{d}x,$$

where we have used the result of the previous section for $\mathrm{d}l$, and its volume is $\mathrm{d}V = \pi y^2\mathrm{d}x$. Summing over all strips in the manner of the previous sections then gives

$$V = \int_a^b \pi y^2\mathrm{d}x \qquad (4.54)$$

for the volume of revolution and

$$A = \int_a^b 2\pi y\left[1 + \left(\frac{\mathrm{d}y}{\mathrm{d}x}\right)^2\right]^{1/2}\mathrm{d}x \qquad (4.55)$$

for the area of revolution. To find the surface area of the whole shape, we need to add the areas of the circular discs at $x = a$ and $x = b$.

Example 4.18

Find the formulas for the volume and surface area of a cone of height h and radius R at the base, as shown in Figure 4.9.

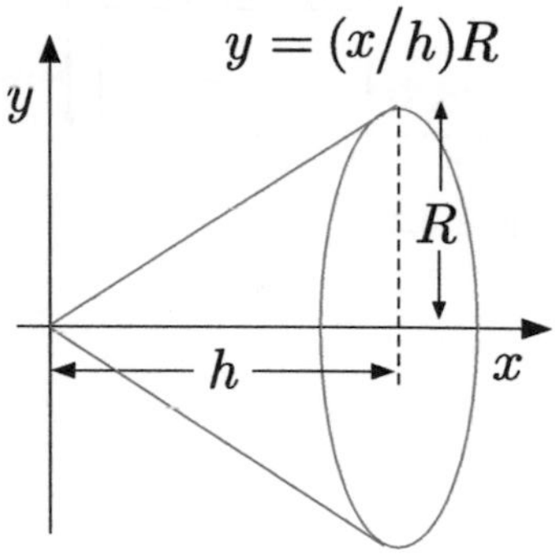

Figure 4.9

Solution

Substituting (4.53) into (4.54) and (4.55) gives

$$V = \frac{\pi R^2}{h^2}\int_0^h x^2\,\mathrm{d}x = \frac{\pi R^2 h}{3} \qquad (4.56a)$$

for the volume of the cone, and

$$A = \int_0^h \frac{2\pi R}{h}\left[\frac{h^2 + R^2}{h^2}\right]^{1/2} x\,\mathrm{d}x = \pi R\left(h^2 + R^2\right)^{1/2} \qquad (4.56b)$$

for the area of revolution. On adding the area of the base, the surface area of the cone is given by

$$A_c = \pi R\left(h^2 + R^2\right)^{1/2} + \pi R^2. \qquad (4.56c)$$

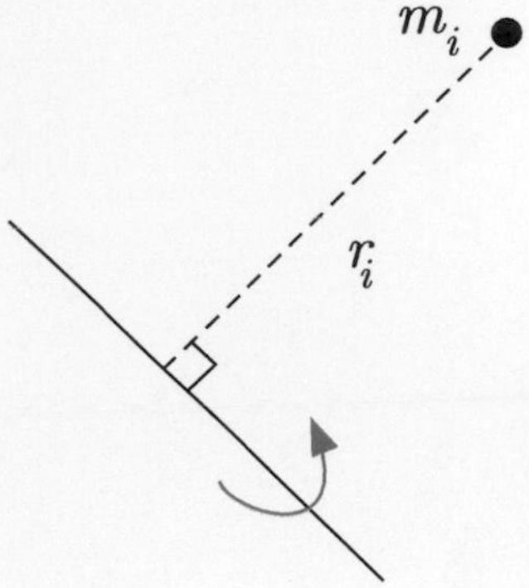

Figure 4.10 Definition of the moment of inertia (4.57) for a point particle.

*4.7.4 Moments of inertia

Moments of inertia play an important role in mechanical problems involving rotations. For a system of point particles, they are defined by

$$I \equiv \sum_i m_i r_i^2, \tag{4.57}$$

where r_i is the perpendicular distance from the mass m_i to the axis of rotation, as illustrated in Figure 4.10. The corresponding moments of inertia for extended objects, which are what is required for most practical applications, are then derived from (4.57) by integration, as we shall illustrate by some simple examples.

Example 4.19

Consider a thin rectangular plate with sides a, b, with $b > a$ and mass M. What is its moment of inertia about an axis passing along one of the short sides?

Solution

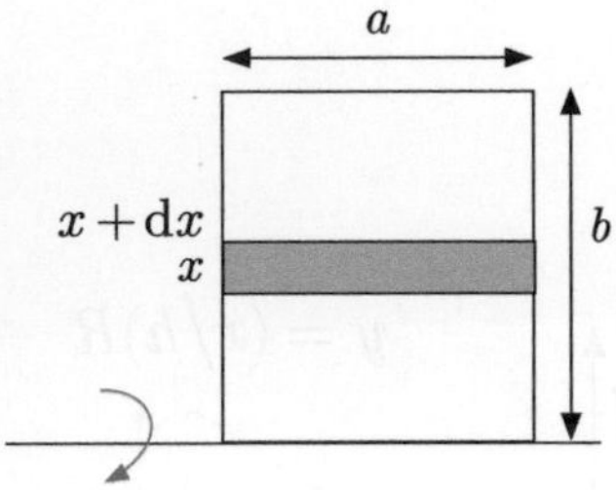

Figure 4.11

Divide the plate into infinitesimal strips of width $\mathrm{d}x$ as shown in Figure 4.11. The mass of a strip is $\sigma a\,\mathrm{d}x$, where $\sigma = M/ab$ is the mass density per unit area; and its distance from the axis is x. Hence its contribution to the total moment of inertia of the plate is

$$\mathrm{d}I = \sigma a x^2 \mathrm{d}x,$$

and summing over the strips gives

$$I = \sigma a \int_0^b x^2\,\mathrm{d}x = \frac{\sigma a b^3}{3} = \frac{M\,b^2}{3} \tag{4.58}$$

for the total moment of inertia.

Example 4.20

Find the moment of inertia of a thin circular disc of radius r and mass m about an axis in the plane of the disc and passing through its centre.

Solution

Consider the thin strip of thickness $\mathrm{d}x$ shown in Figure 4.12, where

$$x = r\sin\theta \quad \text{and} \quad \mathrm{d}x = r\cos\theta\,\mathrm{d}\theta$$

and the length of the strip is $2r\cos\theta$. If the density per unit area is σ, then the mass of the strip is

$$\sigma 2r\cos\theta\,\mathrm{d}x = 2\sigma r^2\cos^2\theta\,\mathrm{d}\theta,$$

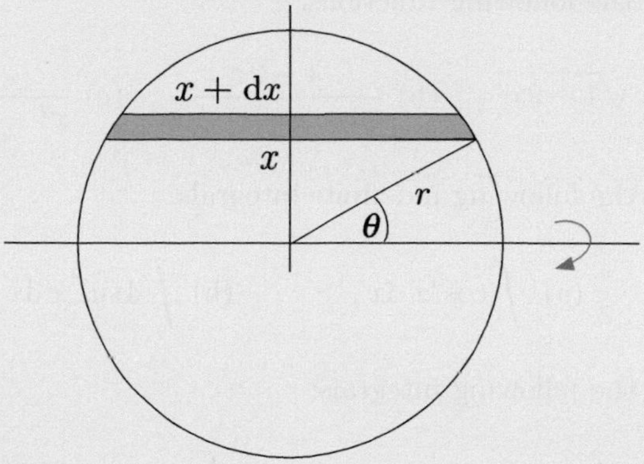

Figure 4.12

and its contribution to the moment of inertia is

$$2\sigma r^2\cos^2\theta\,x^2\,\mathrm{d}\theta = 2\sigma r^4\cos^2\theta\,\sin^2\theta\,\mathrm{d}\theta,$$

so that, on summing over all the strips, the total moment of inertia is

$$I = \sigma r^4 \int_{-\pi/2}^{\pi/2} \cos^2\theta\sin^2\theta\,\mathrm{d}\theta = \frac{\pi\sigma r^4}{8} = \frac{mr^2}{8},$$

where we used the result of Example (4.6) to evaluate the integral.

Problems 4

4.1 Sketch the area between the curves

$$y(x) = 6 - 5x - 2x^2 + x^3 \quad \text{and} \quad y(x) = 1 + 4x + 2x^2 - x^3$$

between the points 1.5 and 3.0 and find its value.

4.2 (a) Prove that for any function $f(x)$ that is differentiable in the interval $a \le x \le b$, there exists a value $x = \zeta$ in the range $a \le \zeta \le b$, such that

$$f(b) - f(a) = (b - a)\,f'(\zeta)$$

is satisfied. This result is called the *first mean value theorem*, as distinct from the first mean value theorem of integration discussed in the text.

(b) Evaluate the mean value of the function $f(x) = x^2/(1 + x^2)$ over the range $0 \le x \le 1$ and verify the first mean value theorem for integration (4.16).

4.3 Use an appropriate substitution to integrate the following functions:

$$\text{(a) } x(2+x^2)^{3/2}, \quad \text{(b) } \frac{x^2}{(3+x)^4}, \quad \text{(c) } \frac{1}{x+2\sqrt{x-1}}.$$

4.4 Integrate the following functions:

$$\text{(a) } \sqrt{1-9x^2}, \quad \text{(b) } \frac{1-2x}{(3+x-x^2)^{1/2}}, \quad \text{(c) } \frac{x}{x^2-4x+4}.$$

4.5 Evaluate the following indefinite integrals:

$$\text{(a) } \int \cos^5 x \, dx, \qquad \text{(b) } \int 4\sin^4 x \, dx.$$

4.6 Evaluate the following integrals:

$$\text{(a) } \int \sinh^6 x \cosh^3 x \, dx, \quad \text{(b) } \int (3+\sin x)^5 \cos^3 x \, dx.$$

4.7 By using partial fractions, evaluate the following integrals:

$$\text{(a) } \int \frac{5x-4}{x^2-x-2} \, dx \quad \text{(b) } \int \frac{1}{(x-1)(x-2)(x-3)} \, dx$$

4.8 Evaluate the following integrals:

$$\text{(a) } \int \frac{2-x^2}{(x+3)(x-4)^2} \, dx, \quad \text{(b) } \int_0^1 \frac{dx}{(x+1)(x^2+1)}$$

4.9 Use appropriate 'tangent substitutions' to integrate the following functions:

$$\text{(a) } \frac{1}{1+\sin x}, \quad \text{(b) } \frac{1}{3-2\sin^2 x}.$$

4.10 Integrate the following functions:

$$\text{(a) } \sec^2 x \ln(\tan x), \quad \text{(b) } \frac{1-\cos x}{1+\cos x}.$$

4.11 Find the following indefinite integrals:

$$\text{(a) } \int \coth(4x) \, dx, \qquad \text{(b) } \int \frac{1}{2+3\cos x} dx.$$

4.12 Find the following indefinite integrals:

$$\text{(a) } \int \frac{dx}{(a^2-x^2)^{3/2}}, \qquad \text{(b) } \int \frac{dx}{ax^2+bx+c} \quad (a \neq 0, \ b^2 < 4ac).$$

4.13 Evaluate the following definite integrals:

$$\text{(a)} \int_1^2 \frac{1}{(10-x)\sqrt{x-1}}\,dx, \quad \text{(b)} \int_0^1 \arcsin x\,dx,$$

$$\text{(c)} \int_0^{\pi/2} \frac{1}{(\sin x - \cos x)^2}\,dx\,.$$

4.14 If

$$I_n = \int_0^{\pi/2} e^{ax}\sin^n x\,dx,$$

where a is a constant, show by repeated integration by parts, that

$$I_n = \frac{ae^{a\pi/2}}{n^2+a^2} + \frac{n(n-1)}{n^2+a^2} I_{n-2}\,, \qquad n \geq 2$$

4.15 If

$$I_n = \int \frac{dx}{(1+x^2)^n},$$

show that

$$2nI_{n+1} = (2n-1)I_n + \frac{x}{(1+x^2)^n}$$

and hence find I_3.

4.16 Find the following indefinite integrals:

$$\text{(a)} \int e^{-x}\sin x\,dx, \quad \text{(b)} \int \frac{\ln x}{(1-x)^2}\,dx, \quad \text{(c)} \int \sin(\ln x)\,dx.$$

4.17 The *Euler gamma function* $\Gamma(x)$ is defined by

$$\Gamma(x) \equiv \int_0^\infty t^{x-1}e^{-t}\,dt, x > 0.$$

(a) Show that $\Gamma(x+1) = x\Gamma(x)$ and hence that $\Gamma(n+1) = n!$ for integer $n \geq 1$.
(b) The relation $\Gamma(x+1) = x\Gamma(x)$ can be used to extend the definition of the gamma function to negative x. Locate the singularities of $\Gamma(x)$ and evaluate $\Gamma(-5/2)$, given that $\Gamma(1/2) = \sqrt{\pi}$.

4.18 A rocket fired vertically into the air at $t = 0$ has an acceleration profile

$$\frac{dv}{dt} = \frac{2g}{1-\alpha t} - g$$

for five seconds, where $\alpha = 0.1\,\text{s}^{-1}$ and $g = 9.8\,\text{ms}^{-2}$ is the acceleration due to gravity. Then the fuel runs out and it subsequently moves freely under gravity. What height does the rocket reach?

4.19 Examine the convergence of the following integrals. Evaluate those that are convergent.

$$\text{(a)} \int_0^\infty x\,e^{-x^2}\,dx, \qquad \text{(b)} \int_0^1 (x-2)\ln x\,dx,$$

$$\text{(c)} \int_1^\infty \frac{\ln x}{x}\,dx\,, \qquad \text{(d)} \int_{-\infty}^\infty \frac{1}{x^2+2x-3}\,dx.$$

4.20 Which of the following integrals are convergent?

$$\text{(a)} \int_0^1 \frac{1-\cos x}{x^{5/2}}\,dx, \quad \text{(b)} \int_1^\infty \frac{\sin x}{x^2}\,dx\,, \quad \text{(c)} \int_0^{\pi/2} \frac{\tan x}{x}\,dx\,.$$

4.21 For what values of α and β (if any) do the following integrals converge?

$$\text{(a)} \int_1^\infty \sin x^\alpha\,dx \quad \text{(b)} \int_1^\infty x^\alpha (\ln x)^\beta\,dx\,.$$

4.22 A particle moves along the x-axis subject to a force given by

$$F(x) = \begin{cases} k/x^2 & a/2 \le x \le a \\ k\,x/a^3 & a \le x \le 2a \\ kx^2/2a^4 & 2a \le x \le 3a \end{cases}$$

where k and a are positive constants. Find the work done W in moving the particle from $a/2$ to $3a$.

4.23 Find the length of the curve $y = (1-x^2)^{1/2}$ between the points 0 and 1.

4.24 Find an integral expression for the length L of that portion of the hyperbola $x^2 - y^2 = 1$ for which $2.5 \ge x \ge 1.5$ and use both the trapezium rule and Simpson's rule with 4 intervals to estimate the value of L.

4.25 Use Simpson's rule with n intervals to calculate the values of pi by integrating the function $(1-x^2)^{-1/2}$ between the limits 0 and 0.5. Find the value of n needed to ensure that the estimated value of π does not change by more than 0.1%.

***4.26** The curve $y = (1-x^2)^{1/2}$ between the points 0 and 1 is rotated about the x-axis. Find The area A and the volume V of rotation and the total surface area S of the resulting shape.

4.27 A *prolate spheroid* is obtained by rotating the ellipse

$$\frac{x^2}{a^2} + \frac{y^2}{b^2} = 1\,, \qquad a > b$$

about the x-axis. Derive a formula for the volume of the spheroid and show that the surface area is given by

$$A = 2\pi b^2 + \frac{2\pi ab}{e}\sin^{-1} e\,,$$

where the *ellipticity* e is defined by $e = \sqrt{1 - b^2/a^2}$.

***4.28** Calculate the moment of inertia of a thin square plate of mass m and sides of length a about (a) a line joining the mid-points of opposite sides, and (b) a diagonal.

***4.29** Find the moment of inertia of a thin circular disc of mass M and radius r about an axis perpendicular to the disc and passing through its centre.

***4.30** Use the result of Problem 4.29 to find the moment of inertia of a cone of mass M, height h and radius R at the base, for rotations about the axis of symmetry.

5

Series and expansions

Perhaps the most important application of the higher derivatives introduced in Section 3.4 is that, provided they exist, they enable functions in the neighbourhood of a given point to be approximated by polynomials in such a way that the accuracy of the approximation increases as the order of the polynomials increases. Such so-called *Taylor expansions* are useful because the resulting polynomials are often much easier to study and evaluate than the original functions themselves, and they have many applications, as we will see. Firstly, however, we must introduce some basic ideas about series and expansions in general.

5.1 Series

A series is the sum $u_0 + u_1 + u_2 + \cdots$ of an ordered sequence $\{u_n\}$ of elements $u_n (n = 1, 2, \ldots)$. The elements may be numbers, for example 0, $\frac{1}{2}$, $\frac{2}{3}$, $\cdots$, obtained from

$$u_n = \frac{n}{n+1}, \qquad n = 0, 1, 2, \ldots \tag{5.1}$$

or functions, such as $u_n = 1, x, x^2, \ldots$, obtained from

$$u_n = x^n, \qquad n = 0, 1, 2, \ldots \tag{5.2}$$

and the sequence may contain a finite number of $(N+1)$ terms,

$$U_N = \sum_{n=0}^{N} u_n, \tag{5.3}$$

or an infinite number of terms

$$U = \sum_{n=0}^{\infty} u_n, \tag{5.4}$$

Mathematics for Physicists, First Edition. B.R. Martin and G. Shaw.

Companion website: www.wiley.com/go/martin/mathsforphysicists

where U_N and U are the sums of the series. In the latter case, we often require the limiting form of u_n as n becomes arbitrarily large. In general, for any sequence $\{u_n\}$, the statement

$$\lim_{n\to\infty} u_n = u, \tag{5.5a}$$

or equivalently

$$u_n \to u \text{ as } n \to \infty, \tag{5.5b}$$

means that for any $\varepsilon > 0$, however small, we can find an integer p such that

$$|u_n - u| < \varepsilon \text{ for all } n > p. \tag{5.5c}$$

Thus for the series (5.1) we have

$$\lim_{n\to\infty} u_n = 1,$$

whereas for the series (5.2) the behaviour depends on the variable x. For example, for $|x| < 1$, $u_n \to 0$ as $n \to \infty$, whereas for $x > 1$, $u_n \to \infty$. For $x \leq -1$ the terms oscillate in sign as n increases and there is no definite limit.

At this point, we note that in writing (5.3) and (5.4) with the same elements u_n, we have implicitly assumed the existence of a well-defined finite limit

$$\lim_{N\to\infty} U_N = U. \tag{5.6}$$

If such a limit does exist, the infinite series (5.4) is said to *converge*, that is, the infinite number of terms yields a finite sum U. On the other hand, if a well-defined finite limit U does not exist, the series does not converge and has no obvious meaning.

The question of whether or not a given sequence $\{u_n\}$ leads to a convergent series will be discussed in general in Section 5.2. Here we will consider just two examples that occur frequently in applications.

(i) *Arithmetic series*

These are any series that can be written in the form

$$A_N = \sum_{n=0}^{N} (a + nx) = a + (a + x) + (a + 2x) + \cdots + (a + Nx) \tag{5.7}$$

for any values of a and x that are independent of n. The series contains $(N + 1)$ terms and since they increase at a steady rate, their average value $\bar{A}$ is given by

$$\bar{A} = \frac{a + (a + Nx)}{2} = \frac{2a + Nx}{2},$$

that is, the average of the first and last terms. The sum of the series is therefore given by

$$A_N = (N+1)\bar{A} = \frac{(N+1)(2a+Nx)}{2}, \tag{5.8}$$

As $N \to \infty$, $A_N \to \infty$, so that the arithmetic series does not lead to a convergent infinite series.

(ii) *Geometric series*
These are series that can be written in the form

$$G_N = \sum_{n=0}^{N} ax^n, \tag{5.9}$$

where again, a and x are independent of n. Explicitly,

$$G_N = a + ax + ax^2 + \cdots + ax^N$$

and

$$xG_N = ax + ax^2 + \cdots + ax^N + ax^{N+1}.$$

Hence

$$G_N - xG_N = a - ax^{N+1}$$

and the geometric series (5.9) sums to

$$G_N = \frac{a(1-x^{N+1})}{(1-x)}. \tag{5.10}$$

In this case, provided $|x| < 1$, $G_N \to a/(1-x)$ as $N \to \infty$, so that

$$G = \sum_{n=0}^{\infty} ax^n = \frac{a}{1-x}, \qquad |x| < 1 \tag{5.11}$$

is a well-defined convergent series. In particular, setting $a = 1$ gives the useful result

$$(1-x)^{-1} = 1 + x + x^2 + \cdots \qquad |x| < 1 \tag{5.12}$$

For $|x| > 1$, however, the limit of (5.10) as $N \to \infty$ is not finite and well-defined, so that

$$\sum_{n=0}^{\infty} ax^n, \qquad |x| > 1 \tag{5.13}$$

is not a convergent series.

Example 5.1

Show that

$$S_N = \sum_{n=0}^{N} \frac{1}{(n+2)(n+3)} = \frac{1}{2} - \frac{1}{N+3}$$

and hence that $S_N \to 1/2$ as $N \to \infty$.

Solution

The sum may be written

$$\begin{aligned}\sum_{n=0}^{N} \frac{1}{(n+2)(n+3)} &= \sum_{n=0}^{N} \left(\frac{1}{n+2} - \frac{1}{n+3}\right)\\ &= \frac{1}{2} - \frac{1}{3} + \frac{1}{3} - \frac{1}{4} + \cdots - \frac{1}{N+3}\\ &= \frac{1}{2} - \frac{1}{N+3},\end{aligned}$$

and so $S_N \to 1/2$ as $N \to \infty$.

Example 5.2

Show that

$$\sum_{n=1}^{N} n = \frac{N(N+1)}{2}. \tag{5.14}$$

Solution

Clearly,

$$\sum_{n=1}^{N} n = \sum_{n=0}^{N} n$$

is an arithmetic series with $a = 0$, $x = 1$. Hence (5.14) follows directly from (5.8).

5.2 Convergence of infinite series

For most series of the form (5.3), the evaluation of the sum U_N as an explicit function of N, enabling the limit (5.6) to be directly taken, is far from easy, if not impossible. Nonetheless, we will often need to know whether such a limit exists, in order to determine whether or not the corresponding infinite series (5.4) converges. In this section we shall introduce two simple tests that will enable this question to be answered in most cases.

The simplest test is that an infinite series can only converge if

$$\lim_{n\to\infty} u_n = 0. \tag{5.15}$$

To see this, we note that the existence of a finite limit (5.6), together with the definition of a limit (5.5), implies

$$\lim_{N\to\infty} (U_N - U_{N-1}) = 0,$$

and substituting (5.3) into this equation leads directly to (5.15) as required. The condition (5.15) is very useful, since we can immediately conclude that any series which does not satisfy (5.15) does not converge. However, we cannot conclude the inverse, that any series satisfying (5.15) does converge, and while many such series do converge, others do not.

The most useful single result on the convergence of infinite series is *d'Alembert's ratio test.* It is formulated in terms of the behaviour of the ratio

$$r_n = \left|\frac{u_{n+1}}{u_n}\right| \tag{5.16}$$

at large n, and can be stated in two forms:

(i) The series (5.4) converges if $r_n < r < 1$, where r is a constant[1], for all $n \geq p$, where p is a finite integer; and does not converge if $r_n > r > 1$ for all $n \geq p$.

(5.17a)

(ii) If r_n has a well-defined limit $r_n \to \rho$ as $n \to \infty$, then if $\rho < 1$, the series (5.4) converges, and does not converge if $\rho > 1$.

(5.17b)

The second form (5.17b) of the test follows directly from the first form (5.17a). This is because if $r_n \to \rho < 1$, for example, then the definition of a limit (5.5a) implies that we can always find an integer p such that $r_n < 1$ for all $n \geq p$. A similar argument applies to the case $r_n > \rho > 1$. Turning to (5.17a) itself, non-convergence for the case $r_n > r > 1$ follows because $r_n = |u_{n+1}|/|u_n| > 1$ for all $n \geq p$ requires $|u_n|$ to increase as n increases, which is clearly incompatible with (5.15). For the case $r_n < r < 1$, however, the proof of convergence is lengthy and complicated. It will be given for completeness

[1]Note that the condition $r_n < r < 1$ for all $n \geq p$ is not the same as $r_n < 1$ for all $n \geq p$, as the latter would allow the case $r_n \to 1$ from below as $n \to \infty$ whereas $r_n < r < 1$ excludes it. If $r_n \to 1$ as $n \to \infty$, the convergence of the series is not guaranteed, as we shall see.

in Section 5.5. Here we simply illustrate its use by example, after first considering its implication for *power series* of the form

$$P = \sum_{n=0}^{\infty} a_n (x - x_0)^n, \tag{5.18}$$

where x_0 and a_n are constants and x is a variable. From the ratio test, if

$$\rho = \lim_{n \to \infty} \left| \frac{a_{n+1}(x - x_0)}{a_n} \right|,$$

the series will converge for $\rho < 1$ and diverge for $\rho > 1$. Since $|ab| = |a|\,|b|$, this in turn implies that the series converges for values of x such that

$$|x - x_0| < R = \lim_{n \to \infty} \left| \frac{a_n}{a_{n+1}} \right|, \tag{5.19}$$

where R is called the *radius of convergence* of the series (5.18). Conversely, the series does not converge outside the radius of convergence, that is, for $|x - x_0| > R$, while the case $|x - x_0| = R$, corresponding to $\rho = 1$, requires special treatment in each case.

Example 5.3

Comment on the possible convergence of the series

$$\text{(a)} \sum_{n=0}^{\infty} \frac{n}{n+1} \quad \text{and} \quad \text{(b)} \sum_{n=0}^{\infty} \frac{1}{n!}.$$

Solution

(a) This violates (5.15) and so does not converge. (b) $r_n = 1/(n+1) \to 0$ as $n \to \infty$, so by d'Alembert's ratio test it converges.

Example 5.4

Show that

$$\lim_{n \to \infty} \left(\frac{x^n}{n!} \right) = 0 \tag{5.20}$$

for all x by considering the series

$$\sum_{n=0}^{\infty} \frac{x^n}{n!}.$$

Solution

We have $r_n = u_{n+1}/u_n = x/(n+1) \to 0$ as $n \to \infty$, so the series converges for all x by d'Alembert's ratio test. Hence the elements of the series must themselves tend to zero by (5.15).

Example 5.5

Does the series

$$\sum_{n=1}^{\infty} \frac{x^n}{3^{n-1}n}$$

converge for the values $x = 1$ and $x = \pi$?

Solution

The radius of convergence is

$$R = \lim_{n\to\infty} \left[\frac{3^n(n+1)}{3^{n-1}n}\right] = 3,$$

so the series converges for $x = 1 < R$, but diverges for $x = \pi > R$.

5.3 Taylor's theorem and its applications

In this section, we introduce a fundamental theorem that enables differentiable functions to be approximated by polynomials in such a way that the approximation becomes more accurate as the order of the polynomials increases. In this section, we will first state and prove the theorem, before discussing some of its simpler applications.

5.3.1 Taylor's theorem

If $f(x)$ is continuous in the range x_0 to $x_0 + h$ inclusive and all its derivatives up to and including $f^{(N+1)}(x)$ exist in the same range, then *Taylor's theorem* states that $f(x)$ can be written in the form

$$f(x_0 + h) = \sum_{n=0}^{N} \frac{h^n}{n!} f^{(n)}(x_0) + R_N, \qquad (5.21a)$$

where the remainder

$$R_N = \frac{h^{N+1}}{(N+1)!} f^{(N+1)}(x_0 + \theta h), \qquad (5.21b)$$

for at least one value of θ in the range $0 < \theta < 1$.

To prove this result, we define a number P for any given h by

$$f(x_0+h) = f(x_0) + hf^{(1)}(x_0) + \cdots + \frac{h^N}{N!}f^{(n)}(x_0) + \frac{h^{N+1}}{(N+1)!}P \tag{5.22}$$

and introduce a function $F(x)$ defined by

$$F(x) \equiv f(x_0+h) - f(x) - (x_0+h-x)f^{(1)}(x) - \cdots - \frac{(x_0+h-x)^N}{N!}f^{(N)}(x) - \frac{(x_0+h-x)^{N+1}}{(N+1)!}P. \tag{5.23}$$

We then have

$$F(x_0+h) = 0, \qquad F(x_0) = 0, \tag{5.24}$$

where the first of these relations follows directly from (5.23) and the second by setting $x = x_0$ in (5.23) and then using (5.22). Since $F(x)$ is continuous in the range x_0 to $x_0 + h$, it follows from (5.24) that there must be at least one value $x = \zeta$ in this range where $F(x)$ is either a maximum or a minimum: and since $F(x)$ is differentiable in this range, we must have[2]

$$F'(\zeta) = 0 \tag{5.25}$$

by (3.43). On substituting (5.23) into this equation, one finds that the resulting terms cancel in pairs, except for the last two, which give

$$-\frac{(x_0+h-\zeta)^N}{N!}f^{(N+1)}(\zeta) + \frac{(x_0+h-\zeta)^N}{N!}P = 0.$$

Hence $P = f^{N+1}(\zeta)$, and writing $\zeta = x_0 + \theta h$, where $0 < \theta < 1$, we obtain the desired result (5.21). We now turn to its applications.

5.3.2 Small changes and l'Hôpital's rule

Most applications of Taylor's theorem rest upon its use to approximate functions by simple polynomials for small values of h. Thus from (5.21) we see that

$$f(x_0+h) = f(x_0) + hf'(x_0) + O(h^2), \tag{5.26}$$

where the notation $O(h^2)$means that terms containing factors of h^2 or higher powers are neglected. (This is referred to as 'neglecting

[2] This is a special case of *Rolle's theorem*, which states that if $f(x)$ is continuous and differentiable in the range $a \le x \le b$ and $f(a) = f(b) = 0$, there must be at least one value of ζ in the range $a \le \zeta \le b$ where $f'(\zeta) = 0$.

terms of order h^2'.) For sufficiently small $h = \delta x$, it is often a good approximation to write

$$f(x_0 + \delta x) \approx f(x_0) + \delta x\, f'(x_0), \tag{5.27}$$

where $\approx$ means 'approximately equal to'. Equation (5.26) also tells us how the limit $f(x) \to f(x_0)$ is approached for small $h = (x - x_0)$. This is particularly useful for taking limits of ratios of two functions $f(x)$ and $g(x)$ that both vanish at $x = x_0$, so that $f(x_0)/g(x_0)$ is indeterminate. In this case,

$$\lim_{x\to x_0}\left[\frac{f(x)}{g(x)}\right] = \lim_{h\to 0}\left[\frac{f(x_0) + hf'(x_0) + O(h^2)}{g(x_0) + hg'(x_0) + O(h^2)}\right] = \frac{f'(x_0)}{g'(x_0)} \tag{5.28a}$$

if

$$f(x_0) = g(x_0) = 0. \tag{5.28b}$$

This result is known as *l'Hôpital's rule.* Of course if $f'(x_0)$ and $g'(x_0)$ also both vanish, i.e.,

$$f(x_0) = f'(x_0) = g(x_0) = g'(x_0) = 0, \tag{5.29a}$$

then (5.28a) is again indeterminate, and repeating the argument gives

$$\lim_{x\to x_0}\left[\frac{f(x)}{g(x)}\right] = \frac{f''(x_0)}{g''(x_0)} \tag{5.29b}$$

and so on.

Example 5.6

Evaluate the limit

$$\lim_{x\to 0}\left[\frac{x \sin x}{1 - \cos x}\right].$$

Solution

Putting $f(x) = x \sin x$ and $g(x) = 1 - \cos x$, leads to $f(0) = g(0) = 0$. Hence by l'Hôpital's rule

$$\lim_{x\to 0}\left[\frac{x \sin x}{1 - \cos x}\right] = \lim_{x\to 0}\left[\frac{x \cos x + \sin x}{\sin x}\right] = 2,$$

since $x/\sin x \to 1$ as $x \to 0$ by (3.15), or by evaluating the latter limit also using l'Hôpital's rule, i.e.

$$\lim_{x\to 0}\frac{\sin x}{x} = \lim_{x\to 0}\left[\frac{\cos x}{1}\right] = 1.$$

5.3.3 Newton's method

At the end of Section 2.1.1, we introduced the bisection method for finding a solution to any desired precision for equations of the form

$$f(x) = 0, \tag{5.30}$$

given an approximate solution $x = x_0$. In Newton's method, solutions are found by substituting the approximation (5.27) into (5.30) to give

$$f(x_0 + \delta x) = f(x_0) + \delta x f'(x_0) = 0$$

and hence a new solution

$$x_1 = x_0 + \delta x = x_0 - f(x_0)/f'(x_0), \tag{5.31}$$

which will be an improvement on x_0 provided the latter is close enough to the exact solution for (5.27) to be a reasonable approximation. This can be iterated in the same way to give a second improved solution

$$x_2 = x_1 - f(x_1)/f'(x_1),$$

and so on until the desired precision is achieved.

Example 5.7

The polynomial $f(x) = x^3 - 3x^2 - 4x + 7$ has a root near $x_0 = 1.1$. (See Figure 2.1.) Use Newton's method to find the root to four significant figures and compare the solution to that obtained by the bisection method of Example 2.5.

Solution

The derivative is given by $f'(x) = 3x^2 - 6x - 4$, so that at $x_0 = 1.1$ we have

$$f(x_0) = 0.301 \text{ and } f'(x_0) = -6.97,$$

giving $x_1 = 1.143$. Iterating a second time gives

$$f(x_1) = -1.92 \times 10^{-3}, \quad f'(x_1) = -6.94,$$

and hence $x_2 = 1.143$ to four significant figures. This is the same answer as that found by the bisection method in Example 2.5 after 11 iterations.

*5.3.4 Approximation errors: Euler's number

When using Taylor's theorem (5.21), the remainder function R_N can often be bounded, enabling the function $f(x) = f(x_0 + h)$ to be evaluated with known accuracy. In this section, we shall illustrate this by evaluating Euler's number e. To do this, we first apply Taylor's theorem to $f(x) = e^x$, using $f^{(n)}(x) = e^x$ for all n. Choosing $x_0 = 0$ in (5.21), we obtain

$$e^h = \sum_{n=0}^{N} \frac{h^n}{n!} + \frac{h^{N+1}}{(N+1)!} e^{\theta h}, \tag{5.32}$$

where $0 < \theta < 1$. Setting $h = 1$ gives

$$e = \sum_{n=0}^{N} \frac{1}{n!} + R_N, \tag{5.33a}$$

where the remainder

$$R_N = \frac{e^\theta}{(N+1)!}, \quad 0 < \theta < 1. \tag{5.33b}$$

From this we see that

$$\frac{1}{(N+1)!} < R_N < \frac{3}{(N+1)!}, \tag{5.34}$$

where we have used the fact that $e < 3$ to bound R_N. Hence if we take $R_N = 2/(N+1)!$, in the middle of the allowed range (5.34), then (5.33a) will give a value of e that is accurate to less than $1/(N+1)!$.

To illustrate this, we will calculate e to six significant figures, which requires $(N+1)! \geq 10^5$, i.e. $N \geq 8$. Taking $N = 8$, we have

$$e = 1 + 1 + \frac{1}{2} + \frac{1}{3!} + \cdots + \frac{1}{8!} + \frac{2}{9!} = 2.71828$$

to six significant figures. This precision increases very rapidly with N, and indeed as $N \to \infty$ we obtain the convergent series

$$e = \sum_{n=0}^{\infty} \frac{1}{n!} = 1 + 1 + \frac{1}{2} + \frac{1}{3!} + \frac{1}{4!} + \cdots \tag{5.35}$$

as $R_N \to 0$.

5.4 Series expansions

In this section, we investigate the existence of power series expansions of the form

$$f(x) = \sum_{n=0}^{\infty} a_n (x - x_0)^n \tag{5.36}$$

for a given function $f(x)$ about a point $x = x_0$.

5.4.1 Taylor and Maclaurin series

Once again, we start from Taylor's theorem (5.21), which we write in the form

$$f(x) = \sum_{n=0}^{N} \frac{f^{(n)}(x_0)}{n!}(x - x_0)^n + R_N, \tag{5.37a}$$

obtained by setting $x = x_0 + h$, where

$$R_N = \frac{(x - x_0)^{N+1}}{(N+1)!} f^{(N+1)}[x_0 + \theta(x - x_0)] \tag{5.37b}$$

and $0 < \theta < 1$. Then taking the limit $N \to \infty$ gives

$$f(x) = \sum_{n=0}^{\infty} \frac{f^{(n)}(x_0)}{n!}(x - x_0)^n, \tag{5.38}$$

provided that: an infinite number of derivatives exist in the range x_0 to x inclusive; the series (5.38) converges; and that[3] $R_N \to 0$ as $N \to \infty$. Equation (5.38) then has the form of the desired series (5.36) with[4]

$$a_n = \frac{f^{(n)}(x_0)}{n!} \tag{5.39}$$

and is called a *Taylor series.* In the special case $x_0 = 0$, it reduces to

$$f(x) = \sum_{n=0}^{\infty} \frac{f^{(n)}(0)}{n!} x^n \tag{5.40}$$

and is called a *Maclaurin series.*

It is important to stress that a Taylor or Maclaurin series does not always exist. For example, $\ln x$ cannot be expanded as a Maclaurin series of the form (5.40) because it is singular at $x = 0$ and no derivatives $f^{(n)}(0)$ exist. Alternatively, $R_N \nrightarrow 0$ as $N \to \infty$, or may do so only for a finite range of x, as we shall see shortly. However, none of these problems arise for $f(x) = \exp(x)$. Then $f^{(n)}(x) = \exp(x)$ for all n, and

$$R_N = \frac{(x - x_0)^{N+1}}{(N+1)!} \exp[x_0 + \theta(x - x_0)] \to 0$$

---

[3] If $f(x)$ is well-defined, then (5.37a) implies the convergence of (5.38) if $R_N \to 0$, whereas the reverse is not the case. Nonetheless, it is usually best to consider convergence first since it is often easier to establish, and if it fails one need not proceed further. In addition, the convergence of (5.38) is sometimes useful in proving $R_N \to 0$, as we shall see in Example 5.9 below.

[4] If we assume that (5.36) is a well-defined convergent expansion, which is not always the case, then (5.39) can be obtained directly by differentiating both sides n times and taking the limit $x \to x_0$, when only the nth term in the series survives.

as $N \to \infty$, since the exponential is always smaller than the larger of e^x or e^{x_0}, and $x^n/n! \to 0$ as $n \to \infty$ by (5.20). The Taylor series (5.38) is

$$e^x = e^{x_0} \sum_{n=0}^{\infty} \frac{(x - x_0)^n}{n!} \tag{5.41}$$

and converges for all finite values of x and x_0, as is easily confirmed using the d'Alembert ratio test. For $x_0 = 0$, it reduces to the Maclaurin series

$$e^x = \sum_{n=0}^{\infty} \frac{x^n}{n!}. \tag{5.42}$$

Equation (5.42) is one of a number of standard Maclaurin series for important elementary functions. Some of the most important of these are listed together in Table 5.1 and two of them are derived as worked examples below. Before doing so, however, some comments are in order. Firstly, in the trigonometric functions the variable x must be measured in radians, as usual. Secondly, the even (odd) nature of some of the functions is reflected in the fact that only even (odd) powers of x appear in their expansions. Finally, the last series in the table is called the *binomial series*, because for positive integers $\alpha = m$ the series terminates and reduces to the binomial

Table 5.1 Standard Maclaurin series

(i) Series valid for all x

$$\sin x = x - x^3/3! + x^5/5! - x^7/7! + \cdots = \sum_{n=0}^{\infty} (-1)^n x^{2n+1}/(2n+1)!$$

$$\cos x = 1 - x^2/2! + x^4/4! - x^6/6! - \cdots = \sum_{n=0}^{\infty} (-1)^n x^{2n}/(2n)!$$

$$e^x = 1 + x + x^2/2! + x^3/3! + x^4/4! + \cdots = \sum_{n=0}^{\infty} x^n/n!$$

$$\sinh x = x + x^3/3! + x^5/5! + x^7/7! + \cdots = \sum_{n=0}^{\infty} x^{(2n+1)}/(2n+1)!$$

$$\cosh x = 1 + x^2/2! + x^4/4! + x^6/6! + \cdots = \sum_{n=0}^{\infty} x^{2n}/(2n)!$$

(ii) Series valid for $-1 < x < 1$

$$\ln(1+x) = x - x^2/2 + x^3/3 - x^4/4 + \cdots = \sum_{n=0}^{\infty} (-1)^n x^{n+1}/(n+1)$$

$$\arctan x = x - x^3/3 + x^5/5 - x^7/7 + \cdots = \sum_{n=0}^{\infty} (-1)^n x^{2n+1}/(2n+1)$$

(iii) Series valid for all α and $-1 < x < 1$

$$(1+x)^\alpha = 1 + \alpha x + \frac{\alpha(\alpha-1)}{2!}x^2 + \frac{\alpha(1-\alpha)(\alpha-2)}{3!}x^3 + \cdots$$

$$= 1 + \sum_{n=1}^{\infty} \frac{[\alpha-(n-1)][\alpha-(n-2)]\cdots\alpha}{n!}x^n$$

theorem (1.23) for $y = 1$, while for $\alpha = -1$ it is identical to the geometric series (5.12).

Example 5.8

Derive the Taylor series expansions of $f(x) = \sin x$ about $x_0 = 0$ and $x_0 = \pi/2$.

Solution

If $f(x) = \sin x$ then from (3.41),

$$f^{(2n)}(x) = (-1)^n \sin x \quad \text{and} \quad f^{(2n+1)}(x) = (-1)^n \cos x.$$

Since the moduli of $\sin x$ and $\cos x$ are both less than or equal to unity for any value of their arguments, from (5.37b) we have

$$|R_N| \leq \frac{(x - x_0)^{N+1}}{(N+1)!} \to 0$$

by (5.20) for all x and x_0. For $x_0 = 0$,

$$f^{(2n)}(0) = 0 \quad \text{and} \quad f^{(2n+1)}(0) = (-1)^n,$$

so that (5.38) gives the Maclaurin series

$$\sin x = \sum_{n=0}^{\infty} \frac{(-1)^n}{(2n+1)!} x^{2n+1}.$$

For $x_0 = \pi/2$,

$$f^{(2n)}(x_0) = \pi/2 \text{ and } f^{(2n+1)}(x_0) = 0,$$

so that (5.38) gives the Taylor series

$$\sin x = \sum_{n=0}^{\infty} \frac{(-1)^n}{(2n)!} \left(x - \frac{\pi}{2}\right)^n.$$

Example 5.9

Derive the Maclaurin series for $f(x) = \ln(1 + x)$ and establish the range of x for which it is valid.

Solution

Differentiating, we have

$$f^{(1)}(x) = (1+x)^{-1}, \quad f^{(2)}(x) = -(1+x)^{-2}, \quad f^{(3)}(x) = 2(1+x)^{-3}$$

and in general

$$f^{(n)}(x) = (-1)^{n-1}(n-1)!(1+x)^{-n}, \quad n \geq 1. \qquad (5.43)$$

Hence, using $f^{(0)}(0) = 0$, the Maclaurin series (5.40) is

$$\ln(1+x) = \sum_{n=1}^{\infty} \frac{(-1)^{n-1} x^n}{n}, \tag{5.44}$$

provided the series converges and

$$R_N = \frac{(-1)^N}{(N1)} \left(\frac{x}{1+\theta x} \right)^{N+1} \tag{5.45}$$

tends to zero as $N \to 0$. Using d'Alembert's ratio test (5.17b), we have

$$|r_n| = \left| \frac{nx}{n+1} \right| \to |x| \quad \text{as} \quad n \to \infty,$$

so that the series converges only if $|x| < 1$ and only this range need be considered.

The proof that $R_N \to 0$ is a little more difficult. Firstly, we note that either $|R_N| \to \infty$ or $|R_N| \to 0$, depending on whether the modulus of the term in brackets is greater than unity or not. So we just have to show that $|R_N| \not\to \infty$ for $-1 < x < 1$. To do this for (5.45), we note that (5.37a) in this case gives

$$\ln(1+x) = \sum_{n=1}^{N} \frac{(-1)^n x^n}{n} + R_N.$$

Then since $\ln(1+x)$ is well-defined and the series converges as $N \to \infty$ for $-1 < x < 1$, R_N must tend to a finite limit in this range. Hence $|R_N| \not\to \infty$ so that $|R_N| \to 0$ from our previous result, as required.

5.4.2 Operations with series

So far we have concentrated on deriving series expansions directly from Taylor's theorem. However, it is often easier to derive them using 'standard series' that have already been obtained, provided that care is taken to confine oneself to regions where the series converges. For example, consider the Maclaurin expansion of $f(x) = \ln(2+x^2)$. Then

$$\ln(2+x^2) = \ln[2(1+x^2/2)] = \ln 2 + \ln(1+z),$$

where $z = x^2/2$. Expanding $\ln(1+z)$ using (5.44) and substituting for z then gives the Maclaurin expansion

$$\ln(2+x^2) = \ln 2 + \sum_{n=1}^{\infty} \frac{(-1)^{n-1} x^{2n}}{n 2^n}. \tag{5.46}$$

However, since the expansion of $\ln(1+z)$ is only valid for $|z| < 1$, the expansion (5.46) only holds for the corresponding range $|x| < \sqrt{2}$.

We next turn to the algebraic manipulation of series. Suppose we have two series

$$f(x) = \sum_{n=0}^{\infty} a_n x^n, \quad g(x) = \sum_{n=0}^{\infty} b_n x^n, \tag{5.47a}$$

which both converge in some given region of x. Then for any number α,

$$\alpha f(x) = \sum_{n=0}^{\infty} (\alpha a_n) x^n \tag{5.48}$$

and

$$f(x) \pm g(x) = \sum_{n=0}^{\infty} (a_n \pm b_n)\, x^n \tag{5.49}$$

both hold and converge in the same region. These results are almost trivial – they follow easily from the definition (5.5) of a convergent series as a limit of a finite series, together with the rules of arithmetic before the limit is taken – but are very useful. For example, from

$$e^x = \sum_{n=0}^{\infty} \frac{x^n}{n!} = 1 + x + \frac{x^2}{2!} + \frac{x^3}{3!} + \cdots$$

and

$$e^{-x} = \sum_{n=0}^{\infty} \frac{(-1)^n x^n}{n!} = 1 - x + \frac{x^2}{2!} - \frac{x^3}{3!} + \cdots,$$

the series

$$\sinh x = \frac{e^x - e^{-x}}{2} = x + \frac{x^3}{3!} + \frac{x^5}{5!} + \cdots \tag{5.50a}$$

and

$$\cosh x = \frac{e^x + e^{-x}}{2} = 1 + \frac{x^2}{2!} + \frac{x^4}{4!} + \cdots \tag{5.50b}$$

given in Table 5.1 follow directly from the definitions (2.57), together with (5.48) and (5.49).

Another useful result, which we will state without proof, is the *Cauchy product.* This states that

$$f(x)g(x) = \sum_{n=0}^{\infty} c_n x^n, \tag{5.51a}$$

where

$$c_n = \sum_{i=0}^{n} a_i b_{n-i}, \tag{5.51b}$$

is the convergent series for the product of the convergent series (5.47a), provided that at least one of the series

$$f(x) = \sum_{n=0}^{\infty} |a_n x^n|, \quad g(x) = \sum_{n=0}^{\infty} |b_n x^n| \tag{5.47b}$$

also converges.[5]

Convergent series can also be differentiated and integrated term by term to give new series that converge in the same region. Suppose we have a series

$$f(x) = \sum_{n=0}^{\infty} a_n x^n. \tag{5.52a}$$

Then differentiation and integration yield

$$f'(x) = \sum_{n=1}^{\infty} n a_n x^{n-1} = \sum_{n=0}^{\infty} (n+1) a_{n+1} x^n, \tag{5.52b}$$

and

$$\int f(x)\mathrm{d}x = c + \sum_{n=0}^{\infty} \frac{a_n x^{n+1}}{n+1} = c + \sum_{n=1}^{\infty} \frac{a_{n-1} x^n}{n}, \tag{5.52b}$$

respectively, where c is an arbitrary constant; and one easily shows, using (5.19), that all three series converge for the same region

$$|x| < R \equiv \lim_{n\to\infty} \left| \frac{a_n}{a_{n+1}} \right| .$$

For example, the series

$$\cos x = \sum_{n=0}^{\infty} \frac{(-1)^n x^{2n}}{(2n)!} = 1 - \frac{x^2}{2!} + \frac{x^4}{4!} - \cdots \tag{5.53}$$

given in Table 5.1 follows directly from the corresponding series for $\sin x$ given in the same table.

Finally, before giving some more examples, we note that while we have for simplicity considered Maclaurin expansions about $x_0 = 0$ in this section, the results extend quite easily to expansions about other values $x_0 \neq 0$.

Example 5.10

Find the Maclaurin series for $\ln[(1+x)/(1-x)]$. For what range of x is it valid?

---

[5] Note that if the convergence of the series in (5.47a) is established by d'Alembert's test, then the series (5.47b) automatically converge.

Solution

We have,

$$\ln[(1+x)/(1-x)] = \ln(1+x) - \ln(1-x).$$

Where from (5.44),

$$\ln(1+x) = x - \frac{x^2}{2} + \frac{x^3}{3} + \cdots$$

and

$$\ln(1-x) = -x - \frac{x^2}{2} - \frac{x^3}{3} + \cdots.$$

Therefore by (5.49),

$$\ln\left(\frac{1+x}{1-x}\right) = 2\left(x + \frac{x^3}{3} + \cdots\right) = \sum_{n=0}^{\infty} \frac{2x^{2n+1}}{(2n+1)}.$$

The series is valid for $|x| < 1$ since the expansions of $\ln(1 \pm x)$ are both valid for $x < 1$.

Example 5.11

Use (5.52a), together with the geometric series (5.12), to derive the Maclaurin series for $f(x) = \arctan x$. What is its range of validity?

Solution

If $f(x) = \arctan x$, then $f(x) \to 0$ as $x \to 0$, so that $a_0 = 0$ in the expansion (5.52a). Further, $x = \tan f$, so that

$$\mathrm{d}x/\mathrm{d}f = \sec^2 f = 1 + \tan^2 f = (1 + x^2)$$

and$f'(x)$ is a geometric series $(1-z)^{-1}$ with $z = -x^2$. Expanding $(1-z)^{-1}$ according to (5.12) gives

$$f'(x) = 1 - x^2 + x^4 - x^6 + \cdots$$

and comparing with the expansion (5.52b) gives

$$a_{2n} = 0, \qquad a_{2n+1} = (-1)^n/(2n+1),$$

so that the series (5.52a) is

$$f(x) = \arctan x = \sum_{n=0}^{\infty} \frac{(-1)^n x^{2n+1}}{(2n+1)}$$
$$= x - x^2/3 + x^5/5 - x^7/7 + \cdots,$$

as given in Table 5.1.

*5.5 Proof of d'Alembert's ratio test

In Section 5.2, we omitted the derivation of d'Alembert's ratio test and any discussion of series to which it cannot be applied. These omissions will be rectified in this and the following section.

*5.5.1 Positive series

We start by considering *positive series*, in which all the terms $u_n \geq 0$, so that U_N, given by (5.3), can only increase as N increases. So either $U_N \to U$ as $N \to \infty$ and the corresponding infinite series converges, or $U_N \to \infty$ and the corresponding series diverges. To decide which, we will obtain two useful results by comparing the series with another series

$$V = \sum_{n=0}^{\infty} v_n, \qquad v_n > 0 \tag{5.54}$$

whose convergence properties are already known; and then obtain d'Alembert's ratio test by choosing (5.54) to be appropriate geometric series.

The first of these results is called the *comparison test* and may be stated as follows:

If $u_n \leq cv_n$ for all $n \geq p$, where c is a positive constant and p is a non-negative integer, then the series U converges if the series V converges; and if $u_n \geq cv_n$, the series U diverges if the series V diverges. (5.55)

We start by proving this for the case $p = 0$, that is, when the u_n are less than or greater than cv_n for all $n \geq 0$. Then in the case $u_n \geq cv_n$, we have

$$0 < U_N = \sum_{n=0}^{N} u_n \leq \sum_{n=0}^{N} cv_n = cV_N$$

where $V_N \to V$ if the series V converges. Hence $U_N \to U < cV$ and the series U also converges. A similar argument shows that U diverges if $u_n > cv_n$ and V diverges. Finally, since the convergence of the series is obviously unaffected by changing the values of the first p terms, the result follows for any finite integer p.

The second result is the *ratio comparison test*:

If $u_{n+1}/u_n < v_{n+1}/v_n$ for all $n \geq p$, then the series U converges if the series V converges; and if $u_{n+1}/u_n \geq v_{n+1}/v_n$ for all $n \geq p$, the series U diverges if the series V diverges.

To begin we prove this for the case $u_{n+1}/u_n < v_{n+1}/v_n$ for all $n \geq p$. Then for $n \geq p$,

$$u_n = \frac{u_n}{u_{n-1}} \frac{u_{n-1}}{u_{n-2}} \cdots \frac{u_{p+1}}{u_p} u_p \leq \frac{v_n}{v_{n-1}} \frac{v_{n-1}}{v_{n-2}} \cdots \frac{v_{p+1}}{v_p} u_p = c v_n,$$

where the constant $c = u_p/v_p > 0$. The result (5.56) then follows directly from the comparison test (5.55), and a similar argument holds for $u_{n+1}/u_n \geq v_{n+1}/v_n$.

We can now use (5.56) to complete the proof of d'Alembert's ratio test for positive series $u_n \geq 0$ by choosing V to be the geometric series (5.12) for positive x. This series has $v_{n+1}/v_n = x$ for all n, and it converges for $x < 1$ and diverges for $x > 1$. Hence if

$$r_n = u_{n+1}/u_n < r < 1, \qquad n \geq p,$$

and we choose V to be a geometric series with $r < x < 1$, then the series V converges and $u_{n+1}/u_n < v_{n+1}/v_n$ for all $n \geq p$. The convergence of the series U then follows directly from the ratio comparison test (5.56), as required. A similar argument establishes that U diverges if $r_n > r > 1$, completing the proof of the d'Alembert ratio test (5.17a) for positive series.

*5.5.2 General series

To generalise the proof of d'Alembert's test to any infinite series

$$U = \sum_{n=0}^{\infty} u_n, \tag{5.56}$$

where the terms u_n may be of either sign, we first consider the related positive series

$$\sum_{n=0}^{\infty} |u_n| \tag{5.57}$$

in which all the terms are replaced by their moduli. The key result is that if this series is convergent, then the original series (5.4) is also convergent. In this case, U is said to be *absolutely convergent*, to distinguish it from *conditional convergence* where (5.4) is convergent, but the related series (5.57) is not convergent.

To show that the convergence of the series (5.57) implies the convergence of (5.4) itself, as stated above, we define

$$w_n = u_n + |u_n|, \qquad n = 0, 1, 2, \ldots$$

and the corresponding series

$$W = \sum_{n=0}^{\infty} w_n = U + \sum_{n=0}^{\infty} |u_n|. \tag{5.58}$$

Then, since (5.57) converges and $0 \leq w_n \leq 2|u_n|$, the series W converges by the comparison test (5.55), and from (5.58), if W and (5.57) converge, we see that U must also converge. We now obtain the desired result by noting that we have already proved that a positive series like (5.57) will converge if $r_n = |u_{n+1}|/|u_n| < \rho < 1$ for all $n \geq p$, where p is a finite integer. Hence, if this condition is satisfied, the related series (5.4) is absolutely convergent, and therefore convergent, as required by the ratio test (5.17a). Since we have already proved, following (5.17a) and (5.17b), that the series (5.4) does not converge if $r_n = |u_{n+1}|/|u_n| > \rho > 1$ for all $n \geq p$, this completes the proof of d'Alembert's ratio test in the form (5.17a). The form (5.17b) then follows directly from (5.17a) using the definition of a limit, as already noted.

*5.6 Alternating and other series

D'Alembert's ratio test enables the convergence properties of most infinite series to be established rather easily, but it says nothing about the convergence of series for which

$$\lim_{n\to\infty} |a_{n+1}/a_n| = \rho = 1. \tag{5.59}$$

Such series must be considered separately, case by case. However, we have already obtained several general results that can be applied in any given instance. To recapitulate:

(i) A series diverges unless its elements $u_n \to 0$ as $n \to \infty$;

(ii) The comparison test (5.55) applies to positive series, with all $u_n \geq 0$;

(iii) Absolute convergence implies convergence.

There are also two important results that apply to *alternating series* of the form

$$U = \sum_{n=0}^{\infty} (-1)^n u_n = u_0 - u_1 + u_2 - u_3 + \cdots . \qquad u_n > 0 \tag{5.60}$$

They are

(iv) an alternating series of the form (5.60) converges if $u_n \to 0$ as $n \to \infty$ and $u_{n+1} \leq u_n$ for all $n \geq p$, where p is a non-negative integer;

(5.61)

(v) the error in curtailing an alternating series in which the magnitude of the terms is monotonically decreasing is less than the magnitude of the first term omitted.

(5.62)

To prove these results, assume that p is even and consider the sum of the first 2 r terms, starting at p. This can be written in the form

$$S_r = (u_p - u_{p+1}) + (u_{p+2} - u_{p+3}) + \cdots + (u_{p+2r-2} - u_{p+2r-1}),$$

and since all the terms in brackets are positive, S_r can only increase as r increases. However, the same sum can also be written in the form

$$S_r = u_p - (u_{p+1} - u_{p+2}) - (u_{p+3} - u_{p+4}) - \cdots$$
$$-(u_{p+2r-3} - u_{p+2r-2}) - u_{p+2r-1},$$

implying $S_r < u_p$ as $r \to \infty$, since all the terms in brackets are again positive. Since S_r increases and remains less than u_p as $r \to \infty$, the original series (5.60) must converge. Furthermore, since

$$U = \sum_{n=0}^{p-1} (-1)^n u_n + \lim_{r\to\infty} S_r,$$

and $0 < S_r < u_p$, (5.62) is also established. A similar argument holds for the case where p is odd.

As well as deriving (5.61) and (5.62), the above proof illustrates the method of *grouping terms* to rewrite a series in such a way that simple arguments and standard results can be used. This technique is frequently used in determining the convergence of series with $\rho = 1$, as is illustrated in Example 5.12.

Example 5.12

Show that the series

$$\sum_{n=1}^{\infty} \frac{(-1)^{n+1}}{n} = \sum_{n=0}^{\infty} \frac{(-1)^n}{n+1} = 1 - \frac{1}{2} + \frac{1}{3} - \frac{1}{4} + \cdots$$

is convergent, but not absolutely convergent.

Solution

The convergence follows directly from (5.61), since this is an alternating series of the form (5.60) with $u_n \to 0$ as $n \to \infty$ and $u_{n+1} < u_n$ for all $n \geq 0$. To prove that it is not absolutely convergent, we must prove that the series

$$\sum_{n=1}^{\infty} \frac{1}{n} = 1 + \frac{1}{2} + \frac{1}{3} + \frac{1}{4} + \cdots$$

is divergent. To do this, we group the terms according to

$$\sum_{n=1}^{\infty} \frac{1}{n} = 1 + \frac{1}{2} + \left(\frac{1}{3} + \frac{1}{4}\right) + \left(\frac{1}{5} + \frac{1}{6} + \frac{1}{7} + \frac{1}{8}\right) + \cdots,$$

where each term in brackets is greater than $\frac{1}{2}$. Since the grouping can be continued indefinitely, the series obviously diverges.

Example 5.13

Show that the series

$$\text{(a)} \sum_{n=1}^{\infty} \left(\frac{2}{2^\alpha}\right)^n \quad \text{and} \quad \text{(b)} \sum_{n=1}^{\infty} \frac{1}{n^\alpha}$$

both converge for all $\alpha > 1$.

Solution

(a) This is a geometric series with $r_n = (2/2^\alpha) < 1$, so $\rho < 1$ and it converges.

(b) For this series, $r_n = [(n+1)/n]^\alpha \to 1$ as $n \to \infty$, so that the ratio test does not help. However, if we group the terms in a way somewhat analogous to that used in Example 5.12, we obtain

$$\begin{aligned}\sum_{n=1}^{\infty} \frac{1}{n^\alpha} &= 1 + \left(\tfrac{1}{2^\alpha} + \tfrac{1}{3^\alpha}\right) + \left(\tfrac{1}{4^\alpha} + \tfrac{1}{5^\alpha} + \tfrac{1}{6^\alpha} + \tfrac{1}{7^\alpha}\right) + \cdots \\ &= 1 + a_1 + a_2 + \cdots,\end{aligned}$$

where

$$a_n \leq \left(\frac{2}{2^\alpha}\right)^n$$

for $\alpha > 1$. Hence by the comparison test (5.55), series (b) converges since series (a) converges. This is called the Riemann series.

Problems 5

5.1 Sum all the odd integers from 23 to 771 inclusive.

5.2 Sum the series

$$S_N = \sum_{n=1}^{N} \ln\left(\frac{nr^n}{n+1}\right)$$

for any $r > 0$. For what values of r, if any, does the series converge as $N \to \infty$?

5.3 Sum the series

$$S_N = \sum_{n=0}^{N} \exp[-(n+1/2)x],$$

where $x > 0$. Does the series converge as $N \to \infty$?

5.4 The *arithmo-geometric series*

$$S_N = \sum_{n=0}^{N} (a + nx)y^n,$$

where a, x and y are constants, may be summed in a similar manner to the geometric series (5.9). Sum the series and show that

$$\lim_{N\to\infty} S_N = S = \frac{a}{1-y} + \frac{xy}{(1-y)^2},$$

provided $|y| < 1$.

5.5 Which of the following series are convergent?

$$\text{(a)} \sum_{n=0}^{\infty} \frac{(2+3n^2)}{3^n} \qquad \text{(b)} \sum_{n=0}^{\infty} \frac{(-1)^n 2^n}{n^2}$$

$$\text{(c)} \sum_{n=1}^{\infty} \frac{(-1)(n+1)(n+2)}{4n^2} \qquad \text{(d)} \sum_{n=0}^{\infty} \frac{n^r}{n!},\ r > 1.$$

5.6 For what range of x values do the following series converge?

$$\text{(a)} \sum_{n=0}^{\infty} n^2 x^n \quad \text{(b)} \sum_{n=0}^{\infty} \frac{2^n (x-1)^n}{(n+3)} \quad \text{(c)} \sum_{n=0}^{\infty} 2^{nx}.$$

5.7 Find the limit as $x \to 1$ of

$$\text{(a)} \frac{x^5 + 3x^2 - 4}{x^2 - 1}, \quad \text{(b)} \frac{\sin \pi x}{\ln x}, \quad \text{(c)} \frac{\arcsin(x-1)}{\sinh(1-x)}, \quad \text{(d)} \frac{1 + \cos \pi x}{\sin^2 \pi x}.$$

5.8 Find the limits of the following functions as $x \to 0$:

$$\text{(a)} \frac{\sqrt{x+5} - \sqrt{5}}{x}, \quad \text{(b)} \frac{\ln^2(1+x)}{x \arcsin x}.$$

5.9 Expand $\cos x$ as a Taylor series about $x = \pi/4$ and establish its region of validity.

5.10 How many terms must be retained in the Maclaurin expansion to evaluate $\sin x$ at $x = 0.6$ rad with an accuracy of 10^{-4}? Confirm your result by comparing it to the more precise value obtained by using a calculator.

5.11 Derive the form of the first three non-vanishing terms in the Maclaurin expansion of (a) $\sec x$, (b) $\tan x$.

5.12 Show that there is no Maclaurin expansion for (a) cotan$x$, (b) $\sqrt{x}$ and (c) $e^{-1/x}$.

5.13 If $f(x) = \exp(-x^2)$, show that $f^{(n)}(0) = 0$ for all $n \geq 0$, and no Maclaurin series exists. Sketch $f(x)$.

5.14 Use the binomial expansion to find the first three terms in the Taylor expansion of $\sqrt{x}$ about $x = 1$ and $x = 2$, respectively. What are the regions of validity of the corresponding expansions?

5.15 Find the following limits:

$$\text{(a)}\ \lim_{x\to 0}\left(\frac{1-\sqrt{1-x}}{x}\right), \quad \text{(b)}\ \lim_{x\to\infty}\{x[(x^2-1)^{1/2}-(x^3-1)^{1/3}]\}.$$

5.16 The polynomial x^5+x^3-1 has a single root in the range $0<x<1$. Use Newton's method to locate this root to 3 significant figures.

5.17 The function $\text{sinc}\,(x)$ is defined by

$$\text{sinc}\,(x)\equiv\frac{\sin x}{x}. \tag{5.63}$$

Sketch the function and show that it has a maximum at $x=0$. For $x\neq 0$, the stationary points of $\text{sinc}\,(x)$ occur approximately at $x=(2n+1)\pi/2$, where $n=\pm 1, \pm 2, \ldots$, as should be clear from the sketch. Show that this approximation becomes increasingly precise as $x\to\infty$. Use Newton's approximation to find the position of the first minimum to an accuracy of one degree.

5.18 Identify the basic rules of elementary algebra (cf. Section 1.2.1) needed to establish the identity

$$\sum_{n=0}^{N} a_n x^n + \sum_{n=0}^{N} b_n x^n = \sum_{n=0}^{N}(a_n+b_n)x^n$$

and hence the identity (5.49) in the limit $N\to\infty$, when the series on the left-hand side converges.

5.19 Show that all three series (5.52a, b, c) converge in the same region.

5.20 Deduce the first three terms in the Maclaurin expansion of $f(x)=\arcsin(x)$. For what values of x is the series valid?

5.21 Deduce the first four terms in the Maclaurin expansion of $f(x)=e^x\cos x$. For what values of x is the series valid?

***5.22** Determine which of the following series are absolutely or conditionally convergent and indicate whether the result depends on the real variable α.

$$\text{(a)}\ \sum_{n=1}^{\infty}\frac{(-1)^n}{\alpha n}, \quad \text{(b)}\ \sum_{n=0}^{\infty}(-1)^n(\cos\alpha)^n, \quad \text{(c)}\ \sum_{n=1}^{\infty}\frac{(-1)^n(\cosh\alpha)^n}{n}.$$

5.23 Determine which of the following series, where $\alpha>0$, are convergent, and state whether they are absolutely or conditionally convergent?

$$\text{(a)}\ \sum_{n=1}^{\infty}\frac{(-1)^n}{\ln(\alpha n)}, \qquad \text{(b)}\ \sum_{n=1}^{\infty}\frac{(-1)^n(n+\alpha)}{(2\alpha n+3)},$$

$$\text{(c)}\ \sum_{n=1}^{\infty}\frac{(-1)^n[\ln(\alpha n)]^n}{n^{n/2}}, \quad \text{(d)}\ \sum_{n=0}^{\infty}\frac{(-1)^n(n+1)\alpha^n}{n!}.$$

6

Complex numbers and variables

In previous chapters we have been discussing real numbers and their algebraic representation. Real numbers are part of a larger set called *complex numbers*. In this chapter we start by showing how the latter arise and then discuss their properties and how they are represented. Complex numbers and complex variables are of great practical importance in a wide range of topics, including vibrations and waves, and quantum theory.

6.1 Complex numbers

Given a positive real number q (not necessarily an integer) we know that its square roots $\pm\sqrt{q}$ are also real numbers. But situations also arise where we meet the square root of a negative number. In Section 2.1.1, for example, we saw that the solution of a general quadratic equation $ax^2 + bx + c = 0$ is of the form

$$x = \frac{-b \pm \sqrt{b^2 - 4ac}}{2a} \tag{6.1}$$

and there is no restriction on the sign of $(b^2 - 4ac)$. Thus we have to face the question: can we find an interpretation of the quantity $\sqrt{-q}$, where $q > 0$? It cannot be the same as $\sqrt{q}$ because squaring would produce a contradiction. A new definition is required. Since

$$\sqrt{-q} = \sqrt{(-1)(q)} = \sqrt{q}\sqrt{-1},$$

it follows that the only new definition needed is for $\sqrt{-1}$. This is denoted by the letter i, with $i^2 = -1$, and is called an *imaginary*

Mathematics for Physicists, First Edition. B.R. Martin and G. Shaw.

Companion website: www.wiley.com/go/martin/mathsforphysicists

number.[1] Thus, $\sqrt{-q} = \pm i\sqrt{q}$ and is also an imaginary number. If x and y are two real numbers, then the number $z = x + iy$ is called a *complex number* and x and y are called its *real* and *imaginary* parts, denoted Re $z = x$ and Im $z = y$. Formally, the quantities Re and Im are functions, whose argument is a complex number z and whose results are the real and imaginary parts of z, respectively. Note that both functions produce real outputs, and in particular the imaginary part y is a *real* number; it is always understood that it is multiplied by i to give an imaginary number.

A first sight it may appear that imaginary numbers have no applications in physical science because all physical measurements yield a real number. In fact the converse is true: complex numbers play a vital role in the mathematical analysis of numerous physical phenomena. We will see that apparently making a problem more complicated by introducing complex variables, can in practice actually simplify analyses by allowing the use of powerful techniques available in the theory of complex quantities.

We can now interpret solutions of equations such as

$$z^2 + z + 1 = 0.$$

Using the standard formula (6.1) gives

$$z = \frac{1}{2}\left[-1 \pm \sqrt{(1-4)}\right] = -\frac{1}{2} \pm \frac{\sqrt{3}}{2}i, \tag{6.2}$$

that is, the roots $z_{1,2}$ are complex numbers, with Re $z_{1,2} = -\frac{1}{2}$, Im $z_1 = \frac{\sqrt{3}}{2}$ and Im $z_2 = -\frac{\sqrt{3}}{2}$. In this case, the two roots only differ by the sign of the imaginary parts. Pairs of complex numbers related in this way are said to be *complex conjugates* of each other. Thus, if a complex number $z = x + iy$, then its complex conjugate, written z^*, is $z^* = x - iy$.[2] It is straightforward to show that complex conjugation has the properties:

$$(z_1 + z_2)^* = z_1^* + z_2^*, \quad (z_1 z_2)^* = z_1^* z_2^* \quad \text{and} \quad (z^*)^* = z,$$

and so on for several complex numbers.

Two complex numbers are defined to be equal only if the real parts of both numbers are equal and the imaginary parts of both numbers are equal. Complex numbers obey the usual rules of addition,

[1] Some authors use the letter j instead of i. This is usually the case in mathematics books for engineers, because engineers use lower case i for electric current.

[2] Other notations, such as $\bar{z}$, are occasionally used for the operation of complex conjugation, that is, changing the sign of the imaginary part of a complex number.

subtraction and multiplication, including the commutative, associative and distributive laws obeyed by real numbers, as discussed in Section 1.1.1. For example,

$$(6+3i)+(-2+5i)=4+8i, \tag{6.3a}$$

and

$$(6+3i)\times(-2+5i)=-12+30i-6i+15i^2=-27+24i, \tag{6.3b}$$

where $i^2=-1$ has been used in (6.3b). Division of a complex number by a real number is straightforward; the real number divides the real and imaginary parts of the complex number separately. Division by a complex number is a little more complicated. If we have two complex numbers p and q, their quotient is in general also a complex number, whose real and imaginary parts are found by rationalisation. In the case of complex numbers, this means multiplying the numerator and denominator by the complex conjugate of the latter to give a new real denominator, which then divides the real and imaginary parts of the numerator. Explicitly, if p and q are two complex numbers,

$$\begin{aligned}\frac{p}{q}&=\frac{(\text{Re }p+i\text{Im }p)}{(\text{Re }q+i\text{Im }q)}=\frac{(\text{Re }p+i\text{Im }p)\,(\text{Re }q-i\text{Im }q)}{(\text{Re }q+i\text{Im }q)\,(\text{Re }q-i\text{Im }q)}\\&=\frac{(\text{Re }p\text{Re }q+\text{Im }p\text{Im }q)+i\,(\text{Im }p\text{Re }q-\text{Re }p\text{Im }q)}{(\text{Re }q)^2+(\text{Im }q)^2}.\end{aligned} \tag{6.4a}$$

The quantity $(\text{Re }q)^2+(\text{Im }q)^2$ that appears in the denominator is the square of the *modulus*, or *absolute value*, of q, written mod$q$, or $|q|$. Thus,

$$|q|^2=(\text{Re }q)^2+(\text{Im }q)^2=qq^*. \tag{6.4b}$$

It follows from (6.4b), that for a general complex number z,

$$z^{-1}=z^*/|z|^2. \tag{6.5}$$

Example 6.1

If $z_1=3+2i$, $z_2=2+3i$ and $z_3=3-i$, find:

(a) $z_1+z_2^*-z_3$, (b) $z_1-(z_2+z_3)^*$, (c) $(z_1+z_2)(z_2+z_3)$, (d) $(z_2-z_1)^*(z_3-z_2)$, (e) $z_1/(z_2z_3)^*$, (f) $(z_1+z_3)/(z_2+z_3)$, (g) $|z_1|$ and z_1^{-1}.

Solution

(a) $z_1 + z_2^* - z_3 = 2$, (b) $z_1 - (z_2 + z_3)^* = -2 + 4i$,

(c) $(z_1 + z_2)(z_2 + z_3) = 15 + 35i$, (d) $(z_2 - z_1)^*(z_3 - z_2) = -5 + 3i$,

(e) $$\frac{z_1}{(z_2 z_3)^*} = \frac{3+2i}{9-7i} = \frac{(3+2i)(9+7i)}{(9-7i)(9+7i)} = \frac{13}{130} + \frac{39}{130}i,$$

(f) $$\frac{z_1 + z_2}{z_2 + z_3} = \frac{6+i}{5+2i} = \frac{(6+i)(5-2i)}{(5+2i)(5-2i)} = \frac{32}{29} - \frac{7}{29}i.$$

(g) $|z_1| = \sqrt{3^2 + 2^2} = \sqrt{13}$ and using $z = z^*/|z|^2$ gives $z_1^{-1} = 3/13 - i2/13$.

Example 6.2

Simplify and rationalise the following expressions:

(a) $\dfrac{(3+2i)(4-2i)}{(3+i)^2(1-3i)}$, (b) $\dfrac{(4+3i)}{(1-i)(2-i)}$, (c) $\dfrac{(3+2i)^2}{(2+i)(3-i)}$.

Solution

(a) $$\frac{(3+2i)(4-2i)}{(3+i)^2(1-3i)} = \frac{16+2i}{(8+6i)(1-3i)} = \frac{16+2i}{26-18i}$$
$$= \frac{(8+i)(13+9i)}{(13-9i)(13+9i)} = \frac{19}{50} + \frac{17}{50}i,$$

(b) $$\frac{(4+3i)}{(1-i)(2-i)} = \frac{(4+3i)}{(1-3i)} = \frac{(4+3i)(1+3i)}{(1-3i)(1+3i)} = -\frac{1}{2} + \frac{2}{3}i,$$

(c) $$\frac{(3+2i)^2}{(2+i)(3-i)} = \frac{5+6i}{7+i} = \frac{(5+6i)(7-i)}{(7+i)(7-i)} = \frac{41}{50} - \frac{37}{50}i.$$

6.2 Complex plane: Argand diagrams

The complex number $z = x + iy$ is an ordered pair of real numbers that can be written (x, y) and these can be viewed as the Cartesian co-ordinates of a point $P(x, y)$ in a plane, called in this context the *complex plane*. The diagram in which complex numbers are represented in this way is called an *Argand diagram*. This is shown in Figure 6.1, with the general point $P(x, y)$ plotted.

An alternative way of representing a complex number is to use two-dimensional polar co-ordinates (r, θ), where r is the positive distance to P from the origin and θ is measured in the counter-clockwise sense from the x-axis. The quantities r and θ are also shown in Figure 6.1, from which we see that

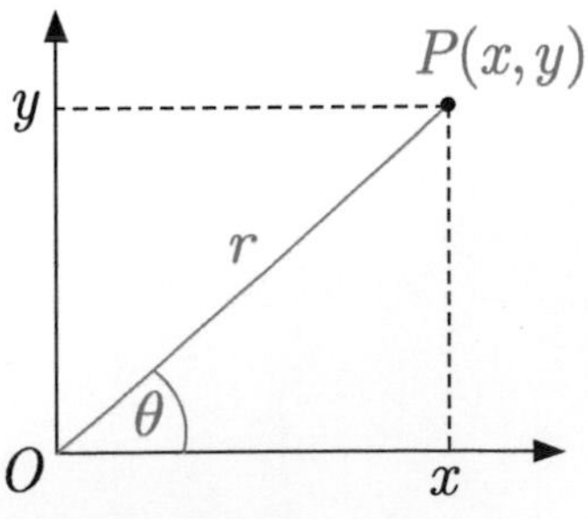

Figure 6.1 Argand diagram.

$$x = r\cos\theta, \quad y = r\sin\theta, \tag{6.6a}$$

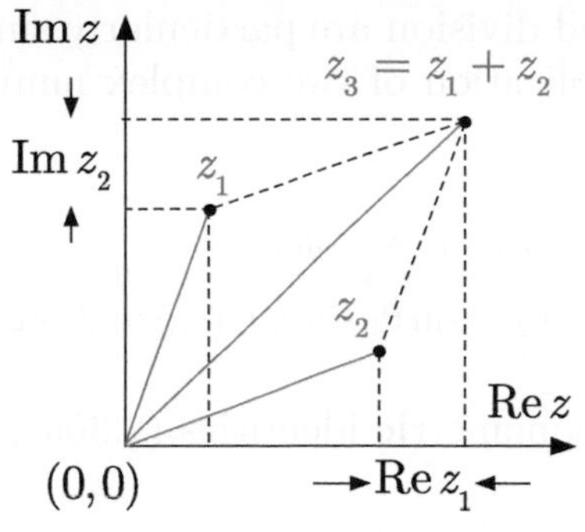

Figure 6.2 Addition of two complex numbers.

so that

$$z = x + i\,y = r\cos\theta + ir\sin\theta. \tag{6.6b}$$

This is called the *polar form* of z. The quantity r, given by $r = +\sqrt{x^2+y^2}$, is the modulus of z, that is, $r = \text{mod}\,z = |z|$. The angle θ is called the *argument* of z and is written $\theta = arg\,z$.[3] It may be found using (6.6b) to be $\theta = \arctan\,(y/x)$, but care must be taken when using this result to take account of the signs of both x and y separately, otherwise not all the values of θ found from their ratio will satisfy (6.6a). As the latter equations only define θ up to an additive integral multiple of 2π, it is usual to quote the so-called *principal value* of θ, that is, the value for which $-\pi < \theta \le \pi$. For example, if $z = x + iy = 2 - 2i$, then $r = 2\sqrt{2}$ and $\theta = \arctan(-1)$, where the latter has solutions

$$\theta = -\pi/4 + 2\pi n \quad \text{and} \quad \theta = 3\pi/4 + 2\pi n,$$

with n any integer. However, only the first of these has $x = r\cos\theta > 0$ and $y = r\sin\theta < 0$, as required, and choosing $n = 0$, we obtain the principal value $\theta = -\pi/4$.

The Argand diagram provides a geometrical interpretation of arithmetical operations involving complex numbers. For example, Figure 6.2 shows two complex numbers z_1 and z_2. It is easy to see from the construction shown that their sum

$$z_1 + z_2 = (x_1, y_1) + (x_2, y_2) = [(x_1 + x_2), (y_1 + y_2)]$$

is the point plotted at z_3. An analogous diagram describes subtraction.

[3] Some authors use Arg for the general argument and reserve arg for the principal value. We will use arg throughout.

Multiplication and division are particularly simple in polar form. In the case of multiplication of two complex numbers z_1 and z_2, we have from (6.6a)

$$\begin{aligned} z_1 z_2 &= (x_1 x_2 - y_1 y_2) + i(x_1 y_2 + x_2 y_1) \\ &= r_1 r_2 \left[(\cos\theta_1 \cos\theta_2 - \sin\theta_1 \sin\theta_2) + i(\cos\theta_1 \sin\theta_2 + \cos\theta_2 \sin\theta_1)\right], \end{aligned}$$

which, using the trigonometric identities (2.36a), is

$$z_1 z_2 = r_1 r_2 \left[\cos(\theta_1 + \theta_2) + i \sin(\theta_1 + \theta_2)\right].$$

Thus the product $z_1 z_2$ is a complex number with modulus $r_1 r_2$ and argument $(\theta_1 + \theta_2)$.

For the of division of two complex numbers z_1/z_2 we can write $z_1/z_2 = (z_1 z_2^*)/|z_2|^2$ and then again use (6.6a), to give

$$\begin{aligned} \frac{z_1}{z_2} &= \frac{1}{r_2^2}\left[(x_1 x_2 + y_1 y_2) + i(x_2 y_1 - x_1 y_2)\right] \\ &= \frac{r_1}{r_2}\left[(\cos\theta_1 \cos\theta_2 + \sin\theta_1 \sin\theta_2) + i(\cos\theta_2 \sin\theta_1 - \cos\theta_1 \sin\theta_2)\right], \end{aligned}$$

which, using the trigonometric identities (2.36a), is

$$\frac{z_1}{z_2} = \frac{r_1}{r_2}\left[\cos(\theta_1 - \theta_2) + i \sin(\theta_1 - \theta_2)\right].$$

Thus the quantity z_1/z_2 is a complex number with modulus r_1/r_2 and argument $(\theta_1 - \theta_2)$. This relation can also be demonstrated on an Argand diagram.

Finally, simple equations define curves in the complex plane. For example, consider the equation

$$\mathrm{Re}(z^2) = 2.$$

Using $z = x + iy$ gives $z^2 = x^2 - y^2 + 2ixy$ and hence

$$\mathrm{Re}(z^2) = x^2 - y^2 = 2.$$

This is a hyperbola, the two branches of which pass through $z = (\pm\sqrt{2},\ 0)$, respectively. Another example is the equation $|z + 3| = 5$, which may be written

$$|x + iy + 3| = 5 \Rightarrow (x + 3)^2 + y^2 = 25.$$

This is the equation of a circle of radius 5 and centre at $(-3, 0)$.

Example 6.3

Find the polar forms of the following complex numbers z, using the principal value of $\arg z$.

(a) $1 - 2i\sqrt{3}$, (b) $\cos\pi + i\sin\pi$, (c) $3 - \sqrt{i}$

Solution

(a) $r = \sqrt{1^2 + (2\sqrt{3})^2} = \sqrt{13}, \theta = \arctan(-2\sqrt{3}/1) = -1.29$ rad,

(b) $r = \sqrt{\cos^2\pi + \sin^2\pi} = 1$, $\theta = \arctan(\sin\pi/\cos\pi) = \pm\pi$ rad,

(c) First write $\sqrt{i} = x + iy$. Then, by squaring and equating real and imaginary parts on both sides, we obtain two solutions $\sqrt{i} = \pm(1+i)/\sqrt{2}$. Choosing the positive sign gives

$$3 - \sqrt{i} = \frac{1}{\sqrt{2}}\left[(3\sqrt{2} - 1) - i\right].$$

Thus

$$r = \frac{1}{\sqrt{2}}\sqrt{\left[(3\sqrt{2}-1)^2 + 1\right]} = \sqrt{10 - 3\sqrt{2}} = 2.40,$$

and

$$\theta = \arctan[-1/(3\sqrt{2} - 1)] = -0.30 \text{ rad}.$$

If instead we take the negative sign for $\sqrt{i}$ and proceed in the same way we obtain

$$3 - \sqrt{i} = \frac{1}{\sqrt{2}}\left[(3\sqrt{2} + 1) + i\right].$$

Thus

$$r = \frac{1}{\sqrt{2}}\sqrt{\left[(3\sqrt{2}+1)^2 + 1\right]} = \sqrt{10 + 3\sqrt{2}} = 3.77,$$

and

$$\theta = \arctan[1/(3\sqrt{2} + 1)] = 0.19 \text{ rad}.$$

Example 6.4

What are the equations of the plane curves represented by the equations: (a) $z^2 - |z| + (z^*)^2 = 0$, (b) $|(z+3)/(z-1)| = 2$, and (c) $\arg[(z+3)/(z-2)] = \pi/4$?

Solution

(a) $z^2 - |z| + (z^*)^2 = (x+iy)^2 - (x+iy)(x-iy) + (x-iy)^2 = x^2 - 3y^2$. Hence $z^2 - |z| + (z^*)^2 = 0$ is a pair of straight lines $y = \pm x/\sqrt{3}$, passing through $(0, 0)$ and with gradients $\pm 1/\sqrt{3}$, respectively.

(b) $|z+3| + \sqrt{(x+3)^2+y^2}$ and $|z-1| = \sqrt{(x-1)^2+y^2}$, so

$$\left(\frac{|z+3|}{|z-1|}\right)^2 = \frac{(x+3)^2+y^2}{(x-1)^2+y^2} = 4,$$

from which one obtains

$$3y^2 + 3x^2 - 2x - 5 = 0.$$

(c) $\dfrac{z+3}{z-2} = \dfrac{(z+3)(z^*-2)}{|z-2|^2} = \dfrac{(|z|^2+x-6)-5iy}{|z-2|^2}$,

so

$$\arg\left(\frac{z+3}{z-2}\right) = \frac{-5y}{(x^2+y^2+x-6)}$$

and

$$\frac{-5y}{(x^2+y^2+x-6)} = \tan\left(\frac{\pi}{4}\right) = 1,$$

from which one obtains

$$x^2+y^2+x+5y-6=0.$$

6.3 Complex variables and series

In Chapter 1, the discussion of real numbers was extended to their algebraic realisation, real variables. In the same way we may extend the current discussion to consider *complex variables* and their associated complex algebra. The rules discussed in Section 1.2 for the algebra of real variables hold, provided we remember that a complex variable is actually a pair of real variables. Thus, for example, the function

$$f(z) = \frac{z}{z+3},$$

written in terms of its real and imaginary parts is

$$f(z) = \frac{(x+iy)}{(x+3)+iy},$$

and may be rationalised by multiplying the numerator and denominator by the complex conjugate of the latter, that is, $[(x+3)-iy]$. This gives

$$f(z) = f_1(x,y) + if_2(x,y),$$

where

$$f_1(x,y) = \frac{x(x+3)+y^2}{(x+3)^2+y^2} \quad \text{and} \quad f_2(x,y) = \frac{3iy}{(x+3)^2+y^2}.$$

Similarly, an equation such as

$$az^2 + bz + c = 0 \tag{6.7}$$

may be written

$$a(x+iy)^2 + b(x+iy) + c = 0,$$

and when expanded is

$$(ax^2 + bx - ay^2 + c) + iy(2ax + b) = 0.$$

Because two complex quantities are only equal if both their real and imaginary parts are equal, (6.7) is equivalent to two equations,

$$ax^2 + bx - ay^2 + c = 0 \quad \text{and} \quad y\,(2ax+b) = 0.$$

We next consider series of the form

$$S = \sum_{n=0}^{\infty} a_n, \tag{6.8}$$

where the individual terms are now complex numbers or expressions. By writing

$$\sum_{n=0}^{\infty} a_n = \sum_{n=0}^{\infty} (\text{Re } a_n) + i \sum_{n=0}^{\infty} (\text{Im } a_n), \tag{6.9}$$

theses series may be expressed in terms of two real series, enabling the results established in Chapter 5 for real series to be easily extended to the complex case. In particular, one can show that d'Alembert's ratio test still holds,[4] so that if

$$r_n = |a_{n+1}|/|a_n| \to \rho, \tag{6.10}$$

as $n \to \infty$, the series converges if $\rho < 1$ and does not converge if $\rho > 1$, while the case $\rho = 1$ requires special treatment. Thus, for example, to test the convergence of the series

$$\sum_{n=1}^{\infty} \frac{(1+i)^n}{2^n},$$

[4] This result is derived in Section 6.3.1.

using the ratio test, we find the quantity

$$\rho = \lim_{n\to\infty}\left|\frac{(1+i)^{n+1}}{2^{n+1}}\frac{2^n}{(1+i)^n}\right| = \lim_{n\to\infty}\left|\frac{(1+i)}{2}\right|$$

$$= \left|\frac{(1+i)}{2}\right| = \frac{1}{\sqrt{2}} < 1,$$

and hence the series is convergent.

We can also extend the previous discussion of power series. These now become series of the form

$$\sum_{n=0}^{\infty} a_n (z-z_0)^n, \tag{6.11}$$

where the variable $z = x + iy$, with $z_0 = x_0 + iy_0$ and the coefficients a_n are complex numbers. Then by the ratio test, the series converges if

$$\rho \equiv \lim_{n\to\infty}\left|\frac{a_{n+1}(z-z_0)}{a_n}\right| < 1,$$

i.e. if

$$|z-z_0| < R \equiv \lim_{n\to\infty}\left|\frac{a_n}{a_{n+1}}\right|, \tag{6.12}$$

as in the case of a real series (5.19). In terms of the real and imaginary parts, this becomes

$$\sqrt{(x-x_0)^2+(y-y_0)^2} < R,$$

i.e.

$$(x-x_0)^2+(y-y_0)^2 < R^2,$$

corresponding to the interior of a circle in the complex plane, centred at $z = z_0$ with radius R. This circle is called the *circle of convergence* and R is called the *radius of convergence*. For example, consider the series

$$\sum_{n=0}^{\infty} (-1)^n \frac{z^n}{n} = 1 - z + \frac{z^2}{2} - \frac{z^3}{3} + \frac{z^4}{4} + \cdots$$

Then by (6.12), the series converges for

$$|z| < R = \lim_{n\to\infty}\left|\frac{n+1}{n}\right| = 1,$$

that is, inside a circle of radius unity centred at the origin $z = 0$ of the complex plane.

One very important power series is the complex exponential series

$$e^z = 1 + z + \frac{z^2}{2!} + \frac{z^3}{3!} + \frac{z^4}{4!} + \cdots = \sum_{n=0}^{\infty}\frac{z^n}{n!}, \tag{6.13}$$

obtained by replacing the real variable x in (5.42) by the complex variable $z = x + iy$. D'Alembert's ratio test (6.10) shows that this series converges for all values of z, so that (6.13) can be used to define the exponential function over the whole complex plane. In the same way, the series for $\sin x$ and $\cos x$ in Table 5.1 can be generalised from real x to complex z, and used to define $\sin z$ and $\cos z$ in the whole complex plane, in which case $\sin z$ and $\cos z$ are no longer real or restricted to the range -1 to $+1$. Other functions can then be defined from these in analogy to the corresponding functions of a real variable. For example $\tan z \equiv \sin z/\cos z$, while the hyperbolic functions are

$$\sinh z = \frac{e^z - e^{-z}}{2} \quad \text{and} \quad \cosh z = \frac{e^z + e^{-z}}{2}, \tag{6.14}$$

in analogy to (2.57) for real $z = x$.

Example 6.5

Use the ratio test to find the circle of convergence for the following infinite series whose general terms R_n are (a) $(n+5)(4iz)^n$, (b) $[(z-1)/2]^2$, (c) $(n-3)(z-3i)^n$.

Solution

By the ratio test, the series convergences if

$$\rho = \lim_{n\to\infty}\left(\frac{R_{n+1}}{R_n}\right) < 1.$$

Applying this condition to each of the series in turn gives

(a) $|z| < 1/4$, that is, a circle of radius 1/4 centred at (0, 0) in the Argand diagram,

(b) $|(z-1)/2| < 1$, that is, a circle of radius 2 centred at (1,0),

(c) $|z - 3i| < 1$, that is, a circle of radius 1 centred at (0, 3).

*6.3.1 Proof of the ratio test for complex series

For the series (6.8) to converge, it is necessary and sufficient for both the real series

$$\sum_{n=0}^{\infty}(\text{Re } a_n) \quad \text{and} \quad \sum_{n=0}^{\infty}(\text{Im } a_n) \tag{6.15}$$

that occur in (6.9) to converge. For $\rho > 1$, this is impossible because (6.10) then implies $|a_n| \nrightarrow 0$, so at least one of the quantities Re a_n

or Im a_n does not tend to zero. Hence the corresponding real series, and by implication (6.8), cannot converge by (5.15).

It remains to prove that (6.8) does converge for $\rho < 1$. To do this, we first consider the series

$$\sum_{n=0}^{\infty} |a_n|.$$

This is a real series, so that d'Alembert's ratio test applies, and it converges if $\rho < 1$. But

$$|a_n| > |\text{Re } a_n|, \quad |\text{Im } a_n| \geq 0,$$

so that the real positive series

$$\sum_{n=0}^{\infty} |\text{Re } a_n| \quad \text{and} \quad \sum_{n=0}^{\infty} |\text{Im } a_n| \tag{6.16}$$

converge by the comparison test (5.55) established in Section 5.5.1*. The series (6.15) are then said to be 'absolutely convergent' and, as shown in Section 5.5.2*, an absolutely convergent series is convergent, as the name implies. Hence both the series (6.16), and thus the complex series (6.8), converge for $\rho < 1$, as required.

6.4 Euler's formula

In this section we introduce an important formula due to Euler and illustrate some of its many applications. To derive this formula, we substitute $z = i\theta$, where θ is real, in the exponential series (6.13). This gives:

$$\begin{aligned} e^{i\theta} &= 1 + i\theta + \frac{(i\theta)^2}{2!} + \frac{(i\theta)^3}{3!} + \frac{(i\theta)^4}{4!} + \cdots \\ &= \left(1 - \frac{\theta^2}{2!} + \frac{\theta^4}{4!} + \cdots\right) + i\left(\theta - \frac{\theta^3}{3!} + \frac{\theta^5}{5!} + \cdots\right). \end{aligned} \tag{6.17}$$

Now from the results given in Table 5.1, the real part of (6.17) is seen to be the series for $\cos\theta$ and the imaginary part is the series for $\sin\theta$. So we have deduced the important result

$$e^{i\theta} = \cos\theta + i\sin\theta. \tag{6.18}$$

This is *Euler's formula,* and enables many useful relations to be derived. For example, from the definition of the hyperbolic functions (2.57) and Euler's formula, we have, for real angles θ,

$$\cosh(i\theta) = \frac{e^{i\theta} + e^{-i\theta}}{2} = \cos\theta,$$

and

$$\sinh(i\theta) = \frac{e^{i\theta} - e^{-i\theta}}{2} = i \sin\theta.$$

Furthermore, using the polar forms (6.6b) together with (6.18), we can write any complex number z in the form

$$z = re^{i\theta}, \tag{6.19}$$

where $r = |z|$ is the modulus and θ is the argument of z as usual. This *exponential form* is very useful in algebraic calculations involving complex variables, particularly multiplication and division. Using the law of exponents discussed in Section 1.1.2, and now extended to complex variables, we have for multiplication

$$z_1 z_2 = r_1 e^{i\theta_1} r_2 e^{i\theta_2} = r_1 r_2 e^{i(\theta_1+\theta_2)}, \tag{6.20a}$$

and for division,

$$\frac{z_1}{z_2} = \frac{r_1 e^{i\theta_1}}{r_2 e^{i\theta_2}} = \frac{r_1}{r_2} e^{i(\theta_1-\theta_2)}. \tag{6.20b}$$

These are the same results that were obtained in Section 6.2, but derived here in a simpler way without using trigonometric identities.

Example 6.6

Use Euler's formula to write the following complex numbers in the form $x + iy$

$$\frac{3+2i}{(1-2i)(3-i)(2+i)}.$$

Solution

We first write each factor in the form $re^{i\theta}$, where $r = \sqrt{\mathrm{Re}^2 z + \mathrm{Im}^2 z}$ and $\theta = \arctan(y/x)$. This gives,

$$(3+2i): r_1 = \sqrt{13},\ \theta_1 = 0.588; \quad (1-2i): r_2 = \sqrt{5},\ \theta_2 = -1.107;$$
$$(3-i): r_3 = \sqrt{10},\ \theta_3 = -0.322; \quad (2+i): r_4 = \sqrt{5},\ \theta_4 = 0.464.$$

Finally,

$$\frac{3+2i}{(1-2i)(3-i)(2+i)} = \frac{r_1}{r_2 r_3 r_4} \exp i(\theta_1 - \theta_2 - \theta_3 - \theta_4)$$
$$= 0.228e^{i1.553} = 0.004 + 0.228i.$$

6.4.1 Powers and roots

The exponential form provides a simple way of finding powers of a complex quantity, since if $z = re^{i\theta}$, then

$$z^n = r^n e^{in\theta} \tag{6.21}$$

by repeated application of (6.20a). For example, to find the cube of $z = (1+i)$, we first convert it to exponential form

$$z = (1+i) = \sqrt{2}\, e^{i\pi/4}$$

to give

$$z^3 = (1+i)^3 = (\sqrt{2}\, e^{i\pi/4})^3 = 2\sqrt{2}\, e^{i3\pi/4} = -2+2i.$$

The nth roots of a complex number z are the solutions w of the equation $w = z^{1/n}$, that is, the complex numbers whose nth power is z. There are always n such roots. To see this, we note that

$$z = re^{i\theta} = re^{i(\theta+2\pi k)}$$

for any integer k. Hence the roots are

$$w_k = z^{1/n} = r^{1/n} \exp[i(\theta + 2\pi k)/n]. \tag{6.22}$$

However, it is easily to see that $w_{k\pm n} = w_k$, so the only roots that are distinct are $w_0, w_1, \ldots, w_{n-1}$, with larger or smaller values of k merely reproducing the roots with $k = 0, 1, \ldots, n-1$. For example, to find the cube roots of $z = (2-2i)$, we use the polar form with $r = 2\sqrt{2}$ and $\theta = -\pi/4$. Then using $k = 0, 1, 2$ gives the three solutions

$$k = 0: \ 1.366 - 0.366i, \quad k = 1: \ -0.366 + 1.366i, \quad k = 2: \ -1 - i.$$

Larger values of k just reproduce the solutions for $k \leq 2$.

Of particular interest are the nth roots of unity. In this case $z^n = 1 = e^{2ik\pi}$, where k is any integer, so $z = e^{2ik\pi/n}$. Hence the solutions are

$$z_{1,2,3,\ldots,\,n} = 1,\ e^{2i\pi/n}, \ldots,\ e^{2i(n-1)\pi/n},$$

corresponding to $k = 0, 1, 2, \ldots, (n-1)$. The solutions for $n = 3$, i.e.

$$z_1 = 1, \quad z_2 = e^{2i\pi/3} = -1/2 + i\sqrt{3}/2, \quad z_3 = e^{4i\pi/3} = -1/2 - i\sqrt{3}/2,$$

are shown plotted on a circle of unit radius in Figure 6.3. Again, larger values of k just reproduce the solutions for $k \leq 2$.

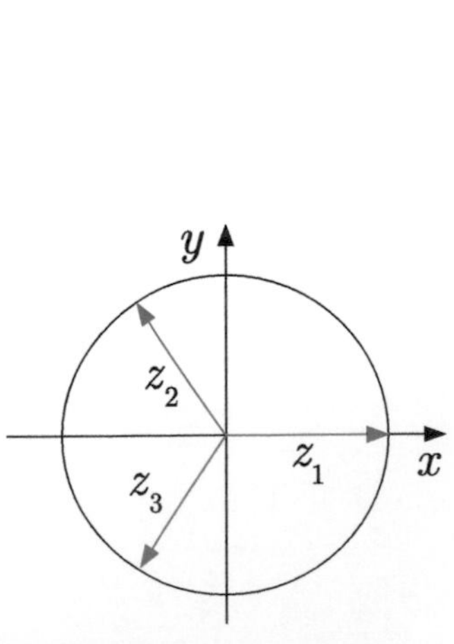

Figure 6.3 The cube roots of unity.

The polar representation of a complex number is also useful when finding the roots of a polynomial equation. To illustrate this, consider the polynomial equation

$$z^6 - 3z^5 + 2z^4 - 7z^3 + 3z^2 - 2z + 6 = 0,$$

which factorises as

$$(z^3 - 1)(z^2 + 2)(z - 3) = 0.$$

Hence solutions are given by $z^3 = 1$, or $z^2 = -2$ or $z = 3$. In the first case we can use (6.21) to give the three solutions obtained given in (6.22)

$$z_1 = 1, \quad z_2 = \left(-\frac{1}{2}, \frac{\sqrt{3}}{2}i\right) \quad \text{and} \quad z_3 = \left(-\frac{1}{2}, -\frac{\sqrt{3}}{2}i\right).$$

The other solutions are $z_4 = i\sqrt{2}$, $z_5 = -i\sqrt{2}$, from $z^2 = -2$, and finally $z_6 = 3$. Thus we have six solutions in accord with the fundamental theorem of algebra. This example also illustrates the general result, that a polynomial equation with real coefficients has roots that occur in complex conjugate pairs. (See Problem 6.4)

Example 6.7

Write the following powers and roots in the form $x + iy$: (a) $(3 + 2i)^6$, (b) $[(1 + i)/(1 - i\sqrt{3})]^4$, (c) $\sqrt[3]{1 - i\sqrt{3}}$. (d) Show that the five fifth roots of unity sum to zero.

Solution

(a) Write $(3 + 2i) = re^{i\theta}$. Then $r = \sqrt{13}$ and $\theta = \arctan(2/3) = 0.588$. Hence

$$(3 + 2i)^6 = (\sqrt{13})^6 e^{6\times 0.588i} = 2197e^{3.528i} = -2034.4 - 832.7i.$$

(b) Using exponential forms we have, $(1 + i) = \sqrt{2}e^{i\pi/4}$ and $(1 - i\sqrt{3}) = 2e^{-i\pi/3}$. So

$$\left(\frac{1+i}{1-i\sqrt{3}}\right)^4 = \left[\frac{1}{\sqrt{2}}\left(e^{i\pi/4+i\pi/3}\right)\right]^4 = \frac{1}{4}e^{7\pi i/3} = \frac{1}{8}\left(1 + i\sqrt{3}\right).$$

(c) To find the three roots, we write

$$1 - i\sqrt{3} = 2e^{-i\pi/3}e^{2\pi ni}, \quad n = 0,\ 1,\ 2,$$

which gives

$$\sqrt[3]{1 - i\sqrt{3}} = 2^{1/3}e^{-i\pi/9} = 1.184 - 0.431i,$$

$$\sqrt[3]{1 - i\sqrt{3}} = 2^{1/3}e^{i5\pi/9} = -0.219 + 1.241i,$$

$$\sqrt[3]{1 - i\sqrt{3}} = 2^{1/3}e^{i11\pi/9} = -0.965 - 0.810i,$$

respectively.

(d) The fifth roots of unity are $\alpha_k = r^{1/5}\exp[i(\theta + 2\pi k)/5]$ for $k = 0, 1, \ldots, 4$, where $r = 1$ and $\theta = 0$. So,

$$\sum_{k=0}^{4} \alpha_k = \sum_{k=0}^{4} e^{2\pi ik/5} = \frac{1 - \exp[10i\pi/5]}{1 - \exp[2\pi i/5]} = 0.$$

6.4.2 Exponentials and logarithms

The exponential function was used in Chapter 2 to define natural logarithms, which we can also generalise for complex arguments. For a complex number z, the natural logarithm is defined by analogy with its definition for a real number. Thus $e^{\ln z} = z$. Substituting $\ln z = \alpha + i\beta$ and $z = re^{i\theta}$, where θ is the principal value, $-\pi < \theta \leq \pi$, we have

$$e^{\alpha} e^{i\beta} = re^{i\theta}$$

and since α and β are both real, $e^{\alpha} = r$, i.e. $\alpha = \ln r$, and $\beta = \theta + 2\pi k(k = 0, 1, \ldots)$, so

$$\ln z = \ln r + i(\theta + 2\pi k), \qquad k = 0, 1, \ldots \tag{6.23}$$

Thus the imaginary part of the logarithm is only defined up to additive multiples of 2π. The *principal value* of the logarithm is defined as the case when $k = 0$, for which[5]

$$\ln z = \ln r + i\theta = \ln|z| + i \arg z. \tag{6.24}$$

It is straightforward to show that the results previously obtained in Chapter 2 for the logarithms of real variables also hold for complex variables. Thus, in general, using (6.19),

$$\ln(z_1 z_2) = \ln z_1 + \ln z_2 = \ln(r_1 r_2) + i(\theta_1 + \theta_2 + 2\pi k)$$

where k is an integer. If $-\pi < \arg z_1 + \arg z_2 < \pi$, this reduces to the result for principal values

$$\ln(z_1 z_2) = \ln z_1 + \ln z_2.$$

Likewise, for division,

$$\ln(z_1/z_2) = \ln z_1 - \ln z_2.$$

[5] Some authors use the notation ln for the natural logarithm of a complex number and Ln when its argument is real, that is, the logarithm corresponding to using the principal value of θ. We will use ln for both cases.

Extending the definition of logarithms to complex arguments enables us to generalise the discussion of Section 6.4.1 to complex powers and roots. For example, to evaluate $(1+i)^z$ where $z = 1 - 2i$, we have

$$(1+i)^z = \left(\sqrt{2}e^{i\pi/4}\right)^{1-2i} = \left(\sqrt{2}e^{i\pi/4}\right)\left(2e^{i\pi/2}\right)^{-i}$$
$$= \sqrt{2}e^{\pi/2}e^{i\pi/4}2^{-i}.$$

Now using logarithms, $2^{-i} = e^{-i\ln 2}$ and so

$$(1+i)^z = \sqrt{2}e^{\pi/2}\exp\left[-i\left(\pi/4\right)+\ln 2\right] = 6.803e^{-1.479i}.$$

So, finally, $(1+i)^z = 0.624 - 6.774i$.

Example 6.8

Express (a) $\ln(1-i)$, (b) $\ln\left[(2+i)/(1-i)\right]$, (c) $\cos(2\pi + i\ln 3)$ in the form $z = x + iy$ using the principal value for arg z.

Solution

(a) The point $z = 1 - i$ in polar form is $re^{i\theta}$, where $r = \sqrt{2}$ and $\theta = -\pi./4$. Therefore,

$$\ln(1-i) = \frac{1}{2}\ln 2 - \frac{i\pi}{4} = 0.3466 - i0.25\pi.$$

(b) In polar form,

$$2+i = \sqrt{5}\exp(0.4636i) \quad \text{and} \quad 1-i = \sqrt{2}\exp(-0.7854i),$$

so

$$(2+i)/(1-i) = \sqrt{2.5}e^{1.249i}$$

and hence

$$\ln\left[(2+i)/(1-i)\right] = 0.458 + 1.249i.$$

(c) Using the trigonometric formula for $\cos(A+B)$ gives $\cos(2\pi + i\ln 3) = \cos(i\ln 3)$. Then using Euler's formula (6.18),

$$\cos(i\ln 3) = -\frac{1}{2}\left[e^{i(i\ln 3)} + e^{-i(i\ln 3)}\right] = -\frac{1}{2}\left[e^{-\ln 3} + e^{\ln 3}\right].$$

6.4.3 De Moivre's theorem

If we substitute Euler's formula into both sides of the simple identity

$$\left(e^{i\theta}\right)^n = e^{in\theta},$$

we immediately obtain *De Moivre's theorem*:

$$(\cos\theta + i\sin\theta)^n = \cos(n\theta) + i\sin(n\theta), \tag{6.25}$$

which is valid for all real values of n, whether integer or not.

De Moivre's theorem provides a very convenient way of obtaining expressions for powers of trigonometric functions, and expansions of these functions for multiple angles. Suppose we wish to express $\cos 4\theta$ and $\sin 4\theta$ in terms of powers of $\sin\theta$ and $\cos\theta$. This could be done for each expansion separately by using the multiple-angle trigonometric formulas of Section 2.2.4, but by apparently making the problem more complicated by introducing complex variables, we can use the result (6.25). This gives

$$\begin{aligned}\cos 4\theta + i\sin 4\theta &= (\cos\theta + i\sin\theta)^4 \\ &= (\cos^4\theta - 6\cos^2\theta\sin^2\theta + \sin^4\theta) \\ &\quad + 4i\sin\theta\cos\theta(\cos^2\theta\sin^2\theta).\end{aligned}$$

Then equating real and imaginary parts of both sides gives

$$\sin 4\theta = 4\sin\theta\cos\theta(\cos^2\theta\sin^2\theta)$$

and

$$\cos 4\theta = \cos^4\theta - 6\cos^2\theta\sin^2\theta + \sin^4\theta.$$

This method may be applied in general to find the forms of $\cos n\theta$ and $\sin n\theta$ for any $n > 0$ by using the general results

$$z^n + \frac{1}{z^n} = 2\cos n\theta \quad \text{and} \quad z^n - \frac{1}{z^n} = 2i\sin n\theta \quad (|z| = 1) \tag{6.26}$$

that follow directly from De Moivre's theorem. In a similar way we can find expressions for $\cos^n\theta$ and $\sin^n\theta$ in terms of simple sines and cosines. For example, consider $\cos^4\theta$. From (6.25) this may be written

$$\cos^4\theta = \frac{1}{16}\left(z + \frac{1}{z}\right)^4 = \frac{1}{16}\left[\left(z^4 + \frac{1}{z^4}\right) + 4\left(z^2 + \frac{1}{z^2}\right) + 6\right],$$

which using (6.25) again is

$$\cos^4\theta = \frac{1}{8}(\cos 4\theta + 4\cos 2\theta + 3).$$

Example 6.9

(a) Express $\sin^5\theta$ as a sum of terms of the form $\sin n\theta$ for $n \leq 5$. (b) Write the expression $(\cos 3\theta + i\sin 3\theta)/(\cos 5\theta - i\sin 5\theta)$ in the form $x + iy$. (c) Show that $\tan 3\theta = t(3 - t^2)/(1 - 3t^2)$, where $t = \tan\theta$. Hence solve the equation $t^3 - 3t^2 - 3t + 1 = 0$.

Solution

(a) Using $2i\sin\theta = z - 1/z$ gives

$$(2i\sin\theta)^5 = \left(z^5 - \frac{1}{z^5}\right) - 5\left(z^3 - \frac{1}{z^3}\right) + 10\left(z - \frac{1}{z}\right),$$

and hence using $2i \sin n\theta = z^n - 1/z^n$,

$$16 \sin^5 \theta = \sin 5\theta - 5 \sin 3\theta + 10 \sin \theta.$$

(b) Use

$$\begin{aligned} 1/(\cos 5\theta - i \sin 5\theta) &= (\cos 5\theta - i \sin 5\theta)^{-1} \\ &= [\cos(-5\theta) - i \sin(-5\theta)] = \cos(5\theta) + i \sin(5\theta). \end{aligned}$$

Then

$$\frac{(\cos 3\theta + i \sin 3\theta)}{(\cos 5\theta - i \sin 5\theta)} = e^{3i\theta} e^{5i\theta} = e^{8i\theta} = (\cos 8\theta + i \sin 8\theta).$$

(c) Use

$$\begin{aligned} (\cos 3\theta + i \sin 3\theta) &= (\cos \theta + i \sin \theta)^3 \\ &= \cos^3 \theta + 3i \cos^2 \theta \sin \theta - 3 \cos \theta \sin^2 \theta - i \sin^3 \theta, \end{aligned}$$

and equate real and imaginary parts on both sides to give

$$\begin{aligned} \cos 3\theta &= \cos^3 \theta - 3 \cos \theta \sin^2 \theta \\ \sin 3\theta &= 3 \cos^2 \theta \sin \theta - \sin^3 \theta \end{aligned}$$

so that

$$\tan 3\theta = \frac{t(3 - t^2)}{1 - 3t^2}.$$

The solutions of the given equation therefore correspond to $\tan 3\theta = 1$, that is, $\theta = \frac{1}{3}\left(\frac{\pi}{4} + n\pi\right)$. Taking $n = 0,\ 1,\ 2$, gives $t = \tan \theta = 0.268,\ 3.732, -1$.

*6.4.4 Summation of series and evaluation of integrals

The Euler formula may also be used to sum many series involving sines and cosines. Consider the series

$$C = \sum_{k=0}^{n} a^k \cos k\theta = 1 + a \cos \theta + a^2 \cos 2\theta + \cdots + a^n \cos n\theta,$$

where a is a real constant. To find C we first form the analogous series for sines,

$$S = \sum_{k=0}^{n} a^k \sin k\theta = 1 + a \sin \theta + a^2 \sin 2\theta + \cdots + a^n \sin n\theta,$$

and then combine them to give the complex series

$$C + iS = \sum_{k=0}^{n} a^k (\cos k\theta + i \sin k\theta) = \sum_{k=0}^{n} a^k e^{ik\theta}.$$

This is a geometric series with a common ratio $R = ae^{i\theta}$ and from Section 4.1 we know that it is

$$C + iS = \frac{1 - R^{n+1}}{1 - R} = \frac{1 - (ae^{i\theta})^{n+1}}{1 - (ae^{i\theta})}. \tag{6.27}$$

Finally, C is given by the real part of the right-hand side and, after some algebra, we find

$$C = \frac{1 + a\cos\theta - a^{n+1}\cos[(n+1)\theta] - a^{n+2}\cos n\theta}{1 + 2a\cos\theta + a^2}.$$

As a bonus, S is given by the imaginary part of the right-hand side of (6.27).

A similar technique may be used for continuous variables, where the analogous quantities are integrals. Consider, for example, the integral

$$C = \int_0^t e^{a\theta}\cos(b\,\theta)\,\mathrm{d}\theta.$$

This could be evaluated directly by integration by parts, but as an illustration of the method we form the analogous integral involving sines,

$$S = \int_0^t e^{a\theta}\sin(b\,\theta)\,\mathrm{d}\theta,$$

and combine these to give

$$C + iS = \int_0^t e^{a\theta}[\cos(b\;\theta) + i\sin(b\;\theta)]\mathrm{d}\theta = \int_0^t e^{a\theta}e^{ib\theta}\,\mathrm{d}\theta = \frac{e^{(a+ib)t} - 1}{a + ib}. \tag{6.28}$$

Finally, C is the real part of the right-hand side of (6.28), that is

$$C = \frac{e^{at}(a\cos bt + b\sin bt) - a}{a^2 + b^2}.$$

and again, as a bonus, S is the imaginary part of the right-hand side.

Problems 6

6.1 If $z_1 = 1 + 4i$, $z_2 = 2 - 3i$ and $z_3 = 4 + 3i$, find:

(a) $z_1 + z_2 - z_3$, (b) $(z_1 - z_2)^* + z_3$, (c) $(z_1 + z_3)(z_2 + z_3)$,
(d) $(z_2 + z_1)(z_3 - z_2)^*$, (e) $z_2/(z_1^* z_3)$, (f) $(z_1 + z_2)/(z_1 + z_3)$,
(g) $|z_1|$ and z_1^{-1}.

6.2 If z_1 and z_2 are two complex numbers, verify that: (a) $(z_1 z_2)^* = z_1^* z_2^*$, (b) $|z_1 z_2| = |z_1||z_2|$, (c) $|z_1 + z_2| \leq |z_1| + |z_2|$ and (d) $|z_1 - z_2| \geq |z_1| - |z_2|$.

6.3 Simplify and rationalise the following expressions:

$$\text{(a)}\ \frac{(1+2i)(3-2i)}{(2+i)^2(2-3i)},\ \text{(b)}\ \frac{(2+i)}{(1-i)(3-i)},\ \text{(c)}\ \frac{(3+i)^2}{(2-i)(3+i)}.$$

6.4 Show that a polynomial equation of order n with real coefficients has roots that are either real, or in complex conjugate pairs.

6.5 Express the following numbers in polar form: (a) $\sqrt{i}$, (b) $(1+i)/(\sqrt{3}-i)$, (c) $(1+i)(\sqrt{3}-i)$, using the principal value of the argument.

6.6 What are the plane curves represented by the equations (a) $|z-1| = 2$, (b) $|z+1| = |z-i|$?

6.7 Use the ratio test to find the circle of convergence for the following infinite series whose general terms R_n are (a) $\left[(n!)^2/(3n)!\right](z-3i)^n$, (b) $(n+3)^3(2iz)^n$, (c) $(-1)^n z^{n+3}/n!$

6.8 What is the modulus and argument of (a) $z = e^{i\pi/2} + \sqrt{2}e^{i\pi/4}$, (b) $z = (1+i)e^{i\pi/6}$, (c) $z = [(2+i)/(i-3)]\,e^{i\pi/3}$.

6.9 Use Euler's formula to write the following complex numbers in the form $x + iy$:

$$\text{(a)}\ \left(\frac{\sqrt{3}}{1+2i}\right)^4,\quad \text{(b)}\ \frac{(2+3i)(1-4i)^*}{(1-2i)^*(3+2i)(4+i)},\quad \text{(c)}\ (1-2i)^{1/2}.$$

6.10 Convert the following complex numbers to the form $(x + iy)$:

$$\text{(a)}\ \left(\frac{i}{1+i}\right)^{10},\quad \text{(b)}\ (6+3i)^{1/3},\quad \text{(c)}\ \left(\frac{\sqrt{2}}{i-1}\right)^6.$$

6.11 Write the following complex numbers in the form $(x + iy)$:

$$\text{(a)}\ i^{1/5},\quad \text{(b)}\ \frac{(2+i)(1-i)^*}{(1+i)^*(3-4i)},\quad \text{(c)}\ \left(\frac{\sqrt{3}i}{i+\sqrt{2}}\right)^7.$$

6.12 Express the following in the form $x + iy$:

$$\text{(a)}\ \operatorname{arcsinh}(i/\sqrt{3}),\quad \text{(b)}\ \sin(\pi - i\ln 2),\quad \text{(c)}\ \ln[(1+i)/(3-i)]$$

6.13 Convert the following complex expressions to the form $x + iy$:

$$\text{(a) } (2i)^{1+i}, \quad \text{(b) } \left(\frac{1+i\sqrt{2}}{1-i\sqrt{2}}\right)^{\sin i}, \quad \text{(c) } \cos\left[i\ln\left(\frac{1-i}{2+i}\right)\right].$$

6.14 Use De Moivre's theorem to (a) write the expression

$$(\cos 2\theta + i\sin 2\theta)^3(\cos 3\theta + i\sin 3\theta)^2$$

in the form $(x + iy)$; (b) show that

$$\sin^7\theta = \frac{1}{64}(35\sin\theta - 21\sin 3\theta + 7\sin 5\theta - \sin 7\theta).$$

6.15 Use De Moivre's theorem to (a) simplify the expression

$$\frac{\cos 2\theta - i\sin 2\theta}{\cos 5\theta + i\sin 5\theta},$$

and (b) show that

$$\tan 4\theta = \frac{4t(1-t^2)}{1-6t^2+t^4}, \text{ where } t = \tan\theta.$$

***6.16** Evaluate the integral

$$I = \int_0^\infty x e^{-x}\cos(2x)\mathrm{d}x.$$

***6.17** Find the sum of the series

$$S(x) = 1 + 2\sin x + \frac{2^2\sin(2x)}{2!} + \frac{2^3\sin(3x)}{3!} + \cdots$$

***6.18** Use the binomial theorem for $(1 + e^{ix})^n$ to show that

$$\sum_{k=o}^{n}\binom{n}{k}\cos kx = 2^n\cos^n\left(\frac{x}{2}\right)\cos\left(\frac{nx}{2}\right),$$

and find the sum of the series

$$\sum_{k=0}^{n}\binom{n}{k}\sin kx,$$

where $\binom{n}{k}$ are the binomial coefficients.

7

Partial differentiation

In this chapter we generalise the discussion of differential calculus in Chapter 3 to functions of more than one variable. Many results will be taken over from Chapter 3 and will be dealt with rather briefly, so that we can focus on the differences between the two cases.

7.1 Partial derivatives

Given a function $f(x_1, x_2, \ldots, x_n)$ of n independent variables, $x_1, x_2, \ldots, x_n$, the *partial derivative* of f with respect to x_1 is defined by

$$\frac{\partial f}{\partial x_1} \equiv \lim_{\delta x_1 \to 0} \left[\frac{f(x_1 + \delta x_1, x_2, \ldots, x_n) - f(x_1, x_2, \ldots, x_n)}{\delta x_1} \right], \quad (7.1)$$

provided the limit exists. In other words, it is obtained by differentiating f with respect to x_1, while treating the other variables $x_2, x_3, \ldots, x_n$ as fixed parameters. Partial derivatives with respect to the other variables are defined in a similar way. For example, if

$$f(x, y) = xy^3 e^x, \quad (7.2)$$

then differentiating with respect to x keeping y fixed gives, using the product rule (3.20),

$$\frac{\partial f}{\partial x} = y^3 e^x + xy^3 e^x, \quad (7.3a)$$

while differentiating with respect to y keeping x fixed gives

$$\frac{\partial f}{\partial y} = 3xy^2 e^x. \quad (7.3b)$$

Mathematics for Physicists, First Edition. B.R. Martin and G. Shaw.

Companion website: www.wiley.com/go/martin/mathsforphysicists

Higher derivatives are obtained by repeated partial differentiation, so that

$$\frac{\partial^2 f}{\partial x_i \partial x_j} \equiv \frac{\partial}{\partial x_i}\left(\frac{\partial f}{\partial x_j}\right), \qquad i, j = 1, 2, \ldots, n \tag{7.4}$$

for the second derivatives. Thus for the function (7.2), using (7.3) one obtains

$$\frac{\partial^2 f}{\partial x^2} = 2y^3 e^x + xy^3 e^x, \qquad \frac{\partial^2 f}{\partial y^2} = 6xye^x,$$

$$\frac{\partial^2 f}{\partial x \partial y} = \frac{\partial}{\partial x}\left(3xy^2 e^x\right) = 3y^2 e^x + 3xy^2 e^x,$$

and

$$\frac{\partial^2 f}{\partial y \partial x} = \frac{\partial}{\partial y}\left(y^3 e^x + xy^3 e^x\right) = 3y^2 e^x + 3xy^2 e^x.$$

From this, one sees that

$$\frac{\partial^2 f}{\partial x \partial y} = \frac{\partial^2 f}{\partial y \partial x}.$$

In general

$$\frac{\partial^2 f}{\partial x_i \partial x_j} = \frac{\partial^2 f}{\partial x_j \partial x_i} \tag{7.5}$$

for any f such that both the derivatives in (7.4) are continuous in x_i and x_j at the point of evaluation.

It is very important when working with partial derivatives to keep track of which variables are kept constant. This can be made explicit by adopting a notation in which the partial derivatives are written in brackets with the fixed variables as subscripts, so that (7.1) becomes

$$\left(\frac{\partial f}{\partial x_1}\right)_{x_2, x_3, \ldots, x_n} = \lim_{\delta x_1 \to 0}\left[\frac{f(x_1 + \delta x_1, x_2, \ldots, x_n) - f(x_1, x_2, \ldots, x_n)}{\delta x_1}\right], \tag{7.6}$$

and (7.3a) and (7.3b) are written

$$\left(\frac{\partial f}{\partial x}\right)_y = y^3 e^x + xy^3 e^x, \qquad \left(\frac{\partial f}{\partial y}\right)_x = 3xy^2 e^x.$$

To emphasise the importance of keeping track of which variables are held constant, we note that if we define $z = xy$, then (7.2) can be written

$$f(x, z) = \frac{z^3}{x^2} e^x,$$

so that

$$\left(\frac{\partial f}{\partial x}\right)_z = \frac{z^3 e^x}{x^2} - 2\frac{z^3 e^x}{x^3} = xy^3 e^x - 2y^3 e^x \neq \left(\frac{\partial f}{\partial x}\right)_y .$$

This notation is widely used in thermal physics, for example, where different choices of variables are often used within the same calculation. Thus, the energy E of a gas at equilibrium is often written both as a function of temperature T and volume V, and also as a function of temperature and pressure P, but

$$\left(\frac{\partial E}{\partial T}\right)_V \neq \left(\frac{\partial E}{\partial T}\right)_P$$

except in the case of a 'perfect gas'. In this chapter, we shall generally use the simpler notation (7.1), resorting to (7.6) only where there is room for ambiguity.

Example 7.1
Verify that

$$\frac{\partial}{\partial x}\left(\frac{\partial f}{\partial y}\right) = \frac{\partial}{\partial y}\left(\frac{\partial f}{\partial x}\right)$$

when $f(x, y) = x \sin y + y \sin^{-1} x$.

Solution
We have [cf. (3.27)],

$$\frac{\partial f}{\partial x} = \sin y + \frac{y}{(1+x)^{1/2}} \frac{\partial f}{\partial y} = x \cos y + \sin^{-1} x,$$

so that

$$\frac{\partial}{\partial y}\left(\frac{\partial f}{\partial x}\right) = \cos y + \frac{1}{(1+x^2)^{1/2}} = \frac{\partial}{\partial x}\left(\frac{\partial f}{\partial y}\right),$$

as required.

7.2 Differentials

For functions $f(x)$ of a single variable x, we are already familiar with the result [cf. (5.27)]

$$\delta f \equiv f(x + \delta x) - f(x) = \delta x \frac{\mathrm{d}f}{\mathrm{d}x} + O[(\delta x)^2], \qquad (7.7)$$

for small changes δx, provided the derivative exists. In the same way, the definition (7.1) implies

$$f(x_1 + \delta x_1, x_2, \ldots, x_n) - f(x_1, x_2, \ldots, x_n) = \delta x_1 \frac{\partial f}{\partial x_1} + O[(\delta x_1)^2], \tag{7.8}$$

since $x_2,\ x_3, \ldots, x_n$ are treated as fixed parameters in defining the partial derivatives. Analogous results are obtained for small changes in the other variables $x_2, x_3, \ldots, x_n$. From this, for a function of two variables $f(x_1,\ x_2)$ one obtains

$$\begin{aligned}\delta f &\equiv f(x_1 + \delta x_1, x_2 + \delta x_2) - f(x_1, x_2) \\ &= [f(x_1 + \delta x_1, x_2 + \delta x_2) - f(x_1 + \delta x_1, x_2)] \\ &\quad + [f(x_1 + \delta x_1, x_2) - f(x_1, x_2)] \\ &= \delta x_2 \frac{\partial f(x_1 + \delta x_1, x_2)}{\partial x_2} + \delta x_1 \frac{\partial f(x_1, x_2)}{\partial x_1} + \ldots\end{aligned}$$

and substituting (7.8) into the first term of this equation gives

$$\delta f = \delta x_1 \frac{\partial f}{\partial x_1} + \delta x_2 \frac{\partial f}{\partial x_2} + \cdots,$$

where the omitted terms are quadratic in $\delta x_1,\ \delta x_2$. On generalising to n variables, this becomes

$$\begin{aligned}\delta f &\equiv f(x_1 + \delta x_1, x_2 + \delta x_2, \ldots, x_n + \delta x_n) - f(x_1, x_2, \ldots, x_n) \\ &= \sum_{i=1}^{n} \delta x_i \frac{\partial f}{\partial x_i} + \cdots, \end{aligned} \tag{7.9}$$

where the omitted terms are again quadratic in δx_i.

At this point, we denote small changes by $\mathrm{d}x$ or $\mathrm{d}x_i$, and define the *differential* $\mathrm{d}f$ by

$$\mathrm{d}f \equiv \frac{\mathrm{d}f}{\mathrm{d}x} \mathrm{d}x \tag{7.10}$$

for the case of single variables, and

$$\mathrm{d}f \equiv \sum_{i=1}^{n} \frac{\partial f}{\partial x_i} \mathrm{d}x_i \tag{7.11}$$

for the case of multi-variables. The important distinction between (7.10, 7.11) and (7.7, 7.9) is that the latter are approximations, with corrections of the order indicated, whereas the former, being definitions, are exact.

Differentials are used repeatedly throughout the rest of this chapter. Here we will show, by an example, how they can be used to obtain partial derivatives when the definition of the relevant function is implicit.

Example 7.2

Find $(\partial y/\partial x)_z$ if $z = (x^2 + y^2)\exp(xy)$.

Solution

Here the function $y(x, z)$ is defined implicitly. However, the partial derivative required may be obtained by using (7.11) to obtain

$$\mathrm{d}z = \left[2x + y(x^2 + y^2)\right] e^{xy}\mathrm{d}x + \left[2y + x(x^2 + y^2)\right] e^{xy}\mathrm{d}y.$$

Keeping z fixed implies $\mathrm{d}z = 0$, so that

$$\left[2y + x(x^2 + y^2)\right] \mathrm{d}y = -\left[2x + y(x^2 + y^2)\right] \mathrm{d}x,$$

on cancelling $e^{xy} \neq 0$, and hence

$$\left(\frac{\partial y}{\partial x}\right)_z = -\frac{[2x + y(x^2 + y^2)]}{[2y + x(x^2 + y^2)]}.$$

Example 7.3

Given that $x^2u - y^2w = 2$ and $x - y = uw$, find $(\partial x/\partial u)_w$.

Solution

Taking differentials, we have

$$x^2\mathrm{d}u + 2xu\,\mathrm{d}x - y^2\mathrm{d}w - 2yw\,\mathrm{d}y = 0$$

and

$$\mathrm{d}x - \mathrm{d}y = u\,\mathrm{d}w + w\,\mathrm{d}u.$$

Since $(\partial x/\partial u)_w \Rightarrow \mathrm{d}w = 0$,

$$x^2\mathrm{d}u + 2xu\,\mathrm{d}x - 2yw\,\mathrm{d}y = 0$$

and

$$\mathrm{d}x - \mathrm{d}y = w\,\mathrm{d}u.$$

Then eliminating $\mathrm{d}y$ from these two equations, we have

$$2(yw - xu)\mathrm{d}x = (x^2 + 2yw^2)\mathrm{d}u,$$

and hence

$$\left(\frac{\partial x}{\partial u}\right)_w = \frac{x^2 + 2yw^2}{2(yw - xu)}.$$

7.2.1 Two standard results

In this subsection we will consider a function of two variables $f(x, y)$ and use differentials to derive the standard results

$$\left(\frac{\partial f}{\partial x}\right)_y = \left[\left(\frac{\partial x}{\partial f}\right)_y\right]^{-1} \tag{7.12}$$

and

$$\left(\frac{\partial f}{\partial x}\right)_y \left(\frac{\partial x}{\partial y}\right)_f \left(\frac{\partial y}{\partial f}\right)_x = -1. \tag{7.13}$$

To do this, we use (7.11) to give

$$\mathrm{d}f = \left(\frac{\partial f}{\partial x}\right)_y \mathrm{d}x + \left(\frac{\partial f}{\partial y}\right)_x \mathrm{d}y \tag{7.14a}$$

and then consider the corresponding function $x(y, f)$ that specifies x in terms of y and f, to obtain the corresponding differential

$$\mathrm{d}x = \left(\frac{\partial x}{\partial y}\right)_f \mathrm{d}y + \left(\frac{\partial x}{\partial f}\right)_y \mathrm{d}f. \tag{7.14b}$$

Substituting (7.14b) into (7.14a) gives

$$\mathrm{d}f = \left(\frac{\partial f}{\partial x}\right)_y \left(\frac{\partial x}{\partial f}\right)_y \mathrm{d}f + \left[\left(\frac{\partial f}{\partial y}\right)_x + \left(\frac{\partial f}{\partial x}\right)_y \left(\frac{\partial x}{\partial y}\right)_f\right] \mathrm{d}y.$$

Since any two of $\mathrm{d}x$, $\mathrm{d}y$, $\mathrm{d}f$ are independent, the coefficient of $\mathrm{d}f$ on the right-hand side must be unity, which gives (7.12), and the square bracket giving the coefficient of $\mathrm{d}y$ must vanish, which gives (7.13), as required.

Finally, we stress again that in using (7.12) and (7.13), it is important to pay attention to the variables being kept fixed in each derivative. In particular,

$$\frac{\partial A}{\partial B} \neq \left(\frac{\partial B}{\partial A}\right)^{-1}$$

in general, and the equality only holds if, as in (7.12), the same variables are kept fixed in each partial derivative.

Example 7.4

At high pressures and/or low temperatures, gases at equilibrium are well-described by the Van der Waal's equation

$$P = \frac{RT}{V - b} - \frac{a}{V^2}, \tag{7.15}$$

where P is the pressure, T the temperature, V the volume, and R, a and b are constants. Use (7.13) to find the coefficient of expansion

$$\alpha = \frac{1}{V}\left(\frac{\partial V}{\partial T}\right)_P$$

as a function of V and T.

Solution

From (7.15), we have

$$\left(\frac{\partial P}{\partial T}\right)_V = \frac{R}{V-b}, \quad \left(\frac{\partial P}{\partial V}\right)_T = \frac{-RT}{(V-b)^2} + \frac{2a}{V^3},$$

But by (7.13),

$$\left(\frac{\partial V}{\partial T}\right)_P \left(\frac{\partial T}{\partial P}\right)_V \left(\frac{\partial P}{\partial V}\right)_T = -1$$

So that by (7.12),

$$\alpha = \frac{1}{V}\left(\frac{\partial V}{\partial T}\right)_P = -\frac{1}{V}\left(\frac{\partial P}{\partial T}\right)_V \left[\left(\frac{\partial P}{\partial V}\right)_T\right]^{-1}$$

$$= \frac{R}{V(V-b)}\left[\frac{RT}{(V-b)^2} - \frac{2a}{V^3}\right]^{-1}.$$

7.2.2 Exact differentials

Given two functions $A(x, y)$ and $B(x, y)$, the quantity

$$A(x, y)\mathrm{d}x + B(x, y)\mathrm{d}y \tag{7.16}$$

is called an *exact* (or *perfect*) *differential* if there exists a function $f(x, y)$ such that

$$\mathrm{d}f = A(x, y)\mathrm{d}x + B(x, y)\mathrm{d}y. \tag{7.17a}$$

If no such function exists, it is called an *inexact differential*. A simple test for whether a differential is exact or not is to note that if it is, (7.17a) implies

$$\frac{\partial f}{\partial x} = A(x, y), \qquad \frac{\partial f}{\partial y} = B(x, y)\,, \tag{7.18}$$

so that, by (7.5),

$$\frac{\partial A}{\partial y} = \frac{\partial B}{\partial x}. \tag{7.19a}$$

The definition of an exact differential may be extended to functions of more than two variables, so that (7.17a) becomes

$$\mathrm{d}f = \sum_{i=1}^{n} C_i(x_1,\ x_2, \ldots, x_n)\mathrm{d}x_i \tag{7.17b}$$

and the condition (7.19a) becomes

$$\frac{\partial C_i}{\partial x_j} = \frac{\partial C_j}{\partial x_i} \quad \text{for all pairs of } i,\ j. \tag{7.19b}$$

Exact differentials are used in solving an important class of differential equations (i.e. equations that contain a function and its derivatives), as we shall see in Section 14.1.4; and in thermal physics, where relations of the form (7.19) are called *Maxwell relations.* In fact, (7.19b) is both a necessary and sufficient condition for (7.16) to be an exact differential. We shall, however, omit the proof of this, and in particular cases where it is satisfied, we shall establish the existence of a suitable function $f(x, y)$ by constructing it, as is shown in the following example.

Example 7.5

Show that

$$3x^2 \sin y \, \mathrm{d}x + (x^3 \cos y + 2y) \, \mathrm{d}y$$

is an exact differential and construct an appropriate function $f(x, y)$.

Solution

On comparing with (7.16) we see that $A = 3x^2 \sin y$ and $B = x^3 \cos y + 2y$, so that

$$\frac{\partial A}{\partial y} = \frac{\partial B}{\partial x} = 3x^2 \cos y$$

and (7.18) is satisfied, as required. Further, using (7.18), we have

$$\frac{\partial f}{\partial x} = 3x^2 \sin y$$

and integrating this, keeping y fixed, gives

$$f(x, y) = x^3 \sin y + g(y) \, ,$$

where the integration constant g may depend on y. The second equation (7.18) then gives

$$\frac{\partial f}{\partial y} = x^3 \cos y + \frac{\mathrm{d}g(y)}{dy} = x^3 \cos y + 2y \, ,$$

so that $g(y) = y^2 + c$, where c is a constant, and therefore

$$f(x, y) = x^3 \sin y + y^2 + c.$$

7.2.3 The chain rule

We next consider a function $f(x_1, \; x_2 \, , \ldots, \; x_n)$ where the variables x_i are themselves functions of another variable t. The rate of change of f with t can then be calculated by substituting the expressions $x_i(t)$ into f and differentiating the result with respect to t.

Alternatively, one can divide the differential (7.11) by dt to obtain the *chain rule*,

$$\frac{df}{dt} = \sum_{i=1}^{n} \frac{\partial f}{\partial x_i}\frac{dx_i}{dt}. \tag{7.20}$$

An important special case is when t is itself one of the arguments of the function, that is, when $f \equiv f(t,\ x_1,\ x_2, \ldots, x_n)$. Equation (7.20), with $n+1$ variables $(x_{n+1} = t)$ then gives

$$\frac{df}{dt} = \frac{\partial f}{\partial t} + \sum_{i=1}^{n} \frac{\partial f}{\partial x_i}\frac{dx_i}{dt}. \tag{7.21}$$

Example 7.6

What is the rate of change of

$$f(x, y, z) = xy + yz + zx$$

at $x = 0$, if $y = x^2 + 1$ and $z = e^x$?

Solution

From (7.21), with $t = x,\ x_1 = y,\ x_2 = z$, we have

$$\frac{df}{dx} = (y + z) + 2x(x + z) + e^z(x + y) = 3,$$

since $y = z = 1$ at $x = 0$.

7.2.4 Homogeneous functions and Euler's theorem

A function $f(x_1,\ x_2, \ldots, x_n)$ is said to be *homogeneous of degree k* if

$$f(\lambda x_1,\ \lambda x_2, \ldots, \lambda x_n) = \lambda^k f(x_1,\ x_2, \ldots, x_n), \tag{7.22}$$

where λ is an arbitrary parameter. For example, the functions

$$f(x, y) = \frac{x}{x^2 + y^2} e^{y/x} + \frac{1}{x + y} \quad \text{and} \quad f(x, y) = x^3 + 2xy^2 - 3y^3$$

are both homogeneous, of degree -1 and 3, respectively. *Euler's theorem* states that if $f(x_1,\ x_2, \ldots, x_n)$ is homogeneous of degree k, then

$$x_1\frac{\partial f}{\partial x_1} + x_2\frac{\partial f}{\partial x_2} + \cdots + x_n\frac{\partial f}{\partial x_n} = kf. \tag{7.23}$$

To derive (7.23) we make the substitutions $x_i = \lambda t_i$ and write

$$f(x_1,\ x_2, \ldots, x_n) = f(\lambda t_1,\ \lambda t_2, \ldots, \lambda t_n) = \lambda^k f(t_1,\ t_2, \ldots, t_n).$$

For any fixed set of $t_1,\ t_2,\ldots,t_n$, this is a function of λ only, and differentiating using the chain rule (7.20) gives

$$\frac{\mathrm{d}f}{\mathrm{d}\lambda}=t_1\frac{\partial f}{\partial x_1}+t_2\frac{\partial f}{\partial x_2}+\cdots+t_n\frac{\partial f}{\partial x_n}=k\lambda^{-1}f.$$

Euler's theorem then follows on multiplying by λ.

Example 7.7

Verify Euler's theorem explicitly for the function $f(x,y)=(x/y)\ln(y/x)$.

Solution

Since $f(\lambda x,\ \lambda y)=f(x,\ y)$, f is homogeneous of degree 0, so that Euler's theorem gives

$$x\frac{\partial f}{\partial x}+y\frac{\partial f}{\partial y}=0\,.$$

Alternatively, using the product and chain rules (3.20) and (7.20), one obtains

$$\frac{\partial f}{\partial x}=\frac{1}{y}\ln\left(\frac{y}{x}\right)-\frac{1}{y},\qquad \frac{\partial f}{\partial y}=-\frac{x}{y^2}\ln\left(\frac{y}{x}\right)+\frac{x}{y^2},$$

so that again

$$x\frac{\partial f}{\partial x}+y\frac{\partial f}{\partial y}=\frac{x}{y}\ln\left(\frac{y}{x}\right)-\frac{x}{y}-\frac{x}{y}\ln\left(\frac{y}{x}\right)+\frac{x}{y}=0,$$

as required by Euler's theorem.

7.3 Change of variables

In this section we address the problem of how to change variables in equations that contain partial derivatives. To do this, we firstly consider a function $f\equiv f(x_1,\ x_2,\ldots,x_n)$ of n variables $x_1,\ x_2,\ldots,x_n$ that are each functions of another n variables $x_i\equiv x_i(t_1,\ t_2,\ldots,t_n)$. Using (7.11) twice then gives

$$\mathrm{d}f=\sum_{i=1}^{n}\frac{\partial f}{\partial x_i}\mathrm{d}x_i=\sum_{i=1}^{n}\sum_{j=1}^{n}\frac{\partial f}{\partial x_i}\frac{\partial x_i}{\partial t_j}\mathrm{d}t_j\,,$$

where we remind the reader that partial differentiation with respect to x_i implies that all the other $x_j (j\neq i)$ are kept constant; and similarly differentiating with respect to t_j means that all the other

$t_i (i \neq j)$ are kept constant. In the same notation, expressing f directly in terms of t_j, $j = 1, 2, \ldots, n$ gives

$$\mathrm{d}f = \sum_{j=1}^{n} \frac{\partial f}{\partial t_j} \mathrm{d}t_j.$$

and comparing these two results gives the relation

$$\frac{\partial f}{\partial t_j} = \sum_{i=1}^{n} \frac{\partial x_i}{\partial t_j} \frac{\partial f}{\partial x_i} \tag{7.24}$$

between partial derivatives with respect to x_i and t_j.

To illustrate the use of this result, consider a function $f(x, y)$ of the Cartesian co-ordinates x, y. We will change variables to the plane polar co-ordinates r, θ of Figure 2.3, where [cf. (2.34) and (2.35)]

$$x = r\cos\theta, \qquad y = r\sin\theta, \tag{7.25a}$$

and conversely

$$r = (x^2 + y^2)^{1/2}, \qquad \theta = \arctan(y/x). \tag{7.25b}$$

From (7.24), setting $(x_1, x_2) = (x, y)$ and $(t_1, t_2) = (r, \theta)$, we obtain

$$\frac{\partial f}{\partial r} = \cos\theta \frac{\partial f}{\partial x} + \sin\theta \frac{\partial f}{\partial y},$$

and

$$\frac{\partial f}{\partial \theta} = -r\sin\theta \frac{\partial f}{\partial x} + r\cos\theta \frac{\partial f}{\partial y},$$

where

$$\frac{\partial f}{\partial r} \equiv \left(\frac{\partial f}{\partial r}\right)_\theta \quad \text{and} \quad \frac{\partial f}{\partial \theta} \equiv \left(\frac{\partial f}{\partial \theta}\right)_r.$$

Using (7.25a), these equations imply

$$r\frac{\partial f}{\partial r} = x\frac{\partial f}{\partial x} + y\frac{\partial f}{\partial y} \tag{7.26a}$$

and

$$\frac{\partial f}{\partial \theta} = -y\frac{\partial f}{\partial x} + x\frac{\partial f}{\partial y}, \tag{7.26b}$$

and conversely[1]

$$\frac{\partial f}{\partial x} = \cos\theta \frac{\partial f}{\partial r} - \frac{\sin\theta}{r}\frac{\partial f}{\partial \theta} \tag{7.27a}$$

and

$$\frac{\partial f}{\partial y} = \sin\theta \frac{\partial f}{\partial r} + \frac{\cos\theta}{r}\frac{\partial f}{\partial \theta}. \tag{7.27b}$$

[1] Note that (7.27a) and (7.27b) can also be obtained directly from (7.24) by setting $(t_1, t_2) = (x, y)$ and $(x_1, x_2) = (r, \theta)$.

Corresponding results involving higher order partial derivatives can be obtained by repeated use of (7.26) and (7.27). As an example, we will transform *Laplace's equation* in two dimensions,

$$\frac{\partial^2 f}{\partial x^2}+\frac{\partial^2 f}{\partial y^2}=0, \tag{7.28}$$

into polar co-ordinates (r,θ). From (7.27a), we have

$$\begin{aligned}\frac{\partial^2 f}{\partial x^2}&=\frac{\partial}{\partial x}\left(\frac{\partial f}{\partial x}\right)=\left(\cos\theta\frac{\partial}{\partial r}-\frac{\sin\theta}{r}\frac{\partial}{\partial\theta}\right)\left(\cos\theta\frac{\partial f}{\partial r}-\frac{\sin\theta}{r}\frac{\partial f}{\partial\theta}\right)\\&=\cos^2\theta\frac{\partial^2 f}{\partial r^2}+\frac{\sin^2\theta}{r^2}\frac{\partial^2 f}{\partial\theta^2}-\frac{2\sin\theta\cos\theta}{r}\frac{\partial^2 f}{\partial r\partial\theta}\\&\quad+\frac{\sin^2\theta}{r}\frac{\partial f}{\partial r}+\frac{2\sin\theta\cos\theta}{r^2}\frac{\partial f}{\partial\theta}.\end{aligned}$$

In a similar way one obtains

$$\begin{aligned}\frac{\partial^2 f}{\partial y^2}&=\sin^2\theta\frac{\partial^2 f}{\partial r^2}+\frac{\cos^2\theta}{r^2}\frac{\partial^2 f}{\partial\theta^2}+\frac{2\sin\theta\cos\theta}{r}\frac{\partial^2 f}{\partial r\partial\theta}\\&\quad+\frac{\cos^2\theta}{r}\frac{\partial f}{\partial r}-\frac{2\sin\theta\cos\theta}{r^2}\frac{\partial f}{\partial\theta}.\end{aligned}$$

Adding these two results and substituting in (7.28) then gives

$$\frac{\partial^2 f}{\partial r^2}+\frac{1}{r}\frac{\partial f}{\partial r}+\frac{1}{r^2}\frac{\partial^2 f}{\partial\theta^2}=0, \tag{7.29}$$

as Laplace's equation expressed in plane polar co-ordinates.

Example 7.8

Consider the change in co-ordinates $(x,y)\to(x',y')$ of a point P, brought about by a rotation through an angle ϕ as shown in Figure 7.1. Show that

$$x'=x\cos\phi+y\sin\phi\,,\quad y'=-x\sin\phi+y\cos\phi \tag{7.30}$$

and hence that Laplace's equation (7.28) is invariant in form under a rotation, that is,

$$\frac{\partial^2 f}{\partial x'^2}+\frac{\partial^2 f}{\partial y'^2}=\frac{\partial^2 f}{\partial x^2}+\frac{\partial^2 f}{\partial y^2}=0. \tag{7.31}$$

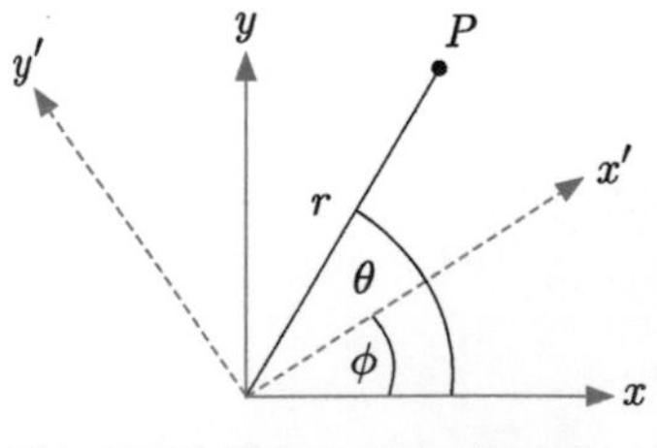

Figure 7.1 The original and rotated co-ordinate systems (7.30).

Solution

If the polar co-ordinates in the un-primed system are $(r\,,\,\theta)$, then in the primed system they are $(r'=r,\theta'=\theta-\phi)$, as can been seen in Figure 7.1. Hence

$$x' = r\cos(\theta - \phi) = r\cos\theta\cos\phi + r\sin\theta\sin\phi = x\cos\phi + y\sin\phi,$$

while

$$y' = r\sin(\theta - \phi) = r\sin\theta\cos\phi - r\cos\theta\sin\phi = -x\sin\phi + y\cos\phi.$$

Then by (7.24), with $(x_1, x_2) = (x', y')$ and $(t_1, t_2) = (x, y)$, one has

$$\frac{\partial f}{\partial x} = \cos\phi\frac{\partial f}{\partial x'} - \sin\phi\frac{\partial f}{\partial y'},$$
$$\frac{\partial f}{\partial y} = \sin\phi\frac{\partial f}{\partial x'} + \cos\phi\frac{\partial f}{\partial y'},$$

and hence

$$\frac{\partial^2 f}{\partial x^2} = \left(\cos\phi\frac{\partial}{\partial x'} - \sin\phi\frac{\partial}{\partial y'}\right)\left(\cos\phi\frac{\partial f}{\partial x'} - \sin\phi\frac{\partial f}{\partial y'}\right)$$
$$= \cos^2\phi\frac{\partial^2 f}{\partial x'^2} + \sin^2\phi\frac{\partial^2 f}{\partial y'^2} - 2\sin\phi\cos\phi\frac{\partial^2 f}{\partial x'\partial y'},$$

and similarly

$$\frac{\partial^2 f}{\partial y^2} = \sin^2\phi\frac{\partial^2 f}{\partial x'^2} + \cos^2\phi\frac{\partial^2 f}{\partial y'^2} + 2\sin\phi\cos\phi\frac{\partial^2 f}{\partial x'\partial y'}.$$

Finally, adding these two results gives

$$\frac{\partial^2 f}{\partial x^2} + \frac{\partial^2 f}{\partial y^2} = \frac{\partial^2 f}{\partial x'^2} + \frac{\partial^2 f}{\partial y'^2},$$

so that (7.31) follows from (7.28) as required.

7.4 Taylor series

The generalisation of Taylor's theorem (5.21) to more than one variable is straightforward. For simplicity, we start by finding an expansion of a function $f(x, y)$ of two variables about $x = x_0,\ y = y_0$ in powers of $h = x - x_0,\ k = y - y_0$. To do this, for any given values of h and k, we define a function

$$F(t) = f(x_0 + ht,\ y_0 + kt),$$

which reduces to $f(x, y)$ when the new variable $t \to 0$. Provided the first $N + 1$ derivatives of $F(t)$ exist over the whole range $0 \le t \le 1$, Taylor's theorem (5.21) gives

$$F(t) = \sum_{n=0}^{\infty} \frac{t^n}{n!} F^{(n)}(0) + R_N, \tag{7.32a}$$

where $F^{(n)}(t)$ is the nth derivative of F with respect to t [cf. (3.40b)], and where the remainder term is

$$R_N = \frac{t^{N+1}}{(N+1)!} F^{(N+1)}(\theta t) \tag{7.32b}$$

for at least one θ in the range $0 \leq \theta \leq 1$. However, from the chain rule (7.20) we also have

$$\frac{\mathrm{d}F}{\mathrm{d}t} = \frac{\partial F}{\partial x}\frac{\mathrm{d}x}{\mathrm{d}t} + \frac{\partial F}{\partial y}\frac{\mathrm{d}y}{\mathrm{d}t} = h\frac{\partial F}{\partial x} + k\frac{\partial F}{\partial y}.$$

Substituting this into (7.32) and setting $t = 1$ then gives

$$f(x_0 + h,\, y_0 + k) = \sum_{n=1}^{N} \left(h\frac{\partial}{\partial x} + k\frac{\partial}{\partial y} \right)^n f(x_0, y_0) + R_N, \tag{7.33a}$$

where

$$\left(h\frac{\partial}{\partial x} + k\frac{\partial}{\partial y} \right)^2 = \left(h\frac{\partial}{\partial x} + k\frac{\partial}{\partial y} \right) \left(h\frac{\partial f}{\partial x} + k\frac{\partial f}{\partial y} \right), \text{ etc.,}$$

and

$$\left(h\frac{\partial}{\partial x} + k\frac{\partial}{\partial y} \right)^n f(x_0, y_0)$$

means the derivatives of $f(x, y)$ are evaluated at $x = x_0,\ \ y = y_0$. The remainder term is

$$R_N = \frac{1}{(N+1)!} \left(h\frac{\partial}{\partial x} + k\frac{\partial}{\partial y} \right)^{N+1} f(x_0 + \theta h,\, y_0 + \theta k), \tag{7.33b}$$

for at least one θ in the range $0 \leq \theta \leq 1$. Assuming $R_N \to 0$ as $N \to \infty$, then leads to the Taylor series

$$f(x_0 + h,\, y_0 + k) = \sum_{n=0}^{\infty} \frac{1}{n!} \left(h\frac{\partial}{\partial x} + k\frac{\partial}{\partial y} \right)^n f(x_0, y_0). \tag{7.34}$$

The above results are easily generalised to more than two variables. For a function $f(x_1, x_2, \ldots, x_k)$ of k variables, (7.34), for example, becomes

$$f(a_1 + h_1, a_2 + h_2, \ldots, a_k + h_k$$
$$= \sum_{n=0}^{\infty} \frac{1}{n!} \left(h_1\frac{\partial}{\partial x_1} + h_2\frac{\partial}{\partial x_2} + \cdots + h_k\frac{\partial}{\partial x_k} \right)^n f(x_1, x_2, \ldots, x_k) \tag{7.35}$$

on expanding about $x_i = a_i (i = 1, 2, \ldots, k)$, where the right-hand side is evaluated at $x_1 = a_1, x_2 = a_2, \ldots, x_k = a_k$. However, expansions such as (7.35) for several variables rapidly become unwieldy, so we will restrict ourselves to explicitly expanding (7.34), when one obtains

$$f(x_0+h,\,y_0+k) = f(x_0,y_0) + h\frac{\partial f}{\partial x} + k\frac{\partial f}{\partial y} + \frac{1}{2!}\left(h^2\frac{\partial^2 f}{\partial x^2} + 2hk\frac{\partial^2 f}{\partial x\partial y} + k^2\frac{\partial^2 f}{\partial y^2}\right) + \cdots, \tag{7.36}$$

where all the derivatives are evaluated at $x = x_0,\ y = y_0$, and we have assumed (7.5). In general, if one assumes that the order of the cross derivatives is unimportant, that is,

$$\frac{\partial^3 f}{\partial x^2\partial y} = \frac{\partial^3 f}{\partial x\partial y\partial x} = \frac{\partial^3 f}{\partial y\partial x^2}\,,\ \text{etc.},$$

as is usually the case[2], (7.34) becomes

$$f(x_0+h,\,y_0+k) = \sum_{n=0}^{\infty}\sum_{m=0}^{n}\frac{h^{n-m}k^m}{m!(n-m)!}\frac{\partial^n f}{\partial x^{n-m}\partial y^m}, \tag{7.37}$$

where we have used the binomial expansion (1.23) and where all the derivatives are again evaluated at $x = x_0,\ y = y_0$.

Example 7.9

Expand the function $f = \sin(xy)$ about $x = 1\,,\ y = \pi/4$, retaining only constant, linear and quadratic terms.

Solution

In the limits $x \to 1\,,\ y \to \pi/4$, we have

$$f = \sin(xy) \to \sin\left(\frac{\pi}{4}\right) = \frac{1}{\sqrt{2}}\,,$$

$$\frac{\partial f}{\partial x} = y\cos(xy) \to \frac{\pi}{4}\frac{1}{\sqrt{2}}, \qquad \frac{\partial f}{\partial y} = x\cos(xy) \to \frac{1}{\sqrt{2}}\,,$$

$$\frac{\partial^2 f}{\partial x^2} = -y^2\sin(xy) \to -\frac{\pi^2}{16}\frac{1}{\sqrt{2}}, \qquad \frac{\partial^2 f}{\partial y^2} = -x^2\sin(xy) \to -\frac{1}{\sqrt{2}}\,,$$

[2] The necessary condition is given by Clairaut's theorem: if $f(x,y)$ is defined in an open region and the derivatives f_{xy} and f_{yx} are continuous in this region, then $f_{xy} = f_{yx}$ at each point in the region. A similar result holds for higher mixed derivatives. See, for example, C. James (1966) *Advanced Calculus,* Wadsworth Publishing Company, Belmont, California.

and

$$\frac{\partial^2 f}{\partial x \partial y} = \cos(xy) - xy\sin(xy) \to \frac{1}{\sqrt{2}}\left(1 - \frac{\pi}{4}\right).$$

Hence, defining $h = x - 1$ and $k = y - \pi/4$, (7.34) gives

$$\sin(xy) = \frac{1}{\sqrt{2}}\left[1 + \left(\frac{\pi}{4}\right)h + k - \frac{\pi^2}{32}h^2 + \left(1 - \frac{\pi}{4}\right)hk - \frac{k^2}{2} + \cdots\right]$$

7.5 Stationary points

The necessary and sufficient conditions for the differential $\mathrm{d}f(x_1, x_2, \ldots, x_n)$ to vanish for arbitrary $\mathrm{d}x_1, \mathrm{d}x_2, \ldots, \mathrm{d}x_n$ are, from (7.11),

$$\frac{\partial f}{\partial x_i} = 0, \qquad i = 1, 2, \ldots, n. \tag{7.38}$$

Points at which (7.38) are satisfied are called *stationary points*, in analogy to those discussed for a function of a single variable in Section 3.4.1. However, determining whether such points are local minima, maxima or saddle points is more complicated than for functions of a single variable. For simplicity, we shall restrict ourselves to functions of two variables $f(x, y)$, which can be regarded as two-dimensional surfaces as shown in Figures 7.2 and 7.3. Suppose that

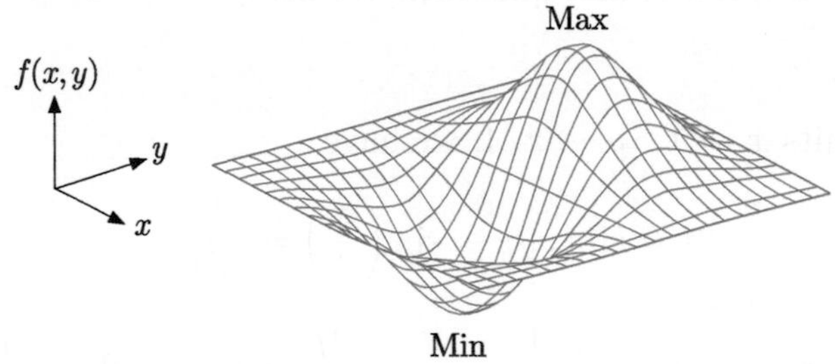

Figure 7.2 A two-dimensional surface $f(x, y)$ showing a maximum (denoted by Max) and a minimum (denoted by Min).

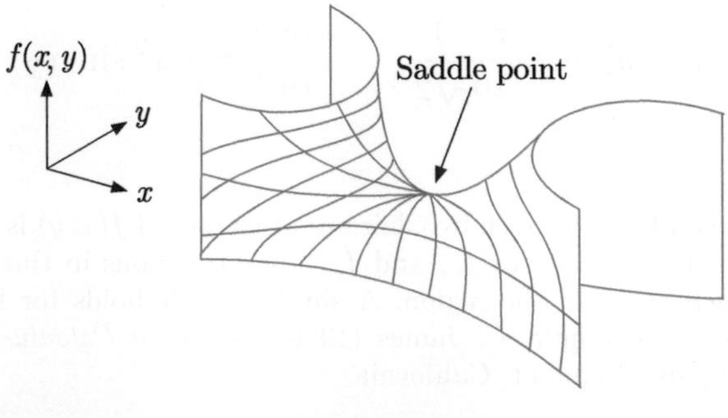

Figure 7.3 A two-dimensional surface $f(x, y)$ showing an example of one type of saddle point.

$f(x, y)$ has a stationary point at $x = x_0$, $y = y_0$, where by (7.38),

$$\frac{\partial f}{\partial x} = 0, \qquad \frac{\partial f}{\partial y} = 0.$$

Then making a Taylor expansion about (x_0, y_0) gives

$$\begin{aligned}\Delta f &= f(x_0 + h,\, y_0 + k) - f(x_0, y_0) \\ &= \frac{1}{2}\left[\frac{\partial^2 f}{\partial x^2}h^2 + 2\frac{\partial^2 f}{\partial x \partial y}hk + \frac{\partial^2 f}{\partial y^2}k^2\right],\end{aligned} \tag{7.39}$$

where we have neglected higher-order terms and assumed $\Delta f \neq 0$ for all values of h and k. Then if $(x_0\,,\, y_0)$ is a minimum (maximum), as opposed to a saddle point, we must have $\Delta f > 0$ $(\Delta f < 0)$ for all non-zero h, k values. For $h \neq 0$, this implies that

$$\Delta\, f = \frac{1}{2}h^2\left(\frac{\partial^2 f}{\partial x^2} + 2\frac{\partial^2 f}{\partial x \partial y}z + \frac{\partial^2 f}{\partial y^2}z^2\right) = 0$$

has no real roots, where $z = k/h$. Since the condition for a quadratic $az^2 + bz + c = 0$ to have no real roots is $b^2 < 4ac$, this implies

$$\left(\frac{\partial^2 f}{\partial x \partial y}\right)^2 < \frac{\partial^2 f}{\partial x^2}\frac{\partial^2 f}{\partial y^2}, \tag{7.40a}$$

and the same condition is obtained if instead we consider $k \neq 0$. Hence (7.40a) is a necessary condition for $f(x_0, y_0)$ to be either a maximum or a minimum, and since the left-hand side is positive definite, this implies that $\partial^2 f/\partial x^2$ and $\partial^2 f/\partial y^2$ are either both positive or both negative. Specifically, if (7.40a) is true and

$$\frac{\partial^2 f}{\partial x^2} > 0\,, \quad \frac{\partial^2 f}{\partial y^2} > 0, \tag{7.40b}$$

then $\Delta\, f > 0$ and $f(x_0, y_0)$ is a maximum; whereas if (7.40a) holds and

$$\frac{\partial^2 f}{\partial x^2} < 0\,, \quad \frac{\partial^2 f}{\partial y^2} < 0, \tag{7.40c}$$

then $\Delta\, f < 0$ and $f(x_0, y_0)$ is a minimum. Examples of a maximum and a minimum in two variables are shown in Figure 7.2.

If on the other hand

$$\left(\frac{\partial^2 f}{\partial x \partial y}\right)^2 > \frac{\partial^2 f}{\partial x^2}\frac{\partial^2 f}{\partial y^2}, \tag{7.41}$$

$f(x_0, y_0)$ is a saddle point. There are several different types of saddle point depending on the behaviour of the second derivative, and one example is shown in Figure 7.3.

Finally, if

$$\frac{\partial^2 f}{\partial x^2} = \frac{\partial^2 f}{\partial x \partial y} = \frac{\partial^2 f}{\partial y^2} = 0,$$

then $\Delta f \equiv 0$ for all h and k, contradicting our earlier assumption, and higher-order terms in the Taylor expansion must be inspected to determine the nature of the stationary point.

Example 7.10

Find the stationary points of the function

$$f(x, y) = 3 + x^2 - y^2 + 2x^2y^2 + y^4.$$

Evaluate the function at each of the stationary points and classify them as maxima, minima or saddle points.

Solution

The conditions for a stationary point are

$$\frac{\partial f}{\partial x} = 2x(2y^2 + 1) = 0 \;\rightarrow\; x = 0\,,$$

$$\frac{\partial f}{\partial y} = 4y^3 + 4x^2y - 2y = 0 \;\rightarrow\; y = 0,\ \pm\tfrac{1}{\sqrt{2}},$$

given that $x = 0$, so that the stationary points are

$$(x, y) = (0, 0)\,,\ (0, \tfrac{1}{\sqrt{2}})\,,\ (0, -\tfrac{1}{\sqrt{2}}).$$

To determine their nature, we require the second derivatives

$$\frac{\partial^2 f}{\partial x^2} = 4y^2 + 2\,, \quad \frac{\partial^2 f}{\partial y^2} = 12y^2 + 4x^2 - 2\,, \quad \frac{\partial^2 f}{\partial x \partial y} = 8xy.$$

Thus at $(x = 0,\ y = 0)$ we have

$$f = 3\,, \quad \frac{\partial^2 f}{\partial x^2} = 2\,, \quad \frac{\partial^2 f}{\partial y^2} = -2\,, \quad \frac{\partial^2 f}{\partial x \partial y} = 0,$$

so that (7.39) is satisfied and (0, 0) is a saddle point. At $(x = 0,\ y = \pm\tfrac{1}{\sqrt{2}})$ we have

$$f = \frac{11}{4}\,, \quad \frac{\partial^2 f}{\partial x^2} = 4\,, \quad \frac{\partial^2 f}{\partial y^2} = 4\,, \quad \frac{\partial^2 f}{\partial x \partial y} = 0,$$

so that (7.40a) and (7.40b) are satisfied and both points are local maxima.

*7.6 Lagrange multipliers

In the preceding section, we discussed how to find the stationary points of a function of two or more variables. However sometimes one needs to find the stationary points of the function when the variables are subject to one or more additional conditions, called 'constraints'. To take a very simple example, one could ask: "What is the maximum area of a rectangular field surrounded by a fence of fixed length, say 200 m?" In other words, if the length and breadth of the field are x and y metres respectively, what is the maximum value of the area $A = xy$ subject to the constraint $x + y = 100\,\text{m}$. In simple problems of this kind, one can use the constraint to eliminate one of the variables. In the above case eliminating y gives

$$A = x(100 - x),$$

which is easily shown to have a maximum value $A = 2500\,\text{m}^2$ for $x = 50\,\text{m}$, corresponding to a square field with $x = y = 50\,\text{m}$. However, in cases where the function and/or the constraint is more complicated, or there are more than two variables and more than one constraint, solving the problem by using each of the constraints to eliminate a variable can become very clumsy and tedious, and it is often easier to use an alternative method due to Lagrange.

Suppose we need to find the stationary points of a function $f(x_1, x_2, \ldots, x_n)$, where the variables are restricted to a limited range of values by k constraints, that we shall assume can be written in the form

$$g_j(x_1, x_2, \ldots, x_n) = 0\,, \qquad j = 1,\, 2, \ldots, k. \tag{7.42}$$

where $k < n$. In this case, the relation

$$\mathrm{d}f = \sum_{i=1}^{n} \left(\frac{\partial f}{\partial x_i}\right) \mathrm{d}x_i = 0 \tag{7.43}$$

no longer leads to the usual conditions

$$\frac{\partial f}{\partial x_i} = 0, \qquad i = 1,\, 2, \ldots, n,$$

because the $\mathrm{d}x_i$ are no longer independent, but are related by conditions of the form

$$\mathrm{d}g_j = \sum_{i=1}^{n} \left(\frac{\partial g_j}{\partial x_i}\right) \mathrm{d}x_i = 0. \qquad j = 1,\, 2, \ldots, k \tag{7.44}$$

This problem can in principle be solved, as in the simple example discussed above, by using conditions (7.42) to eliminate k of the variables, and expressing $f(x_1, x_2, \ldots, x_n)$ as a function of the remaining independent variables, which can then be minimised in the usual way.

However, following Lagrange, it is often more efficient to consider a new function,

$$F(x_1, x_2, \ldots, x_n, \lambda_1, \lambda_2, \ldots, \lambda_k) = f + \sum_{j=1}^{k} \lambda_j g_j, \tag{7.45}$$

where the λ_j are new variables called *undetermined multipliers.* One then determines the stationary points of F by treating $x_1, x_2, \ldots, x_n$ as independent variables to give n conditions

$$\frac{\partial F}{\partial x_i} = \frac{\partial f}{\partial x_i} + \sum_{j=1}^{k} \lambda_j \frac{\partial g_j}{\partial x_i} = 0. \quad i = 1, 2, \ldots, n \tag{7.46}$$

These determine the values of $x_1, x_2, \ldots, x_n$ as functions of the variables $\lambda_j (j = 1, 2, \ldots, k)$, whose values can then be determined by requiring the k conditions (7.42) to be satisfied. In other words, the $n + k$ variables $x_1, \ldots, x_n, \lambda_1, \ldots, \lambda_k$ are determined by the $n + k$ equations (7.46) and (7.42); and since $F \to f$ when (7.42) are satisfied, the x_i values correspond to the stationary points of f subject to the constraints (7.42). This procedure is best illustrated by example.

Example 7.11

Find the largest value of the function $f(x, y) = x + y$ subject to the condition that the point (x, y) lies on the ellipse

$$\frac{x^2}{a^2} + \frac{y^2}{b^2} = 1. \tag{7.47a}$$

Solution

We have the single constraint

$$g(x, y) = \frac{x^2}{a^2} + \frac{y^2}{b^2} - 1 = 0, \tag{7.47b}$$

so that (7.46) becomes

$$\frac{\partial F}{\partial x} = \frac{\partial f}{\partial x} + \lambda \frac{\partial g}{\partial x} = 1 + \frac{2\lambda x}{a^2} = 0,$$

and

$$\frac{\partial F}{\partial y} = \frac{\partial f}{\partial y} + \lambda \frac{\partial g}{\partial y} = 1 + \frac{2\lambda y}{b^2} = 0,$$

with solutions

$$x = -a^2/2\lambda, \quad y = -b^2/2\lambda. \tag{7.48a}$$

The undetermined multiplier λ is determined by substituting (7.48a) into the constraint (7.47b) to give

$$2\lambda = \pm(a^2 + b^2)^{1/2},$$

so that there are four stationary points, which from (7.48a) are

$$x_{\pm} = \pm\frac{a^2}{(a^2+b^2)^{1/2}}, \quad y_{\pm} = \pm\frac{b^2}{(a^2+b^2)^{1/2}}. \qquad (7.48b)$$

These are shown in Figure 7.4. Clearly the largest value of $x+y$ is obtained by choosing positive signs for both x and y, so that

$$(x+y)_{\max} = \frac{a^2+b^2}{(a^2+b^2)^{1/2}} = (a^2+b^2)^{1/2}.$$

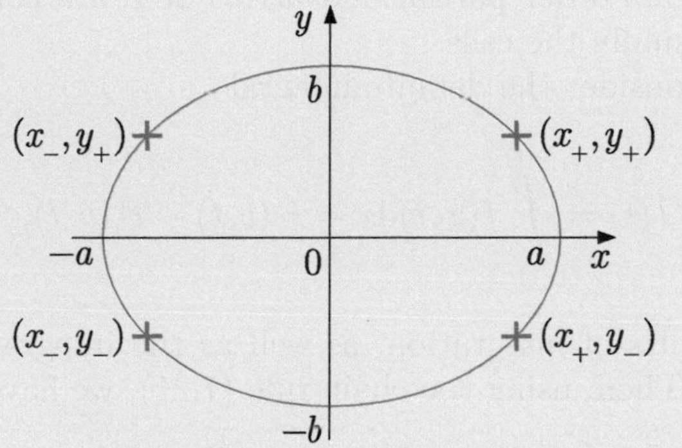

Figure 7.4 The ellipse (7.47a) and the four stationary points (+) given by (7.48b).

*7.7 Differentiation of integrals

We conclude this chapter by using the properties of partial derivatives to deduce the rules for differentiating integrals with respect to a variable parameter, starting with the indefinite integral

$$F(x,t) = \int f(x,t)\mathrm{d}x, \qquad (7.49)$$

where (4.1) together with the definition of partial derivatives implies

$$\frac{\partial F(x,t)}{\partial x} = f(x,t). \qquad (7.50)$$

Then the partial derivative

$$\frac{\partial F(x,t)}{\partial t} \equiv \frac{\partial}{\partial t}\left[\int f(x,t)\mathrm{d}x\right] = \int \frac{\partial f(x,t)}{\partial t}\mathrm{d}x \qquad (7.51)$$

provided that F satisfies (7.5), that is,

$$\frac{\partial^2 F(x,t)}{\partial x \partial t} = \frac{\partial^2 F(x,t)}{\partial t \partial x}. \qquad (7.52)$$

To see this, we note that (7.52) and (7.50) imply

$$\frac{\partial}{\partial x}\left(\frac{\partial F}{\partial t}\right) = \frac{\partial}{\partial t}\left(\frac{\partial F}{\partial x}\right) = \frac{\partial f}{\partial t}.$$

Integrating this equation with respect to x then gives

$$\frac{\partial F}{\partial t} = \int \frac{\partial f(x,t)}{\partial t}\,\mathrm{d}x, \tag{7.53}$$

which together with (7.49) gives (7.51). In other words, we may reverse the order of the differentiation and integration, as in (7.51), provided (7.52) is satisfied. As we saw in Section 7.1, this is so if the first- and second-order partial derivatives of F are continuous in x and t, as is usually the case.

We next consider the definite integral

$$I(t) = \int_{a(t)}^{b(t)} f(x,t)\mathrm{d}x = F(b,t) - F(a,t), \tag{7.54}$$

where the limits of integration, as well as the integrand, may also depend on t. Then, using the chain rule (7.21), we have

$$\frac{\mathrm{d}I(t)}{\mathrm{d}t} = \frac{\partial F(b,t)}{\partial b}\frac{\mathrm{d}b}{\mathrm{d}t} + \frac{\partial F(b,t)}{\partial t} - \frac{\partial F(a,t)}{\partial a}\frac{\mathrm{d}a}{\mathrm{d}t} - \frac{\partial F(a,t)}{\partial t},$$

provided a, b are differentiable functions of t. In addition, (7.53) implies

$$\int_{a(t)}^{b(t)} \frac{\partial f(x,t)}{\partial t}\,\mathrm{d}x = \frac{\partial F(b,t)}{\partial t} - \frac{\partial F(a,t)}{\partial t},$$

so that, using this together with (7.50), one finally obtains *Leibnitz's rule*,

$$\frac{\mathrm{d}I(t)}{\mathrm{d}t} = f(b,t)\frac{\mathrm{d}b}{\mathrm{d}t} - f(a,t)\frac{\mathrm{d}a}{\mathrm{d}t} + \int_{a(t)}^{b(t)} \frac{\partial f(x,t)}{\partial t}\,\mathrm{d}x, \tag{7.55}$$

which reduces to

$$\frac{\mathrm{d}}{\mathrm{d}t}\int_a^b f(x,t)\mathrm{d}x = \int_a^b \frac{\partial f(x,t)}{\partial t}\,\mathrm{d}x \tag{7.56}$$

for fixed limits a, b. Finally, one may allow $b \to \infty$ and/or $a \to -\infty$, provided all the integrals converge.

As well as allowing given integrals to be differentiated, these results can be exploited by using known integrals to evaluate related, unknown integrals. For example, in thermal physics one frequently needs to evaluate integrals of the form

$$I_n = \int_0^\infty x^n e^{-\alpha x} \mathrm{d}x, \tag{7.57}$$

where $n \geq 0$ and $\alpha > 0$. There are no problems with convergence, and for $n = 0$ one easily obtains

$$I_0 = \int_0^\infty e^{-\alpha x} \mathrm{d}x = \frac{1}{\alpha} \quad (\alpha > 0),$$

while differentiating (7.57) with respect to α using (7.56) gives

$$\frac{\mathrm{d}I_n}{\mathrm{d}\alpha} = -I_{n+1},$$

and hence

$$I_n = -\frac{\mathrm{d}I_{n-1}}{\mathrm{d}\alpha} = \frac{\mathrm{d}^2 I_{n-2}}{\mathrm{d}\alpha} = (-1)^n \frac{\mathrm{d}^n I_0}{\mathrm{d}\alpha^n}.$$

Thus, $I_1 = -1/\alpha^2$, $I_2 = 2/\alpha^3$, ... and in general

$$I_n = \int_0^\infty x^n e^{-\alpha x} \mathrm{d}x = \frac{(-1)^n n!}{\alpha^{n+1}} \quad (\alpha > 0). \tag{7.58}$$

Example 7.12

Evaluate

$$\frac{\mathrm{d}}{\mathrm{d}\alpha} \int_\alpha^{\alpha^2} \sin(\alpha x)\, \mathrm{d}x$$

as a function of α.

Solution

Using (7.55) with $t = \alpha$, $b = \alpha^2$, $a = \alpha$ and $f(x, \alpha) = \sin(\alpha x)$ gives

$$\frac{\mathrm{d}}{\mathrm{d}\alpha} \int_\alpha^{\alpha^2} \sin(\alpha x)\, \mathrm{d}x = 2\alpha \sin \alpha^3 - \sin \alpha^2 + \int_\alpha^{\alpha^2} x \cos(\alpha x)\, \mathrm{d}x\,,$$

where

$$\int_\alpha^{\alpha^2} x \cos(\alpha x)\, \mathrm{d}x = \int_\alpha^{\alpha^2} \frac{x}{\alpha} \frac{\mathrm{d}}{\mathrm{d}x} [\sin(\alpha x)]\, \mathrm{d}x$$

$$= \alpha \sin \alpha^3 - \sin \alpha^2 + \frac{\cos \alpha^3}{\alpha^2} - \frac{\cos \alpha^2}{\alpha^2}$$

on integrating by parts. Hence,

$$\frac{\mathrm{d}}{\mathrm{d}\alpha}\int_{\alpha}^{\alpha^2}\sin(\alpha x)\,\mathrm{d}x = 3\alpha\sin\alpha^3 - 2\sin\alpha^2 + \frac{1}{\alpha^2}\left(\cos\alpha^3 - \cos\alpha^2\right).$$

Problems 7

7.1 Show that the relation

$$\frac{\partial}{\partial x}\left(\frac{\partial f}{\partial y}\right) = \frac{\partial}{\partial y}\left(\frac{\partial f}{\partial x}\right)$$

is satisfied for each of the following functions:

$$\text{(a) } x^3 + xy^2 + 2xy + 3x^2, \quad \text{(b) } \frac{x^2+y^2}{xy},$$

$$\text{(c) } (x+y)\ln\left(\frac{y}{x}\right), \qquad \text{(d) } \exp(x^2)\sin^{-1}y.$$

7.2 A function $f(x, y)$ is of the form

$$f(x,y) = \frac{1}{x}g\left(\frac{y}{x}\right),$$

where g is an arbitrary function of y/x. Show that

$$x^2\frac{\partial^2 f}{\partial x^2} + 2xy\frac{\partial^2 f}{\partial x\partial y} + y^2\frac{\partial^2 f}{\partial y^2} + x\frac{\partial f}{\partial x} + y\frac{\partial f}{\partial y} = f.$$

7.3 The plane $z = \alpha x + \beta y + \gamma$ is tangential to the sphere $z^2 = 14 - x^2 - y^2$ at the point $(x, y, z) = (1, 2, 3)$. Find the values of the constants α, β and γ, and hence the equation of the plane.

7.4 F is a function of three independent variables x, y and z, and a, b and k are constants.

(a) If $F = \sin(ax)\sin(by)\sin\left[kz(a^2+b^2)^{1/2}\right]$, show that

$$\frac{\partial F^2}{\partial z^2} = k^2\left(\frac{\partial^2 F}{\partial x^2} + \frac{\partial^2 F}{\partial y^2}\right).$$

(b) If $F = e^{-kz}[\sin(ax) + \cos(by)]$, show that

$$\frac{\partial^2 F}{\partial x^2} + \frac{\partial^2 F}{\partial y^2} = \left(\frac{a^2+b^2}{k}\right)\frac{\partial F}{\partial z}.$$

7.5 Two independent variables u and w are given in terms of two other independent variables x and y, by

$$u + w = x^2 + y^2 - k^2 \quad \text{and} \quad uw = a^2x^2 + b^2y^2 - h^4$$

where a, b, k and h are constants. By using differentials, show that

$$\frac{\partial u}{\partial x} = -2x\left(\frac{a^2-u}{u-w}\right), \quad \frac{\partial w}{\partial y} = 2y\left(\frac{b^2-w}{u-w}\right),$$

and

$$\frac{\partial x}{\partial u} = -\frac{1}{2x}\left(\frac{b^2-w}{a^2-b^2}\right), \quad \frac{\partial y}{\partial w} = \frac{1}{2y}\left(\frac{a^2-u}{a^2-b^2}\right).$$

7.6 A wide class of systems (e.g. a sample of liquid or gas) satisfies the *fundamental thermodynamic identity*

$$\mathrm{d}E = T\mathrm{d}S - P\mathrm{d}V, \tag{7.59}$$

where E is the energy, S is the entropy, and P, V and T are the pressure, volume and temperature of the system, respectively.
(a) Use (7.5) to derive the *Maxwell identity*

$$\left(\frac{\partial T}{\partial V}\right)_S = -\left(\frac{\partial P}{\partial S}\right)_V.$$

(b) Obtain an expression for $\mathrm{d}G$, where $G \equiv E - TS + PV$ and hence derive the second Maxwell identity

$$\left(\frac{\partial V}{\partial T}\right)_P = -\left(\frac{\partial S}{\partial P}\right)_T.$$

7.7 The equilibrium behaviour of a gas at high temperature can be described approximately by *Dieterici's equation*:

$$P(V-b) = RT\exp\left(-\frac{a}{RTV}\right), \tag{7.60}$$

where P, V and T are the pressure, volume and temperature, respectively, R is the gas constant, and a and b are parameters that are characteristic of the particular gas.
(a) Use (7.13) to show that the coefficient of thermal expansion at constant pressure is given by

$$\alpha = \frac{1}{V}\left(\frac{\partial V}{\partial T}\right)_P = \frac{(V-b)}{VT}\left[1+\frac{a}{RTV}\right]\left[1-\frac{a(V-b)}{RTV^2}\right]^{-1} \tag{7.61}$$

in this approximation.
(b) Verify that the same result follows by evaluating $(\partial V/\partial T)_P$ directly from the differential $\mathrm{d}A$, where $A \equiv P(V-b)$.

7.8 Which of the following differentials are exact?
(a) $\mathrm{d}f(x,y) = (3x^2y - xy^3)\mathrm{d}x + (x^3 - \frac{3}{2}x^2y^2)\mathrm{d}y$
(b) $\mathrm{d}f(x,y) = (xy - 3x^2y^2)\mathrm{d}x + (\frac{1}{2}x^2 + 2x^3y)\mathrm{d}y$
(c) $\mathrm{d}f(x,y) = (\sin x \sin y)\,\mathrm{d}x - (\cos x \cos y)\,\mathrm{d}y$
(d) $\mathrm{d}f(x,y,z) = (y^2 + 2xz)\mathrm{d}x + (2xy + yz)\mathrm{d}y + (x^2 + \frac{1}{2}y^2 + z^2)\mathrm{d}z$

7.9 Show that the following are exact differentials $\mathrm{d}f$ of a function $f(x,y)$ and identify the function.

$$\text{(a)}\ \frac{y^2\mathrm{d}x + x^2\mathrm{d}y}{(x+y)^2}, \qquad \text{(b)}\ [2x\ln(xy) + x]\mathrm{d}x + \frac{x^2}{y}\mathrm{d}y.$$

7.10 Find dz/dt when z is given by the following expressions
(a) $z = 2x^2 + 3xy^3 + 4y^4$, where $x = \sin t$ and $y = \cos t$,
(b) $z = \ln(x^{-2} + y^2)$, where $x = e^t$ and $y = e^{-t}$,
(c) $z = xy\ln(x/y) + (x/y)$, where $x = \ln t$ and $y = \ln(1/t)$.

7.11 Which of the following functions $f(x, y, z)$ satisfy the equation

$$x\frac{\partial f}{\partial x} + y\frac{\partial f}{\partial y} + z\frac{\partial f}{\partial z} = kf,$$

and what is the corresponding value of the constant k?

$$\text{(a)}\ \frac{x^2yz + xy^2z}{x+z}, \qquad \text{(b)}\ \frac{xy+z}{x+yz},$$

$$\text{(c)}\ \ln x + 2\ln y - 3\ln z + 4, \quad \text{(d)}\ (x+y+z)^{1/2}.$$

7.12 If $f(x_1, x_2, \ldots, x_n)$ is a homogeneous function of order k, show that

$$\sum_i \sum_j x_i x_j \frac{\partial^2 f}{\partial x_i \partial x_j} = k(k-1)f,$$

so that, for example,

$$x^2\frac{\partial^2 f}{\partial x^2} + 2xy\frac{\partial^2 f}{\partial x \partial y} + y^2\frac{\partial^2 f}{\partial y^2} = k(k-1)f$$

if $f(x, y)$ is homogeneous of degree k.

7.13 If $z = f(x, y)$, where

$$2x = e^u + e^w \text{ and } 2y = e^u - e^w,$$

show that

$$\frac{\partial z}{\partial u} + \frac{\partial z}{\partial w} = x\frac{\partial z}{\partial x} + y\frac{\partial z}{\partial y},$$

and

$$e^{u+w}\frac{\partial z}{\partial x} = e^w\frac{\partial z}{\partial u} + e^u\frac{\partial z}{\partial w}.$$

7.14 A function $f(x, t)$ is given by

$$f(x,t) \doteq \phi_1(x - ct) + \phi_2(x + ct),$$

where ϕ_1 and ϕ_2 are arbitrary differentiable functions, and c is a constant. Show that

$$\frac{\partial^2 f}{\partial x^2} - \frac{1}{c^2}\frac{\partial^2 f}{\partial t^2} = 0.$$

7.15 If the function $f(x, y)$ is transformed to a function $g(u, w)$ by the substitutions

$$3x = u^3 - 3uw^2, \quad 3y = 3u^2w - w^3,$$

show that

$$\left(\frac{\partial^2 g}{\partial u^2} + \frac{\partial^2 g}{\partial w^2}\right) = (u^2 + w^2)^2\left(\frac{\partial^2 f}{\partial x^2} + \frac{\partial^2 f}{\partial y^2}\right).$$

7.16 Use Taylor's theorem to expand $f(x,y) = \exp(x/y)$ to second order about the point $x = 2,\ y = 1$.

7.17 Expand $f(x,y) = [\ln(1+x)]/(1+y)$ as a Taylor series about $x = y = 0$ up to cubic terms.

7.18 Find the maximum and minimum values of the function

$$f(x,y) = \sin x \sin y \sin(x+y)$$

inside the square defined by $0 < x,\ y < \pi$.

7.19 Find the stationary points of the function

$$f(x,y) = e^{x-y}(x^2 + xy + y^2)$$

and classify them as either minima, maxima, or saddle points.

***7.20** Find the stationary points of the function $f(x,y) = x^2 - y^2 - 2$, subject to the constraint $x^2 - 2y = 2$.

***7.21** Find the volume of the largest box with sides parallel to the x, y, z axes that can be fitted into the ellipsoid;

$$\frac{x^2}{a^2} + \frac{y^2}{b^2} + \frac{z^2}{c^2} = 1.$$

***7.22** A set of numbers $x_i\ (i = 1, 2, \ldots, n)$ has a product P. What is the largest value of P, if their sum is equal to N?

***7.23** (a) Evaluate

$$\frac{\mathrm{d}I(x)}{\mathrm{d}x} = \frac{\mathrm{d}}{\mathrm{d}x}\int_1^\infty x e^{-xy}\,\mathrm{d}y,$$

where $x > 0$, and hence find $I(x)$ itself, given that $I(1) = 0$.
(b) If $f(x,t) = 1/\ln(x+t)$, find

$$\frac{\mathrm{d}}{\mathrm{d}t}\int_t^{t^2} f(x,t)\mathrm{d}x.$$

***7.24** (a) Evaluate

$$I(y) = \frac{\mathrm{d}}{\mathrm{d}y}\int_y^{e^y} \frac{\sin(xy)}{x}\,\mathrm{d}x.$$

(b) Show by differentiation with respect to α, that

$$I(\alpha) \equiv \int_0^\infty \frac{e^{-\alpha x}\sin x}{x}\,\mathrm{d}x = \cot^{-1}\alpha, \quad 0 < \alpha < \pi.$$

***7.25** Find an explicit expression for

$$I_k(a) = \int_{-\infty}^{\infty} x^{2k} e^{-ax^2}\,\mathrm{d}x,$$

where $a > 0$ and $k \geq 0$ is an integer, given that $I_0(a) = (\pi/a)^{1/2}$.

8

Vectors

In previous chapters we have been concerned exclusively with quantities that are completely specified by their magnitude. These are called *scalar quantities*, or simply *scalars*. If they have dimensions, then these also must be specified, in appropriate units. Examples of scalars are temperature, electric charge and mass. In physical science one also meets quantities that are specified by both their magnitude (again in appropriate units) and their direction. Provided they obey the particular law of addition specified below, these are called *vector quantities*, or just *vectors*. Examples are force, velocity and magnetic field strength. In this chapter we will be concerned with the algebraic manipulation of vectors, their use in co-ordinate geometry and the most elementary aspects of their calculus. In Chapter 12 we will discuss in more detail the calculus of vectors and vector analysis.

8.1 Scalars and vectors

Because vectors depend on both magnitude and direction, a convenient representation of a vector is by a line with the direction indicated by an arrow anywhere along it, often at its end as shown in Figure 8.1a. The vector represented by the line OA is printed in bold face type $\mathbf{a}$ (or if hand-written, as $\vec{a}$ or $\underline{a}$). The *magnitude* of $\mathbf{a}$ is the length of the line OA and is a scalar. It is written $|\mathbf{a}|$ or a. Vectors are equal if they have the same magnitude *and* are parallel. Thus in Figure 8.1b, all three vectors $\mathbf{a}_1$, $\mathbf{a}_2$ and $\mathbf{a}_3$ have the same magnitude and are parallel and hence are equal. However, in Figure 8.1c the vectors $\mathbf{a}_1$ and $\mathbf{a}_4$ have the same magnitude but are antiparallel and $\mathbf{a}_1 = -\mathbf{a}_4$, that is, reversing the direction of a vector while keeping its magnitude the same changes its sign.

The law of addition is same law that applies to displacements, in which points are moved in the direction of the vector by an

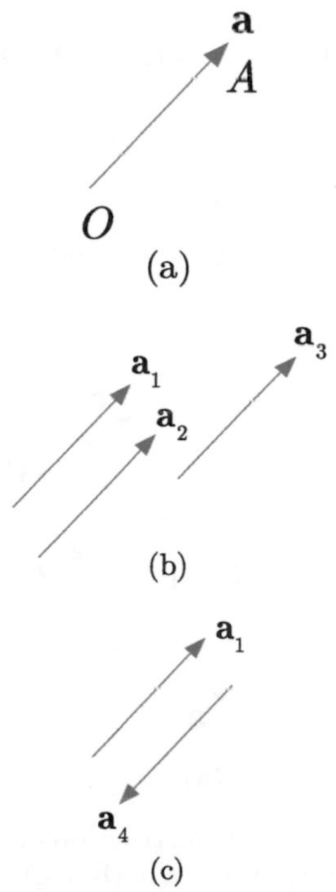

Figure 8.1 Graphical representation of vectors.

Mathematics for Physicists, First Edition. B.R. Martin and G. Shaw.

Companion website: www.wiley.com/go/martin/mathsforphysicists

amount that is equal to its magnitude. This yields the *triangle law of addition,* shown by the construction of Figure 8.2. Here the vector $\mathbf{b}$ is added to the vector $\mathbf{a}$ to yield the vector $\mathbf{s} = \mathbf{a} + \mathbf{b}$. This law is central to the properties of vectors and their usefulness in physical science.

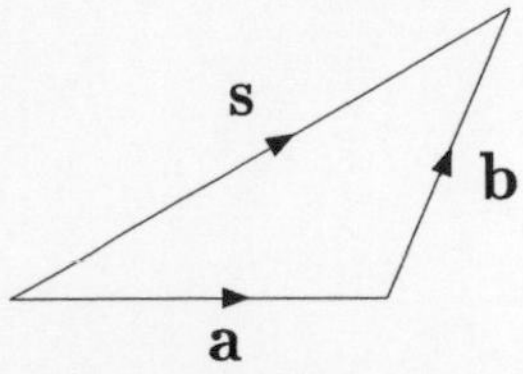

Figure 8.2 Addition of vectors.

8.1.1 Vector algebra

Scalars are manipulated by the rules of ordinary algebra, which we now call *scalar algebra,* that were discussed in Section 1.2.1. Vectors are manipulated by an analogous set of rules known as *vector algebra,* which we will explore in this chapter. We note that the rules differ from those of scalar algebra in several respects: for example, the commutative law of multiplication for scalars does not necessarily hold for vectors; also, if the product of two vectors is zero, this does not imply that one, or both, of them is necessarily zero.

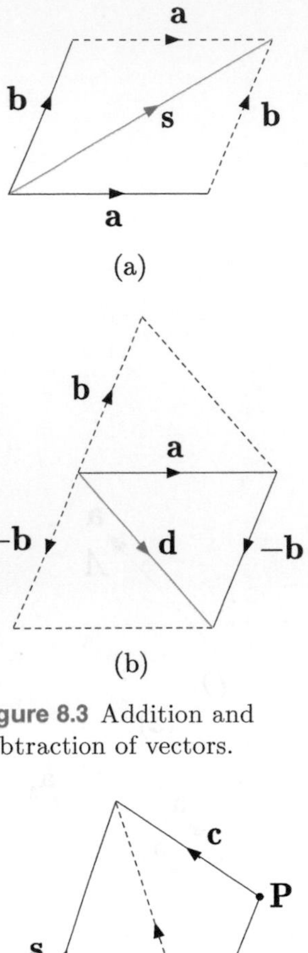

Figure 8.3 Addition and subtraction of vectors.

We begin by looking at addition and subtraction of vectors. In Figure 8.2 we added vector $\mathbf{b}$ to vector $\mathbf{a}$ to obtain the vector $\mathbf{s} = \mathbf{a} + \mathbf{b}$. As can be seen from Figure 8.3a, if we instead add $\mathbf{a}$ to $\mathbf{b}$, we obtain the same result. In other words, addition obeys the commutative law

$$\mathbf{a} + \mathbf{b} = \mathbf{b} + \mathbf{a} \tag{8.1a}$$

The triangle law is also referred to as the *parallelogram law,* because the construction of Figure 8.3a produces a parallelogram. Likewise, Figure 8.3b shows the construction for the difference $\mathbf{d} = \mathbf{a} - \mathbf{b}$, corresponding to adding the vector $-\mathbf{b}$ **to** $\mathbf{a}$. As in the case of scalar algebra, subtraction is not commutative: $\mathbf{a} - \mathbf{b} \neq \mathbf{b} - \mathbf{a}$. Also note that $|\mathbf{s}| \neq |\mathbf{a}| + |\mathbf{b}|$ and $|\mathbf{d}| \neq |\mathbf{a}| - |\mathbf{b}|$. The magnitudes of $\mathbf{s}$ and $\mathbf{d}$ are found from Figure 8.3 by using the cosine rule derived in Chapter 2. These constructions may be extended to more than two vectors and one easily establishes the associative law

$$\mathbf{a} + (\mathbf{b} + \mathbf{c}) = (\mathbf{a} + \mathbf{b}) + \mathbf{c} \tag{8.1b}$$

using the constructions of Figure 8.4.

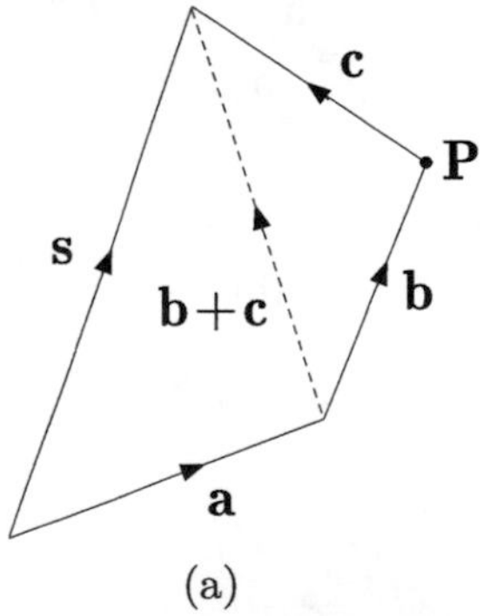

Figure 8.4 Constructions to show that $\mathbf{s} = \mathbf{a} + (\mathbf{b} + \mathbf{c})$ and $\mathbf{s} = (\mathbf{a} + \mathbf{b}) + \mathbf{c}$ are identical. Note that the vertex P is not necessarily in the plane defined by the other three vertices.

Products of vectors will be discussed in Section 8.2. To complete this subsection, we consider the product $\lambda\mathbf{a}$ of a scalar λ with a vector $\mathbf{a}$. If $\lambda > 0$, this is defined to be a vector of magnitude $\lambda|\mathbf{a}|$ in the direction of $\mathbf{a}$; and if $\lambda < 0$, it is defined to be a vector of length $|\lambda||\mathbf{a}|$ in the opposite direction to $\mathbf{a}$, as illustrated in Figure 8.5. Division by a scalar λ is defined as multiplication by λ^{-1} and in both cases the operations are associative and distributive, so that

$$(\lambda\mu)\mathbf{a} = \lambda(\mu\mathbf{a}) = \mu(\lambda\mathbf{a}), \tag{8.2a}$$

$$\lambda(\mathbf{a} + \mathbf{b}) = \lambda\mathbf{a} + \lambda\mathbf{b}, \tag{8.2b}$$

and

$$(\lambda + \mu)\mathbf{a} = \lambda\mathbf{a} + \mu\,\mathbf{a}. \tag{8.2c}$$

Finally, it is often useful to introduce the *null vector* $\mathbf{0}$, which has zero length, so that, for example,

$$0\mathbf{a} = \mathbf{0} \quad \text{and} \quad \mathbf{a} + \mathbf{0} = \mathbf{a}$$

for any vector $\mathbf{a}$. Vectors with unit magnitude, called *unit vectors*, also play a special role and are obtained by dividing a vector by its magnitude, a procedure called *normalisation*. Unit vectors in the same directions as $\mathbf{a}$, $\mathbf{b}, \ldots$ are usually denoted $\hat{\mathbf{a}}$, $\hat{\mathbf{b}}, \ldots$; thus $\hat{\mathbf{a}} = \mathbf{a}|\mathbf{a}|^{-1} = \mathbf{a}/a$.

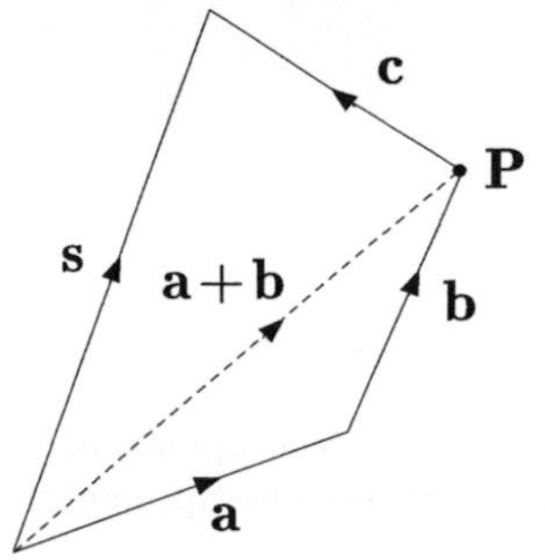

(b)

Figure 8.4 (*Continued*)

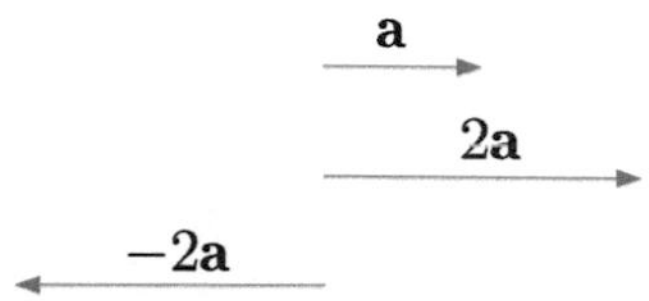

Figure 8.5 The vectors $\mathbf{a}$, $2\mathbf{a}$ and $-2\mathbf{a}$.

Example 8.1

If the mid-points of the consecutive sides of any quadrilateral are joined by straight lines, show that the resulting quadrilateral is a parallelogram.

Solution

Referring to Figure 8.6,

$$\mathbf{PQ} = \tfrac{1}{2}(\mathbf{a}+\mathbf{b}), \quad \mathbf{QR} = \tfrac{1}{2}(\mathbf{b}+\mathbf{c}), \quad \mathbf{RS} = \tfrac{1}{2}(\mathbf{c}+\mathbf{d}),$$
$$\mathbf{SP} = \tfrac{1}{2}(\mathbf{d}+\mathbf{a}),$$

and

$$\mathbf{a}+\mathbf{b}+\mathbf{c}+\mathbf{d} = \mathbf{0},$$

where $\mathbf{0}$ is a null vector. So,

$$\mathbf{PQ} = \tfrac{1}{2}(\mathbf{a}+\mathbf{b}) = -\tfrac{1}{2}(\mathbf{c}+\mathbf{d}) = -\mathbf{RS} = \mathbf{SR},$$

and

$$\mathbf{QR} = \tfrac{1}{2}(\mathbf{b}+\mathbf{c}) = -\tfrac{1}{2}(\mathbf{d}+\mathbf{a}) = -\mathbf{SP} = \mathbf{PS}.$$

Thus the opposite sides are parallel and have the same magnitudes. Hence PQSR is a parallelogram.

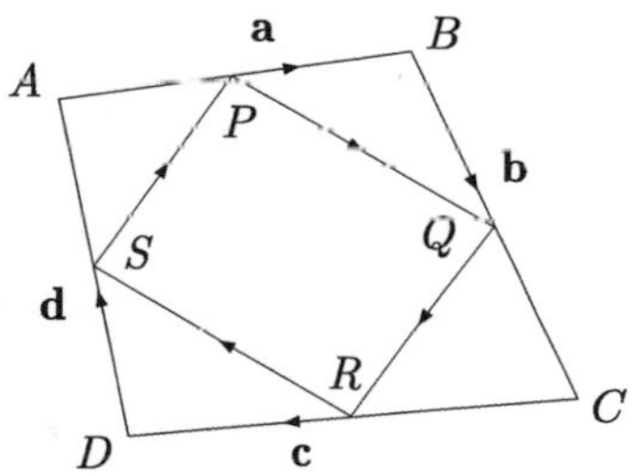

Figure 8.6 Construction to prove the parallelogram relation.

8.1.2 Components of vectors: Cartesian co-ordinates

A useful representation of vectors is obtained by using Cartesian co-ordinates, defined with respect to a right-handed set of axes, as shown in Figure 1.9 and again in Figure 8.7a. Then, using the triangle law of addition, an ordinary vector $\mathbf{a}$ can always be written as the sum

$$\mathbf{a} = \mathbf{a}_x + \mathbf{a}_y + \mathbf{a}_z$$

of three vectors $\mathbf{a}_x$, $\mathbf{a}_y$, $\mathbf{a}_z$ parallel to the x-, y- and z-axes, respectively, as shown in Figure 8.7b. If we now introduce three unit vectors $\mathbf{i}$, $\mathbf{j}$, $\mathbf{k}$ in the x, y and z directions, respectively, then

$$\mathbf{a}_x = |\mathbf{a}_x|\mathbf{i} = a_x\mathbf{i}; \quad \mathbf{a}_y = |\mathbf{a}_y|\mathbf{j} = a_y\mathbf{j}; \quad \mathbf{a}_z = |\mathbf{a}_z|\mathbf{k} = a_z\mathbf{k}$$

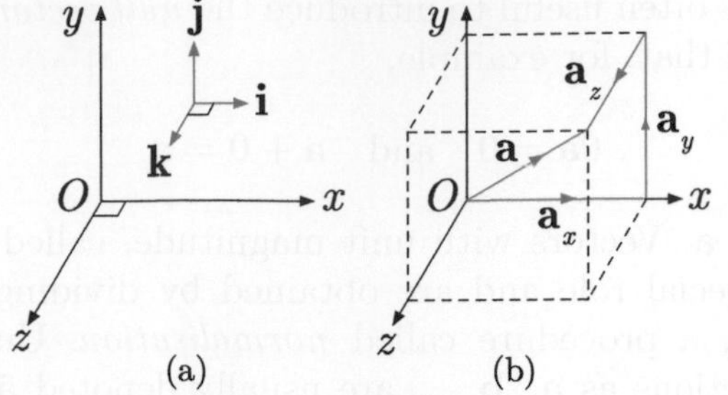

Figure 8.7 Decomposition into Cartesian components.

and

$$\mathbf{a} = a_x\mathbf{i} + a_y\,\mathbf{j} + a_z\mathbf{k}, \tag{8.3a}$$

where

$$|\mathbf{a}|^2 = a_x^2 + a_y^2 + a_z^2 \tag{8.3b}$$

as can be seen by applying Pythagoras' theorem twice to Figure 8.7b.

The vectors **i**, **j** and **k** are called the *basis vectors* of the Cartesian system[1] and a_x, a_y and a_z are called the components of the vector **a** in the x, y and z directions.[2] They are useful because they enable a vector **a** to be defined completely by its components without the necessity of drawing a diagram to specify its direction. They also enable equations involving vectors to be expressed as three equations for their components. This is analogous to the situation we met in Chapter 6 for complex variables, where an equation involving complex quantities could be written as two equations involving real variables. For example, the sum **s** of two vectors **a** and **b**, where

$$\mathbf{a} = a_x\mathbf{i} + a_y\mathbf{j} + a_z\mathbf{k}; \quad \mathbf{b} = b_x\mathbf{i} + b_y\mathbf{j} + b_z\mathbf{k},$$

is, using (8.1) and (8.2),

$$\begin{aligned}\mathbf{s} &= (a_x + b_x)\mathbf{i} + (a_y + b_y)\mathbf{j} + (a_z + b_z)\mathbf{k}\\ &= s_x\mathbf{i} + s_y\mathbf{j} + s_z\mathbf{k},\end{aligned}$$

so the three equations for the components are

$$s_x = (a_x + b_x), \quad s_y = (a_y + b_y), \quad s_z = (a_z + b_z).$$

In Section 1.3.2, a point in a plane was specified by its x and y co-ordinates, defined as its projections onto the x- and y-axes. The description extends directly to three dimensions when a point in space $P(x, y, z)$ is specified by three co-ordinates in the same way. However, in many applications it is useful to describe it using a

[1] Unit basis vectors are usually written without a 'hat'.

[2] We will adopt the convention of representing components of vectors in lower case italic type.

position vector **r**, which translates the origin to a point P, as shown in Figure 8.8. This can be decomposed into its components in exactly the same way as the vector **a** of Figure 8.7, when one obtains

$$\mathbf{r} = x\,\mathbf{i} + y\,\mathbf{j} + z\,\mathbf{k}. \tag{8.4a}$$

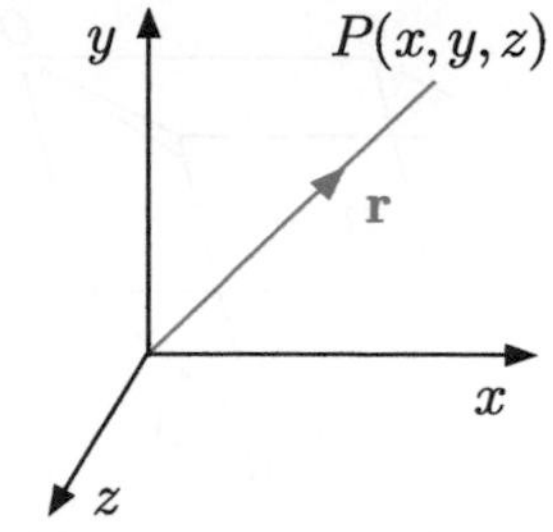

Figure 8.8 Definition of a position vector.

In other words, the components of the position vector are just the Cartesian co-ordinates (x, y, z), and its magnitude, or length,

$$r = \sqrt{x^2 + y^2 + z^2} \tag{8.4b}$$

is just the distance of the point P from the origin of the co-ordinate system.

Another way of representing P is in terms of the angles its position vector makes with the x, y and z-axes. In Figure 8.9a, **r** is the position vector of the point $P(x, y, z)$, with components x, y and z along the three axes and Figure 8.9b shows the three angles α, β, and γ that **r** makes with the x, y and z-axes, respectively. The ratios $x : y : z$ are called the *direction ratios* of **r** and

$$l \equiv \cos\alpha = \frac{x}{r}, \quad m \equiv \cos\beta = \frac{y}{r}, \quad n \equiv \cos\gamma = \frac{z}{r} \tag{8.5a}$$

are its *direction cosines*. It follows from (8.4b) and (8.5a) that

$$l^2 + m^2 + n^2 = 1, \tag{8.5b}$$

and the unit vector is given by

$$\hat{\mathbf{r}} = l\,\mathbf{i} + m\,\mathbf{j} + n\,\mathbf{k} \tag{8.5c}$$

in Cartesian co-ordinates.

So far, we have considered basis vectors **i**, **j**, **k** that are unit vectors at right angles to each other. However, cases sometimes occur, in crystallography for example, where it is advantageous to use basis vectors **a**, **b**, **c**, that do not satisfy these criteria. This is possible

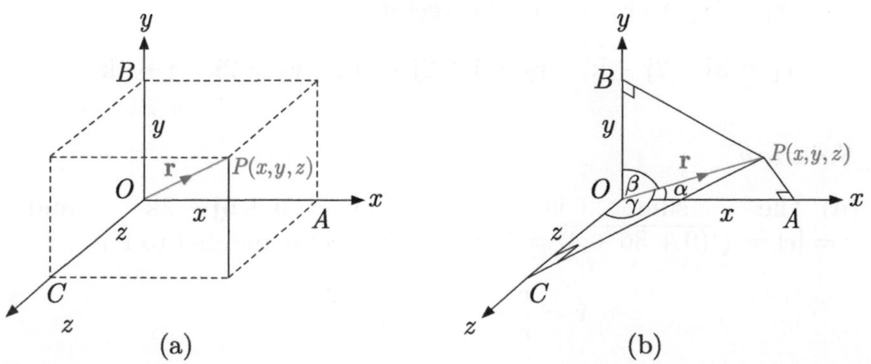

Figure 8.9 Direction ratios and direction cosines.

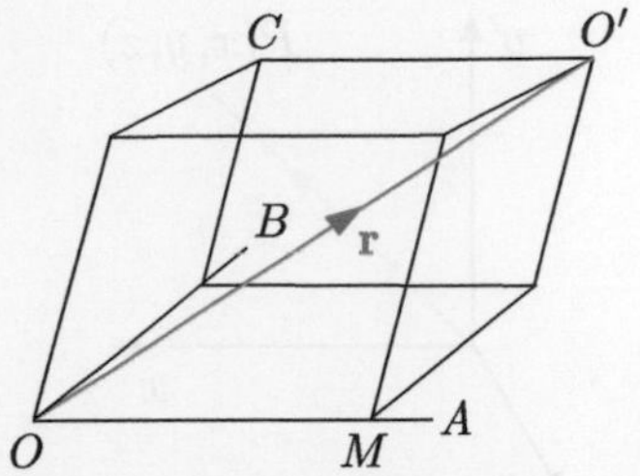

Figure 8.10 Construction to prove the result (8.6a).

because, for any three non-zero vectors **a**, **b**, **c**, which are not parallel to the same plane, one can always write an arbitrary vector **r** in the form

$$\mathbf{r} = \lambda\mathbf{a} + \mu\mathbf{b} + \nu\mathbf{c}, \tag{8.6a}$$

where λ, μ, ν are uniquely determined real numbers.

To prove this result, we use the construction of Figure 8.10. The line OO' has a position vector **r**. The lines OA, OB and OC are parallel to **a**, **b** and **c**, respectively. Taken in pairs, these three lines then define planes parallel respectively to the planes defined by the pairs of vectors $(\mathbf{a}, \mathbf{b})$, $(\mathbf{b}, \mathbf{c})$ and $(\mathbf{c}, \mathbf{a})$. By the law of addition,

$$\mathbf{r} = \mathbf{O\,O'} = \mathbf{O\,M} + \mathbf{M\,C'} + \mathbf{C'O'} = \lambda\mathbf{a} + \mu\mathbf{b} + \nu\mathbf{c},$$

where λ, μ and ν are real numbers. To show they are unique, we assume that **r** can be written

$$\mathbf{r} = \lambda'\,\mathbf{a} + \mu'\,\mathbf{b} + \nu'\,\mathbf{c}. \tag{8.6b}$$

Then subtracting (8.6a) from (8.6b) gives

$$(\lambda' - \lambda)\mathbf{a} = (\mu - \mu')\mathbf{b} + (\nu - \nu')\mathbf{c}.$$

But the left-hand side is a vector parallel to **a**, whereas the right-hand side is a vector parallel to the plane $(\mathbf{b}, \mathbf{c})$. Since **a**, **b** and **c** are all non-zero and not all parallel to the same plane, this result can only be true if

$$(\lambda' - \lambda)\,\mathbf{a} = \mathbf{0}.$$

Thus $\lambda = \lambda'$ and similarly, $\mu = \mu'$ and $\nu = \nu'$.

Example 8.2

(a) Find a unit vector parallel to the sum of $\mathbf{r}_1$ and $\mathbf{r}_2$, where

$$\mathbf{r}_1 = 2\mathbf{i} + 4\mathbf{j} - 5\mathbf{k} \quad \text{and} \quad \mathbf{r}_2 = \mathbf{i} + 2\mathbf{j} + 3\mathbf{k}.$$

(b) Express the vector $\mathbf{v} = 3\mathbf{i} - 3\mathbf{j} + 2\mathbf{k}$ in the form $\mathbf{v} = a\mathbf{r}_1 + b\mathbf{r}_2 + c\mathbf{r}_3$, where the vectors

$$\mathbf{r}_1 = 3\mathbf{i} - 2\mathbf{j} - \mathbf{k}, \quad \mathbf{r}_2 = \mathbf{i} + 2\mathbf{j} - 3\mathbf{k}, \quad \mathbf{r}_3 = 2\mathbf{i} - \mathbf{j} + 4\mathbf{k}.$$

Solution

(a) The sum is $\mathbf{r} = \mathbf{r}_1 + \mathbf{r}_2 = 3\mathbf{i} + 6\mathbf{j} - 2\mathbf{k}$ and $r = |\mathbf{r}| = \sqrt{(9 + 36 + 4)} = 7$. So a unit vector parallel to **r** is

$$\hat{\mathbf{r}} = \frac{\mathbf{r}}{r} = \frac{3}{7}\mathbf{i} + \frac{6}{7}\mathbf{j} - \frac{2}{7}\mathbf{k}.$$

(b) We require

$$3\mathbf{i} - 3\mathbf{j} + 2\mathbf{k} = a(3\mathbf{i} - 2\mathbf{j} - \mathbf{k}) + b(\mathbf{i} + 2\mathbf{j} - 3\mathbf{k}) + c(2\mathbf{i} - \mathbf{j} + 4\mathbf{k}).$$

Since **i**, **j** and **k** are not co-planar, that is, all three vectors are not in the same plane,

$$3 = 3a + b + 2c, \quad -3 = -2a + 2b - c, \quad 2 = -a - 3b + 4c.$$

Solving gives $a = 7/8,\ b = -17/40$ and $c = 2/5$, so that

$$\mathbf{v} = \tfrac{1}{40}(35\mathbf{r}_1 - 17\mathbf{r}_2 + 16\mathbf{r}_3).$$

Example 8.3

A line **OP** is inclined at 60° to the x-axis and 45° to the y-axis. What are its possible inclinations to the z-axis? If the magnitude of **OP** is 4, what are the co-ordinates of the point P?

Solution

From (8.5a) and (8.5b),

$$\cos^2\gamma = 1 - \cos^2 45° - \cos^2 60° = 1/4,$$

so $\cos\gamma = \pm 1/2$ and **OP** is inclined at either 60° or 120° to the z-axis. From (8.5b), the co-ordinates of P are four times the appropriate direction cosine. Thus $P = (2,\ 2\sqrt{2},\ \pm 2)$.

8.2 Products of vectors

We continue the discussion of vector algebra by considering products of vectors. Since in physical science we are mainly concerned with scalar and vector quantities,[3] it is useful to define two sorts of vector products: *scalar products*, which lead to scalars; and *vector products*, which lead to vectors.

8.2.1 Scalar product

Consider a particle that undergoes a linear displacement **d** under the action of a force **F** at an angle θ to the direction of the displacement. The component of **F** in the direction of **d** is $F\cos\theta$ and the product $Fd\cos\theta$ is the work done by the force. The work done is a scalar and $Fd\cos\theta$ is an example of a scalar product. More generally, the scalar product of two vectors **a** and **b** is defined as

$$\mathbf{a} \cdot \mathbf{b} \equiv |\mathbf{a}||\mathbf{b}|\cos\theta = ab\cos\theta, \qquad (8.7)$$

[3] More generally, we are concerned with a class of objects called *tensors*, of which scalars and vectors are the simplest and most commonly occurring examples.

where θ is the angle between the directions of $\mathbf{a}$ and $\mathbf{b}$. Because cosine is an even function, it is irrelevant in which direction the angle θ is measured. Because of the notation on the left-hand side of (8.7), the scalar product is also called the *dot product* of $\mathbf{a}$ and $\mathbf{b}$. From the definition (8.7), we see that the commutative law

$$\mathbf{a} \cdot \mathbf{b} = \mathbf{b} \cdot \mathbf{a} \tag{8.8a}$$

holds for scalar products, and if λ, μ are arbitrary scalars, then

$$(\lambda \mathbf{a}) \cdot (\mu \mathbf{b}) = \lambda\mu(\mathbf{a} \cdot \mathbf{b}). \tag{8.8b}$$

Further, $b\cos\theta$ is the projection of $\mathbf{b}$ onto the axis defined by $\mathbf{a}$ and $a\cos\theta$ is the projection of $\mathbf{a}$ onto the axis defined by $\mathbf{b}$, so that (8.7) can be rewritten in the forms

$$\mathbf{a} \cdot \mathbf{b} = a \times (\text{projection of } \mathbf{b} \text{ onto } \mathbf{a}) = b \times (\text{projection of } \mathbf{a} \text{ onto } \mathbf{b}).$$

Taken together with the triangle law of addition shown in Figure 8.2, this implies the associative law

$$\mathbf{c} \cdot (\mathbf{a} + \mathbf{b}) = \mathbf{c} \cdot \mathbf{a} + \mathbf{c} \cdot \mathbf{b} \tag{8.8c}$$

The algebraic laws (8.8) are similar to those of scalar algebra and

$$\mathbf{a} \text{ or } \mathbf{b} = 0 \Rightarrow \mathbf{a} \cdot \mathbf{b} = 0$$

just as for scalar multiplication. However, the inverse statement is not necessarily true, since $\mathbf{a} \cdot \mathbf{b} = 0$ if $\cos\theta = 0$, that is, if the two vectors are at right angles. So $\mathbf{a} \cdot \mathbf{b} = 0$ does not imply that either $\mathbf{a}$ or $\mathbf{b}$ is necessarily zero, i.e.

$$\mathbf{a} \cdot \mathbf{b} = 0 \not\Rightarrow \text{either } \mathbf{a} \text{ or } \mathbf{b} = 0.$$

This is a fundamental difference between scalar and vector algebra. If

$$\mathbf{a} \cdot \mathbf{b} = 0 \text{ with } \mathbf{a} \neq 0,\ \mathbf{b} \neq 0, \tag{8.9}$$

the vectors $\mathbf{a}$ and $\mathbf{b}$ are said to be *orthogonal*. Also, if $\mathbf{a} = \mathbf{b}$, then

$$\mathbf{a}^2 \equiv \mathbf{a} \cdot \mathbf{a} = |\mathbf{a}|^2 \tag{8.10}$$

is the squared magnitude, or squared 'length', of the vector.

Applying the definition (8.7) to the three unit vectors $\mathbf{i}$, $\mathbf{j}$ and $\mathbf{k}$ in the Cartesian system gives

$$\mathbf{i} \cdot \mathbf{j} = \mathbf{j} \cdot \mathbf{k} = \mathbf{k} \cdot \mathbf{i} = 0 \tag{8.11a}$$

and

$$\mathbf{i}^2 = \mathbf{j}^2 = \mathbf{k}^2 = 1. \tag{8.11b}$$

The unit vectors **i**, **j** and **k** are both orthogonal and normalised and are referred to as an *orthonormal* set of basis vectors. If we now write the vectors **a** and **b** in terms of Cartesian co-ordinates, i.e.

$$\mathbf{a} = a_x\mathbf{i} + a_y\mathbf{j} + a_z\mathbf{k} \quad \text{and} \quad \mathbf{b} = b_x\mathbf{i} + b_y\mathbf{j} + b_z\mathbf{k},$$

then using (8.8) and (8.11) gives

$$\mathbf{a}\cdot\mathbf{b} = a_xb_x + a_yb_y + a_zb_z. \tag{8.12}$$

Example 8.4

Three non-zero vectors **a**, **b** and **c** are such that $(\mathbf{a}+\mathbf{b})$ is perpendicular (symbol $\perp$) to $(\mathbf{a}+\mathbf{c})$ and $(\mathbf{a}-\mathbf{b})$ is $\perp$ to $(\mathbf{a}-\mathbf{c})$. Show that **a** is $\perp$ to $(\mathbf{b}+\mathbf{c})$. If the magnitudes of thc vectors **a**, **b** and **c** are in the ratio 1:2:4, find the angle between **b** and **c**.

Solution

From the definition of a scalar product,

$$(\mathbf{a}+\mathbf{b})\perp(\mathbf{a}+\mathbf{c}) \Rightarrow (\mathbf{a}+\mathbf{b})\cdot(\mathbf{a}+\mathbf{c}) = 0$$

and hence

$$\mathbf{a}\cdot\mathbf{a} + \mathbf{a}\cdot\mathbf{c} + \mathbf{b}\cdot\mathbf{a} + \mathbf{b}\cdot\mathbf{c} = 0. \tag{1}$$

Similarly,

$$(\mathbf{a}-\mathbf{b})\perp(\mathbf{a}-\mathbf{c}) \Rightarrow (\mathbf{a}-\mathbf{b})\cdot(\mathbf{a}-\mathbf{c}) = 0$$

and hence

$$\mathbf{a}\cdot\mathbf{a} - \mathbf{b}\cdot\mathbf{a} - \mathbf{a}\cdot\mathbf{c} + \mathbf{b}\cdot\mathbf{c} = 0 \tag{2}$$

Subtracting (2) from (1) gives

$$\mathbf{a}\cdot(\mathbf{b}+\mathbf{c}) = 0 \Rightarrow \mathbf{a}\perp(\mathbf{b}+\mathbf{c}),$$

as required. Adding (2) and (1) gives

$$2(a^2 + bc\ \cos\theta) = 0, \tag{3}$$

where the magnitudes of the vectors are a, b, c. Using the values $a=\lambda$, $b=2\lambda$, $c=4\lambda$ in (3), so that the ratios are $a:b:c = 1:2:4$, gives

$$\lambda^2 + (2\lambda)(4\lambda)\cos\theta = 0 \Rightarrow \cos\theta = -1/8$$

and so the angle between **b** and **c** is $\theta = 1.70$ radians.

Example 8.5

Two vectors $\hat{\mathbf{a}}$ and $\hat{\mathbf{b}}$ have direction cosines $(1/\sqrt{2}, 1/\sqrt{2}, 0)$ and $(0, \sqrt{3}/2, -1/2)$, respectively. Find the direction cosines of a third vector $\hat{\mathbf{c}}$ that is perpendicular to both $\hat{\mathbf{a}}$ and $\hat{\mathbf{b}}$.

Solution

The two vectors may be written in terms of their direction cosines as

$$\hat{\mathbf{a}} = \frac{1}{\sqrt{2}}\mathbf{i} + \frac{1}{\sqrt{2}}\mathbf{j};\quad \hat{\mathbf{b}} = \frac{\sqrt{3}}{2}\mathbf{j} - \frac{1}{2}\mathbf{k},$$

and so if the direction cosines of c are α, β and γ,

$$\hat{\mathbf{a}}\cdot\hat{\mathbf{c}} = 0 \Rightarrow \frac{\alpha}{\sqrt{2}} + \frac{\beta}{\sqrt{2}} = 0 \quad\text{and}\quad \hat{\mathbf{b}}\cdot\hat{\mathbf{c}} = 0 \Rightarrow \frac{\sqrt{3}\beta}{2} - \frac{\gamma}{2} = 0.$$

In addition,

$$\alpha^2 + \beta^2 + \gamma^2 = 1,$$

and solving these three relations gives the two solutions

$$(\alpha,\ \beta,\ \gamma) = \left(-\frac{1}{\sqrt{5}}, \frac{1}{\sqrt{5}}, \frac{\sqrt{3}}{\sqrt{5}}\right) \quad\text{or}\quad \left(\frac{1}{\sqrt{5}}, -\frac{1}{\sqrt{5}}, -\frac{\sqrt{3}}{\sqrt{5}}\right).$$

8.2.2 Vector product

The *vector product*, also called the *cross product*, of two vectors $\mathbf{a}$ and $\mathbf{b}$ is written $\mathbf{a}\wedge\mathbf{b}$ or $\mathbf{a}\times\mathbf{b}$, (we will use the latter notation) and defined as

$$\mathbf{a}\times\mathbf{b} \equiv a\,b\sin\theta\,\hat{\mathbf{n}}, \tag{8.13}$$

where θ is the angle measured from the direction of $\mathbf{a}$ to that of $\mathbf{b}$ and $\hat{\mathbf{n}}$ is a unit vector perpendicular to the plane containing the two vectors in a direction determined by the 'right-hand screw rule' as shown in Figure 8.11. Because $\sin(-\theta) = -\sin\theta$, it follows that changing the order of the factors in the product changes its sign, that is, the cross product is *anti-commutative*:

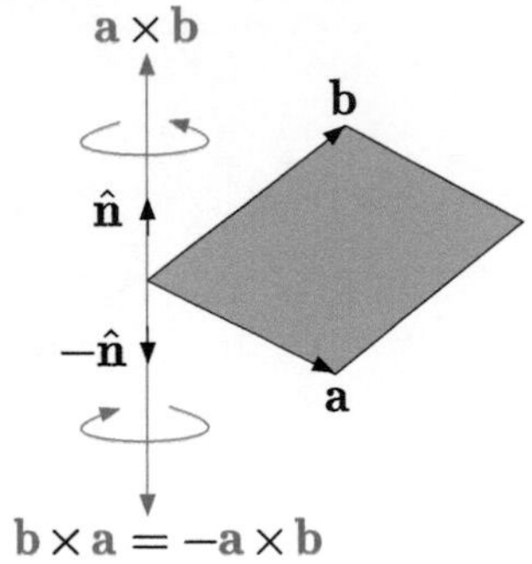

Figure 8.11 Right-hand screw rule for vector products.

$$\mathbf{a}\times\mathbf{b} = -(\mathbf{b}\times\mathbf{a}). \tag{8.14a}$$

Note that

$$\mathbf{a}\times\mathbf{b} = \mathbf{0}, \text{if } \mathbf{a} \text{ or } \mathbf{b} = \mathbf{0},\ \textit{or} \text{ if } \mathbf{a} \text{ and } \mathbf{b} \text{ are parallel,}$$

and

$$\mathbf{a}\times\mathbf{a} = \mathbf{0} \text{ for any vector } \mathbf{a}$$

In addition, the definition (8.13) leads to

$$(\lambda\mathbf{a})\times(\mu\mathbf{b}) = \lambda\mu(\mathbf{a}\times\mathbf{b}) \tag{8.14a}$$

and

$$\mathbf{c}\times(\mathbf{a}+\mathbf{b}) = (\mathbf{c}\times\mathbf{a}) + (\mathbf{c}\times\mathbf{b}), \tag{8.14c}$$

by analogy with (8.8b) (8.8c) for scalar products.

Applying the definition (8.13) to unit Cartesian vectors gives the useful results:

$$\mathbf{i}\times\mathbf{j}=\mathbf{k},\quad \mathbf{j}\times\mathbf{k}=\mathbf{i},\quad \mathbf{k}\times\mathbf{i}=\mathbf{j} \tag{8.15a}$$

and

$$\mathbf{i}\times\mathbf{i}=\mathbf{j}\times\mathbf{j}=\mathbf{k}\times\mathbf{k}=\mathbf{0}. \tag{8.15b}$$

Note the order of the vectors **i**, **j**, **k** in (8.15a). If the order is different from this, a minus sign is required because of the anti-commutative property (8.14a), i.e.

$$\mathbf{j}\times\mathbf{i}=-\mathbf{k},\quad \mathbf{k}\times\mathbf{j}=-\mathbf{i},\quad \mathbf{i}\times\mathbf{k}=-\mathbf{j}.$$

We can now evaluate the vector product of any two vectors in terms of their Cartesian components. Using (8.14) and (8.15) gives

$$\begin{aligned}\mathbf{a}\times\mathbf{b}&=(a_x\mathbf{i}+a_y\,\mathbf{j}+a_z\mathbf{k})\times(b_x\mathbf{i}+b_y\,\mathbf{j}+b_z\mathbf{k})\\&=(a_yb_z-a_zb_y)\mathbf{i}+(a_zb_x-a_xb_z)\mathbf{j}+(a_xb_y-a_yb_x)\mathbf{k}.\end{aligned} \tag{8.16a}$$

The structure of this result is best brought out by relabelling **i**, **j**, **k** as $\hat{\mathbf{e}}_x$, $\hat{\mathbf{e}}_y$, $\hat{\mathbf{e}}_z$, so that the direction with which they are associated is explicit. The vector product (8.16a) then becomes

$$\mathbf{a}\times\mathbf{b}=(a_yb_z-a_zb_y)\hat{\mathbf{e}}_x+(a_zb_x-a_xb_z)\hat{\mathbf{e}}_y+(a_xb_y-a_yb_x)\hat{\mathbf{e}}_z. \tag{8.16b}$$

The sign of each term in (8.16b) is easily memorised by introducing the idea of a *cyclic permutation,* which is also useful in other contexts. A cyclic permutation of any three objects a, b, c in order abc is obtained by removing an object from the end of the sequence and placing it at the beginning, any number of times. Thus abc, cab and bca are cyclic permutations of abc, while acb, bac and cba are non-cyclic permutations. Using this, one sees that (8.16b) is the sum of six terms, each of which is itself the product of three terms, where:

(i) the suffices are all different; and (ii) the sign of the product is +1, or −1, depending on whether the order of the suffices is a cyclic permutation of x, y, z (i.e. xyz, zxy or yzx) or a non-cyclic permutation (i.e. xzy, yxz or zyx).

A physical example of a vector product is provided by considering a rigid body rotating with angular velocity $\boldsymbol{\omega}$, where the direction of $\boldsymbol{\omega}$ corresponds to the axis of rotation. Consider a point P on the body with position vector **r** and angle θ between **r** and $\boldsymbol{\omega}$, as shown in Figure 8.12. The vector product $(\boldsymbol{\omega}\times\mathbf{r})$ is a vector of magnitude $\omega r\sin\theta$ and by the right-hand rule is in the plane perpendicular to the axis of rotation. Since $r\sin\theta$ is the radius of the circle of rotation of P, this has the same magnitude and the same direction as the linear velocity **v** of P, that is, $\mathbf{v}=\boldsymbol{\omega}\times\mathbf{r}$.

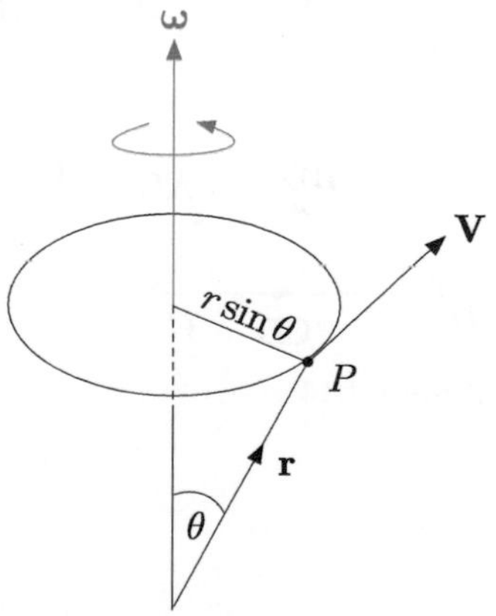

Figure 8.12 Angular velocity.

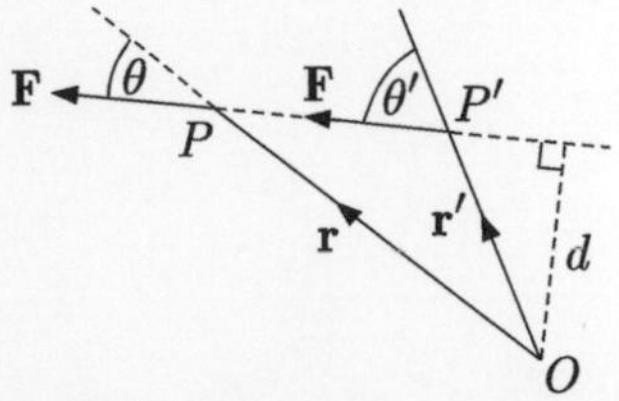

Figure 8.13 Diagram used to calculate the torque about a point O due to a force acting at a point P or a point P' lying on the same line of action.

A second physical example is the *torque,* or *moment,* $\boldsymbol{\tau}$ *about a point* O generated by a force $\mathbf{F}$ acting on an object at a point P, corresponding to a position vector $\mathbf{r}$ relative to O, as shown in Figure 8.13. This is given by

$$\boldsymbol{\tau} = \mathbf{r} \times \mathbf{F} = rF \sin\theta\, \hat{\mathbf{n}}, \tag{8.17}$$

where $\hat{\mathbf{n}} = \hat{\mathbf{r}} \times \hat{\mathbf{F}}$ is a unit vector perpendicular to $\mathbf{r}$ and $\mathbf{F}$ that specifies the direction of $\boldsymbol{\tau}$. The magnitude of the torque is often written in the form $\tau = Fd$ where $d = r\sin\theta$ is the perpendicular distance from O to a straight line through P in the direction of the force, as shown in Figure 8.13. This line is called the *line of action* of $\mathbf{F}$. Now suppose instead that the same force acts at a different point P'. Then the torque is unchanged, provided that P and P' lie on the same line of action, as shown in Figure 8.13 which represents the plane defined by the point O and the line of action of $\mathbf{F}$. This is because

$$\boldsymbol{\tau} = \mathbf{r} \times \mathbf{F} = \mathbf{r}' \times \mathbf{F}$$

by simple geometry, since $\hat{\mathbf{n}} = \hat{\mathbf{r}}' \times \hat{\mathbf{F}}$ remains a unit vector out of the plane and the magnitude of the torque is given by $\tau = Fr'\sin\theta' = Fr\sin\theta = Fd$ in both cases.

Example 8.6

(a) If $\mathbf{d} = \lambda\mathbf{a} + \mu\mathbf{b}$, show that $(\mathbf{a} \times \mathbf{b}) \cdot \mathbf{d} = 0$. (b) A, B, C and D are the consecutive vertices of a parallelogram. Show that

$$(\mathbf{AC})^2 + (\mathbf{BD})^2 = 2[(\mathbf{BC})^2 + (\mathbf{CD})^2].$$

Solution

(a) $(\mathbf{a} \times \mathbf{b}) \cdot \mathbf{d} = (\mathbf{a} \times \mathbf{b}) \cdot (\lambda\mathbf{a} + \mu\mathbf{b}) = \lambda\mathbf{a} \cdot (\mathbf{a} \times \mathbf{b}) + \mu\mathbf{b} \cdot (\mathbf{a} \times \mathbf{b})$, and both terms are zero because $(\mathbf{a} \times \mathbf{b})$ is perpendicular to both $\mathbf{a}$ and $\mathbf{b}$.

(b) From Figure 8.14 below, we see that in triangle BCD, $\mathbf{BD} = \mathbf{BC} + \mathbf{CD}$ and in triangle ABC, $\mathbf{AC} = \mathbf{AB} + \mathbf{BC}$. So,

$$(\mathbf{BD})^2 = (\mathbf{BD} \cdot \mathbf{BD}) = (\mathbf{BC})^2 + 2\mathbf{BC} \cdot \mathbf{CD} + (\mathbf{CD})^2$$

and

$$(\mathbf{AC})^2 = (\mathbf{AC} \cdot \mathbf{AC}) = (\mathbf{AB})^2 + 2\mathbf{AB} \cdot \mathbf{BC} + (\mathbf{BC})^2.$$

Adding these and using the fact that $\mathbf{AB} = -\mathbf{CD}$ gives the result.

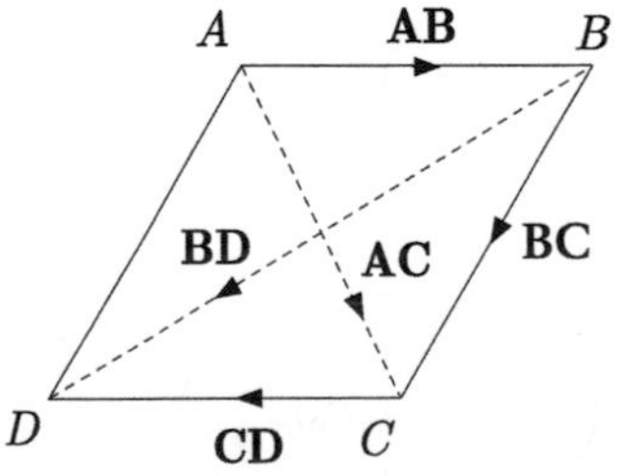

Figure 8.14

Example 8.7

Given that $\mathbf{a} = 2\mathbf{i} - 3\mathbf{j} - \mathbf{k}$ and $\mathbf{b} = \mathbf{i} + 4\mathbf{j} - 2\mathbf{k}$, find $\mathbf{a} \times \mathbf{b}$ and $(\mathbf{a} + \mathbf{b}) \times (\mathbf{a} - \mathbf{b})$.

Solution

Using (8.16) gives

$$\begin{aligned}\mathbf{a}\times\mathbf{b} &= (a_y b_z - a_z b_y)\mathbf{i} - (a_x b_z - a_z b_x)\mathbf{j} + (a_x b_y - a_y b_x)\mathbf{k}\\ &= 10\mathbf{i} + 3\mathbf{j} + 11\mathbf{k}.\end{aligned}$$

Also, $\mathbf{a}+\mathbf{b} = 3\mathbf{i}+\mathbf{j}-3\mathbf{k}$ and $\mathbf{a}-\mathbf{b} = \mathbf{i}-7\mathbf{j}+\mathbf{k}$, so, again using (8.16),

$$(\mathbf{a}+\mathbf{b})\times(\mathbf{a}-\mathbf{b}) = -20\mathbf{i} - 6\mathbf{j} - 22\mathbf{k}.$$

8.2.3 Triple products

The product of a vector $\mathbf{c}$ with a vector product $\mathbf{a}\times\mathbf{b}$ can produce either a scalar $\mathbf{c}\cdot(\mathbf{a}\times\mathbf{b})$ or a vector $\mathbf{c}\times(\mathbf{a}\times\mathbf{b})$. The former is called a *triple scalar product* and the latter a *triple vector product*. We will discuss each in turn.

(i) *Triple scalar product*

If we start by expressing a vector product in terms of the components of its vectors using (8.16a) and then form the scalar product with $\mathbf{c}$ using (8.12), one obtains

$$\mathbf{c}\cdot(\mathbf{a}\times\mathbf{b}) = (a_y b_z - a_z b_y)c_x + (a_z b_x - a_x b_z)c_y + (a_x b_y - a_y b_x)c_z. \tag{8.18}$$

As in (8.16b), the sign of each term is $+1$(or -1) depending on whether the order of the suffices is a cyclic (or non-cyclic) permutation of x, y, z. This in turn implies that if we make a cyclic rearrangement of $\mathbf{a}$, $\mathbf{b}$ and $\mathbf{c}$, for example $\mathbf{c}\cdot(\mathbf{a}\times\mathbf{b}) \to \mathbf{a}\cdot(\mathbf{b}\times\mathbf{c})$, the triple scalar product is unchanged; whereas if we make a non-cyclic rearrangement, for example $\mathbf{c}\cdot(\mathbf{a}\times\mathbf{b}) \to \mathbf{a}\cdot(\mathbf{c}\times\mathbf{b})$, the sign of the triple scalar product is reversed. Introducing the shorthand notation

$$[\mathbf{abc}] \equiv \mathbf{a}\cdot(\mathbf{b}\times\mathbf{c}), \tag{8.19}$$

leads to the results

$$\begin{aligned}[\mathbf{abc}] &= [\mathbf{cab}] = [\mathbf{bca}]\\ &= -[\mathbf{bac}] = -[\mathbf{acb}] = -[\mathbf{cba}].\end{aligned} \tag{8.20}$$

Furthermore, since $\mathbf{a}\cdot(\mathbf{b}\times\mathbf{c}) = (\mathbf{b}\times\mathbf{c})\cdot\mathbf{a}$ by (8.8a), we can use (8.19) to write

$$(\mathbf{a}\times\mathbf{b})\cdot\mathbf{c} = \mathbf{c}\cdot(\mathbf{a}\times\mathbf{b}) = \mathbf{a}\cdot(\mathbf{b}\times\mathbf{c}). \tag{8.21}$$

Thus, provided the order is maintained, the dot and cross are interchangeable.

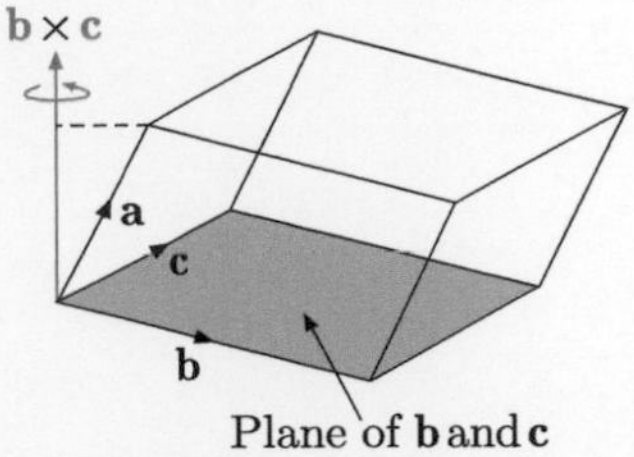

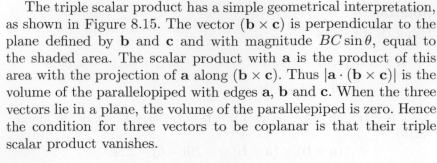

Figure 8.15 Geometrical interpretation of a triple scalar product.

The triple scalar product has a simple geometrical interpretation, as shown in Figure 8.15. The vector $(\mathbf{b}\times\mathbf{c})$ is perpendicular to the plane defined by $\mathbf{b}$ and $\mathbf{c}$ and with magnitude $BC\sin\theta$, equal to the shaded area. The scalar product with $\mathbf{a}$ is the product of this area with the projection of $\mathbf{a}$ along $(\mathbf{b}\times\mathbf{c})$. Thus $|\mathbf{a}\cdot(\mathbf{b}\times\mathbf{c})|$ is the volume of the parallelopiped with edges $\mathbf{a}$, $\mathbf{b}$ and $\mathbf{c}$. When the three vectors lie in a plane, the volume of the parallelepiped is zero. Hence the condition for three vectors to be coplanar is that their triple scalar product vanishes.

$$\mathbf{a}\cdot(\mathbf{b}\times\mathbf{c}) = 0; \quad \mathbf{a},\ \mathbf{b},\ \mathbf{c} \text{ co-planar.} \tag{8.22a}$$

It also vanishes if any two of the vectors are identical.

$$\mathbf{a}\cdot(\mathbf{a}\times\mathbf{b}) = \mathbf{a}\cdot(\mathbf{b}\times\mathbf{a}) = \mathbf{a}\cdot(\mathbf{b}\times\mathbf{b}) = 0. \tag{8.22b}$$

A physical example of a triple scalar product is the *torque, or moment, about an axis* (rather than about a point). Suppose $\mathbf{F}$ is a force acting at a point P and $\mathbf{r}$ is a displacement vector from a point Q to P as shown in Figure 8.16. Then, from (8.17), the torque about the point Q is given by $\boldsymbol{\tau} = (\mathbf{r}\times\mathbf{F})$. However, if the system is constrained so that it can only rotate about an axis OA, specified by the unit vector $\hat{\mathbf{t}}$, as shown in Figure 8.16, then the relevant quantity is the component of $\boldsymbol{\tau}$ in the direction of $\hat{\mathbf{t}}$. This is called the torque about the axis $\hat{\mathbf{t}}$. It is given by the triple scalar product $\hat{\mathbf{t}}\cdot(\mathbf{r}\times\mathbf{F})$ and if we take $\hat{\mathbf{t}}$ to be along the z-axis, then

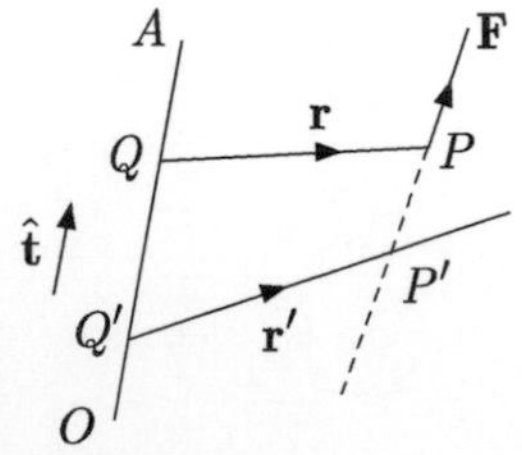

Figure 8.16 Torque as a triple scalar product.

$$\hat{\mathbf{t}}\cdot(\mathbf{r}\times\mathbf{F}) = xF_y - yF_x.$$

Moreover, the result is independent of the choice of the point Q on the axis of rotation and the location of P on the line of action of the force. To see this, suppose $\mathbf{r}'$ is a vector from another point Q' on OA intersecting the line of the force $\mathbf{F}$ through P at P', as shown in Figure 8.16, Then,

$$\hat{\mathbf{t}}\cdot(\mathbf{r}'\times\mathbf{F}) = \hat{\mathbf{t}}\cdot(\mathbf{Q'Q}+\mathbf{r}+\mathbf{PP'})\times\mathbf{F}. \tag{8.23}$$

But $\mathbf{Q'Q}$ is parallel to $\hat{\mathbf{t}}$ and $\mathbf{PP'}$ is parallel to $\mathbf{F}$, so these terms make no contribution to (8.23) and

$$\hat{\mathbf{t}}\cdot(\mathbf{r}'\times\mathbf{F}) = \hat{\mathbf{t}}\cdot(\mathbf{r}\times\mathbf{F}), \tag{8.24}$$

which proves the result.

(ii) *Triple vector product*

Since $(\mathbf{b}\times\mathbf{c})$ is perpendicular to both $\mathbf{b}$ and $\mathbf{c}$ and $\mathbf{a}\times(\mathbf{b}\times\mathbf{c})$ is perpendicular to $(\mathbf{b}\times\mathbf{c})$, it follows that the triple vector product $\mathbf{a}\times(\mathbf{b}\times\mathbf{c})$ must be co-planar with $\mathbf{b}$ and $\mathbf{c}$. Hence

$$\mathbf{a}\times(\mathbf{b}\times\mathbf{c}) = \alpha\,\mathbf{b} + \beta\,\mathbf{c}, \tag{8.25}$$

where the unknown scalars α and β are conveniently determined by considering components, for example

$$[\mathbf{a} \times (\mathbf{b} \times \mathbf{c})]_x = \alpha\ b_x + \beta\ c_x.$$

To do this, we introduce

$$\mathbf{F} \equiv (\mathbf{b} \times \mathbf{c}) = (b_y c_z - b_z c_y)\mathbf{i} + (b_z c_x - b_x c_z)\mathbf{j} + (b_x c_y - b_y c_x)\mathbf{k} \tag{8.26}$$

by (8.16). Using (8.16) again gives

$$\begin{aligned} \alpha\ b_x + \beta\ c_x = [\mathbf{a} \times \mathbf{F}]_x &= a_z F_y - a_y F_z \\ &= (a_y c_y + a_z c_z) b_x - (a_y b_y + a_z b_z) c_x, \end{aligned}$$

on substituting from (8.26) and rearranging. Adding and subtracting $a_x b_x c_x$ from this equation then gives

$$\begin{aligned} \alpha\ b_x + \beta\ c_x &= (a_x c_x + a_y c_y + a_z c_z) b_x - (a_x b_x + a_y b_y + a_z b_z) c_x \\ &= (\mathbf{a} \cdot \mathbf{c}) b_x - (\mathbf{a} \cdot \mathbf{b}) c_x. \end{aligned}$$

This can only hold for arbitrary vectors $\mathbf{a}$, $\mathbf{b}$, $\mathbf{c}$ if $\alpha = \mathbf{a} \cdot \mathbf{c}$ and $\beta = -\mathbf{a} \cdot \mathbf{b}$, so that (8.25) gives

$$\mathbf{a} \times (\mathbf{b} \times \mathbf{c}) = (\mathbf{a} \cdot \mathbf{c})\ \mathbf{b} - (\mathbf{a} \cdot \mathbf{b})\ \mathbf{c}. \tag{8.27}$$

Equation (8.27) not only facilitates the evaluation of triple vector products, but it is also extremely useful in deriving general results. For example, using it, one easily verifies that

$$\mathbf{a} \times (\mathbf{b} \times \mathbf{c}) + \mathbf{c} \times (\mathbf{a} \times \mathbf{b}) + \mathbf{b} \times (\mathbf{c} \times \mathbf{a}) = \mathbf{0}, \tag{8.28}$$

while setting $\mathbf{a} = \mathbf{c}$ in (8.27) gives

$$\mathbf{a} \times (\mathbf{b} \times \mathbf{a}) = a^2 \mathbf{b} - (\mathbf{a} \cdot \mathbf{b})\mathbf{a}.$$

Rearranging the latter gives

$$\mathbf{b} = \frac{(\mathbf{a} \cdot \mathbf{b})\mathbf{a}}{a^2} + \frac{\mathbf{a} \times (\mathbf{b} \times \mathbf{a})}{a^2}, \tag{8.29}$$

which enables an arbitrary vector $\mathbf{b}$ to be expressed as a sum of vectors parallel to and perpendicular to a given vector $\mathbf{a}$.

Finally, we consider triple vector products of the form $(\mathbf{a} \times \mathbf{b}) \times \mathbf{c}$. Since vector products anti-commute, this is given by

$$(\mathbf{a} \times \mathbf{b}) \times \mathbf{c} = -\mathbf{c} \times (\mathbf{a} \times \mathbf{b}) = (\mathbf{c} \cdot \mathbf{a}) \times \mathbf{b} - (\mathbf{c} \cdot \mathbf{b}) \times \mathbf{a}. \tag{8.30}$$

On comparing (8.30) with (8.27), we see that

$$\mathbf{a} \times (\mathbf{b} \times \mathbf{c}) \neq (\mathbf{a} \times \mathbf{b}) \times \mathbf{c} \tag{8.31}$$

in general. In contrast to vector addition (8.1a) and (8.1b), vector products are not associative and in triple vector products the positions of the brackets are important.

A physical example of a triple vector product is the angular momentum of a particle of mass m fixed to the point P on a rigid body, as shown in Figure 8.12. By definition, the angular momentum $\mathbf{L}$ is given by

$$\mathbf{L} = \mathbf{r} \times (m\mathbf{v}) = m\,\mathbf{r} \times \mathbf{v}.$$

We have seen that the velocity is related to the angular velocity $\boldsymbol{\omega}$ by $\mathbf{v} = \boldsymbol{\omega} \times \mathbf{r}$, so $\mathbf{L}$ is given by the triple vector product

$$\mathbf{L} = m\,\mathbf{r} \times (\boldsymbol{\omega} \times \mathbf{r}). \tag{8.32}$$

Example 8.8

If $\mathbf{a} = \mathbf{i} + 2\mathbf{j} + \mathbf{k}$, $\mathbf{b} = 2\mathbf{i} + \mathbf{j} + 3\mathbf{k}$ and $\mathbf{c} = 3\mathbf{i} - 2\mathbf{j} - 2\mathbf{k}$, find the triple vector products

$$\mathbf{a} \times (\mathbf{b} \times \mathbf{c}) \text{ and } (\mathbf{a} \times \mathbf{b}) \times \mathbf{c}.$$

Solution

We have, $\mathbf{a} \cdot \mathbf{b} = 7$, $\mathbf{b} \cdot \mathbf{c} = -2$, $\mathbf{a} \cdot \mathbf{c} = -3$, so that

$$\mathbf{a} \times (\mathbf{b} \times \mathbf{c}) = -3\mathbf{b} - 7\mathbf{c} = -27\mathbf{i} + 11\mathbf{j} + 5\mathbf{k}$$

by (8.27) and

$$(\mathbf{a} \times \mathbf{b}) \times \mathbf{c} = -3\mathbf{b} + 2\mathbf{a} = -4\mathbf{i} + \mathbf{j} - 7\mathbf{k}$$

by (8.30).

Example 8.9

Show that if $\mathbf{a}$, $\mathbf{b}$, $\mathbf{c}$ and $\mathbf{d}$ are four arbitrary vectors:

$$(\mathbf{a} \times \mathbf{b}) \cdot (\mathbf{c} \times \mathbf{d}) = (\mathbf{a} \cdot \mathbf{c})(\mathbf{b} \cdot \mathbf{d}) - (\mathbf{a} \cdot \mathbf{d})(\mathbf{b} \cdot \mathbf{c}), \tag{8.33a}$$

$$(\mathbf{a} \times \mathbf{b}) \times (\mathbf{c} \times \mathbf{d}) = [\mathbf{abd}]\mathbf{c} - [\mathbf{abc}]\mathbf{d}, \tag{8.33b}$$

and

$$(\mathbf{a} \times \mathbf{b}) \cdot (\mathbf{b} \times \mathbf{c}) \times (\mathbf{c} \times \mathbf{a}) = [\mathbf{a} \cdot (\mathbf{b} \times \mathbf{c})]^2. \tag{8.33c}$$

Equation (8.33a) is called the *Lagrange identity.*

Solution

(a) Defining $\mathbf{p} \equiv (\mathbf{a} \times \mathbf{b})$ and using (8.27), we have

$$\begin{aligned}(\mathbf{a} \times \mathbf{b}) \cdot (\mathbf{c} \times \mathbf{d}) &= \mathbf{p} \cdot (\mathbf{c} \times \mathbf{d}) = (\mathbf{p} \times \mathbf{c}) \cdot \mathbf{d} \\ &= [(\mathbf{a} \times \mathbf{b}) \times \mathbf{c}] \cdot \mathbf{d} = -[\mathbf{c} \times (\mathbf{a} \times \mathbf{b})] \cdot \mathbf{d} \\ &= -[(\mathbf{c} \cdot \mathbf{b})\mathbf{a} - (\mathbf{c} \cdot \mathbf{a})\mathbf{b}] \cdot \mathbf{d} = (\mathbf{a} \cdot \mathbf{c})(\mathbf{b} \cdot \mathbf{d}) - (\mathbf{a} \cdot \mathbf{d})(\mathbf{b} \cdot \mathbf{c}).\end{aligned}$$

(b) Defining $\mathbf{p} \equiv (\mathbf{a} \times \mathbf{b})$ and using (8.27), we have

$$(\mathbf{a} \times \mathbf{b}) \times (\mathbf{c} \times \mathbf{d}) = \mathbf{p} \times (\mathbf{c} \times \mathbf{d}) = (\mathbf{p} \cdot \mathbf{d})\mathbf{c} - (\mathbf{p} \cdot \mathbf{c})\mathbf{d}$$
$$= [(\mathbf{a} \times \mathbf{b}) \cdot \mathbf{d}]\mathbf{c} - [(\mathbf{a} \times \mathbf{b}) \cdot \mathbf{c}]\mathbf{d} = [\mathbf{abd}]\mathbf{c} - [\mathbf{abc}]\mathbf{d}.$$

(c) Using (8.33b),

$$(\mathbf{b} \times \mathbf{c}) \times (\mathbf{c} \times \mathbf{a}) = [\mathbf{bca}]\mathbf{c} - [\mathbf{bcc}]\mathbf{a} = [\mathbf{bca}]\mathbf{c}$$

because $[\mathbf{bcc}] = \mathbf{b} \cdot (\mathbf{c} \times \mathbf{c}) = 0$. So,

$$(\mathbf{a} \times \mathbf{b}) \cdot (\mathbf{b} \times \mathbf{c}) \times (\mathbf{c} \wedge \mathbf{a}) = [(\mathbf{a} \times \mathbf{b}) \cdot \mathbf{c}][\mathbf{bca}]$$
$$= [\mathbf{abc}]^2 = [\mathbf{a} \cdot (\mathbf{b} \times \mathbf{c})]^2.$$

Example 8.10

Three points A, B and C lie of the surface of a sphere of unit radius centred at the origin, as shown in Figure 8.17. Let the vectors **OA**, **OB** and **OC** be denoted **a**, **b** and **c**, respectively. By considering the vector product $(\mathbf{a} \times \mathbf{b}) \times (\mathbf{a} \times \mathbf{c})$, show that

$$\frac{\sin\alpha}{\sin a} = \frac{\sin\beta}{\sin b} = \frac{\sin\gamma}{\sin c},$$

where the α, β, γ and a, b, c, are defined in Figure 8.17.

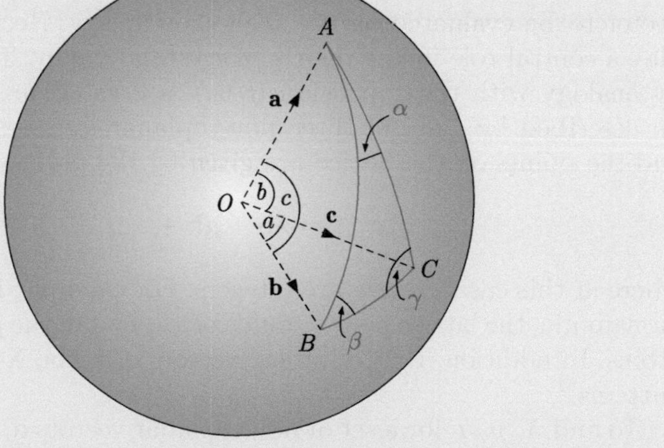

Figure 8.17 Definitions of the angles on the sphere.

Solution

The angle α is the angle between the planes OAB and OAC, where OAB is perpendicular to the vector $(\mathbf{a} \times \mathbf{b})$ and OAC is perpendicular to the vector $(\mathbf{a} \times \mathbf{c})$. Thus,

$$\sin\alpha = \frac{|(\mathbf{a} \times \mathbf{b}) \times (\mathbf{a} \times \mathbf{c})|}{|(\mathbf{a} \times \mathbf{b})||(\mathbf{a} \times \mathbf{c})|}.$$

Now using the result from Example 8.9(b),

$$(\mathbf{a}\times\mathbf{b})\times(\mathbf{a}\times\mathbf{c}) = [(\mathbf{a}\times\mathbf{b})\cdot\mathbf{c}]\mathbf{a},$$

and since $\mathbf{a}$ is a unit vector,

$$|(\mathbf{a}\times\mathbf{b})\times(\mathbf{a}\times\mathbf{c})| = |(\mathbf{a}\times\mathbf{b})\cdot\mathbf{c}|$$

so that

$$\sin\alpha = \frac{|(\mathbf{a}\times\mathbf{b})\cdot\mathbf{c}|}{|(\mathbf{a}\times\mathbf{b})||(\mathbf{a}\times\mathbf{c})|}.$$

From Figure 8.17, $\sin a = |\mathbf{b}\times\mathbf{c}|$ and so

$$\frac{\sin\alpha}{\sin a} = \frac{|(\mathbf{a}\times\mathbf{b})\cdot\mathbf{c}|}{|(\mathbf{a}\times\mathbf{b})||(\mathbf{a}\times\mathbf{c})||(\mathbf{b}\times\mathbf{c})|}.$$

Finally, using the permutation properties of the triple scalar product from (8.20), it follows that

$$\frac{\sin\alpha}{\sin a} = \frac{\sin\beta}{\sin b} = \frac{\sin\gamma}{\sin c}.$$

*8.2.4 Reciprocal vectors

In this section, we will use the properties of products of vectors described above to introduce the concept of *reciprocal vectors.* They enable the coefficients $\lambda,\ \mu,\ \nu$ in the expansion (8.6a) of an arbitrary vector to be evaluated, as we shall show below. Reciprocal vectors play a central role in the theory of crystallography. This is because, by analogy with the expansion (8.6a), a crystalline substance may be described by a set of three non-coplanar *lattice vectors* **a, b, c** and the points on the lattice are given by the position vectors

$$\mathbf{r}_{\lambda\mu v} = \lambda\mathbf{a} + \mu\mathbf{b} + \nu\mathbf{c},$$

where in this case $\lambda,\ \mu,\ \nu$ are integers. For example, if the crystal is monatomic, the lattice points could correspond to the positions of the atoms. In addition, the reciprocal vectors describe X-ray diffraction patterns.[4]

To find $\lambda,\ \mu,\ \nu$ for a set of non-coplanar vectors **a**, **b**, **c**, we define a set of corresponding reciprocal vectors $\mathbf{a}',\ \mathbf{b}',\ \mathbf{c}'$ by

$$\mathbf{a}\cdot\mathbf{a}' = \mathbf{b}\cdot\mathbf{b}' = \mathbf{c}\cdot\mathbf{c}' = 1, \qquad (8.34a)$$

and

$$\mathbf{a}'\cdot\mathbf{b} = \mathbf{a}'\cdot\mathbf{c} = \mathbf{b}'\cdot\mathbf{a} = \mathbf{b}'\cdot\mathbf{c} = \mathbf{c}'\cdot\mathbf{a} = \mathbf{c}'\cdot\mathbf{b} = 0. \qquad (8.34b)$$

[4]See, for example, Chapter 11 of J.R. Hook and H.E. Hall (1991) *Solid State Physics*, 2nd edn., John Wiley & Sons, Chichester.

Assuming such vectors can be identified, then on multiplying both sides of (8.6a) by $\mathbf{a}'$, we immediately obtain

$$\mathbf{a}' \cdot \mathbf{r} = \lambda \mathbf{a}' \cdot \mathbf{a} + \mu \mathbf{a}' \cdot \mathbf{b} + \nu \mathbf{a}' \cdot \mathbf{c} = \lambda$$

by (8.34a) and (8.34b). Similar results applying for μ and ν, so that (8.6a) can be rewritten in the form

$$\mathbf{r} = (\mathbf{a}' \cdot \mathbf{r})\mathbf{a} + (\mathbf{b}' \cdot \mathbf{r})\mathbf{b} + (\mathbf{c}' \cdot \mathbf{r})\mathbf{c}. \tag{8.35}$$

It remains to obtain explicit expressions for $\mathbf{a}'$, $\mathbf{b}'$, $\mathbf{c}'$. To do this, we note that since $\mathbf{a}' \cdot \mathbf{b} = \mathbf{a}' \cdot \mathbf{c} = 0$, $\mathbf{a}'$ must be perpendicular to both $\mathbf{b}$ and $\mathbf{c}$ and hence parallel or antiparallel to $\mathbf{b} \times \mathbf{c}$. This implies that

$$\mathbf{a}' = \alpha(\mathbf{b} \times \mathbf{c}),$$

where α is a constant. Substituting in the requirement $\mathbf{a}'\mathbf{a} = 1$ then gives $\alpha = [\mathbf{abc}]^{-1}$ and hence

$$\mathbf{a}' = \frac{\mathbf{b} \times \mathbf{c}}{[\mathbf{abc}]} = \frac{\mathbf{b} \times \mathbf{c}}{\mathbf{a} \cdot (\mathbf{b} \times \mathbf{c})}. \tag{8.36a}$$

Similar arguments, together with (8.21), give

$$\mathbf{b}' = \frac{\mathbf{c} \times \mathbf{a}}{\mathbf{a} \cdot (\mathbf{b} \times \mathbf{c})} \quad \text{and} \quad \mathbf{c}' = \frac{\mathbf{a} \times \mathbf{b}}{\mathbf{a} \cdot (\mathbf{b} \times \mathbf{c})}, \tag{8.36b}$$

where $\mathbf{a} \cdot (\mathbf{b} \times \mathbf{c}) \neq 0$, since the vectors are non-coplanar.

Example 8.11

Find the reciprocal vectors corresponding to

$$\mathbf{a} = \mathbf{i}, \quad \mathbf{b} = 2\mathbf{j} + \mathbf{k}, \quad \mathbf{c} = \mathbf{i} + 2\mathbf{k}$$

and hence expand the vector $\mathbf{r} = 2\mathbf{i} + 2\mathbf{j} - \mathbf{k}$ as $\mathbf{r} = \lambda\mathbf{a} + \mu\mathbf{b} + \nu\mathbf{c}$.

Solution

Using (8.18) and (8.21) gives

$$\mathbf{a} \cdot (\mathbf{b} \times \mathbf{c}) = \mathbf{i} \cdot (2\mathbf{j} + \mathbf{k}) \times (\mathbf{i} + 2\mathbf{k}) = 4,$$

while (8.36) and (8.16) give the reciprocal vectors

$$\begin{aligned}
\mathbf{a}' &= \tfrac{1}{4}(2\mathbf{j} + \mathbf{k}) \times (\mathbf{i} + 2\mathbf{k}) = \tfrac{1}{4}(4\mathbf{i} + \mathbf{j} - 2\mathbf{k}),\\
\mathbf{b}' &= \tfrac{1}{4}(\mathbf{i} + 2\mathbf{k}) \times \mathbf{i} = \tfrac{1}{2}\mathbf{j},\\
\mathbf{c}' &= \tfrac{1}{4}\mathbf{i} \times (2\mathbf{j} + \mathbf{k}) = -\tfrac{1}{4}(-\mathbf{j} + 2\mathbf{k}).
\end{aligned}$$

Hence, $\mathbf{r} \cdot \mathbf{a}' = 3$, $\mathbf{r} \cdot \mathbf{b}' = 1$, $\mathbf{r} \cdot \mathbf{c}' = -1$, so that $\mathbf{r} = 3\mathbf{a} + \mathbf{b} - \mathbf{c}$ by (8.35).

8.3 Applications to geometry

We have discussed some simple uses of vectors in describing geometry in earlier sections. We will continue that discussion here by considering straight lines and planes in more detail.

8.3.1 Straight lines

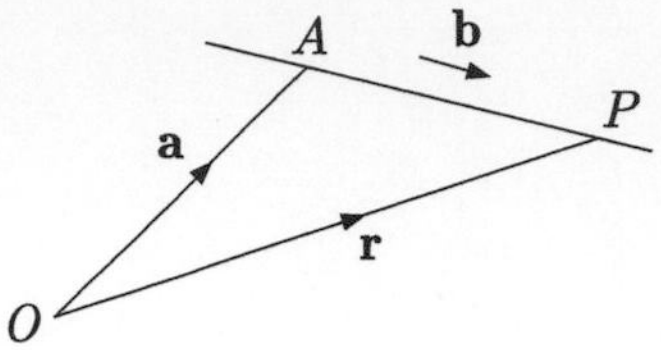

Figure 8.18 Construction of equation for a straight line.

We will find the vector equation of a line through a given point A, with position vector $\mathbf{a}$, in the direction of a given vector $\mathbf{b}$, by reference to Figure 8.18. The position vector of any point P on the line is given by

$$\mathbf{r} = \mathbf{OP} = \mathbf{OA} + \mathbf{AP} = \mathbf{a} + s\,\mathbf{b},$$

where s is the length of the line AP in units of $|\mathbf{b}|$. By allowing the scale parameter to vary over all values $-\infty < s < \infty$, we obtain the position vectors for all points on the line. Thus, as the parameter s varies, the equation

$$\mathbf{r} = \mathbf{a} + s\,\mathbf{b} \tag{8.37a}$$

defines a straight line through the point A with position vector $\mathbf{a}$ in a direction of the vector $\mathbf{b}$. It is often written in the standard form

$$\mathbf{r} = \mathbf{a} + \lambda\hat{\mathbf{b}}, \tag{8.37b}$$

where $\lambda = s|\mathbf{b}|$ and $\hat{\mathbf{b}}$ is a unit vector in the $\mathbf{b}$-direction as usual. More generally, one can choose any point A on the line and $\mathbf{b}$ can be replaced by any vector that is parallel or antiparallel to $\hat{\mathbf{b}}$. For example, consider the equation for a straight line through the two points C and D, with vectors $\mathbf{c}$ and $\mathbf{d}$, respectively. Then we could take $A = C$ and use $\mathbf{b} = \mathbf{d} - \mathbf{c}$ to specify the direction of the line, so that (8.37) becomes

$$\mathbf{r} = \mathbf{c} + s(\mathbf{d} - \mathbf{c}); \tag{8.38a}$$

or we could take $A = D$ and use $\mathbf{b} = \mathbf{c} - \mathbf{d}$, yielding

$$\mathbf{r} = \mathbf{d} + s(\mathbf{c} - \mathbf{d}). \tag{8.38b}$$

These two equations (and others) describe the same straight line through C and D, although any given point on the line will correspond to a different value of s, depending on which form is used.

We turn now to Cartesian co-ordinates, in which

$$\mathbf{r} = x\,\mathbf{i} + y\,\mathbf{j} + z\,\mathbf{k}$$

and

$$\hat{\mathbf{b}} = \cos\alpha\,\mathbf{i} + \cos\beta\,\mathbf{j} + \cos\gamma\,\mathbf{k}, \tag{8.39}$$

where $\cos\alpha, \cos\beta, \cos\gamma$ are the direction cosines of $\mathbf{b} = |\mathbf{b}|\hat{\mathbf{b}}$. Substituting into (8.37a) then gives

$$x = a_x + sb_x, \quad y = a_y + sb_y, \quad z = a_z + sb_z$$

in an obvious notation. Eliminating s gives the two equations

$$\frac{x - a_x}{b_x} = \frac{y - a_y}{b_y} = \frac{z - a_z}{b_z} \tag{8.40}$$

that are required to define a straight line in Cartesian co-ordinates in three dimensions.

As an example we will find the shortest distance from a point P to the line (8.37). This distance is the length d of the perpendicular from P to the line as shown in Figure 8.19a. If the position vector of P is $\mathbf{p}$, then from the right-angled triangle,

$$d = |\mathbf{p} - \mathbf{a}| \sin\theta = |(\mathbf{p} - \mathbf{a}) \times \hat{\mathbf{b}}|, \tag{8.41a}$$

where $\hat{\mathbf{b}}$ is a unit vector in the direction of the line and we have used the definition of the vector product. We can also find the shortest distance d' between two arbitrary lines in the directions of the vectors $\mathbf{a}$ and $\mathbf{b}$. Referring to Figure 8.19b, the line normal to both $\mathbf{a}$ and $\mathbf{b}$ is $\mathbf{a} \times \mathbf{b}$ and so a unit vector normal to both lines is

$$\hat{\mathbf{n}} = \frac{\mathbf{a} \times \mathbf{b}}{|\mathbf{a} \times \mathbf{b}|}.$$

Then if P is a point on the line in the direction $\mathbf{a}$ with position vector $\mathbf{p}$ and Q is a point on the line in the direction $\mathbf{b}$ with a position vector $\mathbf{q}$, the line $\mathbf{QP} = (\mathbf{p} - \mathbf{q})$ and the minimum distance between the lines is the component of QP along the unit normal $\hat{\mathbf{n}}$, i.e.

$$d' = |(\mathbf{p} - \mathbf{q}) \cdot \hat{\mathbf{n}}| = \frac{|(\mathbf{p} - \mathbf{q}) \cdot (\mathbf{a} \times \mathbf{b})|}{|\mathbf{a} \times \mathbf{b}|}. \tag{8.41b}$$

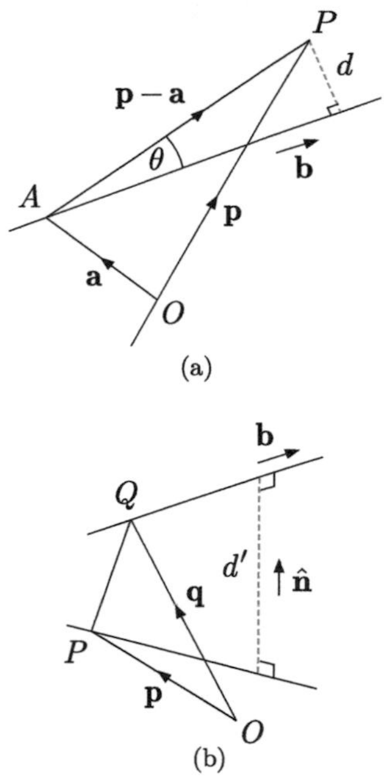

Figure 8.19 Shortest distance between (a) a point and a line, (b) two lines.

Example 8.12

Show that the lines defined by

$$\mathbf{r} = \mathbf{a} + s\mathbf{b}, \quad \mathbf{r}' = \mathbf{a}' + s\mathbf{b}'$$

are the same if

$$\mathbf{a} = 2\mathbf{i} + 2\mathbf{j} - \mathbf{k}, \quad \mathbf{a}' = \mathbf{i} - \mathbf{j} + 2\mathbf{k}, \quad \mathbf{b}' = 2\mathbf{b} = \mathbf{i} + 3\mathbf{j} - 3\mathbf{k}.$$

Solution

The vectors $\mathbf{b}$ and $\mathbf{b}'$ are obviously parallel, so that it is only necessary to show that the running vector $\mathbf{r}'$ passes through the point A with position vector $\mathbf{a}$. That is, there exists a value of s such that $\mathbf{a} = \mathbf{a}' + s\mathbf{b}'$. In terms of components, this condition is

$$2 = 1 + s, \quad 2 = -1 + 3s, \quad -1 = 2 - 3s,$$

which are all satisfied for $s = 1$. So $\mathbf{r}$ and $\mathbf{r}'$ represent the same line.

Example 8.13

Find the shortest distance (a) from a point P with co-ordinates $(1, 2, 3)$ to the line $\mathbf{r} = \mathbf{a} + s\mathbf{b}$, where $\mathbf{a} = \mathbf{i} - 2\mathbf{j} + \mathbf{k}$ and $\mathbf{b} = 2\mathbf{i} + \mathbf{j} - 2\mathbf{k}$, (b) between the line $\mathbf{r}_1$ that passes through the points $(1, 2, 3)$ and $(1, 0, 1)$ and the line $\mathbf{r}_2$ that passes through the points $(3, 2, 1)$ and $(0, 1, 0)$.

Solution

(a) Using the notation in (8.39),

$$\hat{\mathbf{b}} = \tfrac{1}{3}(2\mathbf{i} + \mathbf{j} - 2\mathbf{k})$$

and the position vector of P is

$$\mathbf{p} = \mathbf{i} + 2\mathbf{j} + 3\mathbf{k}.$$

So, using (8.16),

$$(\mathbf{p} - \mathbf{a}) \times \hat{\mathbf{b}} = 2(-5\mathbf{i} + 2\mathbf{j} - 4\mathbf{k})/3$$

and (8.41a) gives

$$d = |(\mathbf{p} - \mathbf{a}) \times \hat{\mathbf{b}}| = 2\sqrt{5}.$$

(b) Using the notation of (8.37a),

$$\mathbf{r}_1 = (\mathbf{i} + \mathbf{k}) + s(2\mathbf{j} + 2\mathbf{k}) \quad \text{and} \quad \mathbf{r}_2 = \mathbf{j} + t(3\mathbf{i} + \mathbf{j} + \mathbf{k}).$$

Hence, again using (8.16),

$$\mathbf{n} = (2\mathbf{j} + 2\mathbf{k}) \times (3\mathbf{i} + \mathbf{j} + \mathbf{k}) = 6\mathbf{j} - 6\mathbf{k},$$

and $\hat{\mathbf{n}} = (\mathbf{j} - \mathbf{k})\big/\sqrt{2}$. A vector between the two lines is, for example, one that joins (1, 0, 1) to (0, 1, 0), that is, $(\mathbf{i} - \mathbf{j} + \mathbf{k})$, so the shortest distance between the lines is, from (8.41b),

$$d = \frac{1}{\sqrt{2}}|(\mathbf{i} - \mathbf{j} + \mathbf{k}) \cdot (\mathbf{j} - \mathbf{k})| = \sqrt{2}.$$

8.3.2 Planes

The above ideas can be extended to find the equation for a plane. Consider a plane defined by three non-collinear points A, B, C, with position vectors $\mathbf{a}$, $\mathbf{b}$, $\mathbf{c}$. Then any other point R with position vector $\mathbf{r}$ that lies in the plane can be written in the form

$$\mathbf{r} = \mathbf{a} + s(\mathbf{b} - \mathbf{a}) + s'(\mathbf{c} - \mathbf{a}), \tag{8.42}$$

as shown in Figure 8.20, where s and s' are real parameters in the range $-\infty < s,\ s' < \infty$. This expression can then be re-arranged in the more symmetric form

$$\mathbf{r} = \alpha\mathbf{a} + \beta\mathbf{b} + \gamma\mathbf{c}, \tag{8.43a}$$

where $\alpha = 1 - s - s'$, $\beta = s$, $\gamma = s'$ satisfy

$$\alpha + \beta + \gamma = 1. \tag{8.43b}$$

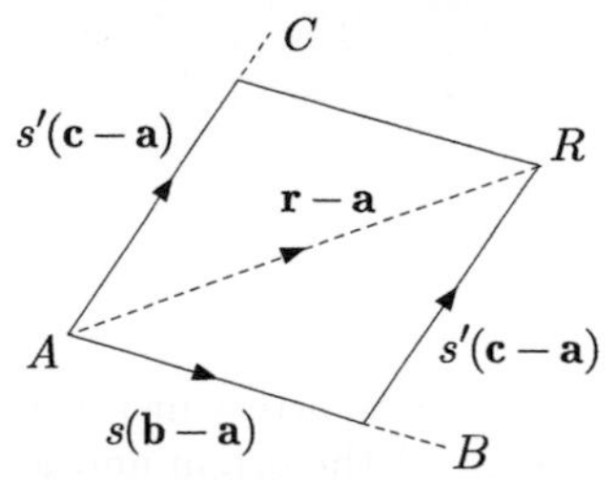

Figure 8.20 A point R in the plane defined by the points A, B, C, having position vectors $\mathbf{a}$, $\mathbf{b}$, $\mathbf{c}$, as given in (8.42). In the figure we have chosen the point R so that s and s' are both less than 1.

Equation (8.42) is similar in form to (8.38) used to describe a straight line, except that two parameters s and s' are needed to define a given point on a plane, as opposed to one for a straight line. However, a very useful alternative description is obtained by considering a plane through the point A that is perpendicular to a given vector $\mathbf{n}$. Then a vector $\mathbf{r} - \mathbf{a}$ will be perpendicular to $\mathbf{n}$ if, and only if, the point $\mathbf{r}$ also lies in the plane. Hence

$$(\mathbf{r} - \mathbf{a}) \cdot \mathbf{n} = 0 \tag{8.44a}$$

is the condition that $\mathbf{r}$ lies in the plane and (8.44a) corresponds to the same plane as (8.42) if we choose $\mathbf{n} = (\mathbf{b} \times \mathbf{c})$ or $\mathbf{n} = -(\mathbf{b} \times \mathbf{c})$. In Cartesian co-ordinates, this becomes

$$\alpha x + \beta y + \gamma z = d, \tag{8.45a}$$

where α, β, γ are the direction ratios of $\mathbf{n}$, i.e.

$$\mathbf{n} = \alpha\mathbf{i} + \beta\mathbf{j} + \gamma\mathbf{k}, \tag{8.45b}$$

and $d = \mathbf{r} \cdot \mathbf{a}$. The latter parameter has a simple interpretation in the case that $\mathbf{n}$ is chosen to be a unit vector $\hat{\mathbf{n}}$ in the same direction as $\mathbf{n}$. Then α, β, γ reduce to the direction cosines of $\hat{\mathbf{n}}$ (or $\mathbf{n}$), satisfying [cf. Eq. (8.5)]

$$\alpha^2 + \beta^2 + \gamma^2 = 1, \tag{8.46a}$$

and

$$|d| = |\mathbf{a} \cdot \hat{\mathbf{n}}| \tag{8.46b}$$

is the perpendicular distance from the origin to the plane, as can be seen from Figure 8.21a. The sign of d depends on which side of

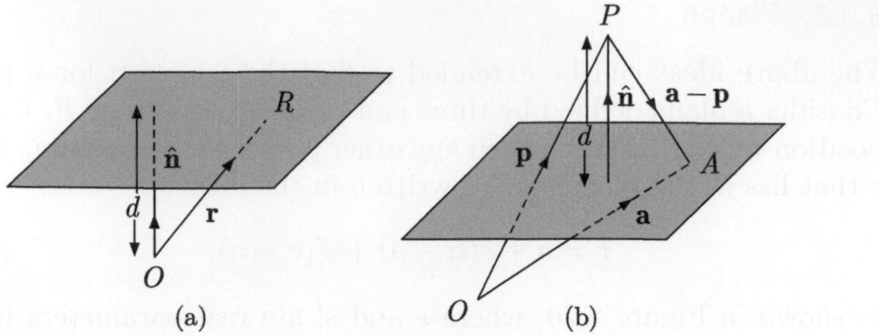

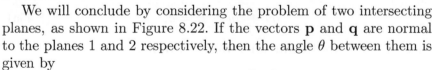

Figure 8.21 Shortest distance between (a) the origin and a plane; and (b) an arbitrary point P and a plane.

the plane the origin is situated relative to the direction of $\hat{\mathbf{n}}$, where Figure 8.21a corresponds to the case $d > 0$.

To illustrate the use of (8.44), we will find the shortest distance from a point P, with position vector $\mathbf{p}$, to the plane described by (8.44). Consider the vector $(\mathbf{a} - \mathbf{p})$, where $\mathbf{a}$ is the position vector of any point A in the plane, as shown in Figure 8.21b. Its component normal to the plane, which is the distance to the plane, is

$$d = (\mathbf{a} - \mathbf{p}) \cdot \hat{\mathbf{n}}, \tag{8.47}$$

where the sign will depend on the direction of $(\mathbf{a} - \mathbf{p})$ relative to $\hat{\mathbf{n}}$. The same result can be extended to a line $\mathbf{r} = \mathbf{c} + s\,\mathbf{b}$ that is parallel to the plane, that is, for which $\mathbf{b} \cdot \hat{\mathbf{n}} = 0$. Then the perpendicular distance from a point on the line to the plane will be independent of which point is chosen and so will again be given by (8.47), where P is now any point on the line and A is any point in the plane. However, if the line and the plane are not parallel, that is, if $\mathbf{b} \cdot \hat{\mathbf{n}} \neq 0$, the line will pass through the plane and the minimum distance is zero.

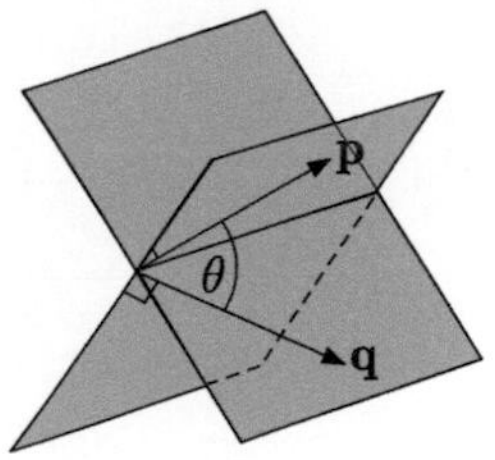

Figure 8.22 Intersection of two planes.

We will conclude by considering the problem of two intersecting planes, as shown in Figure 8.22. If the vectors $\mathbf{p}$ and $\mathbf{q}$ are normal to the planes 1 and 2 respectively, then the angle θ between them is given by

$$\cos\theta = \hat{\mathbf{p}} \cdot \hat{\mathbf{q}} = \frac{\mathbf{p} \cdot \mathbf{q}}{|\mathbf{p}||\mathbf{q}|}, \tag{8.48}$$

which is also the angle of intersection of the two planes. For example, if the planes are

$$2x + y + z = 2, \quad x - 2y - z = 4, \tag{8.49}$$

then we can take

$$\mathbf{p} = 2\mathbf{i} + \mathbf{j} + \mathbf{k}, \quad \mathbf{q} = \mathbf{i} - 2\mathbf{j} - \mathbf{k},$$

giving $\theta = 99.6^\circ$. Finally, the line of intersection $\mathbf{r}$ of the two planes is contained in both planes, so that both equations (8.49) must be simultaneously satisfied, from which one obtains

$$x = \frac{y+6}{3} = \frac{8-z}{5} = s,$$

where s is a scalar parameter. So the co-ordinates of any point on the line are $x = s,\ y = 3s + 6,\ z = 8 - 5s$ and the vector equation of the line of intersection is

$$\mathbf{r} = (6\mathbf{j} + 8\mathbf{k}) + s(\mathbf{i} + 3\mathbf{j} - 5\mathbf{k}).$$

Example 8.14

A line is given by

$$\mathbf{r} = \mathbf{a} + s\mathbf{b} \text{ with } \mathbf{a} = \mathbf{i} + 2\mathbf{j} - 2\mathbf{k} \text{ and } \mathbf{b} = 2\mathbf{i} - \mathbf{j} + 3\mathbf{k}.$$

(a) Show that the line intersects the plane $x + 2y - z = 4$; (b) find the co-ordinates of the point of intersection.

Solution

(a) A vector that is perpendicular to the plane is $\mathbf{n} = \mathbf{i} + 2\mathbf{j} - \mathbf{k}$ and since $\mathbf{b} \cdot \mathbf{n} = -3 \neq 0$, the line does intersect the plane.

(b) The point of intersection is found by first substituting the co-ordinate values of the line into the equation of the plane, i.e.

$$(1 + 2s) + 2(2 - s) - (-2 + 3s) = 4$$

to give $s = 1$ and then using this value of s in the equation of the line. This gives the co-ordinates as $(x,\ y,\ z) = (3,\ 1,\ 1)$.

8.4 Differentiation and integration

In this section we extend the ideas of differentiation and integration to vectors that depend on a continuously varying parameter s or t.

Differentiation of a vector $\mathbf{a}$ is defined in the same way as for a scalar. Thus, if $\mathbf{a}$ is a function of the scalar variable t, then

$$\frac{\mathrm{d}\mathbf{a}}{\mathrm{d}t} = \lim_{\delta t \to 0} \frac{\delta \mathbf{a}}{\delta t},$$

assuming that the limit exists. If this is the case, then just as for the differentiation of a scalar function, $\mathbf{a}$ is said to be differentiable at the point. Similarly, the differential $\mathrm{d}\mathbf{a}$ of a vector is given by

$$\mathrm{d}\mathbf{a} = \frac{\mathrm{d}\mathbf{a}}{\mathrm{d}t}\mathrm{d}t.$$

Both the derivative and the differential are vectors. In particular, they are *not* in general in the same direction as **V** itself. This is easily seen by writing **a** in components. For example, using rectangular coordinates,

$$\mathbf{a} = a_x\mathbf{i} + a_y\,\mathbf{j} + a_z\mathbf{k}, \tag{8.50}$$

and so if **a** is a function of a variable t,

$$\frac{\mathrm{d}\mathbf{a}}{\mathrm{d}t} = \frac{\mathrm{d}a_x}{\mathrm{d}t}\mathbf{i} + \frac{\mathrm{d}a_y}{\mathrm{d}t}\mathbf{j} + \frac{\mathrm{d}a_z}{\mathrm{d}t}\mathbf{k} \tag{8.51}$$

and because the coefficients of the basis vectors are in general different in (8.50) and (8.51), the two vectors will have different directions. Similar results hold for higher derivatives.

Scalar and vector products are also differentiated by using the product rule. Thus,

$$\frac{\mathrm{d}(\mathbf{a}\cdot\mathbf{b})}{\mathrm{d}t} = \frac{\mathrm{d}\mathbf{a}}{\mathrm{d}t}\cdot\mathbf{b} + \mathbf{a}\cdot\frac{\mathrm{d}\mathbf{b}}{\mathrm{d}t}$$

and

$$\frac{\mathrm{d}(\mathbf{a}\times\mathbf{b})}{\mathrm{d}t} = \frac{\mathrm{d}\mathbf{a}}{\mathrm{d}t}\times\mathbf{b} + \mathbf{a}\times\frac{\mathrm{d}\mathbf{b}}{\mathrm{d}t}.$$

In the latter case, the order of the vectors must be strictly maintained. Triple products are dealt with by applying the product rule a second time. Thus, for the triple scalar product,

$$\frac{\mathrm{d}[\mathbf{a}\cdot(\mathbf{b}\times\mathbf{c})]}{\mathrm{d}t} = \frac{\mathrm{d}\mathbf{a}}{\mathrm{d}t}\cdot(\mathbf{b}\times\mathbf{c}) + \mathbf{a}\cdot\left(\frac{\mathrm{d}\mathbf{b}}{\mathrm{d}t}\times\mathbf{c}\right) + \mathbf{a}\cdot\left(\mathbf{b}\times\frac{\mathrm{d}\mathbf{c}}{\mathrm{d}t}\right),$$

and for the triple vector product,

$$\frac{\mathrm{d}[\mathbf{a}\times(\mathbf{b}\times\mathbf{c})]}{\mathrm{d}t} = \frac{\mathrm{d}\mathbf{a}}{\mathrm{d}t}\times(\mathbf{b}\times\mathbf{c}) + \mathbf{a}\times\left(\frac{\mathrm{d}\mathbf{b}}{\mathrm{d}t}\times\mathbf{c}\right) + \mathbf{a}\times\left(\mathbf{b}\times\frac{\mathrm{d}\mathbf{c}}{\mathrm{d}t}\right).$$

Again the order of the vectors must be maintained.

Integration of a vector, or an expression involving vectors (which may be itself be a vector or a scalar) with respect to a scalar is the inverse of differentiation. As for differentiation, it is important to remember that integrating an expression involving vectors (including the differential) does not change the character (either scalar or vector) of the expression and that the order of terms in a vector integrand must be maintained. In addition, in indefinite integrals the integration constant is a constant vector. That is, if

$$\mathbf{b}(t) = \mathrm{d}\mathbf{a}(t)/\mathrm{d}t \tag{8.52a}$$

then

$$\int \mathbf{b}(t)\,\mathrm{d}t = \mathbf{a}(t) + \mathbf{a}_0, \tag{8.52b}$$

where $\mathbf{a}_0$ is an arbitrary vector, independent of t.

Example 8.15

If the vectors **a** and **b** may be written in the parametric form

$$\mathbf{a} = 3t^3\mathbf{i} - 2t\mathbf{j} + t^2\mathbf{k} \quad \text{and} \quad \mathbf{b} = 3\sin t\ \mathbf{i} + 2\cos t\ \mathbf{k}$$

find (a) $\mathrm{d}(\mathbf{a}\cdot\mathbf{b})/\mathrm{d}t$ and (b) $\mathrm{d}(\mathbf{a}\times\mathbf{b})/\mathrm{d}t$.

Solution

(a) Either one can use the result

$$\frac{\mathrm{d}}{\mathrm{d}t}(\mathbf{a}\cdot\mathbf{b}) = \mathbf{a}\cdot\frac{\mathrm{d}\mathbf{b}}{\mathrm{d}t} + \frac{\mathrm{d}\mathbf{a}}{\mathrm{d}t}\cdot\mathbf{b}$$

and evaluate both terms, or, more directly one can calculate $\mathbf{a}\cdot\mathbf{b} = 9t^3\sin t + 2t^2\cos t$ and then differentiate, giving

$$\frac{\mathrm{d}}{\mathrm{d}t}(\mathbf{a}\cdot\mathbf{b}) = 25t^2\sin t + t(9t^2+4)\cos t.$$

(b) Again, one can either use the result

$$\frac{\mathrm{d}}{\mathrm{d}t}(\mathbf{a}\times\mathbf{b}) = \mathbf{a}\times\frac{\mathrm{d}\mathbf{b}}{\mathrm{d}t} + \frac{\mathrm{d}\mathbf{a}}{\mathrm{d}t}\times\mathbf{b},$$

or first evaluate $\mathbf{a}\times\mathbf{b}$ and then differentiate. Using (8.16),

$$\mathbf{a}\times\mathbf{b} = -4t\ \cos t\ \mathbf{i} - 3t^2(2t\ \cos t - \sin t)\ \mathbf{j} + 6t\sin t\ \mathbf{k},$$

which on differentiating gives

$$\begin{aligned}\frac{\mathrm{d}(\mathbf{a}\times\mathbf{b})}{\mathrm{d}t} &= 4(t\sin t - \cos t)\ \mathbf{i} + [6t(t^2+1)\sin t - 15t^2\cos t]\mathbf{j} \\ &\quad + 6(\sin t + t\ \cos t)\ \mathbf{k}.\end{aligned}$$

Example 8.16

The acceleration of a particle for times $t \geq 0$ is given by

$$\mathbf{a} = 4\cos t\ \mathbf{i} + 12\sin 2t\ \mathbf{j} - 10t\ \mathbf{k},$$

Find expressions for the velocity $\mathbf{v}(t)$ and the displacement $\mathbf{r}(t)$ if both zero at $t = 0$.

Solution

The velocity is given by

$$\begin{aligned}\mathbf{v} &= \int \mathbf{a}(t)\ \mathrm{d}t = \int (4\cos t\ \mathbf{i} + 12\sin 2t\ \mathbf{j} - 10t\ \mathbf{k})\ \mathrm{d}t \\ &= 4\sin t\ \mathbf{i} - 6\cos 2t\ \mathbf{j} - 5t^2\ \mathbf{k} + \mathbf{c}_1,\end{aligned}$$

where $\mathbf{c}_1$ is a constant vector. Using the boundary conditions at $t = 0$, gives $\mathbf{c}_1 = 6\mathbf{j}$ and hence

$$\mathbf{v} = 4\sin t\ \mathbf{i} - 6(\cos 2t - 1)\ \mathbf{j} - 5t^2\ \mathbf{k}.$$

Also,

$$\mathbf{r} = \int \mathbf{v}(t)\ \mathrm{d}t = \int (4\sin t\ \mathbf{i} - 6(\cos 2t - 1)\ \mathbf{j} - 5t^2\ \mathbf{k})\mathrm{d}t$$

$$= -4\cos t\ \mathbf{i} - (3\sin 2t - 6t)\ \mathbf{j} - \tfrac{5}{3}t^3\ \mathbf{k} + \mathbf{c}_2,$$

where $\mathbf{c}_2$ is a constant vector. Again using the boundary conditions at $t = 0$ gives $\mathbf{c}_2 = 4\mathbf{i}$ and hence

$$\mathbf{r} = 4(1 - \cos t)\ \mathbf{i} - (3\sin 2t - 6t)\ \mathbf{j} - \tfrac{5}{3}t^3\ \mathbf{k}.$$

Problems 8

8.1 Three vectors of lengths a, $2a$ and $3a$ meet at a point and are directed along the diagonals of the faces of a cube meeting at the point. Determine their sum in the form $\mathbf{r} = x\,\mathbf{i} + y\,\mathbf{j} + z\,\mathbf{k}$ and find its magnitude.

8.2 Use the vector law of addition to prove that the diagonals of a parallelogram bisect one another.

8.3 If $\mathbf{a}_0$ and $\mathbf{b}_0$ are the position vectors of the points (1,2,3) and (3,2,1) relative to the origin, show that the lines corresponding to the vectors

$$\mathbf{a} = \mathbf{a}_0 + \lambda(\mathbf{i} - 2\mathbf{j} + 3\mathbf{k}) \quad \text{and} \quad \mathbf{b} = \mathbf{b}_0 + \mu(-3\mathbf{i} + 2\mathbf{j} - \mathbf{k})$$

intersect and find the co-ordinates of the point of intersection.

8.4 (a) Two vectors $\mathbf{a}_1$ and $\mathbf{a}_2$ passing through the origin and with an angle θ between them, have direction cosines a, b, c and α, β, γ, respectively. Show that $\cos\theta = a\alpha + b\beta + c\gamma$.

(b) Find the angle between the position vectors $(3, -4, 0)$ and $(-2, 1, 0)$, and the direction cosines of a vector perpendicular to both.

8.5 If AB is the diameter of a circle centre O and P is any point on the circumference, use vector methods to show that the angle subtended at P by the lines AP and BP is a right-angle.

8.6 (a) Show that the vectors

$$\mathbf{a} = 3\mathbf{i} - 2\mathbf{j} + \mathbf{k}, \quad \mathbf{b} = \mathbf{i} - 3\mathbf{j} + 5\mathbf{k}, \quad \mathbf{c} = 2\mathbf{i} + \mathbf{j} - 4\mathbf{k}$$

form a right-angled triangle.

(b) Use the vector law of addition for the triangle of Figure 8.2 to prove the cosine law for a plane triangle.

8.7 Find a unit vector that is perpendicular to both the vectors

$$\mathbf{a} = \mathbf{i} + \mathbf{j} - \mathbf{k} \quad \text{and} \quad \mathbf{b} = 3\mathbf{i} - 2\mathbf{j} + \mathbf{k}.$$

If $\mathbf{a}$ and $\mathbf{b}$ are two sides of a triangle, what is its area?

8.8 If $\mathbf{v}_i (i = 1, 2, 3, 4)$ are four vectors whose magnitudes are equal to the areas of the faces of a tetrahedron and whose directions are perpendicular to the faces in the outward direction, show that $\sum_i \mathbf{v}_i = \mathbf{0}$.

8.9 The volume of a tetrahedron is given by

$$\tfrac{1}{3} \times \text{area of base} \times \text{height}.$$

Find the volume of the tetrahedron whose four vertices have the Cartesian co-ordinates $A(1, 2, 3)$, $B(-2, 3, 1)$, $C(2, 0, -3)$ and $D(-2, -1, 0)$.

8.10 (a) Evaluate the triple vector product $\mathbf{a} \times (\mathbf{b} \times \mathbf{c})$, where

$$\mathbf{a} = \mathbf{i} - 2\mathbf{j} + \mathbf{k}, \quad \mathbf{b} = 2\mathbf{i} + \mathbf{j} + 2\mathbf{k}, \quad \text{and} \quad \mathbf{c} = 2\mathbf{i} + \mathbf{j} - \mathbf{k}.$$

(b) A vector $\mathbf{r}$ satisfies the equations

$$\mathbf{r} \times \mathbf{a} = \mathbf{b} \times \mathbf{a}, \quad \mathbf{r} \cdot \mathbf{c} = 0, \quad (1)$$

where $\mathbf{a}$, $\mathbf{b}$, $\mathbf{c}$ are fixed vectors, with $\mathbf{a} \cdot \mathbf{c} \neq 0$. Solve for $\mathbf{r}$ in terms of $\mathbf{a}$, $\mathbf{b}$, $\mathbf{c}$.

8.11 A particle of mass m is attached to a rigid body and is rotating as shown in Figure 8.12. If $\mathbf{v}$ is the velocity of $\mathbf{r}$, and $\mathbf{r}$ and $\boldsymbol{\omega}$ are perpendicular, use the results of Section 8.2.3 to show: (a) the magnitude of its angular momentum L is given by $L = mvr$, where $v = |\mathbf{v}|$; and (b) its acceleration $\mathbf{a}$, given by $\mathbf{a} = \boldsymbol{\omega} \times (\boldsymbol{\omega} \times \mathbf{r})$, is directed towards the centre of the circle with magnitude $a = v^2/r$.

8.12 A force $\mathbf{F} = 2\mathbf{i} + 3\mathbf{j}$ acts at a point $P(3, 1, 3)$. Find the moment $\boldsymbol{\tau}$: (a) about the origin $(0, 0, 0)$; (b) about the point $A(1, -1, 2)$; (c) about the z-axis; (d) about the line from the origin to the point $(0, 1, 1)$.

***8.13** Show that if $\mathbf{a}$, $\mathbf{b}$ and $\mathbf{c}$ are an orthogonal set of vectors, then the reciprocal vectors $\mathbf{a}'$, $\mathbf{b}'$ and $\mathbf{c}'$ are also an orthogonal set.

***8.14** If $\mathbf{a}'$, $\mathbf{b}'$ and $\mathbf{c}'$ are reciprocal to the vectors $\mathbf{a}$, $\mathbf{b}$ and $\mathbf{c}$, then the definition (8.34) implies that $\mathbf{a}$, $\mathbf{b}$ and $\mathbf{c}$ are reciprocal to $\mathbf{a}'$, $\mathbf{b}'$ and $\mathbf{c}'$. Confirm this by showing that

$$\mathbf{a}'' \equiv \frac{\mathbf{b}' \times \mathbf{c}'}{[\mathbf{a}'\mathbf{b}'\mathbf{c}']} = \mathbf{a}.$$

***8.15** (a) Construct the reciprocal vectors of $\mathbf{a} = 3\mathbf{i} - \mathbf{k}$, $\mathbf{b} = \mathbf{j} + \mathbf{k}$ and $\mathbf{c} = \mathbf{i} + 2\mathbf{j}$.

(b) In crystals that have the so-called 'face-centred cubic' structure, the positions of the atoms are often specified using the basis vectors

$$\mathbf{a} = \frac{a}{2}(\mathbf{i} + \mathbf{j}), \quad \mathbf{b} = \frac{a}{2}(\mathbf{j} + \mathbf{k}), \quad \mathbf{c} = \frac{a}{2}(\mathbf{i} + \mathbf{k}),$$

where a is a constant depending on the interatomic spacing. Find the corresponding reciprocal vectors.

8.16 If $\mathbf{a}$, $\mathbf{b}$, $\mathbf{c}$ and $\mathbf{d}$ are constant vectors, and λ and μ are scalar parameters, show that the lines $\mathbf{v}_1 = \mathbf{a} + \lambda \mathbf{c}$ and $\mathbf{v}_2 = \mathbf{b} + \mu \mathbf{d}$ intersect if $\mathbf{d} \cdot (\mathbf{b} \times \mathbf{c}) = \mathbf{d} \cdot (\mathbf{a} \times \mathbf{c})$. Find the parameters λ and μ at the point of intersection in terms of $\mathbf{a}$, $\mathbf{b}$, $\mathbf{c}$ and $\mathbf{d}$.

8.17 Find the shortest distance from the point $P(3, 2, 4)$ and the line that passes through the points $A(1, 0, -1)$ and $B(-1, 2, 1)$. Find the point D on this line that is closest to P.

8.18 Find the forms of the surfaces whose equations are: (a) $|\mathbf{r} - \mathbf{a}| = \lambda$, (b) $(\mathbf{r} - \mathbf{a}) \cdot \mathbf{a} = 0$, (c) $|\mathbf{r} - (\mathbf{r} \cdot \hat{\mathbf{n}})\hat{\mathbf{n}}| = \lambda$, where $\mathbf{a}$ is the position vector of a fixed point (a_x, a_y, a_z), $\mathbf{r}$ is the position vector of a variable point (x, y, z), $\hat{\mathbf{n}}$ is a fixed unit vector and λ is a scalar constant.

8.19 Find the equation in Cartesian co-ordinates for the plane perpendicular to the vector $\mathbf{n} = \mathbf{i} - 2\mathbf{j} - 3\mathbf{k}$ and passing through a point A whose position vector is $\mathbf{a} = \mathbf{i} + 3\mathbf{j} - 4\mathbf{k}$. Find also the shortest distance from the origin to the plane.

8.20 Find an equation in Cartesian co-ordinates for the plane determined by the points $P_1(1, -2, 2)$, $P_2(3, 2, -1)$, $P_3(-1, 3, -1)$.

8.21 Find the angle of intersection of the planes $x + 2y + 3z + 4 = 0$ and $2x + 3y + 4z + 5 = 0$ and the equation of their line of intersection in vector form.

8.22 Find a unit tangent vector to the curve defined by

$$x = t^2 + 1, \quad y = 4t - 3, \quad z = 2t^2 - 6t$$

at the point where $t = 2$.

8.23 Evaluate the integral

$$\mathbf{I} = \int \left(\mathbf{a} \times \frac{\mathrm{d}^2\mathbf{a}}{\mathrm{d}t^2} \right) \mathrm{d}t.$$

9

Determinants, Vectors and Matrices

In Chapter 8, we introduced vectors as objects associated with a direction in everyday three-dimensional space and showed how they can be discussed using equations for their three components in a given reference frame. Here we shall show how to extend the number of components to define vectors in spaces of more than three dimensions. This leads to the introduction of matrices, which are two-dimensional arrays that enable vectors to be transformed into other vectors. The properties of matrices are discussed in detail and their uses illustrated in, for example, solving simultaneous linear equations. In the following chapter we continue the discussion of matrices, with applications to vibrating systems and to geometry. Firstly, however, we study related quantities called determinants, which will play a crucial role in this development.

9.1 Determinants

These occur in many contexts and we have already met examples in the discussion of vectors in Chapter 8. From (8.16b), the vector product of two vectors **a** and **b** in Cartesian co-ordinates has an x-component $(a_y b_z - a_z b_y)$. Any four quantities $a_{ij}(i, j = 1, 2)$ combined in this way can be written in the form of a square array, denoted by Δ_2, called a *determinant.* This is written in the form

$$\Delta_2 = \begin{vmatrix} a_{11} & a_{12} \\ a_{21} & a_{22} \end{vmatrix} \equiv a_{11}a_{22} - a_{12}a_{21}, \tag{9.1}$$

where the quantities $a_{ij}(i, j = 1, 2)$ are called the *elements* of the determinant. For example,

$$\begin{vmatrix} 2 & 3 \\ -1 & -4 \end{vmatrix} = 2 \times (-4) - 3 \times (-1) = -5.$$

Mathematics for Physicists, First Edition. B.R. Martin and G. Shaw.

Companion website: www.wiley.com/go/martin/mathsforphysicists

The result, in this case -5, is called the value of the determinant. It is important to note that the vertical bars in (9.1) do not mean that a modulus is to be taken, as this example confirms. Although we have used real numbers for the elements in this example, in general they can be algebraic expressions, real or complex, so the value of the determinant may also be real or complex expressions or numbers.

Determinants of larger dimensionality can also be constructed. Thus the 3×3 determinant

$$\Delta_3 = \begin{vmatrix} a_{11} & a_{12} & a_{13} \\ a_{21} & a_{22} & a_{23} \\ a_{31} & a_{32} & a_{33} \end{vmatrix} \tag{9.2a}$$

is defined as

$$\Delta_3 \equiv a_{11}(a_{22}a_{33} - a_{23}a_{32}) - a_{12}(a_{21}a_{33} - a_{23}a_{31}) + a_{13}(a_{21}a_{32} - a_{22}a_{31}). \tag{9.2b}$$

Comparing this with (8.18), we see that the triple scalar product of three vectors **a**, **b**, **c**

$$\mathbf{a} \cdot (\mathbf{b} \times \mathbf{c}) = \begin{vmatrix} a_x & a_y & a_z \\ b_x & b_y & b_z \\ c_x & c_y & c_z \end{vmatrix} \tag{9.3a}$$

is a determinant whose elements are the Cartesian components of the vectors. Likewise, comparing (9.2a) with (8.16a) shows that the vector product of two vectors **a** and **b** can also be written as a 3×3 determinant

$$\mathbf{a} \times \mathbf{b} = \begin{vmatrix} \mathbf{i} & \mathbf{j} & \mathbf{k} \\ a_x & a_y & a_z \\ b_x & b_y & b_z \end{vmatrix}. \tag{9.3b}$$

The two compact forms (9.3a) (9.3b) are probably the easiest way of remembering the expressions (8.18) and (8.16a) for the triple scalar product and vector product, respectively.

Returning to (9.2b), we see that the terms in brackets on the right-hand side are themselves 2×2 determinants. Hence we can write

$$\Delta_3 = a_{11} \begin{vmatrix} a_{22} & a_{23} \\ a_{32} & a_{33} \end{vmatrix} - a_{12} \begin{vmatrix} a_{21} & a_{23} \\ a_{31} & a_{33} \end{vmatrix} + a_{13} \begin{vmatrix} a_{21} & a_{22} \\ a_{31} & a_{32} \end{vmatrix}$$

where the determinants that occur on the right-hand side are examples of *minors*. In general, the minor m_{ij} of any element a_{ij} of Δ_3 is the 2×2 determinant obtained by deleting all the elements in the ith row and jth column of Δ_3. Therefore (9.2b) can be written

$$\Delta_3 = \sum_{j=1}^{3} (-1)^{1+j} a_{1j} m_{1j} = \sum_{j=1}^{3} a_{1j} A_{1j}, \tag{9.4a}$$

where the *co-factor* of any element a_{ij} is defined by

$$A_{ij} \equiv (-1)^{i+j} m_{ij}. \tag{9.4b}$$

Equation (9.4a) is called the *Laplace expansion* along the first row of Δ_3. For example, the minors of the elements along the first row of the determinant

$$\Delta = \begin{vmatrix} 1 & 3 & -1 \\ 2 & 0 & 4 \\ 2 & -2 & 1 \end{vmatrix} \tag{9.5}$$

are

$$m_{11} = \begin{vmatrix} 0 & 4 \\ -2 & 1 \end{vmatrix}, \quad m_{12} = \begin{vmatrix} 2 & 4 \\ 2 & 1 \end{vmatrix}, \quad m_{13} = \begin{vmatrix} 2 & 0 \\ 2 & -2 \end{vmatrix},$$

so that (9.4a) gives

$$\begin{aligned} \Delta &= 1\begin{vmatrix} 0 & 4 \\ -2 & 1 \end{vmatrix} - 3\begin{vmatrix} 2 & 4 \\ 2 & 1 \end{vmatrix} - 1\begin{vmatrix} 2 & 0 \\ 2 & 2 \end{vmatrix} \\ &= 1(0+8) - 3(2-8) - 1(-4-0) = 30. \end{aligned}$$

Laplace expansions can be made along any row or column. For example, the expression in (9.2b) can be rearranged to give

$$\begin{aligned} \Delta_3 = &-a_{21}(a_{12}a_{33} - a_{13}a_{32}) + a_{22}(a_{11}a_{33} - a_{13}a_{31}) \\ &- a_{23}(a_{11}a_{32} - a_{12}a_{31}), \end{aligned}$$

which is the Laplace expansion

$$\Delta_3 = \sum_{j=1}^{3} (-1)^{2+j} a_{2j} m_{2j} = \sum_{j-1}^{3} a_{2j} A_{2j}$$

along the second row. Using this expansion for the determinant (9.5) gives

$$\begin{aligned} \begin{vmatrix} 1 & 3 & -1 \\ 2 & 0 & -4 \\ 2 & -2 & 1 \end{vmatrix} &= -2\begin{vmatrix} 3 & -1 \\ -2 & 1 \end{vmatrix} + 0\begin{vmatrix} 1 & -1 \\ 2 & 1 \end{vmatrix} = 4\begin{vmatrix} 1 & 3 \\ 2 & -2 \end{vmatrix} \\ &= -2(3-2) + 0(1+2) + 4(-2-6) = 30, \end{aligned}$$

in agreement with the value obtained by expanding along the first row. Alternatively (9.2b) can be written in the form

$$\Delta_3 = \sum_{i=1}^{3} (-1)^{i+3} a_{i3} m_{i3} = \sum_{i=1}^{3} a_{i3} A_{i3},$$

which is a Laplace expansion along the third column.

The definition of a determinant can now be extended to integers $n > 3$ by generalising the Laplace expansion (9.4a) to any n. To do this, we first write an $n \times n$ array

$$\Delta_n = \begin{vmatrix} a_{11} & a_{12} & \cdots & a_{1n} \\ a_{21} & a_{22} & \cdots & a_{2n} \\ \vdots & \vdots & \ddots & \vdots \\ a_{n1} & a_{n2} & \cdots & a_{nn} \end{vmatrix}, \tag{9.6a}$$

where the elements are a_{ij} $(i, j = 1, 2, \ldots, n)$ and the indices i and j again label the rows and columns, respectively. Then, by analogy with the expansion (9.4a) for 3×3 determinants, we define

$$\Delta_n \equiv \sum_{j-1}^{n} (-1)^{1+j} a_{1j} m_{1j} = \sum_{j=1}^{n} a_{1j} A_{1j}, \tag{9.6b}$$

where the minors m_{ij} are again the determinants obtained by deleting all the elements of the ith row and jth column, and the co-factors are given by (9.4b). Since the minors associated with the elements of an $n \times n$ determinant are $(n-1) \times (n-1)$ determinants, (9.6b) defines 4×4 determinants in terms of a sum of 3×3 determinants, and so on. Such higher order determinants are required in, for example, the solution of n simultaneous linear equations, as we shall see in Sections 9.1.2 and 9.4.4.

Example 9.1

Find the vector product $\mathbf{b} \times \mathbf{c}$ and the triple scalar product $\mathbf{a} \cdot (\mathbf{b} \times \mathbf{c})$, where

$$\mathbf{a} = \mathbf{i} + 4\mathbf{j} + \mathbf{k}, \quad \mathbf{b} = -\mathbf{i} + 2\mathbf{j} + 2\mathbf{k}, \quad \mathbf{c} = 2\mathbf{i} - \mathbf{k}.$$

Solution

Using (9.3b),

$$\mathbf{b} \times \mathbf{c} = \begin{vmatrix} \mathbf{i} & \mathbf{j} & \mathbf{k} \\ -1 & 2 & 2 \\ 2 & 0 & -1 \end{vmatrix},$$

and using the Laplace expansion for the first row gives

$$\mathbf{b} \times \mathbf{c} = \mathbf{i} \begin{vmatrix} 2 & 2 \\ 0 & -1 \end{vmatrix} - \mathbf{j} \begin{vmatrix} -1 & 2 \\ 2 & -1 \end{vmatrix} + \mathbf{k} \begin{vmatrix} -1 & 2 \\ 2 & 0 \end{vmatrix} = -2\mathbf{i} + 3\mathbf{j} - 4\mathbf{k}.$$

Taking the scalar product with $\mathbf{a}$ gives

$$\mathbf{a} \cdot (\mathbf{b} \times \mathbf{c}) = -2 + 12 - 4 = 6.$$

Alternatively, the same result may be obtained by using (9.3a), that is, directly evaluating the determinant

$$\mathbf{a}\cdot(\mathbf{b}\times\mathbf{c}) = \begin{vmatrix} 1 & 4 & 1 \\ -1 & 2 & 2 \\ 2 & 0 & -1 \end{vmatrix}.$$

Example 9.2

Evaluate the determinant

$$\Delta^{(a)} = \begin{vmatrix} 1 & 2 & 3 \\ 4 & 5 & 6 \\ 7 & 8 & 9 \end{vmatrix}$$

by using the Laplace expansion along (i) the second row and (ii) the first column.

Solution

(i) The Laplace expansion about the second row gives

$$\begin{aligned}\Delta^{(a)} &= -4\begin{vmatrix} 2 & 3 \\ 8 & 9 \end{vmatrix} + 5\begin{vmatrix} 1 & 3 \\ 7 & 9 \end{vmatrix} - 6\begin{vmatrix} 1 & 2 \\ 7 & 8 \end{vmatrix} \\ &= -4(18-24) + 5(9-21) - 6(8-14) = 0.\end{aligned}$$

(ii) About the first column, the Laplace expansion gives

$$\begin{aligned}\Delta^{(a)} &= 1\begin{vmatrix} 5 & 6 \\ 8 & 9 \end{vmatrix} - 4\begin{vmatrix} 2 & 3 \\ 8 & 9 \end{vmatrix} + 7\begin{vmatrix} 2 & 3 \\ 5 & 6 \end{vmatrix} \\ &= (45-48) - 4(18-24) + 7(12-15) = 0.\end{aligned}$$

9.1.1 General properties of determinants

The evaluation of determinants using the Laplace expansion involves the arithmetical operations of addition, subtraction and multiplication, the number of which increases rapidly as the dimensionality of the determinant increases. The work involved can sometimes be reduced by exploiting a number of general properties of determinants that are given below.

Although these results hold in general, here we will only consider the case for 3×3 determinants. In this case it is convenient to define the totally antisymmetric symbol ε_{ijk} as follows:

$$\varepsilon_{ijk} \equiv \begin{cases} 1 & \text{if } ijk \text{ is a cyclic permutation of 1, 2, 3} \\ -1 & \text{if } ijk \text{ is a non-cyclic permutation of 1, 2, 3} \\ 0 & \text{if two or more indices are equal} \end{cases} \tag{9.7}$$

where cyclic permutations were defined following (8.16b). Using (9.7), Eqs. (9.2a) and (9.2b) may be written

$$\Delta_3 = \sum_{i,j,k} \varepsilon_{ijk}\, a_{1i}\, a_{2j}\, a_{3k}, \tag{9.8}$$

where we have used a shorthand notation for a sum over three dummy indices i, j and k, which may each take the values 1, 2 and 3, i.e.

$$\sum_{i,j,k} \varepsilon_{ijk}\, a_{1i}\, a_{2j}\, a_{3k} = \sum_{i=1}^{3}\sum_{j=1}^{3}\sum_{k=1}^{3} \varepsilon_{ijk}\, a_{1i}\, a_{2j}\, a_{3k}$$

The theorems are as follows.

(i) *The value of a determinant is unchanged by interchanging (called transposing) its rows and columns.*

This corresponds to the transformation $a_{ij} \to a_{ji}$ for i, j equal to 1, 2 and 3. Using the notation in (9.8) and denoting the new determinant by Δ_3^T, this gives

$$\begin{aligned}\Delta_3^T &= \sum_{i,j,k} \varepsilon_{ijk}\, a_{i1}\, a_{j2}\, a_{k3}\\ &= a_{11}(a_{22}a_{33} - a_{23}a_{32}) - a_{21}(a_{12}a_{33} - a_{13}a_{32})\\ &\quad + a_{31}(a_{12}a_{23} - a_{13}a_{22}).\end{aligned}$$

Rearranging the right-hand side gives

$$\begin{aligned}\Delta_3^T &= a_{11}(a_{22}a_{33} - a_{23}a_{32}) - a_{12}(a_{21}a_{33} - a_{23}a_{31})\\ &\quad + a_{13}(a_{21}a_{32} - a_{22}a_{31}) = \Delta_3.\end{aligned}$$

It follows that theorems about rows also apply to columns, so it is sufficient to prove them only for the former.

(ii) *The sign of a determinant is reversed by interchanging any two of its rows (or columns).*

This result again follows directly from (9.8). For example, interchanging the first row and second column gives

$$\Delta_3 = \sum_{i,j,k} \varepsilon_{ijk}\, a_{1i}\, a_{2j}\, a_{3k} \to \sum_{i,j,k} \varepsilon_{ijk}\, a_{2i}\, a_{1j}\, a_{3k},$$

and using the definition (9.7),

$$\sum_{i,j,k} \varepsilon_{ijk}\, a_{2i}\, a_{1j}\, a_{3k} = -\sum_{i,j,k} \varepsilon_{jik}\, a_{1j}\, a_{2i}\, a_{3k} = -\Delta_3.$$

(iii) *The value of a determinant is zero if any two rows (or columns) are identical.*

This follows immediately from the preceding result, because this interchange gives $\Delta_3 = -\Delta_3$ and hence $\Delta_3 = 0$.

(iv) *If the elements of any one row (or column) are multiplied by a common factor, the value of the determinant is multiplied by this factor.*

This follows trivially, because each term in (9.8) contains a single element from each row (or column).
Using these theorems, a number of other useful results may be established as follows.

(v) *If any two rows (or columns) have proportional elements, the value of the determinant is zero.*

(vi) *If the elements of any row (or column) are the sums or differences of two or more terms, the determinant may be written as the sum or difference of two or more determinants.*

(vii) *The value of a determinant is unchanged if equal multiples of the elements of any row (or column) are added to the corresponding elements of any other row (or column).*

These properties can often be used to manipulate a determinant into a form that is easier to evaluate. For example, consider the determinant

$$\Delta = \begin{vmatrix} 99 & 18 & 63 \\ 15 & -1 & 9 \\ 4 & -3 & 2 \end{vmatrix}.$$

The elements of the first row are all multiples of 9, which can therefore be factored out to give

$$\Delta = 9\begin{vmatrix} 11 & 2 & 7 \\ 15 & -1 & 9 \\ 4 & -3 & 2 \end{vmatrix}.$$

Then by property (vii) we can add row 3 to row 1 without changing the value of the determinant, when we obtain

$$r_1 \to r_1 + r_3, \quad \Delta = 9\begin{vmatrix} 15 & -1 & 9 \\ 15 & -1 & 9 \\ 4 & -3 & 2 \end{vmatrix} = 0$$

because a determinant with two equal rows has a value zero [property (iii)].

In other cases, property (vii) can often be used to manipulate a determinant into a form where it has one or more zeros in a given row or column. Then if this row or column is used in the Laplace expansion, the number of arithmetic operations can be reduced considerably. Consider the evaluation of the determinant

$$\Delta = \begin{vmatrix} 1 & 2 & 1 & -1 \\ 1 & 3 & 5 & -3 \\ -3 & 2 & -7 & -5 \\ 1 & 4 & 4 & 5 \end{vmatrix}.$$

In this case, one way of proceeding is to add column 4 to each of columns 1 and 3, and add twice column 4 to column 2, when we obtain

$$\left.\begin{array}{l} c_1 \to c_1 + c_4 \\ c_2 \to c_2 + 2c_4 \\ c_3 \to c_3 + c_4 \end{array}\right\} \quad \Delta = \begin{vmatrix} 0 & 0 & 0 & -1 \\ -2 & -3 & 2 & -3 \\ -8 & -8 & -12 & -5 \\ 6 & 14 & 9 & 5 \end{vmatrix}$$

Making a Laplace expansion along the first row gives

$$\Delta = \begin{vmatrix} -2 & -3 & 2 \\ -8 & -8 & -12 \\ 6 & 14 & 9 \end{vmatrix} = 4 \begin{vmatrix} -2 & -3 & 2 \\ -2 & -2 & -3 \\ 6 & 14 & 9 \end{vmatrix}.$$

Then subtracting row 2 from row 1 gives

$$\begin{aligned} r_1 \to r_1 - r_2, \quad \Delta &= 4 \begin{vmatrix} 0 & -1 & 5 \\ -2 & -2 & -3 \\ 6 & 14 & 9 \end{vmatrix} \\ &= 4 \begin{vmatrix} -2 & -3 \\ 6 & 9 \end{vmatrix} + 20 \begin{vmatrix} -2 & -2 \\ 6 & 14 \end{vmatrix} = -320. \end{aligned}$$

The Laplace expansion is most suited for determinants of low dimensionality (i.e., small values of n) and where in numerical calculations the elements do not differ much in magnitude. For large-dimensional determinants, the final result may still be formed from the addition and subtraction of many terms, each of which is itself the product of several elements. In these cases there is a significant probability of inaccuracies being introduced in numerical calculations due to rounding errors, particularly if the elements differ considerably in magnitude. Special computer programs exist[1] that address this problem, and are capable of evaluating determinants exactly.

[1] Such as *Mathematica* – see for example: *mathworld.wolfram.com*.

Example 9.3

Evaluate the determinant

$$\Delta^{(b)} = \begin{vmatrix} 7 & 5 & 14 \\ -2 & 1 & 6 \\ 9 & 8 & 4 \end{vmatrix}.$$

Solution

Labelling the rows and columns by r and c, we have

$$r_1 \to r_1 - r_2 - r_3 \text{ gives } \Delta^{(b)} = \begin{vmatrix} 0 & -4 & 4 \\ -2 & 1 & 6 \\ 9 & 8 & 4 \end{vmatrix},$$

and then

$$c_2 \to c_2 + c_3 \text{ gives } \Delta^{(b)} = \begin{vmatrix} 0 & 0 & 4 \\ -2 & 7 & 6 \\ 9 & 12 & 4 \end{vmatrix} = 4\begin{vmatrix} 0 & 0 & 1 \\ -2 & 7 & 6 \\ 9 & 12 & 4 \end{vmatrix},$$
$$= 4(-24 - 63) = -348.$$

9.1.2 Homogeneous linear equations

We have seen that determinants appear naturally when manipulating vectors. They also appear in the theory of simultaneous linear equations. If there are n simultaneous linear equations in n unknowns $x_i (i = 1, 2, \ldots, n)$, they may be written in the general form,

$$\begin{aligned} a_{11}x_1 + a_{12}x_2 + \cdots + a_{1n}x_n &= b_1 \\ a_{21}x_1 + a_{22}x_2 + \cdots + a_{2n}x_n &= b_2 \\ \vdots \quad\quad \vdots \quad\quad\quad\quad \vdots \quad &\quad \vdots \\ a_{n1}x_1 + a_{n2}x_2 + \cdots + a_{nn}x_n &= b_n \end{aligned} \tag{9.9}$$

where the a_{ij} $(i = 1, n;\ j = 1, n)$ and b_j $(j = 1, n)$ are constants. These equations are not necessarily compatible. In the general case where the b_j are not all zero, the equations are called *inhomogeneous,* and their solution will be discussed in Section 9.4.4. In the simpler *homogeneous* case, where all the constants b_j are zero, the equations are never inconsistent, because they always have a so-called *trivial solution* where all the x_i are zero. But they may also have *non-trivial solutions*, where not all the x_i are zero. Because the equations are linear and homogeneous, it follows that if a non-trivial solution exists for a particular set of values $x_i (i = 1, 2, \ldots, n)$, then the set $cx_i (i = 1, 2, \ldots, n)$, where c is a constant, is also a solution. Thus non-trivial solutions are characterised by the ratios $x_1 : x_2 : x_3 : \cdots : x_n$, rather than by unique values.

We will examine below how to find non-trivial solutions, using initially the example of $n = 3$, that is, the set of equations

$$\begin{aligned} a_{11}x_1 + a_{12}x_2 + a_{13}x_3 &= 0 \\ a_{21}x_1 + a_{22}x_2 + a_{23}x_3 &= 0 \\ a_{31}x_1 + a_{32}x_2 + a_{33}x_3 &= 0 \end{aligned} \tag{9.10}$$

which has an associated determinant of coefficients

$$\Delta = \begin{vmatrix} a_{11} & a_{12} & a_{13} \\ a_{21} & a_{22} & a_{23} \\ a_{31} & a_{32} & a_{33} \end{vmatrix}.$$

The value of this determinant determines whether or not a non-trivial solution exists.

An obvious way to proceed is to use the third equation in (9.10) to give an expression for x_3 in terms of x_2 and x_1, then substitute this into the other two equations and examine the two resulting equations in x_1 and x_2 to see if they have compatible solutions. However, this is algebraically rather cumbersome and rapidly becomes very tedious if one considers more than three equations.

Instead, we will use another method, in which the key result follows from the equation

$$\begin{aligned} &(a_{11}A_{11} + a_{21}A_{21} + a_{31}A_{31})x_1 \\ &\quad + (a_{12}A_{11} + a_{22}A_{21} + a_{32}A_{31})x_2 \\ &\qquad + (a_{13}A_{11} + a_{23}A_{21} + a_{33}A_{31})x_3 = 0 \end{aligned} \tag{9.11a}$$

obtained by multiplying the first equation in (9.10) by the co-factor A_{11}, the second by A_{21}, and the third by A_{31}, and adding the three resulting equations together. The first term in brackets in (9.11a) is seen to be the Laplace expansion of Δ using the first column, and so has the value Δ. On comparing the second bracket with the first, we see that it is the Laplace expansion of a determinant in which the first column a_{11}, a_{21}, a_{31} of Δ has been replaced by a_{12}, a_{22}, a_{32}. Hence

$$(a_{12}A_{11} + a_{22}A_{21} + a_{32}A_{31}) = \begin{vmatrix} a_{12} & a_{12} & a_{13} \\ a_{22} & a_{22} & a_{23} \\ a_{32} & a_{32} & a_{33} \end{vmatrix} = 0$$

because two columns are identical. The third bracket in (9.11a) vanishes for a similar reason, so that (9.11a) reduces to

$$x_1\Delta = 0 \tag{9.11b}$$

and therefore $x_1 = 0$ unless $\Delta = 0$. Analogous arguments show that $x_2\Delta = x_3\Delta = 0$, so a necessary condition for a non-trivial solution to (9.10) is

$$\Delta = \begin{vmatrix} a_{11} & a_{12} & a_{13} \\ a_{21} & a_{22} & a_{23} \\ a_{31} & a_{32} & a_{33} \end{vmatrix} = 0. \tag{9.12}$$

Furthermore, if we substitute

$$x_1 : x_2 : x_3 = A_{11} : A_{12} : A_{13} \tag{9.13a}$$

into (9.10), we see that the left-hand sides of the three equations (9.10) equal the three terms in brackets in (9.11a), which have all been shown to vanish for $\Delta = 0$. Hence (9.13a) is the desired non-trivial solution and (9.12) is both a necessary and sufficient condition for it to exist. A similar argument shows that the solution can equally well be expressed in the form

$$x_1 : x_2 : x_3 = A_{21} : A_{22} : A_{23} = A_{31} : A_{32} : A_{33}. \tag{9.13b}$$

In contrast to the direct method of solution, the above chain of reasoning can be extended in a straightforward way to solve n homogeneous linear equations for any integer n. The condition for a non-trivial solution then becomes

$$\begin{vmatrix} a_{11} & a_{12} & \cdots & a_{1n} \\ a_{21} & a_{22} & \cdots & a_{2n} \\ \vdots & \vdots & \ddots & \vdots \\ a_{n1} & a_{n2} & \cdots & a_{nn} \end{vmatrix} = 0, \tag{9.14}$$

and provided this is satisfied, the non-trivial solution is given by the co-factors, i.e.,

$$x_1 : \ x_2 : \ldots : x_n = A_{11} : A_{12} : \ldots : A_{1n}. \tag{9.15}$$

Finally, we note that for the case $n = 3$, the homogeneous equations (9.10) have a simple geometrical interpretation if we interpret x_1, x_2 and x_3 as Cartesian co-ordinates x, y and z. On comparing to (1.51), we see that the three equations (9.10) are those of three planes passing through the origin. Hence the line of intersection of two of these planes, assuming they are not identical, will also pass through the origin. If this line lies in the plane described by the third equation, then any point on it is a solution to all three equations (9.10). In this case, there is a non-trivial solution given by (9.13a), which is indeed the equation of a straight line through the origin, as can be seen by comparing with (8.40). On the other hand, if it does not lie in the plane described by the third equation, then it just passes through that plane at the origin and there is no non-trivial solution to all three equations.

Example 9.4

Find the values of λ for which the simultaneous equations

$$\begin{aligned} 2y + z &= 0 \\ (1+\lambda)x - y - 2z &= 0 \\ 4x + \lambda y - z &= 0 \end{aligned}$$

have a non-trivial solution, and express y and z in terms of x in each case.

Solution

The necessary condition is

$$\begin{vmatrix} 0 & 2 & 1 \\ 1+\lambda & -1 & -2 \\ 4 & \lambda & -1 \end{vmatrix} = 0,$$

which expanding the determinant gives

$$-2\begin{vmatrix} 1+\lambda & -2 \\ 4 & -1 \end{vmatrix} + \begin{vmatrix} 1+\lambda & -1 \\ 4 & \lambda \end{vmatrix} = \lambda^2 + 3\lambda - 10 = 0,$$

with solutions $\lambda = 2$ and $\lambda = -5$. Using $\lambda = 2$ in (9.13a) gives

$$x : y : z = +\begin{vmatrix} -1 & -2 \\ 2 & -1 \end{vmatrix} : -\begin{vmatrix} 3 & -2 \\ 4 & -1 \end{vmatrix} : +\begin{vmatrix} 3 & -1 \\ 4 & 2 \end{vmatrix} = 5 : -5 : 10,$$

so the solution is $y = -x, z = 2x$ for any x. Similarly, using $\lambda = 5$ gives

$$\begin{aligned} x : y : z &= +\begin{vmatrix} -1 & -2 \\ -5 & -1 \end{vmatrix} : -\begin{vmatrix} -4 & -2 \\ 4 & -1 \end{vmatrix} : +\begin{vmatrix} -4 & -1 \\ 4 & -5 \end{vmatrix} \\ &= -9 : -12 : 24, \end{aligned}$$

so the solution is $y = 4x/3,\ z = -8x/3$ for any x.

9.2 Vectors in *n* Dimensions

In Chapter 8, three-dimensional vectors were defined as mathematical quantities having magnitude and direction and satisfying the parallelogram law of addition. This approach is a geometrical one and is independent of the co-ordinate system. We also developed an algebraic approach using basis vectors $(\mathbf{i}, \mathbf{j}, \mathbf{k})$ in the directions of the x, y, z axes of a three-dimensional Cartesian co-ordinate system.

Any vector $\mathbf{a}$ could then be specified by its components a_x, a_y, a_z along the directions of the basis vectors, i.e.

$$\mathbf{a} = a_x\mathbf{i} + a_y\mathbf{j} + a_z\mathbf{k},$$

or equivalently $\mathbf{a} = (a_x, a_y, a_z)$. The basis vectors are not unique (for example, we could rotate the three axes through a fixed angle and use these new directions to define new basis vectors) but they are *linearly independent.* This means that there is no linear combination of them that vanishes, unless the coefficients are all zero. That is,

$$\alpha\mathbf{i} + \beta\mathbf{j} + \gamma\mathbf{k} = \mathbf{0},$$

only if $\alpha = \beta = \gamma = 0$.

In the physical sciences it is common to encounter ordered sets of n quantities $\mathbf{a} = (a_1, a_2, \ldots, a_n), \mathbf{b} = (b_1, b_2, \ldots, b_n)$ etc., whose elements satisfy the same algebraic properties as the components of vectors. In particular, if we define their sums by

$$\mathbf{a} + \mathbf{b} = (a_1 + b_1, a_2 + b_2, \ldots,\ a_n + b_n) \tag{9.16a}$$

and multiplication by a scalar λ by

$$\lambda\mathbf{a} = (\lambda a_1,\ \lambda a_2, \ldots,\ \lambda a_n), \tag{9.16b}$$

then they obey all the general rules (8.1), (8.2) deduced for vectors in Chapter 8. For this reason $(a_1, a_2, \ldots, a_n)$ and $(b_1, b_2, \ldots, b_n)$ are referred to as the components of vectors $\mathbf{a}$ and $\mathbf{b}$ in an *n-dimensional vector space.* In addition, we can define a null vector $\mathbf{0}$, whose n components are all zero, so that for any vector $\mathbf{a}$,

$$0\mathbf{a} = \mathbf{0} \quad \text{and} \quad \mathbf{a} + \mathbf{0} = \mathbf{a}.$$

9.2.1 Basis vectors

Implicit in the choice of the word 'component' to describe $(a_1, a_2, \ldots, a_n), (b_1, b_2, \ldots, b_n)$, etc. is the existence of a set of basis vectors, for example,

$$\begin{aligned} \mathbf{e}_1 &\equiv (1, 0, 0, \ldots, 0), \\ \mathbf{e}_2 &\equiv (0, 1, 0, \ldots 0), \ldots, \mathbf{e}_n \equiv (0, 0, 0, \ldots 1), \end{aligned} \tag{9.17}$$

so that

$$\mathbf{a} = a_1\mathbf{e}_1 + a_2\mathbf{e}_2 + \cdots + a_n\mathbf{e}_n \tag{9.18}$$

in analogy to $\mathbf{a} = a_x\mathbf{i} + a_y\mathbf{j} + a_z\mathbf{k}$ for ordinary three-dimensional vectors. As for the case of ordinary vectors, the choice of basis vectors is not unique, and we can equally well expand the vector $\mathbf{a}$ in terms of any set of basis vectors $\mathbf{e}_i (i = 1, 2, \ldots, n)$, providing the latter are linearly independent, that is, provided that

$$\sum_{i=1}^{n} \mu_i \, \mathbf{e}_i = \mu_1 \mathbf{e}_1 + \mu_2 \mathbf{e}_2 + \cdots + \mu_n \mathbf{e}_n = 0 \tag{9.19a}$$

has no solutions for the constants μ_i except

$$\mu_i = 0, \qquad i = 1, 2, \ldots, n. \tag{9.19b}$$

This ensures that none of the basis vectors can be expressed in terms of the others, and, in general, a vector space is said to be n-dimensional if it contains no linearly independent set of vectors within it with more than n members. Such a set of n linearly independent vectors is called a *complete set*. It also guarantees the uniqueness of the expansion (9.18). This is easily seen by writing

$$\mathbf{a} = \sum_{i=1}^{n} \tilde{a}_i \mathbf{e}_i$$

and equating this to (9.18) gives

$$\sum_{i=1}^{n} (a_i - \tilde{a}_i) \, \mathbf{e}_i = \mathbf{0},$$

which from (9.19) has no solution other than $a_i = \tilde{a}_i$ for all $i = 1, 2, \ldots, n$. Of course the components $(a_1, a_2, \ldots, a_n)$ will depend on the particular basis vectors chosen, and $(a_1, a_2, \ldots, a_n)$ is said to be a *representation* of $\mathbf{a}$ in the basis $\mathbf{e}_i (i = 1, 2, \ldots, n)$.

In what follows, we will need to relate the components a_i in a given representation (9.18) to the components a_i' in a representation

$$\mathbf{a} = a_1' \mathbf{e}'_1 + a_2' \mathbf{e}'_2 + \cdots + a_n' \mathbf{e}'_n \tag{9.20}$$

defined with respect to a different set of basis vectors where $\mathbf{e}'_i$ $(i = 1, 2, \cdots, n)$. To do this, we note that any vector in the space can be written in the form (9.18), including the new basis vectors $\mathbf{e}'_i$. Hence we can write

$$\mathbf{e}'_j = \sum_{i=1}^{n} p_{ij} \mathbf{e}_i, \qquad i = 1, 2, \ldots, n \tag{9.21a}$$

where p_{ij} are numerical constants. On substituting (9.21a) into (9.20), we obtain

$$\mathbf{a} = \sum_{i=1}^{n} a_i' \mathbf{e}'_i = \sum_{i=1}^{n} \sum_{j=1}^{n} a_j' p_{ij} \mathbf{e}_i.$$

This is only compatible with (9.18) for arbitrary vectors **a** if

$$a_i = \sum_{j=1}^{n} p_{ij}\, a'_j, \qquad i = 1, 2, \ldots, n, \tag{9.21b}$$

which is the desired relation.

9.2.2 Scalar products

The components of vectors need not be restricted to real quantities. Complex vectors in an arbitrary number of dimensions play an important role in, for example, quantum mechanics. Generalising the vectors and scalar variables to complex quantities does not alter any of the equations (8.1), (8.2) or (9.16)–(9.18), but does affect the definition of the scalar product. To distinguish this from the scalar product defined in Chapter 8 for three-dimensional vectors, we will use the notation $(\mathbf{a}, \mathbf{b})$ (also called the *inner product* in this context).

For the moment, we restrict ourselves to the basis (9.17), when the inner product of two vectors $\mathbf{a} = (a_1, a_2, \ldots, a_n)$ and $\mathbf{b} = (b_1, b_2, \ldots, b_n)$ is defined to be

$$(\mathbf{a}, \mathbf{b}) \equiv \sum_{i=1}^{n} a_i^*\, b_i. \tag{9.22}$$

It reduces to the scalar (dot) product defined in Chapter 8 for the case of real coefficients and ensures that the squared length

$$(\mathbf{a}, \mathbf{a}) \equiv |\mathbf{a}\,|^2 = |a_1|^2 + |a_2|^2 + \cdots + |a_n|^2$$

remains real and positive. This leads to the basic properties

$$(\mathbf{a}, [\mathbf{b} + \mathbf{c}]) = (\mathbf{a}, \mathbf{b}) + (\mathbf{a}, \mathbf{c}) \text{ distributive law of addition,} \tag{9.23a}$$

$$(\mathbf{a}, \lambda\mathbf{b}) = \lambda(\mathbf{a}, \mathbf{b}) \text{ scalar multiplication,} \tag{9.23b}$$

$$(\mathbf{a}, \mathbf{b}) = (\mathbf{b}, \mathbf{a})^* \text{ complex conjugation,} \tag{9.23c}$$

from which it follows that

$$([\lambda\mathbf{a} + \mu\mathbf{b}], \mathbf{c}) = \lambda^*(\mathbf{a}, \mathbf{c}) + \mu^*(\mathbf{b}, \mathbf{c}), \tag{9.23d}$$

and

$$(\lambda\mathbf{a}, \mu\mathbf{b}) = \lambda^*\mu(\mathbf{a}, \mathbf{b}), \tag{9.23e}$$

where λ and μ are both in general complex constants. Note that these relations reduce to the corresponding relations (8.8a), (8.8b) and (8.8c) for the real vectors discussed in Chapter 8 when λ, μ and the vectors themselves are real. In particular, we see from (9.23c) that the scalar product is only commutative for real vectors.

We can now apply the general properties (9.23a)–(9.23e) to a general basis (9.18). In doing so, we will assume that the chosen basis satisfies the orthonormality relations [cf. (8.11)]

$$(\mathbf{e}_i, \mathbf{e}_j) = \delta_{ij}, \tag{9.24a}$$

where δ_{ij} is the *Kronecker delta symbol*, defined by

$$\delta_{ij} \equiv \begin{cases} 1 & i = j \\ 0 & i \neq j \end{cases}. \tag{9.24a}$$

Then using (9.23) repeatedly we have

$$\begin{aligned}(\mathbf{a}, \mathbf{b}) &= ([a_1\mathbf{e}_1 + a_2\mathbf{e}_2 + \cdots + a_n\mathbf{e}_n], [b_1\mathbf{e}_1 + b_2\mathbf{e}_2 + \cdots + b_n\mathbf{e}_n]) \\ &= \sum_{i,j} a_i^* b_j (\mathbf{e}_i, \mathbf{e}_j) = \sum_i a_i^* b_i,\end{aligned}$$

using (9.24). Thus the expression (9.22) holds in all bases (9.18) provided the orthonormality relations (9.24) are satisfied. Furthermore, using (9.18) and (9.24) we have

$$(\mathbf{e}_i, \mathbf{a}) = \sum_{j=1}^{n} (\mathbf{e}_i, a_j\mathbf{e}_j) = a_i,$$

i.e. the vector $\mathbf{a}$ is given by

$$\mathbf{a} = \sum_i (\mathbf{e}_i, \mathbf{a})\mathbf{e}_i. \tag{9.25}$$

Example 9.5

Prove the *Schwarz inequality,*

$$|(\mathbf{a}, \mathbf{b})| \leq |\mathbf{a}|\,|\mathbf{b}|, \tag{9.26}$$

where $\mathbf{a}$ and $\mathbf{b}$ are two arbitrary vectors.

Solution

If $\mathbf{b} = \alpha\,\mathbf{a}$, where α is a constant, then one easily shows by direct substitution that

$$|(\mathbf{a}, \mathbf{b})| = |\mathbf{a}|\,|\mathbf{b}|,$$

so that (9.26) is satisfied. If $\mathbf{b} \neq \alpha\,\mathbf{a}$ for any constant α, then

$$\mathbf{c} = \mathbf{a} - \lambda(\mathbf{b}, \mathbf{a})\mathbf{b}$$

cannot be a null vector, where λ is any constant. Hence

$$\begin{aligned}(\mathbf{c}, \mathbf{c}) &= ([\mathbf{a} - \lambda(\mathbf{a}, \mathbf{b})\mathbf{b}], [\mathbf{a} - \lambda(\mathbf{b}, \mathbf{a})\mathbf{b}]) \\ &= (\mathbf{a}, \mathbf{a}) - 2\lambda(\mathbf{a}, \mathbf{b})(\mathbf{b}, \mathbf{a}) + \lambda^2(\mathbf{a}, \mathbf{b})(\mathbf{b}, \mathbf{a})(\mathbf{b}, \mathbf{b}) > 0,\end{aligned}$$

implying that

$$|\mathbf{a}|^2 - 2\lambda\,|(\mathbf{a},\mathbf{b})|^2 + \lambda^2\,|(\mathbf{a},\mathbf{b})|^2\,|\mathbf{b}|^2 = 0$$

has no solutions for real λ. Using (2.40) and (2.6) we see that the condition for this is

$$|(\mathbf{a},\mathbf{b})|^2 < |\mathbf{a}|^2\,|\mathbf{b}|^2 .$$

Finally, taking square roots on both sides gives

$$|(\mathbf{a},\mathbf{b})| < |\mathbf{a}|\,|\mathbf{b}| .$$

9.3 Matrices and linear transformations

In this section we introduce matrices and discuss their role in transforming vectors into other vectors.

9.3.1 Matrices

Consider the set of linear simultaneous equations

$$\begin{array}{ccccccc} a_{11}x_1 & + a_{12}x_2 & + & \cdots & + a_{1n}x_n & = & y_1 \\ a_{21}x_1 & + a_{22}x_2 & + & \cdots & + a_{2n}x_n & = & y_2 \\ \vdots & \vdots & & & \vdots & & \vdots \\ a_{m1}x_1 & + a_{m2}x_2 & + & \cdots & + a_{mn}x_n & = & y_m \end{array} \tag{9.27}$$

where the coefficients $a_{ij}(i = 1, 2, \ldots, m;\ j = 1, 2, \ldots, n)$ are constants. These equations determine m variables $y_i(i = 1, 2, \ldots, m)$ in terms of n given variables $x_j(j = 1, 2, \ldots, n)$, where the integers m and n are not necessarily equal. It is convenient to write (9.27) in a form that separates the variables x_j from the coefficients a_{ij} as follows:

$$\begin{pmatrix} a_{11} & a_{12} & \cdots & a_{1n} \\ a_{21} & a_{22} & \cdots & a_{2n} \\ \vdots & \vdots & \ddots & \vdots \\ a_{m1} & a_{m2} & \cdots & a_{mn} \end{pmatrix} \begin{pmatrix} x_1 \\ x_2 \\ \vdots \\ x_n \end{pmatrix} = \begin{pmatrix} y_1 \\ y_2 \\ \vdots \\ y_m \end{pmatrix}. \tag{9.28}$$

This array of coefficients is called a *matrix* and the quantities a_{ij} are called the *elements* of the matrix. It is said to be of *order* $m \times n$ because it has m rows and n columns. The vertical arrays $y_i(i = 1, 2, \ldots, m)$ and $x_j(j = 1, 2, \ldots, n)$ are also matrices, in this case of order $m \times 1$ and $n \times 1$. They are referred to as *column matrices,* or *column vectors.* Likewise, matrices of order $1 \times n$ are referred to as *row matrices*, or *row vectors*. On comparing (9.28) with (9.27), we see that each of the $y_i(i = 1, 2, \ldots, m)$ is obtained by multiplying the element in the ith row of the $m \times n$ matrix by the numbers $x_j(j = 1, 2, \ldots, n)$ in turn and adding, so that

$$y_i = \sum_{j=1}^{n} a_{ij}x_j, \quad j = 1, 2, \ldots, m. \tag{9.29}$$

For example, if

$$a_{ij} = \begin{pmatrix} 2 & -1 & 1 \\ 1 & 3 & 2 \end{pmatrix} \quad \text{and} \quad x_j = \begin{pmatrix} -2 \\ 4 \\ 1 \end{pmatrix},$$

then

$$y_1 = 2(-2) - 1(4) + 1(1) = -7 \text{ and } y_2 = 1(-2) + 3(4) + 2(1) = 12.$$

So far we have merely rewritten (9.27) in the different, but equivalent, form (9.28). The usefulness of this form results from developing rules for manipulating matrices directly. In doing this, it is convenient to denote matrices by upper-case bold Roman letters **A**, **B**, **C**, etc., with the exception that both row and column vectors are denoted by lower-case bold Roman letters **a**, **b**, **c**, etc. Thus, (9.28) may be written in the compact form

$$\mathbf{y} = \mathbf{A}\,\mathbf{x}. \tag{9.30}$$

Matrix algebra is then defined by the following rules.

(i) *Equality*
Two matrices **A**, with elements a_{ij}, and **B**, with elements b_{ij}, are equal, if, and only if, they are of the same order $m \times n$, and $a_{ij} = b_{ij}$ for all $i = 1, 2, \ldots, m$ and $j = 1, 2, \ldots, n$.

(ii) *Addition*
The sum **S** of two matrices **A** and **B** is defined if, and only if, they have the same order. The elements of **S** are then given by

$$(\mathbf{S})_{ij} = (\mathbf{A} + \mathbf{B})_{ij} = a_{ij} + b_{ij}, \quad i = 1, 2, \ldots, m; \quad j = 1, 2, \ldots, n. \tag{9.31}$$

This leads directly to the commutative and associative laws

$$\mathbf{A} + \mathbf{B} = \mathbf{B} + \mathbf{A} \tag{9.32a}$$

and

$$\mathbf{A} + (\mathbf{B} + \mathbf{C}) = (\mathbf{A} + \mathbf{B}) + \mathbf{C}, \tag{9.32b}$$

respectively.

(iii) *Scalar multiplication*
If a matrix **A** is multiplied by a scalar quantity λ, then every element of **A** is multiplied by λ, i.e.

$$(\lambda \mathbf{A})_{ij} = \lambda\, a_{ij}. \tag{9.33}$$

If λ and μ are arbitrary constants, (9.31)–(9.33) lead to the associative and distributive laws

$$(\lambda\mu)\mathbf{A} = \lambda(\mu\mathbf{A}) = \mu(\lambda\mathbf{A}), \tag{9.34a}$$

$$\lambda(\mathbf{A}+\mathbf{B}) = \lambda\mathbf{A} + \lambda\mathbf{B}, \tag{9.34b}$$

and

$$(\lambda+\mu)\mathbf{A} = \lambda\mathbf{A} + \mu\mathbf{A}, \tag{9.34c}$$

provided again that **A** and **B** are of the same order. In addition, we define *null matrices* **0** of any dimension, whose elements are all zero, so that

$$0\mathbf{A} = \mathbf{0} \text{ and } \mathbf{A} + \mathbf{0} = \mathbf{A}. \tag{9.34d}$$

(iv) *Matrix multiplication*
The product of two matrices **AB** is defined if, and only if, the number of columns in **A** is the same as the number of rows in **B**. Then, if **A** is an $l \times m$ matrix and **B** is an $m \times n$ matrix, the product **AB** is an $l \times n$ matrix whose elements are defined by

$$(\mathbf{AB})_{ik} \equiv \sum_{j=1}^{m} a_{ij}\, b_{jk} \tag{9.35}$$

for all $i = 1, 2, \ldots, l$; $j = 1, 2, \ldots, n$. In other words, the element $(\mathbf{AB})_{ik}$ is obtained by multiplying each element of row i of **A** by the corresponding element of column k of **B**, and adding. For example, if

$$\mathbf{A} = \begin{pmatrix} 2 & 1 & 1 \\ 1 & -2 & 3 \end{pmatrix} \quad \text{and} \quad \mathbf{B} = \begin{pmatrix} -1 & 1 \\ -3 & 2 \\ 1 & -1 \end{pmatrix}, \tag{9.36a}$$

then **AB** is the 2×2 matrix

$$\begin{aligned} \mathbf{AB} &= \begin{pmatrix} 2 & 1 & 1 \\ 1 & -2 & 3 \end{pmatrix} \begin{pmatrix} -1 & 1 \\ -3 & 2 \\ 1 & -1 \end{pmatrix} \\ &= \begin{pmatrix} -2-3+1 & 2+2-1 \\ -1+6+3 & 1-4-3 \end{pmatrix} = \begin{pmatrix} -4 & 3 \\ 8 & -6 \end{pmatrix}. \end{aligned} \tag{9.36b}$$

It is worth noting that, just as for the scalar products of ordinary three-dimensional vectors, $\mathbf{AB} = \mathbf{0} \not\Rightarrow$ either $\mathbf{A} = \mathbf{0}$ or $\mathbf{B} = \mathbf{0}$. For example, if

$$\mathbf{A} = \begin{pmatrix} a & a \\ b & b \end{pmatrix} \text{ and } \mathbf{B} = \begin{pmatrix} -1 & 1 \\ 1 & -1 \end{pmatrix},$$

then

$$\mathbf{AB} = \begin{pmatrix} 0 & 0 \\ 0 & 0 \end{pmatrix} = \mathbf{0},$$

but neither $\mathbf{A}$ nor $\mathbf{B}$ is a null matrix.

To motivate the definition (9.35) and to derive another important relation, let us suppose the n-component column vector $\mathbf{x}$ in (9.30) is related to a p-component column vector $\mathbf{z}$ by

$$\mathbf{x} = \mathbf{B}\,\mathbf{z}, \tag{9.37a}$$

where $\mathbf{B}$ is an $n \times p$ matrix, so that

$$x_j = \sum_{k=1}^{p} b_{jk}\, z_k, \quad j = 1, 2, \ldots, n. \tag{9.37b}$$

Substituting (9.37a) into (9.30) gives

$$\mathbf{y} = \mathbf{A}(\mathbf{B}\,\mathbf{z}). \tag{9.38a}$$

On the other hand, substituting (9.37b) into (9.29), gives

$$y_i = \sum_{j=1}^{n} a_{ij}\, x_j = \sum_{j=1}^{n} \sum_{k=1}^{p} a_{ij}\, b_{jk} x_k,$$

which, on comparing with (9.35), is seen to be

$$y_i = \sum_{k=1}^{p} (\mathbf{AB})_{ik} z_k, \quad i = 1, 2, \ldots, m.$$

Hence $\mathbf{y} = (\mathbf{AB})\mathbf{z}$ and on comparing this with (9.38a), we finally obtain

$$\mathbf{y} = (\mathbf{AB})\mathbf{z} = \mathbf{A}(\mathbf{Bz}). \tag{9.38b}$$

From this we see that the position of the brackets is immaterial and we can write $\mathbf{y} = \mathbf{ABz}$ without ambiguity. By a similar argument one can show that

$$\mathbf{A}(\mathbf{BC}) = (\mathbf{AB})\mathbf{C} = \mathbf{ABC}, \tag{9.39}$$

and so on. However, while the position of brackets in matrix products is not important, the order is crucial, since matrix multiplication is not in general commutative, that is, $\mathbf{AB} \neq \mathbf{BA}$. This is obvious for the multiplication of a $n \times m$ matrix $\mathbf{A}$ and a $m \times n$ matrix $\mathbf{B}$, because the products **AB** and **BA** have different dimensionalities, but it is also true even if $n = m$. Matrix multiplication is however distributive with respect to addition, i.e.

$$\mathbf{A}(\mathbf{B} + \mathbf{C}) = \mathbf{AB} + \mathbf{AC} \tag{9.40a}$$

and

$$(\mathbf{B} + \mathbf{C})\mathbf{A} = \mathbf{BA} + \mathbf{CA}, \tag{9.40b}$$

but (9.40a) and (9.40b) are not in general identical.

Example 9.6

Consider the matrices

$$\mathbf{A} = \begin{pmatrix} 1 & 2 \\ 3 & 4 \end{pmatrix} \qquad \mathbf{B} = \begin{pmatrix} 2 & 1 \\ 4 & 3 \end{pmatrix}$$

$$\mathbf{C} = \begin{pmatrix} 1 & -1 & 4 \\ 2 & 3 & 5 \end{pmatrix} \quad \mathbf{D} = \begin{pmatrix} 2 & 2 \\ 1 & -1 \\ 0 & 4 \end{pmatrix}.$$

(a) Which of the following additions are defined: $\mathbf{A} + \mathbf{B}, \mathbf{A} + \mathbf{C}, \mathbf{C} + \mathbf{D}$?

(b) Evaluate the matrix $\mathbf{A} + 2\mathbf{B}$.

(c) Which of the following products are defined: $\mathbf{AB}, \mathbf{BA}, \mathbf{AC}, \mathbf{CA}, \mathbf{AD}, \mathbf{DA}$?

(d) Evaluate **BA**, where **A** and **B** are the matrices (9.36a) and compare it with the product **AB** given in (9.36b).

Solution

(a) The addition of two matrices is only defined if the number of rows and columns of the two matrices are equal. Thus only $\mathbf{A} + \mathbf{B}$ is defined; $\mathbf{A} + \mathbf{C}$ and $\mathbf{C} + \mathbf{D}$ are undefined.

(b)

$$\mathbf{A} + 2\mathbf{B} = \begin{pmatrix} 1 & 2 \\ 3 & 4 \end{pmatrix} + 2\begin{pmatrix} 2 & 1 \\ 4 & 3 \end{pmatrix}$$

$$= \begin{pmatrix} 1 & 2 \\ 3 & 4 \end{pmatrix} + \begin{pmatrix} 4 & 2 \\ 8 & 6 \end{pmatrix} = \begin{pmatrix} 5 & 4 \\ 11 & 10 \end{pmatrix}.$$

(c) The products of two matrices is only defined if the number of columns in the first matrix is equal to the number of rows in the second matrix. Thus only the products $\mathbf{AB}, \mathbf{BA}, \mathbf{AC}, \mathbf{DA}$ are defined; **CA** and **AD** are undefined.

(d) The number of rows in **B** matches the number of columns in **A** so the product **BA** is defined and is given by

$$\mathbf{BA} = \begin{pmatrix} -1 & 1 \\ -3 & 2 \\ 1 & -1 \end{pmatrix} \begin{pmatrix} 2 & 11 \\ 1 & -23 \end{pmatrix} = \begin{pmatrix} 1 & -3 & 2 \\ -4 & -7 & 3 \\ 1 & 3 & -2 \end{pmatrix}.$$

Comparing with (9.36b), we see that **BA** does not even have the same dimensions as **AB**.

9.3.2 Linear transformations

Column matrices are special cases of $m \times n$ matrices with $n = 1$ and are written with the second index suppressed, that is, we write them with a single row index. For example,

$$\mathbf{a} = \begin{pmatrix} a_1 \\ a_2 \\ \vdots \\ a_n \end{pmatrix} \text{ and } \mathbf{b} = \begin{pmatrix} b_1 \\ b_2 \\ \vdots \\ b_n \end{pmatrix}. \tag{9.41}$$

With this convention, for any two column matrices **a** and **b**, (9.31) and (9.33) reduce to

$$(\mathbf{a} + \mathbf{b})_i = a_i + b_i \tag{9.42a}$$

and

$$(\lambda\, a_i) = \lambda\, a_i. \tag{9.42b}$$

These relations are identical to (9.16a) and (9.16b) used to characterise the components of an n-dimensional vector in Section 9.2. Similarly, the matrix relations (9.32)–(9.34) reduce to the vector relations (8.1) and (8.2) when applied to column matrices. Hence column matrices are with justification referred to as column vectors. The scalar product of a vector **a** with a vector **b** is also easily expressed in matrix notation, since the product of a row vector and a column vector of the same order n is given by

$$(a_1, a_2, \ldots, a_n) \begin{pmatrix} b_1 \\ b_2 \\ \vdots \\ b_n \end{pmatrix} = \sum_{i=1}^{n} a_i b_i.$$

Comparing this with (9.22), we see that in an orthogonal basis, the scalar product is

$$(\mathbf{a}, \mathbf{b}) = \mathbf{a}^\dagger\, \mathbf{b}, \tag{9.43}$$

where the row vector $\mathbf{a}^\dagger$ corresponding to the column vector $\mathbf{a}$ is defined by

$$\mathbf{a}^\dagger \equiv (a_1^*, a_2^*, \cdots, a_n^*) \tag{9.44}$$

and is called the Hermitian conjugate of $\mathbf{a}$ for reasons that will become clear in Section 9.3.3.

Returning to (9.30), we now interpret the matrix $\mathbf{A}$ as a matrix *operator* that transforms an n-dimensional vector $\mathbf{x}$ into an m-dimensional vector $\mathbf{y}$. By an operator we mean anything that acts on the object to its right, called the *operand*, to give a new object. Furthermore, it is easy to show, using (9.29) and (9.42) that

$$\mathbf{A}(\lambda\,\mathbf{a} + \mu\mathbf{b}) = \lambda\mathbf{A}\mathbf{a} + \mu\mathbf{A}\mathbf{b}, \tag{9.45}$$

where λ and μ are arbitrary constants and $\mathbf{a}$, $\mathbf{b}$ are arbitrary vectors. Any operator that satisfies an equation of the form (9.45) is called a *linear operator* and, correspondingly, (9.30) is called a *linear transformation.* Another linear operator, which we will meet in Chapter 10, is the *differential operator* $D \equiv \mathrm{d}/\mathrm{d}x$, which transforms a function $f(x)$ into its derivative. Thus,

$$D\,f(x) = \frac{\mathrm{d}f(x)}{\mathrm{d}x} = f'(x), \tag{9.46a}$$

where the linearity condition

$$D\,[\lambda f_1(x) + \mu f_2(x)] = \lambda D\,f_1(x) + \mu D\,f_2(x) \tag{9.46b}$$

follows directly from (3.19).

Linear operators and transformations are widely used in mathematics and physical science. Here we shall confine ourselves to matrix operators. A simple example is provided by considering a position vector in two dimensions,

$$\mathbf{r} = x\,\mathbf{i} + y\,\mathbf{j} = r\cos\phi\,\mathbf{i} + r\sin\phi\,\mathbf{j}. \tag{9.47}$$

When rotated through an angle θ, this gives a new position vector

$$\mathbf{r}' = x'\,\mathbf{i} + y'\mathbf{j} = r\cos(\phi+\theta)\mathbf{i} + r\sin(\phi+\theta)\mathbf{j}$$

of the same length r, as shown in Figure 9.1. Using the trigonometric identities (2.36), we have

$$\begin{aligned} x' &= r\cos(\phi+\theta) = r\cos\phi\cos\theta - r\sin\phi\sin\theta \\ &= x\cos\theta - y\sin\theta, \end{aligned}$$

and similarly

$$y' = x\sin\theta + y\cos\theta.$$

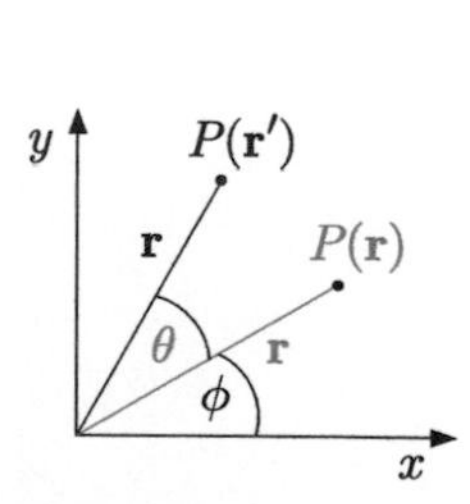

Figure 9.1 The rotation of the two-dimensional vector (9.47) through an angle θ.

Hence in matrix notation,

$$\begin{pmatrix} x' \\ y' \end{pmatrix} = \begin{pmatrix} \cos\theta & -\sin\theta \\ \sin\theta & \cos\theta \end{pmatrix} \begin{pmatrix} x \\ y \end{pmatrix}, \tag{9.48}$$

or equivalently,

$$\mathbf{r}' = \mathbf{R}(\theta)\mathbf{r}, \tag{9.49}$$

where the *rotation matrix*

$$\mathbf{R}(\theta) = \begin{pmatrix} \cos\theta & -\sin\theta \\ \sin\theta & \cos\theta \end{pmatrix}. \tag{9.50}$$

Finally, we consider the product of two transformation matrices **A** and **B**. Equation (9.38b) implies

$$\mathbf{ABz} = \mathbf{A}(\mathbf{Bz})$$

so that the transformation **AB** is equivalent to the operator **B** acting first, followed by the operator **A**. In other words, the operator on the right acts first, and if **A** acts before **B**, the appropriate operator is $\mathbf{BA} \neq \mathbf{AB}$, since in general matrices do not commute.

Example 9.7

A two-dimensional vector $\mathbf{r}\,(x, y)$ undergoes an expansion of its x component represented by a matrix **E**, i.e.

$$\mathbf{Er} \equiv \begin{pmatrix} \lambda & 0 \\ 0 & 1 \end{pmatrix} \begin{pmatrix} x \\ y \end{pmatrix} = \begin{pmatrix} \lambda x \\ y \end{pmatrix},$$

followed by a rotation $\mathbf{R}(\theta)$ given by (9.50). Find the matrix operator describing the overall transformation, and the corresponding matrix for which the order of the transformations is reversed.

Solution

If the expansion occurs before the rotation, the overall transformation matrix is

$$\mathbf{RE} = \begin{pmatrix} \cos\theta & -\sin\theta \\ \sin\theta & \cos\theta \end{pmatrix} \begin{pmatrix} \lambda & 0 \\ 0 & 1 \end{pmatrix} = \begin{pmatrix} \lambda\cos\theta & -\sin\theta \\ \lambda\sin\theta & \cos\theta \end{pmatrix}.$$

In contrast, if the vector is first rotated, the overall transformation matrix is

$$\mathbf{ER} = \begin{pmatrix} \lambda & 0 \\ 0 & 1 \end{pmatrix} \begin{pmatrix} \cos\theta & -\sin\theta \\ \sin\theta & \cos\theta \end{pmatrix} = \begin{pmatrix} \lambda\cos\theta & -\lambda\sin\theta \\ \sin\theta & \cos\theta \end{pmatrix}.$$

Hence the result depends on the order in which the two operations occur.

9.3.3 Transpose, complex, and Hermitian conjugates

Given a matrix $\mathbf{A}$ with elements a_{ij}, it is useful to define three related matrices, as follows.

(i) The *transpose* of $\mathbf{A}$, denoted $\mathbf{A}^{\mathrm{T}}$, is obtained by interchanging rows and columns. An example is

$$\mathbf{A} = \begin{pmatrix} 1 & -1 & 3 \\ 2 & 1 & 4 \end{pmatrix} \;\Rightarrow\; \mathbf{A}^{\mathrm{T}} = \begin{pmatrix} 1 & 2 \\ -1 & 1 \\ 3 & 4 \end{pmatrix},$$

while the general relation is

$$a_{ij}^{\mathrm{T}} = a_{ji}. \tag{9.51}$$

It follows from this that

$$(\mathbf{AB})^{\mathrm{T}} = \mathbf{B}^{\mathrm{T}}\mathbf{A}^{\mathrm{T}} \tag{9.52}$$

since

$$(\mathbf{AB})_{ij}^{\mathrm{T}} = (\mathbf{AB})_{ji} = \sum_k a_{jk}b_{ki} = \sum_k b_{ik}^{\mathrm{T}}\, a_{kj}^{\mathrm{T}} = \mathbf{B}^{\mathrm{T}}\mathbf{A}^{\mathrm{T}}.$$

In general, the transpose of a product of matrices is the product of the individual transposed matrices taken in reverse order. Thus,

$$(\mathbf{ABC})^{\mathrm{T}} = \mathbf{C}^{\mathrm{T}}(\mathbf{AB})^{\mathrm{T}} = \mathbf{C}^{\mathrm{T}}\mathbf{B}^{\mathrm{T}}\mathbf{A}^{\mathrm{T}},$$

and so on, which follows by repeated application of (9.52).

(ii) The *complex conjugate* of a matrix $\mathbf{A}$ is denoted $\mathbf{A}^*$ and has elements a_{ij}^*. Complex conjugation has no effect on the order in products, i.e.

$$(\mathbf{ABC}\ldots)^* = \mathbf{A}^*\mathbf{B}^*\mathbf{C}^*\ldots.$$

The *Hermitian conjugate*[2] of a matrix $\mathbf{A}$, written $\mathbf{A}^\dagger$, is defined as the transpose of the complex conjugate matrix, or vice versa, i.e.

$$\mathbf{A}^\dagger \equiv (\mathbf{A}^*)^{\mathrm{T}} = (\mathbf{A}^{\mathrm{T}})^* \tag{9.53a}$$

so that[3]

$$a_{ij}^\dagger = a_{ji}^*. \tag{9.53b}$$

[2] The Hermitian conjugate is sometimes called the *adjoint*. We reserve the latter term for a different matrix, to be defined in Section 9.4.3.

[3] We have already met (9.53b) for the special case of a column matrix $\mathbf{a}$ in Section 9.3.2 [cf. (9.44)].

Since Hermitian conjugation involves a transpose, it also reverses the order of products, i.e.

$$(\mathbf{ABC}\ldots)^\dagger = \ldots \mathbf{C}^\dagger\mathbf{B}^\dagger\mathbf{A}^\dagger \tag{9.54}$$

For a real matrix, the Hermitian conjugate is just the transpose.

Example 9.8
Given the matrices

$$\mathbf{A} = \begin{pmatrix} 1 & 2 & 3 \\ 2 & 3 & 2 \\ 1 & 0 & 1 \\ 2 & 1 & 0 \end{pmatrix}, \quad \mathbf{C} = \begin{pmatrix} 1 & 2 & 3 \\ 2 & 2 & 0 \end{pmatrix},$$

find $\mathbf{AC}^{\mathrm{T}}$.

Solution

$$\mathbf{C}^{\mathrm{T}} = \begin{pmatrix} 1 & 2 \\ 2 & 2 \\ 3 & 0 \end{pmatrix} \quad \text{and so} \quad \mathbf{AC}^{\mathrm{T}} = \begin{pmatrix} 14 & 6 \\ 14 & 10 \\ 4 & 2 \\ 4 & 6 \end{pmatrix}.$$

9.4 Square Matrices

Matrices with the same number of rows and columns are called *square matrices,* and their dimension $n = m$ is called their *order*. We discuss here some of the most important types of square matrices that will be required in later sections.

9.4.1 Some special square matrices

(i) *Diagonal matrix*
A matrix $\mathbf{A}$ is *diagonal* if its elements a_{ij} are zero unless they lie on the *leading diagonal* $i = j$, so that $a_{ij} = a_i\delta_{ij}$, where δ_{ij} is the Kronecker delta symbol of (9.24b). The sum of the elements along this diagonal is called the *trace*, denoted Tr. As an exception to the general rule, diagonal matrices of the same order commute under multiplication, that is, $\mathbf{AB} = \mathbf{BA}$ if $\mathbf{A}$ and $\mathbf{B}$ are both diagonal. An important example of a diagonal matrix is the *unit matrix* $\mathbf{I}$ defined by

$$\mathbf{I} \equiv \begin{pmatrix} 1 & 0 & \cdots & 0 \\ 0 & 1 & \cdots & 0 \\ \vdots & \vdots & \ddots & \vdots \\ 0 & 0 & \cdots & 1 \end{pmatrix} \tag{9.55}$$

which has the property

$$\mathbf{IA} = \mathbf{AI} = \mathbf{A} \tag{9.56}$$

for any matrix $\mathbf{A}$ (not necessarily diagonal) of the same order.

(ii) *Symmetric and anti-symmetric matrices*
A matrix is *symmetric* if it satisfies the condition $\mathbf{A} = \mathbf{A}^{\mathrm{T}}$, i.e. $a_{ij} = a_{ji}$, and *anti-symmetric* (or *skew symmetric*) if $\mathbf{A} = -\mathbf{A}^{\mathrm{T}}$, i.e. $a_{ij} = -a_{ji}$, where $\mathbf{A}^{\mathrm{T}}$ is the transpose of $\mathbf{A}$. Any matrix $\mathbf{A}$ may be expressed as the sum of a symmetric and an anti-symmetric matrix, by analogy with the decomposition of functions as the sum of symmetric and anti-symmetric functions, as discussed in Section 1.3.1. Thus

$$\mathbf{A} = \tfrac{1}{2}(\mathbf{A} + \mathbf{A}^{\mathrm{T}}) + \tfrac{1}{2}(\mathbf{A} - \mathbf{A}^{\mathrm{T}}),$$

where by construction the first bracket is a symmetric matrix and the second is anti-symmetric.

(iii) *Hermitian matrix*
A matrix is *Hermitian*, if it satisfies $\mathbf{A} = \mathbf{A}^{\dagger}$, where the dagger indicates the combined operation of complex conjugation and transposition, carried out in either order, that is, if $a^{\dagger}_{ij} = (a_{ji})^* = a_{ij}$. If $\mathbf{A}^{\dagger} = -\mathbf{A}$, the matrix $\mathbf{A}$ is said to be *anti-Hermitian* (or *skew Hermitian*). Any complex matrix can be expressed as the sum of a Hermitian matrix and an anti-Hermitian matrix. Thus,

$$\mathbf{A} = \tfrac{1}{2}(\mathbf{A} + \mathbf{A}^{\dagger}) + \tfrac{1}{2}(\mathbf{A} - \mathbf{A}^{\dagger})$$

where by construction the first bracket is a Hermitian matrix and the second is anti-Hermitian. A real, symmetric matrix is automatically Hermitian, because $\mathbf{A}^{\dagger} = \mathbf{A}^{\mathrm{T}}$ in this case.

(iv) *Unitary matrix*
A matrix $\mathbf{U}$ is said to be *unitary* if it satisfies

$$\mathbf{UU}^{\dagger} = \mathbf{U}^{\dagger}\mathbf{U} = \mathbf{I} \quad \Rightarrow \quad \mathbf{U}^{-1} = \mathbf{U}^{\dagger}. \tag{9.57a}$$

If we make the *unitary transformation*

$$\mathbf{x}' = \mathbf{Ux},$$

on a vector $\mathbf{x}$, then by (9.43) and (9.57a),

$$(\mathbf{x}', \mathbf{x}') = (\mathbf{Ux}, \mathbf{Ux}) = \mathbf{x}^{\dagger}\mathbf{U}^{\dagger}\mathbf{Ux} = \mathbf{x}^{\dagger}\mathbf{x} = (\mathbf{x}, \mathbf{x}),$$

so that the length of the vector is unchanged.

(v) *Orthogonal matrix*
An *orthogonal* matrix $\mathbf{O}$ is a real unitary matrix. It therefore also leaves the length of a vector unchanged and (9.57a) becomes

$$\mathbf{OO}^{\mathrm{T}} = \mathbf{O}^{\mathrm{T}}\mathbf{O} = \mathbf{I} \quad \Rightarrow \quad = \mathbf{O}^{-1} = \mathbf{O}^{\mathrm{T}}. \tag{9.57b}$$

Example 9.9

Find the transpose, complex conjugate and Hermitian conjugate of the matrix

$$\mathbf{A} = \begin{pmatrix} 1+i & 2-i \\ 3i & 2 \end{pmatrix},$$

and hence decompose **A** into a sum of symmetric and antisymmetric matrices, and a sum of Hermitian and anti-Hermitian matrices.

Solution

The transpose, complex conjugate and Hermitian conjugate matrices are

$$\mathbf{A}^T = \begin{pmatrix} 1+i & 3i \\ 2-i & 2 \end{pmatrix}, \quad \mathbf{A}^* = \begin{pmatrix} 1-i & 2+i \\ -3i & 2 \end{pmatrix}$$

$$\mathbf{A}^\dagger = \begin{pmatrix} 1-i & -3i \\ 2+i & 2 \end{pmatrix}.$$

Therefore, using the decomposition into symmetric and antisymmetric matrices

$$\mathbf{A} = \mathbf{A}_S + \mathbf{A}_{AS} = \tfrac{1}{2}(\mathbf{A} + \mathbf{A}^\mathrm{T}) + \tfrac{1}{2}(\mathbf{A} - \mathbf{A}^\mathrm{T})$$

gives

$$\mathbf{A} = \begin{pmatrix} 1+i & 1+i \\ 1+i & 2 \end{pmatrix} + \begin{pmatrix} 0 & 1-2i \\ -1+2i & 0 \end{pmatrix}.$$

Similarly, using the decomposition into Hermitian and anti-Hermitian matrices

$$\mathbf{A} = \mathbf{A}_H + \mathbf{A}_{AH} = \tfrac{1}{2}(\mathbf{A} + \mathbf{A}^\dagger) + \tfrac{1}{2}(\mathbf{A} - \mathbf{A}^\dagger)$$

gives

$$\mathbf{A} = \begin{pmatrix} 1 & 1-2i \\ 1+2i & 2 \end{pmatrix} + \begin{pmatrix} i & 1+i \\ -1+i & 0 \end{pmatrix}.$$

9.4.2 The determinant of a matrix

Given a square matrix **A** of order n, we can define an associated determinant by

$$\det \mathbf{A} = |\mathbf{A}\,| \equiv \begin{vmatrix} a_{11} & a_{12} & \cdots & a_{1n} \\ a_{21} & a_{22} & \cdots & a_{2n} \\ \vdots & \vdots & \ddots & \vdots \\ a_{n1} & a_{n2} & \cdots & a_{nn} \end{vmatrix}. \tag{9.58}$$

If $\det \mathbf{A} = 0$, the matrix is said to be *singular*; if $\det \mathbf{A} \neq 0$, then **A** is *non-singular*.

The properties of determinants have been summarised in Section 9.1. Since interchanging rows and columns leaves the value of the determinant unchanged, it follows that

$$\det(\mathbf{A}^{\mathrm{T}}) = \det \mathbf{A}. \tag{9.59a}$$

Similarly, since $\det \mathbf{A}^* = (\det \mathbf{A})^*$, we have

$$\det \mathbf{A}^{\dagger} = \det(\mathbf{A}^*)^{\mathrm{T}} = (\det \mathbf{A})^* \tag{9.59b}$$

for the Hermitian conjugate matrix $\mathbf{A}^{\dagger}$. Multiplying a matrix by a scalar constant λ multiplies every element a_i by λ, but since only one member of each row occurs in the determinant, we have

$$\det(\lambda\mathbf{A}) = \lambda^n \det \mathbf{A} \tag{9.60a}$$

for a square matrix of order n. The determinant of a product of matrices is equal to the product of the determinants.

$$\det(\mathbf{AB}) = (\det \mathbf{A})(\det \mathbf{B}). \tag{9.60b}$$

The proof of (9.60b) is rather lengthy and will not be reproduced here[4]. However, it follows from it that

$$\det(\mathbf{AB}) = \det(\mathbf{BA}) \tag{9.60c}$$

and repeated application of (9.60b) leads to

$$\det(\mathbf{ABC}\ldots) = \det \mathbf{A} \det \mathbf{B} \det \mathbf{C}\ldots, \tag{9.60d}$$

for any number of matrices, independent of their order.

Equation (9.60b) also leads to useful results for unitary and orthogonal matrices. Specifically, from (9.57a) and (9.60b), we obtain

$$|\det \mathbf{U}|^2 = 1. \tag{9.61}$$

Hence the determinant of a unitary matrix is either +1 or −1, and since an orthogonal matrix $\mathbf{O}$ is just a real unitary matrix, the same result applies to orthogonal matrices.

A simple example of an orthogonal matrix is the rotation matrix in two dimensions $\mathbf{R}(\theta)$ described in (9.50). One sees that

$$\det \mathbf{R}(\theta) = \cos^2\theta + \sin^2\theta = 1,$$

consistent with (9.61). In contrast, a matrix that generates a reflection in a given axis, for example

$$\begin{pmatrix} x' \\ y' \end{pmatrix} = \begin{pmatrix} -1 & 0 \\ 0 & 1 \end{pmatrix} \begin{pmatrix} x \\ y \end{pmatrix}$$

[4] It may be found in, for example, G. Strang, *Linear Algebra and its Applications*, 3rd edn., Harcourt, Brace, Jovanovich, San Diego, California, 1988, p215.

so that $x' = -x$, $y' = y$, has determinant -1. This behaviour is characteristic of rotations and reflections about any given axis.

Example 9.10

Verify the general relations (9.60b) (9.60c) for the matrices

$$\mathbf{A} = \begin{pmatrix} 0 & 1 \\ 1 & 2 \end{pmatrix} \quad \text{and} \quad \mathbf{B} = \begin{pmatrix} 1 & 3 \\ 2 & 4 \end{pmatrix}$$

by explicitly evaluating the various determinants.

Solution

The determinants of $\mathbf{A}$ and $\mathbf{B}$ are $|\mathbf{A}| = (0 \times 2) - (1 \times 1) = -1$ and $|\mathbf{B}| = (1 \times 4) - (3 \times 2) = -2$, so that $|\mathbf{A}|\,|\mathbf{B}| = 2$. Similarly,

$$\mathbf{AB} = \begin{pmatrix} 0 & 1 \\ 1 & 2 \end{pmatrix}\begin{pmatrix} 1 & 3 \\ 2 & 4 \end{pmatrix} = \begin{pmatrix} 2 & 4 \\ 5 & 11 \end{pmatrix},$$

so that $|\mathbf{AB}| = (2 \times 11) - (4 \times 5) = 2$ and

$$\mathbf{BA} = \begin{pmatrix} 1 & 3 \\ 2 & 4 \end{pmatrix}\begin{pmatrix} 0 & 1 \\ 1 & 2 \end{pmatrix} = \begin{pmatrix} 3 & 7 \\ 4 & 10 \end{pmatrix},$$

so that $|\mathbf{BA}| = (3 \times 10) - (7 \times 4) = 2$, and therefore $|\mathbf{AB}| = |\mathbf{BA}| = |\mathbf{A}|\,|\mathbf{B}|$, as required.

9.4.3 Matrix inversion

We can now complete the discussion of matrix algebra. The operation of division by a matrix is not defined. However, if we can find a matrix $\mathbf{D}$ such that $\mathbf{AD} = \mathbf{DA} = \mathbf{I}$, then $\mathbf{D}$ is called the *inverse* of $\mathbf{A}$ and is written $\mathbf{A}^{-1}$, so that

$$\mathbf{AA}^{-1} = \mathbf{A}^{-1}\mathbf{A} = \mathbf{I}. \tag{9.62}$$

The analogy with division is then multiplication by $\mathbf{A}^{-1}$, so that, for example,

$$\mathbf{AB} = \mathbf{C} \quad \Rightarrow \quad \mathbf{B} = \mathbf{A}^{-1}\mathbf{C}.$$

Equation (9.62) can only be satisfied if $\mathbf{A}$ and $\mathbf{A}^{-1}$ are square matrices of the same order, while (9.60b) then implies

$$\det(\mathbf{A}^{-1}) = (\det \mathbf{A})^{-1},$$

so that a singular matrix (one having $\det \mathbf{A} = 0$) has no inverse, whereas a non-singular matrix does have an inverse. To find the inverse of a matrix $\mathbf{A}$, we need a new matrix called the *adjoint*, denoted $\operatorname{adj}\mathbf{A}$. This is defined as the transpose matrix of the co-factors of $\mathbf{A}$. Thus for the $n \times n$ matrix $\mathbf{A}$, with co-factors A_{ij} corresponding to the element a_{ij}, the adjoint matrix is

$$\text{adj}\mathbf{A} \equiv \begin{pmatrix} A_{11} & A_{21} & \cdots & A_{n1} \\ A_{12} & A_{22} & \cdots & A_{n2} \\ \vdots & \vdots & \ddots & \vdots \\ A_{1n} & A_{2n} & \cdots & A_{nn} \end{pmatrix}, \tag{9.63}$$

from which it follows that

$$(\mathbf{A}\text{adj}\mathbf{A})_{ij} = \sum_k a_{ik}A_{jk} = \delta_{ij} \det \mathbf{A}. \tag{9.64}$$

To see this, we note that for $i = j$, (9.64) is just the Laplace expansion of $\det \mathbf{A}$ along row i; while for $i \neq j$, it is the Laplace expansion of a matrix $\mathbf{A}'$ which differs from $\mathbf{A}$ in that the jth row is replaced by the ith row. Thus we have arrived at the result that the matrix defined by $\mathbf{D} \equiv \text{adj}\mathbf{A}/|\mathbf{A}|$ has the property that $\mathbf{AD} = \mathbf{I}$ and hence $\mathbf{D}$ can be identified with the inverse matrix $\mathbf{A}^{-1}$, i.e.

$$\mathbf{A}^{-1} = \frac{1}{|\mathbf{A}|}\text{adj}\mathbf{A}, \tag{9.65}$$

and $\mathbf{AA}^{-1} = \mathbf{I}$. A similar argument gives $\mathbf{A}^{-1}\mathbf{A} = \mathbf{I}$, and hence (9.62) is satisfied.

Using this result, it is easy to prove that

$$(\mathbf{A}^{-1})^{-1} = \mathbf{A} \tag{9.66a}$$

and

$$(\mathbf{A}^{\mathrm{T}})^{-1} = (\mathbf{A}^{-1})^{\mathrm{T}}, \tag{9.66b}$$

while

$$(\mathbf{ABC}\ldots)^{-1} = (\ldots\mathbf{C}^{-1}\mathbf{B}^{-1}\mathbf{A}^{-1}). \tag{9.66c}$$

For a 2×2 matrix $\mathbf{A}$, (9.65) reduces to

$$\mathbf{A}^{-1} = \mathbf{A}^{\mathrm{T}}(\det \mathbf{A})^{-1}, \tag{9.67}$$

but the evaluation of the inverses of matrices with higher dimensionality can be somewhat tedious. However the computational work needed can be reduced by a process called *row reduction*, or *Gaussian elimination*.

The three elementary operations used in row reductions are:

(i) Multiply any row by a non-zero constant;

(ii) Interchange any two rows;

(iii) Replace any row by the sum (or difference) of itself and any multiple of another row.

Since by the law of matrix multiplication, the identity $\mathbf{A}\mathbf{A}^{-1} = \mathbf{I}$ involves only the rows of $\mathbf{A}$ and the columns of $\mathbf{A}^{-1}$, it follows that the equality is preserved if one applies the same row reductions to $\mathbf{A}$ and the unit matrix; hence if a set of row reductions can be found which transform $\mathbf{A}$ to $\mathbf{I}$, the same set will transform $\mathbf{I}$ to $\mathbf{A}^{-1}$. For example, if

$$\mathbf{A} = \begin{pmatrix} 1 & 0 & 2 \\ 1 & 1 & 0 \\ 0 & 0 & 1 \end{pmatrix},$$

then the row reduction $r_1 \to r_1 - 2r_3$ transforms the first row of $\mathbf{A}$ to (1, 0, 0), and when followed by the reduction $r_2 \to r_2 - r_1$ yields a unit matrix, as follows:

$$\underset{\mathbf{A}}{\begin{pmatrix} 1 & 0 & 2 \\ 1 & 1 & 0 \\ 0 & 0 & 1 \end{pmatrix}} \xrightarrow{r_1 \to r_1 - 2r_3} \begin{pmatrix} 1 & 0 & 0 \\ 1 & 1 & 0 \\ 0 & 0 & 1 \end{pmatrix} \xrightarrow{r_2 \to r_2 - r_1} \underset{\mathbf{I}}{\begin{pmatrix} 1 & 0 & 0 \\ 0 & 1 & 0 \\ 0 & 0 & 1 \end{pmatrix}}$$

Applying the same sequence of reductions to the unit matrix $\mathbf{I}$ gives

$$\underset{\mathbf{I}}{\begin{pmatrix} 1 & 0 & 0 \\ 0 & 1 & 0 \\ 0 & 0 & 1 \end{pmatrix}} \xrightarrow{r_1 \to r_1 - 2r_3} \begin{pmatrix} 1 & 0 & -2 \\ 0 & 1 & 0 \\ 0 & 0 & 1 \end{pmatrix} \xrightarrow{r_2 \to r_2 - r_1} \underset{\mathbf{A}^{-1}}{\begin{pmatrix} 1 & 0 & -2 \\ -1 & 1 & 2 \\ 0 & 0 & 1 \end{pmatrix}}$$

so that

$$\mathbf{A}^{-1} = \begin{pmatrix} 1 & 0 & -2 \\ -1 & 1 & 2 \\ 0 & 0 & 1 \end{pmatrix}$$

The calculations involved in manipulating matrices of large dimensionality can be very tedious and in these cases useful computer programs exist, such as that referenced in footnote 1 in Section 9.1.1. Simpler, but effective, free programs may also be found on the internet.

Example 9.11

(a) Use equation (9.65) to find the inverse of the matrix

$$\mathbf{A} = \begin{pmatrix} 1 & 4 & 1 \\ -1 & 2 & 2 \\ 2 & 0 & -1 \end{pmatrix}.$$

(b) Use the Gaussian elimination method to find the inverse of the matrix

$$\mathbf{B} = \begin{pmatrix} 1 & 0 & -1 \\ -2 & 1 & 1 \\ 0 & -1 & 2 \end{pmatrix}.$$

Solution

(a) From the definition,

$$\mathbf{A}^{-1} \equiv \frac{\text{adj}\mathbf{A}}{\det \mathbf{A}},$$

we have

$$\det \mathbf{A} = 1\begin{vmatrix} 2 & 2 \\ 0 & -1 \end{vmatrix} - 4\begin{vmatrix} -1 & 2 \\ 2 & -1 \end{vmatrix} + 1\begin{vmatrix} -1 & 2 \\ 2 & 0 \end{vmatrix} = 6,$$

and

$$\text{adj}\mathbf{A} = \begin{pmatrix} A_{11} & A_{21} & A_{31} \\ A_{12} & A_{22} & A_{23} \\ A_{13} & A_{23} & A_{33} \end{pmatrix} = \begin{pmatrix} -2 & 4 & 6 \\ 3 & -3 & -3 \\ -4 & 8 & 6 \end{pmatrix},$$

where A_{ij} is the co-factor associated with the element a_{ij} of $\mathbf{A}$. Then

$$\mathbf{A}^{-1} = \frac{1}{6}\begin{pmatrix} -2 & 4 & 6 \\ 3 & -3 & -3 \\ -4 & 8 & 6 \end{pmatrix}.$$

(b) The sequence of row reductions is as follows:

$$\begin{array}{ccccccc}
 & r_3 \to r_3 + 2r_1 + r_2 & & r_2 \to r_2 + 2r_1 + r_3 & & r_1 \to r_1 + r_3 & \\
\mathbf{B} & & & & & & \mathbf{I} \\
\begin{pmatrix} 1 & 0 & -1 \\ -2 & 1 & 1 \\ 0 & -1 & 2 \end{pmatrix} & \to & \begin{pmatrix} 1 & 0 & -1 \\ -2 & 1 & 1 \\ 0 & 0 & 1 \end{pmatrix} & \to & \begin{pmatrix} 1 & 0 & -1 \\ 0 & 1 & 0 \\ 0 & 0 & 1 \end{pmatrix} & \to & \begin{pmatrix} 1 & 0 & 0 \\ 0 & 1 & 0 \\ 0 & 0 & 1 \end{pmatrix} \\
\downarrow & & & & & & \downarrow \\
\mathbf{I} & & & & & & \mathbf{B}^{-1} \\
\begin{pmatrix} 1 & 0 & 0 \\ 0 & 1 & 0 \\ 0 & 0 & 1 \end{pmatrix} & \to & \begin{pmatrix} 1 & 0 & 0 \\ 0 & 1 & 0 \\ 2 & 1 & 1 \end{pmatrix} & \to & \begin{pmatrix} 1 & 0 & 0 \\ 4 & 2 & 1 \\ 2 & 1 & 1 \end{pmatrix} & \to & \begin{pmatrix} 3 & 1 & 1 \\ 4 & 2 & 1 \\ 2 & 1 & 1 \end{pmatrix}
\end{array}$$

from which we deduce that

$$\mathbf{B}^{-1} = \begin{pmatrix} 3 & 1 & 1 \\ 4 & 2 & 1 \\ 2 & 1 & 1 \end{pmatrix}.$$

9.4.4 Inhomogeneous simultaneous linear equations

The n simultaneous linear equations in n unknowns $x_i (i=1, 2, \ldots, n)$ given in (9.9) are conveniently written in matrix form

$$\mathbf{A}\mathbf{x} = \mathbf{b}, \tag{9.68a}$$

where

$$\mathbf{A} = \begin{pmatrix} a_{11} & a_{12} & \cdots & a_{1n} \\ a_{21} & a_{22} & \cdots & a_{2n} \\ \vdots & \vdots & \ddots & \vdots \\ a_{n1} & a_{n2} & \cdots & a_{nn} \end{pmatrix}, \mathbf{x} = \begin{pmatrix} x_1 \\ x_2 \\ \vdots \\ x_n \end{pmatrix}, \mathbf{b} = \begin{pmatrix} b_1 \\ b_2 \\ \vdots \\ b_n \end{pmatrix}. \tag{9.68b}$$

The solution of (9.68) for the *homogeneous* case $\mathbf{b} = \mathbf{0}$ was discussed in Section 9.1.3. Here we consider the *inhomogeneous case*, when $\mathbf{b} \neq \mathbf{0}$. We will also start by assuming that $\mathbf{A}$ is non-singular so that $\mathbf{A}^{-1}$ exists. Then the solution of (9.68) is

$$\mathbf{x} = \mathbf{A}^{-1}\,\mathbf{b}, \;(\det \mathbf{A} \neq 0), \tag{9.69}$$

and the solution is unique. The latter statement follows from assuming there are two solutions, $\mathbf{x}^{(1)}$ and $\mathbf{x}^{(2)}$, so that $\mathbf{A}\mathbf{x}^{(i)} = \mathbf{b}_i (i = 1, 2)$. Then $\mathbf{A}\mathbf{x}^{(1)} = \mathbf{A}\mathbf{x}^{(2)}$, and since $\mathbf{A}$ has an inverse, we may multiple by $\mathbf{A}^{-1}$ to obtain $\mathbf{x}^{(1)} = \mathbf{x}^{(2)}$, as required for the solution to be unique.

The solution of linear simultaneous equations by finding the inverse matrix $\mathbf{A}^{-1}$ can be tedious and it is sometimes simpler to use an alternative method based on *Cramer's rule*, which we now discuss. We will again consider the set of equations (9.68a), which we will write in the form

$$\sum_{k=1}^{n} a_{ik} x_k = b_i, \qquad (i = 1, 2, \ldots, n). \tag{9.70}$$

Multiplying the equation for b_i by A_{ij} and summing over i using (9.64), gives

$$\begin{aligned} \sum_{i=1}^{n} b_i A_{ij} &= \sum_{i=1}^{n}\sum_{k=1}^{n} a_{ik} A_{ij} x_k \\ &= \sum_{k=1}^{n} \det(\mathbf{A})\delta_{kj} x_k = \det(\mathbf{A}) x_j. \end{aligned} \tag{9.71}$$

Hence, provided $\det \mathbf{A} \neq 0$, and setting $\Delta = \det \mathbf{A}$, (9.71) becomes

$$x_j = \frac{1}{\Delta}\sum_{i=1}^{n} b_i A_{ij}, \quad (j = 1, 2, \ldots, n), \tag{9.72a}$$

or equivalently,

$$x_j = \Delta_j/\Delta\,, \qquad (j = 1, 2, \ldots, n), \tag{9.72b}$$

where Δ_j is the determinant obtained by replacing the elements in the jth column of Δ by the elements of the column vector $\mathbf{b}$. Equations (9.72a) and (9.72b) are the combined statement of Cramer's rule.

We now briefly consider the cases where $\mathbf{A}^{-1}$ does not exist, that is, when $\det \mathbf{A} = 0$. There are two possibilities:

(i) If any of the determinants in the numerators of (9.72) are non-zero, then since the determinant in the denominator is $\Delta = \det \mathbf{A} = 0$, no finite solution to the set of equations exists. The equations are said to be *inconsistent*, or *incompatible*.

(ii) If $\Delta = \det(\mathbf{A}) = 0$, but all the determinants in the numerators of (9.72) are also zero, then in general one can show that an infinity of solutions exists.

In the case of three simultaneous equations, these results have a simple geometrical interpretation. For $n = 3$, (9.68b) reduces to the three equations

$$\begin{aligned}
a_{11}x_1 + a_{12}x_2 + a_{13}x_3 &= b_1,\\
a_{21}x_1 + a_{22}x_2 + a_{23}x_3 &= b_2,\\
a_{31}x_1 + a_{32}x_2 + a_{33}x_3 &= b_3,
\end{aligned}$$

and if we interpret x_1, x_2 and x_3 as Cartesian co-ordinates x, y and z, on comparing to (1.51) we see that these are the equations of three planes. Assuming they are not identical, the first two planes will intersect in a straight line. There are then three possibilities. If the line lies in the plane described by the third equation, then any point on it is a solution to all three equations so that there is an infinite number of solutions. This corresponds to case (ii) above. Alternatively, if the line of intersection is parallel to, but not in, the third plane, there is no solution. This corresponds to case (i) above. Finally, if the line of intersection is not parallel to the third plane, it will pass through it at a single point, corresponding to a unique solution.

Example 9.12

Solve the equations

$$\begin{aligned}
x + 2y + z &= 3\\
x - y + 2z &= 2\\
-2y + z &= 4
\end{aligned}$$

by using (a) matrix inversion, and (b) Cramer's rule.

Solution

(a) In the notation of (9.68a),

$$\mathbf{A} = \begin{pmatrix} 1 & 2 & 1 \\ 1 & -1 & 2 \\ 0 & -2 & 1 \end{pmatrix}, \quad \mathbf{x} = \begin{pmatrix} x \\ y \\ z \end{pmatrix}, \quad \mathbf{b} = \begin{pmatrix} 3 \\ 2 \\ 4 \end{pmatrix}.$$

Then,

$$\mathbf{x} = \mathbf{A}^{-1}\mathbf{b}\,, \text{where } \mathbf{A}^{-1} = \begin{pmatrix} -3 & 4 & -5 \\ 1 & -1 & 1 \\ 2 & -2 & 3 \end{pmatrix},$$

so

$$\mathbf{x} = \begin{pmatrix} -3 & 4 & -5 \\ 1 & -1 & 1 \\ 2 & -2 & 3 \end{pmatrix} \begin{pmatrix} 3 \\ 2 \\ 4 \end{pmatrix},$$

and hence $x = -21$, $y = 5$ and $z = 14$.

(b) In the notation of (9.72b),

$$x_j = \Delta_j/\Delta \qquad (j = 1, 2, 3),$$

where $x_1 = x$, $x_2 = y$, $x_3 = z$ and $\Delta = \det \mathbf{A}$, i.e.

$$\Delta = \begin{vmatrix} 1 & 2 & 1 \\ 1 & -1 & 2 \\ 0 & -2 & 1 \end{vmatrix} = -1.$$

The terms Δ_j are found by replacing the elements in the jth column of Δ by the elements of the column vector $\mathbf{b}$. Thus,

$$\Delta_1 = \begin{vmatrix} 3 & 2 & 1 \\ 2 & -1 & 2 \\ 4 & -2 & 1 \end{vmatrix}, \quad \Delta_2 = \begin{vmatrix} 1 & 3 & 1 \\ 1 & 2 & 2 \\ 0 & 4 & 1 \end{vmatrix}, \quad \Delta_3 = \begin{vmatrix} 1 & 2 & 3 \\ 1 & -1 & 2 \\ 0 & -2 & 4 \end{vmatrix},$$

i.e. $\Delta_1 = 21$, $\Delta_2 = -5$, $\Delta_3 = -14$, and so $x = -21$, $y = 5$, $z = 14$.

Problems 9

9.1 The vectors **a**, **b**, **c**, are given by

$$\begin{aligned} \mathbf{a} &= \mathbf{i} + 2\mathbf{j} + 3\mathbf{k}, \\ \mathbf{b} &= 4\mathbf{i} - \mathbf{j} + 2\mathbf{k}, \\ \mathbf{c} &= 5\mathbf{i} - 6\mathbf{j} + 3\mathbf{k}. \end{aligned}$$

Use determinants to evaluate $\mathbf{a} \times \mathbf{b}$ and $\mathbf{b} \cdot \mathbf{a} \times \mathbf{c}$.

9.2 (a) Evaluate the determinant

$$\begin{vmatrix} 2 & 2-i & 3 \\ 1 & i & 4 \\ 1+i & 1 & -2i \end{vmatrix}$$

by using the Laplace expansion about (i) the third column and (ii) the first row.
(b) Use the general properties of a determinant, as stated in Section 9.1.2, to show that the determinant

$$\begin{vmatrix} 27 & 14 & 5 \\ 8 & 3 & -1 \\ 13 & 7 & 3 \end{vmatrix}$$

may be written

$$\begin{vmatrix} 1 & 0 & 0 \\ 5 & 3 & 4 \\ 6 & 7 & 9 \end{vmatrix}$$

and find its value.

9.3 Simplify and hence evaluate the determinant

$$\Delta = \begin{vmatrix} 3 & 7 & 12 \\ 6 & 11 & -11 \\ 8 & 18 & 23 \\ -2 & -7 & 42 \end{vmatrix}.$$

9.4 (a) Solve the equation

$$\Delta_1 = \begin{vmatrix} -1+x & -1 & 1 \\ 1 & 1+x & 1 \\ 1 & 1 & 1+x \end{vmatrix} = 0.$$

(b) Write the determinant

$$\Delta_2 = \begin{vmatrix} 1 & 1 & 1 \\ \alpha & \beta & \gamma \\ \alpha^3 & \beta^3 & \gamma^3 \end{vmatrix}$$

as the product of factors that are linear in α, β, γ.

9.5 The $n \times n$ determinant Δ_n is given by

$$\Delta_n = \begin{vmatrix} -2 & 1 & 0 & 0 & \cdots & 0 \\ 1 & -2 & 1 & 0 & \cdots & 0 \\ 0 & 1 & -2 & 1 & \cdots & 0 \\ \vdots & \vdots & \vdots & \vdots & \vdots & \vdots \\ 0 & 0 & 0 & 0 & 1 & -2 \end{vmatrix}.$$

Establish a recurrence relation for $S_n \equiv \Delta_n + \Delta_{n-1}$ and hence find an explicit formula for Δ_n

9.6 Consider the two sets of homogeneous equations

$$\begin{aligned} \text{(a) } 2x - 5y + 5z &= 0 \\ 4x + y - 2z &= 0 \\ x - 3y + 3z &= 0 \end{aligned} \qquad \begin{aligned} \text{(b) } 3x + 5y + 4z &= 0 \\ x + y + 2z &= 0 \\ 3x + 7y + 2z &= 0 \end{aligned}$$

Determine whether these sets have non-trivial solutions for x, y, z and, if so, find them.

9.7 Find the values of α for which the equations

$$\begin{aligned}\alpha x + 3y - 2 &= 0\\ -3x + \alpha y + (\alpha + 4) &= 0\\ -x + 3y + 4 &= 0\end{aligned}$$

have a unique consistent solution and solve the equations for the larger of these values.

9.8 Given two vectors $\mathbf{a}$ and $\mathbf{b}$ in an arbitrary number of dimensions, use the properties of the inner product and the Cauchy–Schwarz inequality, (9.26), to prove:

(a) the *parallelogram equality*

$$|\mathbf{a}+\mathbf{b}|^2 + |\mathbf{a}-\mathbf{b}|^2 = 2\left[|\mathbf{a}|^2 + |\mathbf{b}|^2\right],$$

(b) the *triangle inequality* $|\mathbf{a}+\mathbf{b}| \leq |\mathbf{a}| + |\mathbf{b}|$.

9.9 Consider the matrices

$$\mathbf{A} = \begin{pmatrix} 1 & -2 & 0\\ 3 & 2 & 5\\ -1 & 3 & 1\end{pmatrix}, \quad \mathbf{B} = \begin{pmatrix} 3 & 1 & -2\\ 1 & 0 & 2\\ -2 & 4 & 3\end{pmatrix},$$

$$\mathbf{C} = \begin{pmatrix} 7 & -1 & 3\\ 1 & 6 & -2\end{pmatrix}, \quad \mathbf{D} = \begin{pmatrix} 5 & 2\\ 1 & -2\\ -3 & 3\end{pmatrix}.$$

(a) Find $\mathbf{A} - 3\mathbf{B}$, $\mathbf{AB}$ and $\mathbf{BA}$.

(b) State which of the products $\mathbf{AC}$, $\mathbf{CA}$, $\mathbf{AD}$, $\mathbf{DA}$, $\mathbf{CD}$ and $\mathbf{DC}$ are defined and evaluate those that are.

9.10 (a) The three matrices

$$\sigma_x = \begin{pmatrix} 0 & 1\\ 1 & 0\end{pmatrix}, \quad \sigma_y = \begin{pmatrix} 0 & -i\\ i & 0\end{pmatrix}, \quad \sigma_z = \begin{pmatrix} 1 & 0\\ 0 & -1\end{pmatrix},$$

called the *Pauli spin matrices*, form a 'vector' $\boldsymbol{\sigma}$. Show that $(\boldsymbol{\sigma}\cdot\mathbf{a})^2 = a^2\,\mathbf{I}$, where $\mathbf{a}$ is an arbitrary real vector $\mathbf{a} = (a_x,\ a_y,\ a_z)$ and $\mathbf{I}$ is the 2×2 unit matrix.

(b) If the matrices $\mathbf{M}_\pm$ are defined by $\mathbf{M}_\pm \equiv \mathbf{M}_x \pm i\mathbf{M}_y$, where

$$\mathbf{M}_x = \frac{1}{\sqrt{2}}\begin{pmatrix} 0 & 1 & 0\\ 1 & 0 & 1\\ 0 & 1 & 0\end{pmatrix}, \quad \mathbf{M}_y = \frac{1}{\sqrt{2}}\begin{pmatrix} 0 & -i & 0\\ i & 0 & -i\\ 0 & i & 0\end{pmatrix},$$

$$\mathbf{M}_z = \frac{1}{\sqrt{2}}\begin{pmatrix} 1 & 0 & 0\\ 0 & 0 & 0\\ 0 & 0 & -1\end{pmatrix},$$

show that the *commutator* $[\mathbf{M}_+,\ \mathbf{M}_-] \equiv \mathbf{M}_+\mathbf{M}_- - \mathbf{M}_-\mathbf{M}_+ = 2\mathbf{M}_z$.

9.11 Write down the matrix operator corresponding to a rotation $\mathbf{R}(\theta)$ through an angle θ about the z-axis in three dimensions, where positive θ corresponds to the x-axis moving towards the original y-axis. Use the form of this matrix to verify explicitly that

$$\mathbf{R}(\theta_1)\mathbf{R}(\theta_2) = \mathbf{R}(\theta_1 + \theta_2) = \mathbf{R}(\theta_2)\mathbf{R}(\theta_1),$$

and that

$$\mathbf{R}^{-1}(\theta) = \mathbf{R}(-\theta) = \mathbf{R}^T(\theta).$$

9.12 The matrix operators corresponding to rotations $\mathbf{R}_x(\theta)$ and $\mathbf{R}_y(\theta)$ through an angle θ about the x and y axes are given by

$$\mathbf{R}_x(\theta) = \begin{pmatrix} 1 & 0 & 0 \\ 0 & \cos\theta & -\sin\theta \\ 0 & \sin\theta & \cos\theta \end{pmatrix} \text{ and } \mathbf{R}_y(\theta) = \begin{pmatrix} \cos\theta & 0 & \sin\theta \\ 0 & 1 & 0 \\ -\sin\theta & 0 & \cos\theta \end{pmatrix}.$$

(a) Show that the matrix corresponding to a rotation through θ_1 about the x-axis, followed by a rotation through θ_2 about the y-axis, is given by

$$\mathbf{R}(\theta_1,\, \theta_2) = \begin{pmatrix} \cos\theta_2 & \sin\theta_1\sin\theta_2 & \sin\theta_2\cos\theta_1 \\ 0 & \cos\theta_1 & -\sin\theta_2 \\ -\sin\theta_2 & \sin\theta_1\cos\theta_2 & \cos\theta_1\cos\theta_2 \end{pmatrix}.$$

Do $\mathbf{R}_x(\theta_1)$ and $\mathbf{R}_y(\theta_2)$ commute?
(b) Write an expression for the inverse matrix $\mathbf{R}^{-1}(\theta_1, \theta_2)$ in terms of $\mathbf{R}_x(\theta)$ and $\mathbf{R}_y(\theta)$ and hence confirm explicitly the relation $\mathbf{R}^{-1} = \mathbf{R}^T$, which holds for any orthogonal matrix and show that $\det[\mathbf{R}(\theta_1)\mathbf{R}(\theta_2)] = 1$ in this case.

9.13 The powers of a matrix $\mathbf{X}$ are defined by $\mathbf{X}^2 \equiv \mathbf{XX}, \mathbf{X}^3 \equiv \mathbf{XXX}$ etc., while its exponential is defined as

$$\exp(\mathbf{X}) \equiv \sum_{n=0}^{\infty} \frac{\mathbf{X}^n}{n!}.$$

If $\mathbf{A}$ and $\mathbf{B}$ are square matrices: (a) find an expression for $(\mathbf{A}+\mathbf{B})^3$ in terms of the products of $\mathbf{A}$ and $\mathbf{B}$ and their powers; (b) derive a condition for the relation

$$e^{(\mathbf{A}+\mathbf{B})} = e^{\mathbf{A}}e^{\mathbf{B}}$$

to be valid.

9.14 Find the transpose, complex conjugate and Hermitian conjugate of the matrix

$$\mathbf{A} = \begin{pmatrix} i & 2 & -3+i \\ 2i & 1 & 3 \\ 2 & 1+i & 2 \end{pmatrix}.$$

9.15 (a) Verify that the matrix

$$\mathbf{A} = \begin{pmatrix} 0 & 1 & 0 \\ (-1+i)/\sqrt{6} & 0 & (1-i)/\sqrt{3} \\ 2/\sqrt{6} & 0 & 1/\sqrt{3} \end{pmatrix}$$

is unitary.

(b) Express the matrix

$$\mathbf{A} = \begin{pmatrix} 1 & 2 \\ 3 & 2 \end{pmatrix}$$

in the form $\mathbf{A}_S + \mathbf{A}_{AS}$, where $\mathbf{A}_S$ is a symmetric matrix and $\mathbf{A}_{AS}$ is an anti-symmetric matrix.

9.16 Which of the matrices below are: (i) symmetric, (ii) orthogonal, (iii) unitary or (iv) Hermitian? Use the matrix that has none of these properties to construct (v) an anti-symmetric matrix and (vi) an anti-Hermitian matrix.

$$\mathbf{A} = \begin{pmatrix} 1 & -2 & i \\ -2 & -1 & -i \\ i & -i & 0 \end{pmatrix} \quad \mathbf{B} = \begin{pmatrix} 2 & 1+2i & 1-2i \\ 1-2i & 0 & 3 \\ 1+2i & 3 & 6 \end{pmatrix}$$

$$\mathbf{C} = \begin{pmatrix} 1 & 1-i & 2i \\ -3i & -2i & 1 \\ 1+3i & 2 & 3 \end{pmatrix} \quad \mathbf{D} = \frac{1}{\sqrt{2}}\begin{pmatrix} 1 & i & 0 \\ -i & -1 & 0 \\ 0 & 0 & \sqrt{2} \end{pmatrix}$$

$$\mathbf{E} = \begin{pmatrix} \cos\theta & -\sin\theta & 0 \\ \sin\theta & \cos\theta & 0 \\ 0 & 0 & 1 \end{pmatrix}.$$

9.17 (a) If $\mathbf{S}$ is a symmetric matrix and $\mathbf{A}$ is an anti-symmetric matrix, show that $\mathrm{Tr}\,(\mathbf{SA}) = 0$.
(b) Prove that diagonal matrices commute with each other.

9.18 Find the inverse of the matrix

$$\mathbf{A} = \begin{pmatrix} 3 & -2 & 2 \\ 1 & -2 & -3 \\ -4 & 1 & 2 \end{pmatrix}$$

and check the answer by direct multiplication.

9.19 Find the inverse of the matrix

$$\mathbf{A} = \begin{pmatrix} -2 & 1 & 2 \\ -1 & 1 & 1 \\ 0 & 2 & -1 \end{pmatrix}$$

and hence solve the matrix equation

$$\mathbf{AX} = \begin{pmatrix} 6 & -2 \\ 4 & 0 \\ 1 & 3 \end{pmatrix}.$$

9.20 Find by matrix inversion the solution of the equations

$$\begin{aligned} 3x + y - z &= 1 \\ x - y + z &= 2 \\ -2x + 2y + 2z &= 3 \end{aligned}$$

9.21 Find the solution of the equations

$$\begin{aligned} 2x + 3y - z &= 0 \\ x - y + z &= 1 \\ -x + y + 2z &= 2 \end{aligned}$$

by Cramer's rule.

9.22 The half-life τ of a radioactive atom is defined as the time it takes for half of a given quantity of atoms to decay. A sample consists of just two radioactive components A and B, both of which decay to gaseous products that rapidly disperse. The sample is weighed after 8 and 12 hours and is found to weigh 90 and 30 grams, respectively. If the half-lives of A and B are $\tau_a = 2\,\mathrm{h}$ and $\tau_b = 4\,\mathrm{h}$, respectively, use Cramer's rule to calculate the amounts of A and B initially in the sample.

9.23 (a) For what values of the constants α and β do the simultaneous equations

$$\begin{aligned} 4x + 2y + \alpha z &= \beta, \\ 7x + 3y + 4z &= 8, \\ x + y + 2z &= 4, \end{aligned}$$

have a unique solution?

(b) Solve the equations for the case $\alpha = 2,\ \beta = 3$ by inverting the appropriate matrix.

(c) Comment on both the existence and uniqueness of solutions in the cases: (i) $\alpha = 3, \beta = 6$; (ii) $\alpha = 3, \beta = 2$.

10

Eigenvalues and eigenvectors

Given a square matrix **A**, it is often required to find scalar constants λ and vectors **x** such that

$$\mathbf{A}\mathbf{x} = \lambda\mathbf{x} \tag{10.1}$$

is satisfied. This equation only has non-trivial solutions $\mathbf{x} \neq \mathbf{0}$ for particular values of λ. These values are called *eigenvalues* and the corresponding vectors **x** are called *eigenvectors*.[1] In physical applications the eigenvalues often correspond to the allowed values of observable quantities. In what follows, we shall firstly consider the solutions of (10.1) in general, before specialising to Hermitian matrices, which are the most important in physical applications. We then show how knowledge of the eigenvalues can be used to transform the matrix **A** to diagonal form, with applications to the theory of small vibrations and geometry.

10.1 The eigenvalue equation

The *eigenvalue equation* (10.1) may be written in the form

$$(\mathbf{A} - \lambda\mathbf{I})\mathbf{x} = \mathbf{0}. \tag{10.2}$$

This is a set of homogeneous linear simultaneous equations in the components x_i ($i = 1, 2, \ldots, n$) of the type discussed in Section 9.1.2 and has non-trivial solutions if, and only if,

$$\det(\mathbf{A} - \lambda\mathbf{I}) = 0, \tag{10.3}$$

[1] These hybrid words come from the German *eigenwert* and *eigenvektor*, where 'eigen' means 'characteristic'.

Mathematics for Physicists, First Edition. B.R. Martin and G. Shaw.

Companion website: www.wiley.com/go/martin/mathsforphysicists

which is called the *characteristic equation* of the matrix **A**. The determinant is given by

$$\det(\mathbf{A}-\lambda\mathbf{I}) = \begin{vmatrix} a_{11}-\lambda & a_{12} & \cdots & a_{1n} \\ a_{21} & a_{22}-\lambda & \cdots & a_{2n} \\ \vdots & \vdots & \ddots & \vdots \\ a_{n1} & a_{n2} & \cdots & a_{nn}-\lambda \end{vmatrix} = f(\lambda), \qquad (10.4a)$$

where

$$f(\lambda) = (-1)^n(\lambda^n + \alpha_1\lambda^{n-1} + \cdots + \alpha_n) \qquad (10.4b)$$

is a polynomial in $f(\lambda)$ in λ of degree n, called the *characteristic polynomial*, whose coefficients α_i $(i = 1, 2, \ldots, n)$ depend on the matrix elements a_{ij}. Solving (10.3) is equivalent to finding the roots of this polynomial. In general, any polynomial of order n has n roots when complex values are allowed,[2] so (10.4b) may be written in the form

$$f(\lambda) = (-1)^n \prod_{i=1}^{n} (\lambda - \lambda_i), \qquad (10.4c)$$

and thus (10.3) gives rise to n eigenvalues λ_i $(i = 1, 2, \ldots, n)$. However, not all these eigenvalues are necessarily distinct, that is, two or more may have the same numerical value.

Once the eigenvalues have been determined, each value of $\lambda = \lambda_i$ may be substituted into (10.2). In each case this yields a set of n simultaneous homogeneous linear equations in the components $[\mathbf{x}^{(i)}]_j$ of the corresponding eigenvector $\mathbf{x}^{(i)}$, which may be solved by the methods discussed Section 9.1.2, as we shall shortly illustrate.[3] However, this does not uniquely determine the eigenvectors, because if **x** is a solution of (10.2), then so is $\alpha\mathbf{x}$, where α is any constant. We will usually exploit this to choose normalised eigenvectors **x** of unit modulus, that is, such that $(\mathbf{x}, \mathbf{x}) = |\mathbf{x}|^2 = 1$.

Example 10.1

Find the eigenvalues and eigenvectors of the matrices:

$$\text{(a) } \mathbf{A} = \begin{pmatrix} 4 & 4 & 0 \\ 4 & 4 & 0 \\ 0 & 0 & 1 \end{pmatrix} \quad \text{and} \quad \text{(b) } \mathbf{B} = \begin{pmatrix} 1 & 1+i \\ 1-i & 2 \end{pmatrix}.$$

[2] This is the fundamental theorem of algebra mentioned in Section 2.1.1.

[3] To clarify the notation: x_i are the components of a vector **x**; $\mathbf{x}^{(i)}$ is an eigenvector belonging to the eigenvalue λ_i; its components are written $[\mathbf{x}^{(i)}]_j$ where $j = 1, 2, \ldots, n$.

Solution

(a) The characteristic equation is

$$\begin{vmatrix} 4-\lambda & 4 & 0 \\ 4 & 4-\lambda & 0 \\ 0 & 0 & 1-\lambda \end{vmatrix} = (4-\lambda)^2(1-\lambda) - 16(1-\lambda) = 0,$$

with eigenvalue solutions $\lambda_1 = 0$, $\lambda_2 = 1$ and $\lambda_3 = 8$. The corresponding eigenvectors are found by using these values in the eigenvalue equation $\mathbf{Ax} = \lambda\mathbf{x}$. Thus, for $\lambda_1 = 0$, we have

$$4x_1 + 4x_2 = 0 \text{ and } x_3 = 0 \quad \Rightarrow \quad x_1 = -x_2 = \alpha \text{ and } x_3 = 0.$$

Normalising the eigenvectors to have unit norm, gives $\alpha = 1/\sqrt{2}$ and so

$$\mathbf{x}^{(1)} = \frac{1}{\sqrt{2}}\begin{pmatrix} 1 \\ -1 \\ 0 \end{pmatrix}.$$

Proceeding in the same way for the other two eigenvalues $\lambda_2 = 1$, $\lambda_3 = 8$ gives

$$\mathbf{x}^{(2)} = \begin{pmatrix} 0 \\ 0 \\ 1 \end{pmatrix} \text{ and } \mathbf{x}^{(3)} = \frac{1}{\sqrt{2}}\begin{pmatrix} 1 \\ 1 \\ 0 \end{pmatrix}.$$

(b) The characteristic equation is

$$\begin{vmatrix} 1-\lambda & 1+i \\ 1-i & 2-\lambda \end{vmatrix} = (1-\lambda)(2-\lambda) - (1+i)(1-i) = 0,$$

with eigenvalue solutions $\lambda_1 = 0$, $\lambda_2 = 3$. The corresponding eigenvectors are found by using these values in the eigenvalue equation $\mathbf{Ax} = \lambda\mathbf{x}$. Thus, for $\lambda_1 = 0$, we have

$$x_1 + (1+i)x_2 = 0 \quad \Rightarrow \quad x_1 = -(1+i)x_2.$$

If we normalise the eigenvectors to have unit norm, then

$$\mathbf{x}^{(1)} = \frac{1}{\sqrt{3}}\begin{pmatrix} -(1+i) \\ 1 \end{pmatrix}.$$

Proceeding in the same way for the other eigenvalue gives

$$\mathbf{x}^{(2)} = \frac{1}{\sqrt{6}}\begin{pmatrix} (1+i) \\ 2 \end{pmatrix}.$$

10.1.1 Properties of eigenvalues

In this section we will derive some useful properties of eigenvalues that follow directly from (10.3).

Firstly, if $\mathbf{A}$ is singular, that is, $\det \mathbf{A} = 0$, then it follows from (10.3) that it has an eigenvalue $\lambda = 0$; conversely, if $\mathbf{A}$ has an eigenvalue $\lambda = 0$, then it is singular. Secondly, it follows from (10.4a) and (10.4c) that

$$\det(\mathbf{A} - \lambda\mathbf{I}) = (-1)^n \prod_{i=1}^{n} (\lambda - \lambda_i).$$

Setting $\lambda = 0$ then gives

$$\det \mathbf{A} = \prod_{i=1}^{n} \lambda_i, \tag{10.5}$$

that is, the determinant of any matrix is equal to the product of its eigenvalues. Similarly, as we shall show, the sum of the eigenvalues is given by

$$\mathrm{Tr}\, \mathbf{A} = \sum_{i=1}^{n} \lambda_i, \tag{10.6}$$

where the *trace* is

$$\mathrm{Tr}\, \mathbf{A} \equiv \sum_{i=1}^{n} a_{ii}. \tag{10.7}$$

Together with (10.6), Equation (10.7) is very useful in checking that the eigenvalues of a given matrix have been computed correctly. It is proved by computing the coefficient of λ^{n-1} in (10.4b) using (10.4a) and (10.4c) in turn, and comparing the results. In (10.4a), the co-factors of $a_{12}, a_{13}, \ldots, a_{1n}$ are polynomials of order λ^{n-2}. Hence terms of order λ^{n-1} can only occur in the product of the diagonal elements in (10.4a), giving

$$\alpha_1 = (-1)^n (a_{11} + a_{22} + \cdots + a_{nn}).$$

On the other hand, expanding (10.4c) gives

$$\alpha_1 = (-1)^n (\lambda_1 + \lambda_2 + \cdots + \lambda_n),$$

and comparing the two expressions yields the desired result.

Finally, suppose that an $n \times n$ matrix $\mathbf{A}$ has $k \leq n$ distinct eigenvalues $\lambda_1, \lambda_2, \ldots, \lambda_k$, that is, $\lambda_i \neq \lambda_j$ for $i \neq j$ and $i, j \leq k$. Then the following related matrices also have a total of k distinct eigenvalues, as specified below.

(a) The transpose matrix $\mathbf{A}^{\mathrm{T}}$ has the same eigenvalues λ_i.

(b) The matrix $\alpha\mathbf{A}$ has eigenvalues $\alpha\lambda_i$, where α is a scalar constant.

(c) The Hermitian conjugate matrix $\mathrm{A}^{\dagger}$ has eigenvalues λ_i^*.

(d) The inverse matrix $\mathbf{A}^{-1}$, if it exists, has eigenvalues λ_i^{-1}.

Here we will prove (iii) and leave the others as exercises for the reader. Since λ_i is an eigenvalue of $\mathbf{A}$,

$$\det(\mathbf{A} - \lambda_i \mathbf{I}) = 0,$$

which by (9.59b) implies

$$\det(\mathrm{A} - \lambda_i \mathbf{I})^\dagger = [\det(\mathrm{A} - \lambda_i \mathbf{I})]^* = 0.$$

From (9.33) and (9.55), we have

$$(\lambda_i \mathbf{I})^\dagger = \lambda_i^* \mathbf{I}^\dagger = \lambda_i^* \mathbf{I},$$

so that

$$\det(\mathbf{A} - \lambda_i \mathbf{I})^\dagger = \det(\mathbf{A}^\dagger - \lambda_i^* \mathbf{I}) = 0,$$

and hence λ_i^* is an eigenvalue of $\mathrm{A}^\dagger$ for all $i = 1, 2, \ldots, k$. That they are the only distinct eigenvalues of $\mathrm{A}^\dagger$, even if $k < n$, follows by using the argument in reverse. Suppose $\mathrm{A}^\dagger$ had an extra eigenvalue $\lambda_i \neq \lambda_i^*$, $i = 1, 2, \ldots, k$. Then since $(\mathbf{A}^\dagger)^\dagger = \mathbf{A}$, this would imply that $\mathbf{A}$ had a distinct eigenvalue $\lambda \neq \lambda_i$, $i = 1, 2, \ldots, k$, in contradiction to the requirement that k is the total number of distinct eigenvalues of $\mathbf{A}$.

Example 10.2

(a) If $\mathbf{A}$ and $\mathbf{B}$ are both $n \times n$ matrices, show that

$$\mathrm{Tr}(\mathbf{AB}) = \mathrm{Tr}(\mathbf{BA}). \tag{10.8}$$

(b) Use (10.5) and (10.6) to find the eigenvalues of the Pauli matrices σ_i $(i = 1, 2, 3)$, defined by

$$\sigma_1 = \begin{pmatrix} 0 & 1 \\ 1 & 0 \end{pmatrix} \quad \sigma_2 = \begin{pmatrix} 0 & i \\ -i & 0 \end{pmatrix} \quad \sigma_3 = \begin{pmatrix} 1 & 0 \\ 0 & -1 \end{pmatrix}.$$

Solution

(a) We have

$$\mathrm{Tr}(\mathbf{AB}) = \sum_i (\mathbf{AB})_{ii} = \sum_{i,j} a_{ij} b_{ji}$$
$$= \sum_{i,j} b_{ji} a_{ij} = \sum_j (\mathbf{BA})_{jj} = \mathrm{Tr}(\mathbf{BA}).$$

(b) For all three matrices, we have $|\sigma_i| = -1$ and $\mathrm{Tr}\,\sigma_i = 0$, so that

$$\lambda_1 \lambda_2 = -1 \quad \text{and} \quad \lambda_1 + \lambda_2 = 0$$

by (10.6) and (10.7), respectively. Therefore the eigenvalues are $\lambda_1 = 1, \lambda_2 = -1$, or equivalently, $\lambda_1 = -1, \lambda_2 = 1$, which is just a relabeling. These values can also be obtained by solving the characteristic equations $\det(\sigma_i - \lambda \mathbf{I}) = 0$.

10.1.2 Properties of eigenvectors

If $\mathbf{x}^{(i)}$ $(i = 1, 2, \ldots, k)$ is a set of eigenvectors corresponding to k different eigenvalues λ_i $(i = 1, 2, \ldots, k)$, then $\mathbf{x}^{(i)}$ are linearly independent. That is, there is no linear relationship of the type

$$c_1\mathbf{x}^{(1)} + c_2\mathbf{x}^{(2)} + \ldots + c_k\mathbf{x}^{(k)} = \mathbf{0}, \tag{10.9}$$

where the c_i are constants, except the trivial case $c_i = 0$ where $i = 1, 2, \ldots, k$. The proof is as follows.

Since $\mathbf{A}\mathbf{x}^{(i)} = \lambda_i\mathbf{x}^{(i)}$ $(i = 1, 2, \ldots, k)$,

$$(\mathbf{A} - \lambda_j\mathbf{I})\mathbf{x}^{(i)} = \mathbf{A}\mathbf{x}^{(i)} - \lambda_j\mathbf{I}\,\mathbf{x}^{(i)} = (\lambda_i - \lambda_j)\mathbf{x}^{(i)}. \tag{10.10}$$

Suppose now that a condition of the form (10.9) does exist and we operate on it by $(\mathbf{A} - \lambda_j\mathbf{I})$, with the result

$$(\mathbf{A} - \lambda_j\mathbf{I})[c_1\mathbf{x}^{(1)} + c_2\mathbf{x}^{(2)} + \cdots + c_k\mathbf{x}^{(k)}] = \mathbf{0}. \tag{10.11}$$

For $j = 2$, using (10.10) and (10.11) gives

$$c_1(\lambda_1 - \lambda_2)\mathbf{x}^{(1)} + c_3(\lambda_3 - \lambda_2)\mathbf{x}^{(3)} + \cdots + c_k(\lambda_k - \lambda_2)\mathbf{x}^{(k)} = \mathbf{0}, \tag{10.12}$$

where the term in $\mathbf{x}^{(2)}$ is absent. If this operation is now repeated on (10.12) using $j = 3$, an additional bracket $(\lambda_1 - \lambda_3)$ multiplying each term will be generated and the term in $\mathbf{x}^{(3)}$ will be eliminated. Repeating the operation for the remaining values of j successively, eventually yields the result

$$c_1(\lambda_1 - \lambda_2)(\lambda_1 - \lambda_3)\cdots(\lambda_1 - \lambda_k)\,\mathbf{x}^{(1)} = \mathbf{0},$$

and since all the λ_i are assumed to be different, this implies that $c_1 = 0$. The same method can be used to show that $c_2 = 0$, and so on. Hence if all the values of λ_i are different, only the trivial solution $c_i = 0$ $(i = 1, 2, 3, \ldots, k)$ exists, and so the eigenvectors are linearly independent.

We next consider the implications of this for an $n \times n$ matrix $\mathbf{A}$. If all the eigenvalues λ_i $(i = 1, 2, \ldots, n)$ are distinct, then $k = n$ above and there are n linearly independent eigenvectors $\mathbf{x}^{(1)}, \mathbf{x}^{(2)}, \ldots, \mathbf{x}^{(n)}$. Since an n-dimensional space cannot contain more than n linearly independent vectors, the eigenvectors form a complete set of linearly independent vectors, as defined in Section 9.2.1. Hence an arbitrary vector $\mathbf{x}$ can always be written as a sum of eigenvectors of the form

$$\mathbf{x} = \alpha_1\mathbf{x}^{(1)} + \alpha_2\mathbf{x}^{(2)} + \cdots + \alpha_n\mathbf{x}^{n}, \tag{10.13}$$

where the numerical constants α_i depend on $\mathbf{x}$.

It remains to consider the case where $k < n$, that is, when there are less than n distinct eigenvalues. To illustrate this, suppose the characteristic polynomial is of the form

$$\det(\mathbf{A} - \lambda\mathbf{I}) = f(\lambda) = (-1)^n(\lambda - \lambda_1)(\lambda - \lambda_2)\cdots(\lambda - \lambda_{n-1})^2,$$

so that there are $k = n - 1$ distinct eigenvalues. Nonetheless, one can usually find two linearly independent eigenvectors $\mathbf{x}^{(n-1)}$, $\mathbf{x}^{(n)}$ that both have eigenvalue λ_{n-1}. Hence there are still n linearly independent eigenvectors, and an arbitrary vector $\mathbf{x}$ can still be expanded in the form (10.13). However, sometimes, as we shall illustrate by an example below, there is only a single eigenvector $\mathbf{x}^{(n-1)}$ corresponding to λ_{n-1}. Hence there are only $n - 1$ linearly independent eigenvectors. Matrices like these, which have fewer independent eigenvectors than dimension of the matrix, are called *defective matrices.* For such matrices, an arbitrary vector in the n dimensional space cannot be expanded in terms of its eigenvectors.

Example 10.3

Find the eigenvalues and eigenvectors of the matrices

$$\text{(a) } \mathbf{A} = \begin{pmatrix} -2 & -1 & -1 \\ 6 & 3 & 2 \\ 0 & 0 & 1 \end{pmatrix} \quad \text{and} \quad \text{(b) } \mathbf{B} = \begin{pmatrix} 1 & 1 & 2 \\ 2 & 2 & 2 \\ -1 & -1 & -1 \end{pmatrix},$$

and hence determine if either is defective.

Solution

(a) The characteristic equation is

$$\det(\mathbf{A} - \lambda\mathbf{I}) = \begin{vmatrix} -2-\lambda & -1 & -1 \\ 6 & 3-\lambda & 2 \\ 0 & 0 & 1-\lambda \end{vmatrix} = -\lambda(1-\lambda)^2 = 0,$$

as is easily seen by expanding the determinant along row 3. The eigenvalues are thus $\lambda = 0$ and $\lambda = 1$. For $\lambda = 0$, the eigenvalue equation $(\mathbf{A} - \lambda\mathbf{I})\mathbf{x} = \mathbf{0}$ becomes

$$\begin{pmatrix} -2 & -1 & -1 \\ 6 & 3 & 2 \\ 0 & 0 & 1 \end{pmatrix}\begin{pmatrix} x \\ y \\ z \end{pmatrix} = \begin{pmatrix} 0 \\ 0 \\ 0 \end{pmatrix},$$

which on expanding becomes

$$\begin{aligned} -2x - y - z &= 0, \\ 6x + 3y + 2z &= 0, \\ z &= 0, \end{aligned}$$

with solution $z = 0,\ y = -2x$. A normalised eigenvector is therefore

$$\mathbf{x}^{(1)} = \frac{1}{\sqrt{5}}\begin{pmatrix}1\\-2\\0\end{pmatrix}.$$

For $\lambda = 1$, the corresponding eigenvalue equation is

$$\begin{pmatrix}-3 & -1 & -1\\ 6 & 2 & 2\\ 0 & 0 & 0\end{pmatrix}\begin{pmatrix}x\\y\\z\end{pmatrix} = \begin{pmatrix}0\\0\\0\end{pmatrix},$$

leading to the same single condition $3x + y + z = 0$, and thus $3x$ is fixed in terms of y and z. Choosing $y = 0$ and $z = 0$ in turn gives the normalised eigenvectors

$$\mathbf{x}^{(2)} = \frac{1}{\sqrt{10}}\begin{pmatrix}1\\0\\-3\end{pmatrix} \quad \text{and} \quad \mathbf{x}^{(3)} = \frac{1}{\sqrt{10}}\begin{pmatrix}1\\-3\\0\end{pmatrix},$$

so that, although there are only two distinct eigenvalues, there are three linearly independent eigenvectors, and thus the matrix is not defective. The choice of eigenvectors is not unique, since any linear combination $\alpha\mathbf{x}^{(2)} + \beta\mathbf{x}^{(3)}$ is also an eigenvector with eigenvalue $\lambda = 1$, so that other choices are possible. For example, instead of $\mathbf{x}^{(2)}$ and $\mathbf{x}^{(3)}$ we could choose the normalised eigenvalues

$$\mathbf{x}_+ = \frac{1}{\sqrt{11}}(\mathbf{x}^{(2)} + \mathbf{x}^{(3)}) = \frac{1}{\sqrt{22}}\begin{pmatrix}2\\-3\\-3\end{pmatrix}$$

and

$$\mathbf{x}_- = \frac{1}{\sqrt{9}}(\mathbf{x}^{(2)} - \mathbf{x}^{(3)}) = \frac{1}{\sqrt{18}}\begin{pmatrix}0\\3\\-3\end{pmatrix}.$$

(b) The characteristic equation is now

$$\begin{vmatrix}1-\lambda & 1 & 2\\ 2 & 2-\lambda & 2\\ -1 & -1 & -1-\lambda\end{vmatrix} = -\lambda(\lambda-1)^2 = 0,$$

so that again we have only two eigenvalues $\lambda = 0, 1$. For $\lambda = 0$, the eigenvalue equation $(\mathbf{B} - \lambda\mathbf{I})\mathbf{x} = \mathbf{0}$ becomes

$$\begin{pmatrix}1 & 1 & 2\\ 2 & 2 & 2\\ -1 & -1 & -1\end{pmatrix}\begin{pmatrix}x\\y\\z\end{pmatrix} = \begin{pmatrix}0\\0\\0\end{pmatrix},$$

i.e.,

$$\begin{aligned} x+y+2z&=0,\\ 2x+2y+2z&=0,\\ -x-y-z&=0, \end{aligned}$$

with solution $z=0$, $y=-x$. The corresponding normalised eigenvector is

$$\mathbf{x}^{(1)}=\frac{1}{\sqrt{2}}\begin{pmatrix}1\\-1\\0\end{pmatrix}.$$

For $\lambda=1$, the corresponding equation is

$$\begin{pmatrix}0&1&2\\2&1&2\\-1&-1&-2\end{pmatrix}\begin{pmatrix}x\\y\\z\end{pmatrix}=\begin{pmatrix}0\\0\\0\end{pmatrix},$$

i.e.,

$$\begin{aligned} y+2z&=0,\\ 2x+y+2z&=0,\\ -x-y-2z&=0, \end{aligned}$$

with solution $x=0$, $y=-2z$, yielding the normalised eigenvector

$$\mathbf{x}^{(2)}=\frac{1}{\sqrt{5}}\begin{pmatrix}0\\-2\\1\end{pmatrix}.$$

Thus, in this case there are only two independent eigenvectors, and the matrix is therefore defective.

10.1.3 Hermitian matrices

In most physical applications, and especially in quantum mechanics, the eigenvalues and eigenvectors of interest are those of Hermitian matrices. This is because the eigenvalues are real and so can correspond to measurable quantities. In addition, the eigenvectors corresponding to different eigenvalues are not only linearly independent, but also orthogonal. In particular, these results apply to real, symmetric matrices, which are automatically Hermitian.

To prove these properties, consider a Hermitian matrix $\mathbf{A}$ and an eigenvector $\mathbf{a}$, corresponding to an eigenvalue λ_a, so that

$$\mathbf{A}\mathbf{a}=\lambda_a\mathbf{a}. \tag{10.14a}$$

Taking the Hermitian conjugate, we obtain

$$(\mathbf{A}\mathbf{a})^\dagger=\mathbf{a}^\dagger\mathbf{A}^\dagger=\mathbf{a}^\dagger\mathbf{A}=\lambda_a^*\mathbf{a}^\dagger, \tag{10.14b}$$

where we have used $\mathbf{A}=\mathbf{A}^\dagger$ and the relation

$$(\lambda_a\mathbf{a})^\dagger=\lambda_a^*\mathbf{a}^\dagger$$

which follows from (9.33) and (9.53). Then multiplying (10.14a) on the left by $\mathbf{a}^\dagger$ and (10.14b) on the right by $\mathbf{a}$, we obtain

$$\mathbf{a}^\dagger\mathbf{A}\mathbf{a} = \lambda_a\mathbf{a}^\dagger\mathbf{a} = \lambda_a(\mathbf{a},\mathbf{a})$$

and

$$\mathbf{a}^\dagger\mathbf{A}\mathbf{a} = \lambda_a^*\mathbf{a}^\dagger\mathbf{a} = \lambda_a^*(\mathbf{a},\mathbf{a}).$$

Since $(\mathbf{a},\mathbf{a}) \neq 0$, these equations can only be satisfied if $\lambda_a = \lambda_a^*$, that is, the eigenvalue is real, as required.

Next we consider a second eigenvector $\mathbf{b}$ satisfying

$$\mathbf{A}\mathbf{b} = \lambda_b\mathbf{b}, \qquad \lambda_b \neq \lambda_a. \tag{10.14c}$$

On multiplying (10.14c) on the left by $\mathbf{a}^\dagger$ and (10.14b) on the right by $\mathbf{b}$, we obtain

$$\mathbf{a}^\dagger\mathbf{A}\mathbf{b} = \lambda_b\mathbf{a}^\dagger\mathbf{b} = \lambda_b(\mathbf{a},\mathbf{b})$$

and

$$\mathbf{a}^\dagger\mathbf{A}\mathbf{b} = \lambda_a\mathbf{a}^\dagger\mathbf{b} = \lambda_a(\mathbf{a},\mathbf{b}),$$

where in the second equation we have used the result $\lambda^* = \lambda$ proved above. Since $\lambda_a \neq \lambda_b$, these two equations are only compatible if

$$(\mathbf{a},\mathbf{b}) = 0, \quad \lambda_a \neq \lambda_b \tag{10.15}$$

that is, the eigenvectors are orthogonal.

An $n \times n$ Hermitian matrix $\mathbf{A}$ always has n linearly independent eigenvectors[4] $\mathbf{x}^{(i)}$. Hence an arbitrary n-dimensional vector can always be expanded in the form (10.13), that is,

$$\mathbf{x} = \sum_{i=1}^{n} \alpha_i \hat{\mathbf{x}}^{(i)}, \tag{10.16a}$$

where

$$\mathbf{A}\hat{\mathbf{x}}_i = \lambda_i \hat{\mathbf{x}}^{(i)}, \quad i = 1, 2, \cdots, n \tag{10.16b}$$

and we have chosen unit eigenvectors $\hat{\mathbf{x}}^{(i)}$. If the eigenvalues λ_i are all different, then the eigenvectors are orthonormal, that is,

$$(\hat{\mathbf{x}}^{(i)}, \hat{\mathbf{x}}^{(j)}) = [\hat{\mathbf{x}}^{(i)}]^\dagger \hat{\mathbf{x}}^{(j)} = \delta_{ij}, \tag{10.17a}$$

[4]This result applies not only to Hermitian matrices but to any matrix $\mathbf{A}$ that commutes with its Hermitian conjugate, that is, for which $\mathbf{A}\mathbf{A}^\dagger = \mathbf{A}^\dagger\mathbf{A}$. Such matrices are called *normal matrices*, and automatically include Hermitian, anti-Hermitian, and unitary matrices. See p. 311 of G. Strang (1988) *Linear Algebra and its Applications*, 3rd edn., Harcourt, Brace, Jovanovich, San Diego, California.

where δ_{ij} is the kronecker delta symbol defined in (9.24b). Multiplying (10.16a) by $[\hat{\mathbf{x}}^{(j)}]^\dagger$ and using (10.17a) then gives

$$\alpha_j = (\mathbf{x}^{(j)}, \mathbf{x}), \quad j = 1, 2, \ldots, n \tag{10.17b}$$

for the coefficients α_j.

Equations (10.16) and (10.17) are very convenient in applications, but are only *automatically* valid if the eigenvalues λ_i are all different. If this is not so, the eigenvectors (10.16b) are not uniquely defined. However, one may always choose a complete set of linearly independent eigenvectors (10.16a) and (10.16b) that do satisfy (10.17a) and (10.17b). To see this, let us suppose there are k linearly independent eigenvectors $\mathbf{u}^{(1)}, \mathbf{u}^{(2)}, \ldots, \mathbf{u}^{(k)}$ corresponding to a given eigenvalue $\bar{\lambda}$, that is,

$$\mathbf{A}\mathbf{u}^{(i)} = \bar{\lambda}\mathbf{u}^{(i)}, \quad i = 1, 2, \ldots, k.$$

Then the eigenvalue $\bar{\lambda}$ is said to be *k-fold degenerate* and any linear combination of the form

$$\mathbf{x} = \sum_{i=1}^{k} \alpha_i \mathbf{u}^{(i)}, \quad i = 1, 2, \ldots, k, \tag{10.18}$$

where the α_i are arbitrary constants, is also an eigenvector. In particular, it is possible to choose a sequence of eigenvectors

$$\begin{aligned}
\mathbf{x}^{(1)} &= \mathbf{u}^{(1)}, \\
\mathbf{x}^{(2)} &= \mathbf{u}^{(2)} - (\hat{\mathbf{x}}^{(1)}, \mathbf{u}^{(2)})\hat{\mathbf{x}}^{(1)}, \\
\mathbf{x}^{(3)} &= \mathbf{u}^{(3)} - (\hat{\mathbf{x}}^{(1)}, \mathbf{u}^{(3)})\hat{\mathbf{x}}^{(1)} - (\hat{\mathbf{x}}^{(2)}, \mathbf{u}^{(3)})\hat{\mathbf{x}}^{(2)}, \\
&\vdots \\
\mathbf{x}^{(k)} &= \mathbf{u}^{(k)} - \sum_{j=1}^{k-1} (\hat{\mathbf{x}}^{(j)}, \mathbf{u}^{(k)})\hat{\mathbf{x}}^{(j)},
\end{aligned} \tag{10.19a}$$

in which each $\mathbf{x}^{(i)}$, $i \leq k$, is chosen to be orthogonal to all $\mathbf{x}^{(j)}$ with $j < i$. These can then be normalised:

$$\hat{\mathbf{x}}^{(i)} = \mathbf{x}^{(i)}/|\mathbf{x}^{(i)}|, \quad i = 1, 2, \ldots, k. \tag{10.19b}$$

This procedure is called *Gram-Schmidt orthogonalisation*, and the resulting eigenvectors $\mathbf{x}^{(i)}$ satisfy (10.17a), as required. They are, however, not unique and other choices of linearly independent eigenvectors satisfying (10.17a) are also possible.

Example 10.4
Show that the Hermitian matrix

$$\begin{pmatrix} 0 & 0 & i \\ 0 & 1 & 0 \\ -i & 0 & 0 \end{pmatrix}$$

has only two real eigenvalues and find an orthonormal set of three eigenvectors.

Solution
The characteristic equation is

$$\begin{vmatrix} -\lambda & 0 & i \\ 0 & 1-\lambda & 0 \\ -i & 0 & -\lambda \end{vmatrix} = -(\lambda+1)(\lambda-1)^2,$$

so that the eigenvalues are $\lambda = -1$ and $\lambda = 1$. For $\lambda = -1$, the eigenvalue equation gives

$$x + iz = 0, \quad 2y = 0, \quad -ix + z = 0,$$

yielding the unit vector

$$\mathbf{x}^{(1)} = \frac{1}{\sqrt{2}}\begin{pmatrix} 1 \\ 0 \\ i \end{pmatrix}.$$

For $\lambda = 1$, the corresponding equations are

$$-x + iz = 0, \quad 0(x+y+z) = 0, \quad -ix - z = 0,$$

so that $z = -ix$ and y is undetermined. Suitable orthonormal eigenvectors are

$$\mathbf{x}^{(2)} = \begin{pmatrix} 0 \\ 1 \\ 0 \end{pmatrix} \quad \text{and} \quad \mathbf{x}^{(3)} = \frac{1}{\sqrt{2}}\begin{pmatrix} 1 \\ 0 \\ -i \end{pmatrix}.$$

Note that these eigenvectors are already orthonormal, so that further orthogonalisation is not required.

*10.2 Diagonalisation of matrices

In Section 9.2.1, we emphasised that the components of a vector depend on the choice of basis vectors. To find the corresponding dependence of a linear operator $\mathbf{A}$, we first note that (9.21b) can be written in the matrix form $\mathbf{a} = \mathbf{P}\mathbf{a}'$ on transforming from the primed to unprimed basis. Re-labeling the vector $\mathbf{a}$ as $\mathbf{x}$ for convenience, this becomes

$$\mathbf{x} = \mathbf{P}\,\mathbf{x}' \tag{10.20}$$

on transforming from the primed to unprimed basis. Furthermore, if we write the reverse transformation in the form $\mathbf{x}' = \mathbf{P}'\mathbf{x}$, then we have

$$\mathbf{x} = \mathbf{P}\,\mathbf{x}' = \mathbf{P}\mathbf{P}'\mathbf{x},$$

and since this must hold for any vector $\mathbf{x}$, we must have $\mathbf{P}' = \mathbf{P}^{-1}$ and hence

$$\mathbf{x}' = \mathbf{P}^{-1}\mathbf{x}. \tag{10.21}$$

The corresponding transformation for a matrix $\mathbf{A}$ is then obtained by applying (10.21) to a vector $\mathbf{y} = \mathbf{A}\mathbf{x}$ and using (10.20) to give

$$\mathbf{y}' = \mathbf{P}^{-1}\mathbf{y} = \mathbf{P}^{-1}\mathbf{A}\mathbf{x} = \mathbf{P}^{-1}\mathbf{A}\mathbf{P}\,\mathbf{x}' = \mathbf{A}'\mathbf{x}',$$

where

$$\mathbf{A}' = \mathbf{P}^{-1}\mathbf{A}\mathbf{P}. \tag{10.22}$$

Equations of the type (10.22) are called *similarity transformations* and two matrices $\mathbf{A}$ and $\mathbf{A}'$ related in this way are said to be *similar*. In geometrical problems we know that a suitable choice of coordinates can often simplify calculations and likewise problems involving linear transformations can often be simplified by a judicious choice of basis. In particular, any n-dimensional matrix with n linearly independent eigenvectors[5] can be transformed to diagonal form by means of a similarity transformation. To see this, set

$$p_{ij} = [\mathbf{x}^{(j)}]_i \tag{10.23a}$$

i.e. the columns of $\mathbf{P}$ are the eigenvectors of $\mathbf{A}$. Then from (10.22),

$$\begin{aligned}
\mathbf{A}'_{ij} \equiv a'_{ij} &= (\mathbf{P}^{-1}\mathbf{A}\mathbf{P})_{ij} \\
&= \sum_k \sum_l (\mathbf{P}^{-1})_{ik} a_{kl} p_{lj} = \sum_k \sum_l (\mathbf{P}^{-1})_{ik} a_{kl} [\mathbf{x}^{(j)}]_l \\
&= \sum_k (\mathbf{P}^{-1})_{ik} \lambda_j [\mathbf{x}^{(j)}]_k = \lambda_j \sum_k (\mathbf{P}^{-1})_{ik} \mathbf{P}_{kj} \\
&= \lambda_j \delta_{ij}.
\end{aligned}$$

The matrix $\mathbf{A}'$ is thus diagonal with elements that are the eigenvalues of $\mathbf{A}$, that is,

$$\mathbf{A}' = \mathbf{P}^{-1}\mathbf{A}\mathbf{P} = \begin{pmatrix} \lambda_1 & 0 & \cdots & 0 \\ 0 & \lambda_2 & \cdots & 0 \\ \vdots & \vdots & \ddots & \vdots \\ 0 & 0 & \cdots & \lambda_n \end{pmatrix}. \tag{10.23b}$$

[5] In other words, any non-defective matrix.

Using this expression, together with (9.60b) and (10.8), it follows that

$$\det \mathbf{A} = \det \mathbf{A}' = \prod_{i=1}^{n} \lambda_i \quad \text{and} \quad \operatorname{Tr} \mathbf{A} = \operatorname{Tr} \mathbf{A}' = \sum_{i=1}^{n} \lambda_i,$$

in accordance with (10.6) and (10.9). In addition, with this transformation, the basis vectors with respect to which $\mathbf{A}'$ is defined are just the eigenvectors, since

$$\begin{pmatrix} \lambda_1 & 0 & \cdots & 0 \\ 0 & \lambda_2 & \cdots & 0 \\ \vdots & \vdots & \ddots & \vdots \\ 0 & 0 & \cdots & \lambda_n \end{pmatrix} \begin{pmatrix} 1 \\ 0 \\ 0 \\ \vdots \end{pmatrix} = \lambda_1 \begin{pmatrix} 1 \\ 0 \\ 0 \\ \vdots \end{pmatrix},$$

i.e. $\mathbf{x}'^{(1)} = \mathbf{e}^{(1)}$ and so on.

Finally, we note that for Hermitian operators $\mathbf{A}$, and some other types of matrices,[6] the eigenvectors can always be chosen to be an orthonormal set. We then have

$$(\mathbf{P}^\dagger \mathbf{P})_{ij} = \sum_k (\mathbf{P}^\dagger)_{ik} (\mathbf{P})_{kj} = \sum_k [\mathbf{x}^{(i)}]_k^* [\mathbf{x}^{(j)}]_k$$
$$= [\mathbf{x}^{(i)}]^\dagger [\mathbf{x}^{(j)}] = \delta_{ij}.$$

Hence $\mathbf{P}$ is unitary, that is, $\mathbf{P}^{-1} = \mathbf{P}^\dagger$ and so the original matrix can be diagonalised by

$$\mathbf{A}' = \mathbf{P}^{-1}\mathbf{A}\mathbf{P} = \mathbf{P}^\dagger \mathbf{A}\mathbf{P}, \tag{10.24}$$

which is easier to evaluate.

Example 10.5

Diagonalise the matrix

$$\mathbf{A} = \begin{pmatrix} 1 & 1 \\ 0 & 2 \end{pmatrix}.$$

Solution

The eigenvalues are found as usual from

$$\begin{vmatrix} 1-\lambda & 1 \\ 0 & 2-\lambda \end{vmatrix} = (\lambda - 1)(\lambda - 2) = 0,$$

that is, $\lambda_1 = 1$ and $\lambda_2 = 2$. The corresponding eigenvectors $\mathbf{u}^{(1,2)}$ are found from

$$(\mathbf{A} - \lambda_{1,2}\mathbf{I})\mathbf{u}^{(1,2)} = \mathbf{0}.$$

---

[6] This applies to all normal matrices $\mathbf{A}$, defined by the condition $\mathbf{A}^\dagger\mathbf{A} = \mathbf{A}\mathbf{A}^\dagger$, as noted in Section 10.1.3, footnote 4.

This gives the eigenvectors

$$\mathbf{u}^{(1)} = \begin{pmatrix} 1 \\ 0 \end{pmatrix} \quad \text{and} \quad \mathbf{u}^{(2)} = \begin{pmatrix} 1 \\ 1 \end{pmatrix}$$

and hence a diagonalising matrix $\mathbf{N}$ is [cf. (10.23)]

$$\mathbf{N} = \begin{pmatrix} 1 & 1 \\ 0 & 1 \end{pmatrix} \quad \text{with} \quad \mathbf{N}^{-1} = \begin{pmatrix} 1 & -1 \\ 0 & 1 \end{pmatrix}.$$

Finally, the diagonal matrix $\mathbf{A}'$ is given by

$$\mathbf{A}' = \mathbf{N}^{-1}\mathbf{A}\mathbf{N} = \begin{pmatrix} 1 & -1 \\ 0 & 1 \end{pmatrix}\begin{pmatrix} 1 & 1 \\ 0 & 2 \end{pmatrix}\begin{pmatrix} 1 & 1 \\ 0 & 1 \end{pmatrix} = \begin{pmatrix} 1 & 0 \\ 0 & 2 \end{pmatrix},$$

and, as expected, the diagonal elements are the eigenvalues of $\mathbf{A}$.

*10.2.1 Normal modes of oscillation

In physical applications, diagonalisation of a matrix often enables one to choose a set of variables that decouple from each other. A typical application in mechanics is that of coupled oscillations. An example is given in Figure 10.1. This shows two equal masses m that are joined by a spring and suspended from fixed points by strings of equal length l. We will analyse the motion of the system when the weights are displaced small distances from their equilibrium positions, as shown.

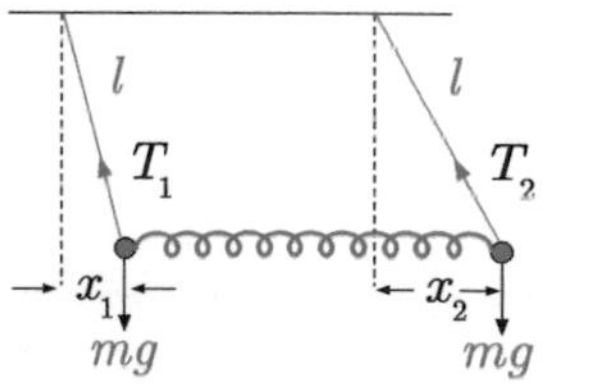

Figure 10.1 An example of coupled motion, showing the coupling of two weights via a spring.

If the instantaneous displacements are x_1 and x_2, then the force due to the spring pulling the two masses together is $mk(x_2 - x_1)$, where mk is the spring constant. The tension T_i in the string produces a horizontal restoring force of magnitude mgx_i/l, for small displacements, and so the equations of motion of the system are

$$m\frac{\mathrm{d}^2 x_1}{\mathrm{d}t^2} = -\frac{mg}{l}x_1 + mk(x_2 - x_1) \tag{10.25a}$$

and

$$m\frac{\mathrm{d}^2 x_2}{\mathrm{d}t^2} = -\frac{mg}{l}x_2 - mk(x_2 - x_1). \tag{10.25b}$$

These coupled equations may be written in the matrix form

$$\frac{\mathrm{d}^2 \mathbf{x}}{\mathrm{d}t^2} = \mathbf{A}\mathbf{x}, \tag{10.26a}$$

where

$$\mathbf{x} = \begin{pmatrix} x_1 \\ x_2 \end{pmatrix} \quad \text{and} \quad \mathbf{A} = \begin{pmatrix} \alpha & \beta \\ \beta & \alpha \end{pmatrix} = \begin{pmatrix} -g/l - k & k \\ k & -g/l - k \end{pmatrix}. \tag{10.26b}$$

We now look for a transformation $\mathbf{P}$ such that

$$\mathbf{x} = \mathbf{P}\,\mathbf{x}'$$

and

$$\mathbf{A}' = \mathbf{P}^{-1}\mathbf{A}\mathbf{P} = \begin{pmatrix} \lambda_1 & 0 \\ 0 & \lambda_2 \end{pmatrix}.$$

Since $\mathbf{P}$ is independent of t, the equations of motion become

$$\frac{\mathrm{d}^2(\mathbf{P}\,\mathbf{x}')}{\mathrm{d}t^2} = \mathbf{A}(\mathbf{P}\,\mathbf{x}') \quad \Rightarrow \quad \frac{\mathrm{d}^2\,\mathbf{x}'}{\mathrm{d}t^2} = (\mathbf{P}^{-1}\,\mathbf{A}\mathbf{P})\,\mathbf{x}' = \mathbf{A}'\mathbf{x}',$$

so that in terms of x_1' and x_2', the equations of motion decouple

$$\frac{\mathrm{d}^2 x_1'}{\mathrm{d}t^2} = \lambda_1 x_1' \quad \text{and} \quad \frac{\mathrm{d}^2 x_2'}{\mathrm{d}t^2} = \lambda_2 x_2'. \tag{10.27}$$

The eigenvalues are obtained using the characteristic equation

$$\begin{vmatrix} \alpha - \lambda & \beta \\ \beta & \alpha - \lambda \end{vmatrix} = 0 \quad \Rightarrow \quad \lambda_{1,2} = (\alpha \pm \beta),$$

that is

$$\lambda_1 = \alpha + \beta = -g/l \quad \text{and} \quad \lambda_2 = \alpha - \beta = -g/l - 2k.$$

The solution of the equations of motion (10.27) are then

$$x_1' = a_1 \sin(\omega_1 t) + b_1 \cos(\omega_1 t) \tag{10.28a}$$

and

$$x_2' = a_2 \sin(\omega_2 t) + b_2 \cos(\omega_2 t), \tag{10.28b}$$

where $\omega_1 = \sqrt{g/l}$, $\omega_2 = \sqrt{g/l + 2k}$, and where a_1, b_1, a_2, b_2 are arbitrary constants. If the latter are chosen such that $x_1' = 0$ (or $x_2' = 0$), the system vibrates with a single frequency ω_1 (or ω_2) and the motion is called a *normal mode* of the system. In general the actual motion will be a linear combination of its normal modes.

To express the motion (10.28) in terms of the original variables x_1, x_2, we need to find the matrix $\mathbf{P}$. To do this, we first have to find the eigenvectors $\mathbf{u}^{(1)}$ and $\mathbf{u}^{(2)}$. Using the techniques discussed previously, we find the two eigenvectors

$$\mathbf{u}^{(1)} = \begin{pmatrix} 1 \\ 1 \end{pmatrix} \quad \text{and} \quad \mathbf{u}^{(2)} = \begin{pmatrix} 1 \\ -1 \end{pmatrix} \quad \Rightarrow \quad \mathbf{P} = \begin{pmatrix} 1 & 1 \\ 1 & -1 \end{pmatrix}.$$

Thus, from $\mathbf{x} = \mathbf{P}\,\mathbf{x}'$,

$$x_1 = (x_1' + x_2') \quad \text{and} \quad x_2 = (x_1' - x_2'), \tag{10.29}$$

which, together with (10.28), completes the matrix analysis of solution. Specific motions depend on the values of the constants a_1, b_1, a_2, b_2, as shown in Example 10.6 below.

Finally, we note that coupled oscillations occur in a wide variety of contexts in physical science, which include compound pendulums, electrical circuits and infra-red spectroscopy. Provided the oscillations are small,[7] as in the example above, they are always described by equations of the form (10.26a), where **A** can in general be a real $n \times n$ matrix with $n \geq 2$. As in the example, these are solved by diagonalising the matrix to obtain a set of n decoupled equations analogous to (10.27), with solutions of the form (10.28) for each of the new variables. Further examples, from classical mechanics, are explored in the problems at the end of this chapter.

Example 10.6

Find the resulting motions of the system shown in Figure 10.1b discussed above for the following conditions: (a) $a_2 = b_2 = 0$, (b) $a_1 = b_1 = 0$, (c) both masses initially at rest and hanging vertically, then ball 2 moved to a point $x_2 = A$ and released.

Solution

(a) If $a_2 = b_2 = 0$, only the normal mode x'_1 is excited. Then from (10.29)

$$x_1 = x_2 = a_1 \sin \omega_1 t + b_1 \cos \omega_1 t.$$

So the two masses move in phase with the spring unstretched. This is shown in Figure 10.2a.

(b) If $a_1 = b_1 = 0$, only the normal mode x'_2 is excited. Then from (10.29),

$$x_1 = -x_2 = a_2 \sin \omega_2 t + b_2 \cos \omega_2 t.$$

The two masses are out of phase and the spring is alternately stretched and compressed. This is shown in Figure 10.2b.

(c) From (10.28) and (10.29),

$$x_1(t) = a_1 \sin \omega_1 t + b_1 \cos \omega_1 t + a_2 \sin \omega_2 t + b_2 \cos \omega_2 t$$

and

$$x_2(t) = a_1 \sin \omega_1 t + b_1 \cos \omega_1 t - a_2 \sin \omega_2 t - b_2 \cos \omega_2 t.$$

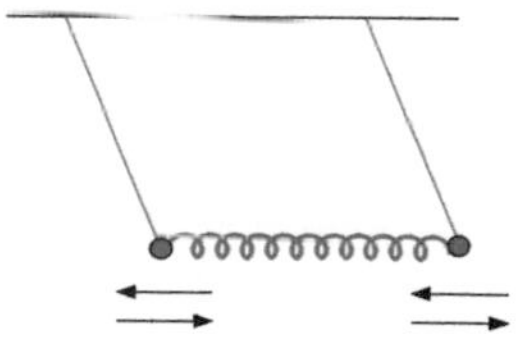
(a)

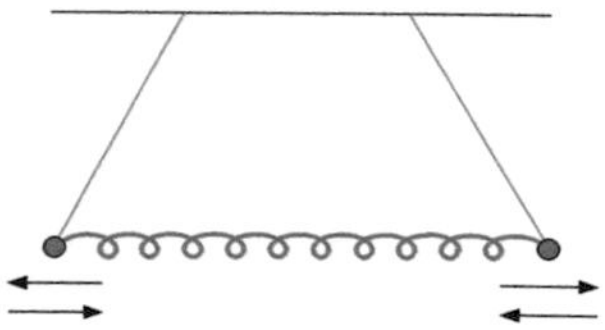
(b)

Figure 10.2 The normal modes of the system shown in Figure 10.1.

[7] The restriction to small oscillations is important in real applications, because if the oscillations increase, a point will usually be reached where additional terms occur on the right-hand side of (10.26a) and the problem then becomes much more difficult to solve. For example, if the quantity $(x_2 - x_1)$ in Figure 10.1b becomes large, the spring will cease to be perfectly elastic and the right-hand side of (10.26a) will cease to be a good approximation.

At $t = 0$, we have $x_1 = 0, \mathrm{d}x_1/\mathrm{d}t = \mathrm{d}x_2/\mathrm{d}t = 0, x_2 = A$, so

$$\begin{aligned} x_1(0) &= b_1 + b_2 = 0 \quad \Rightarrow \quad b_2 = -b_1, \\ x_2(0) &= b_1 - b_2 = A \quad \Rightarrow \quad b_1 = A + b_2, \end{aligned}$$

so that $b_1 = -b_2 = A/2$ and

$$\begin{aligned} \mathrm{d}x_1/\mathrm{d}t = 0 &\Rightarrow a_1\omega_1 + a_2\omega_2 = 0, \\ \mathrm{d}x_2/\mathrm{d}t = 0 &\Rightarrow a_1\omega_1 - a_2\omega_2 = 0 \end{aligned}$$

so that $a_1 = a_2 = 0$. Thus,

$$x_1(t) = \frac{A}{2}(\cos\omega_1 t - \cos\omega_2 t) \quad \text{and}$$

$$x_2(t) = \frac{A}{2}(\cos\omega_1 t + \cos\omega_2 t),$$

which, defining $\omega \equiv \frac{1}{2}(\omega_1 + \omega_2)$ and $\Omega \equiv \frac{1}{2}(\omega_1 - \omega_2)$, may be written

$$x_1(t) = A\sin\omega t \sin\Omega\, t \quad \text{and} \quad x_2(t) = A\cos\omega t \cos\Omega\, t.$$

*10.2.2 Quadratic forms

Another example of matrix diagonalisation occurs in the theory of *quadratic forms.* These are expressions of the type

$$Q = \sum_{i=1}^{n}\sum_{j=1}^{n} a_{ij}x_i x_j, \tag{10.30}$$

where the quantities x_i and the coefficients a_{ij} are real. The latter form an $n \times n$ square matrix $\mathbf{A}$, so (10.30) may be written

$$Q = \mathbf{x}^{\mathrm{T}}\mathbf{A}\,\mathbf{x}, \tag{10.31}$$

where $\mathbf{x}^{\mathrm{T}} = (x_1, x_2, \cdots, x_n)$. Furthermore, it can be seen from (10.30) that Q is the sum of terms of the form $(a_{ij} + a_{ji})x_i x_j$, which may be written $(c_{ij}x_i x_j + c_{ji}x_j x_i)$, where

$$c_{ij} = c_{ji} = \tfrac{1}{2}(a_{ij} + a_{ji}).$$

Hence the quadratic form (10.31) can always be written in the form

$$Q = \mathbf{x}^{\mathrm{T}}\mathbf{C}\mathbf{x} = \sum_{i=1}^{n}\sum_{j=1}^{n} c_{ij}x_i x_j,$$

where $\mathbf{C}$ is a real symmetric matrix. Therefore, in considering the quadratic forms (10.30), we may, without loss of generality, consider only cases where $\mathbf{A}$ is a real symmetric matrix. If $Q > 0$, it is said to be *positive definite.*

One application of quadratic forms is in analytic geometry. For example, suppose a surface in three-dimensional space is described by the equation

$$k = a_{11}x^2 + a_{22}y^2 + a_{33}z^2 + 2a_{12}xy + 2a_{23}yz + 2a_{31}zx, \quad (10.32)$$

where x, y, z are Cartesian co-ordinates and k is a constant. Because of the cross terms in xy, etc., it is not obvious what is the geometrical nature of the surface. Its visualisation would be simpler if the surface could be expressed in co-ordinates such that the cross terms were absent. This may be done by using the technique of diagonalisation. We start by writing (10.32) in the matrix form

$$\mathbf{x}^{\mathrm{T}}\mathbf{A}\mathbf{x} = k, \quad (10.33)$$

where $\mathbf{x} = (x, y, z)^{\mathrm{T}}$ and $\mathbf{A}$ is a real symmetric matrix. Since $\mathbf{A}$ is Hermitian it can be diagonalised by a unitary matrix $\mathbf{P}$, where $\mathbf{P}^{-1} = \mathbf{P}^{\dagger}$; and since it is also real, it can be chosen to be a real orthogonal matrix, with $\mathbf{P}^{-1} = \mathbf{P}^{\mathrm{T}}$, so that

$$\mathbf{P}^{-1}\mathbf{A}\mathbf{P} = \mathbf{P}^{\mathrm{T}}\mathbf{A}\mathbf{P} = \begin{pmatrix} \lambda_1 & 0 & 0 \\ 0 & \lambda_2 & 0 \\ 0 & 0 & \lambda_3 \end{pmatrix} \equiv \Lambda,$$

where λ_i $(i = 1, 2, 3)$ are the eigenvalues of $\mathbf{A}$. Given $\mathbf{P}$, we can define new co-ordinates $\mathbf{x}' = (x', y', z')$ in terms of which (10.32) becomes simpler. The equation for the surface in these new co-ordinates may be found by writing

$$\mathbf{x}^{\mathrm{T}}\mathbf{A}\,\mathbf{x} = \mathbf{x}^{\mathrm{T}}\mathbf{P}\mathbf{P}^{\mathrm{T}}\mathbf{A}\mathbf{P}\mathbf{P}^{\mathrm{T}}\mathbf{x} = (\mathbf{P}^{\mathrm{T}}\mathbf{x})^{\mathrm{T}}\Lambda(\mathbf{P}^{\mathrm{T}}\mathbf{x}),$$

so that (10.33) becomes

$$(\mathbf{x}')^{\mathrm{T}}\Lambda\,\mathbf{x}' = k, \quad (10.34)$$

where $\mathbf{x}' = \mathbf{P}^{\mathrm{T}}\mathbf{x} = (x', y', z')$. Writing this in terms of the new Cartesian co-ordinates gives

$$\frac{x'^2}{(k/\lambda_1)} + \frac{y'^2}{(k/\lambda_2)} + \frac{z'^2}{(k/\lambda_3)} = 1, \quad (10.35)$$

which is the equation of the quadratic surface where the eigenvectors of $\mathbf{A}$ define the direction the new co-ordinate axes x', y', z', called the *principal axes*. They are related to the original axes x, y, z by rotations about, and possibly a reflection in, the origin.

The geometrical interpretation depends on the signs of the denominators in (10.35). If all three are positive, then (10.35) describes an *ellipsoid*, as shown in Figure 10.3. In this case the principal axis x', for example, cuts the quadratic surface where $y' = z' = 0$,

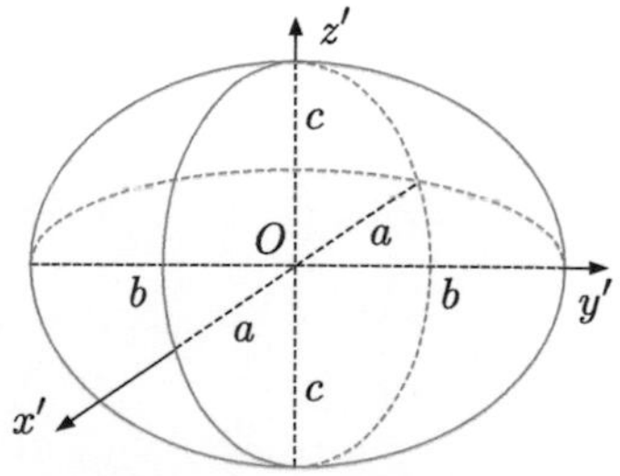

Figure 10.3 An ellipsoid, showing the principal axes x', y', z' and the lengths of the semi-axes a, b, c.

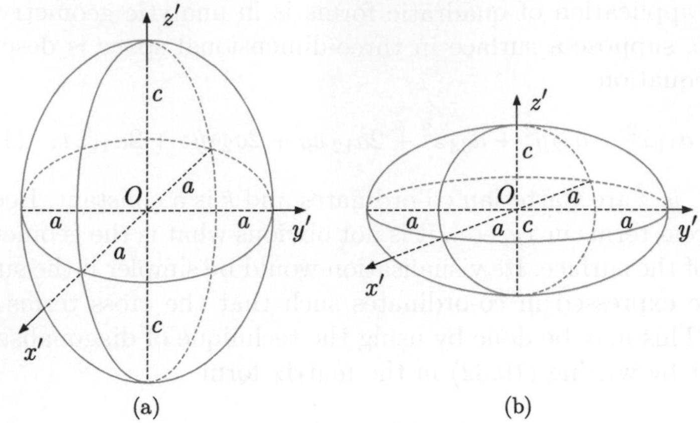

Figure 10.4 (a) Prolate spheroid resulting when the lengths of the semi-axes satisfy $a = b < c$. (b) Oblate spheroid resulting when the lengths of the semi-axes satisfy $a = b > c$.

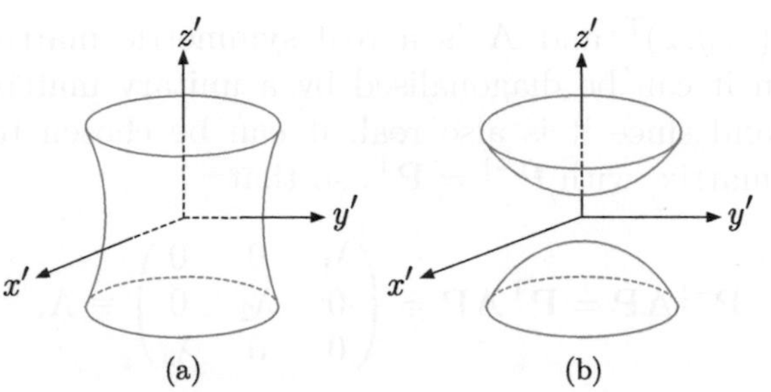

Figure 10.5 (a) Hyperboloid of one sheet, (b) hyperboloid of two sheets.

which from (10.35) is where $x' = \pm(k/\lambda_1)^{1/2}$. Thus the distance along the x' axis from the origin to the point of intersection is $a = (k/\lambda_1)^{1/2}$. This is called the length of the semi-axis. The lengths of the other semi-axes are similarly given by $b = (k/\lambda_2)^{1/2}$ and $c = (k/\lambda_3)^{1/2}$, as shown in Figure 10.3.[8]

If all three denominators are different, then the ellipsoid is said to be *triaxial*. More familiar shapes are obtained when two of the denominators are equal. For example, if $a = b > c$, the ellipsoid reduces to an *oblate spheroid*, as shown in Figure 10.4b; while if $a = b < c$, it reduces to a *prolate spheroid*, as shown in Figure 10.4a. A familiar example of the former is the shape of earth, which is to a good approximation an oblate spheroid; while a rugby (or American) football is roughly a prolate spheroid. If $a = b = c$, the spheroid reduces to a sphere.

Finally, if one of the denominators in (10.35) is negative, the shape is a *hyperboloid* of one sheet, while if two are negative, it corresponds to a hyperboloid of two sheets, as shown in Figure 10.5a

[8] If the ellipsoid is drawn using the original x, y, z, co-ordinates, it has the same shape but is oriented so that its principal axes lie along the x', y', z' directions, that is, the directions of the eigenvalues of **A**.

and Figure 10.5b, respectively. Examples of the former are the large cooling towers seen at power stations.

Example 10.7

Show that the curve $5x^2 + 5y^2 + 6xy = 8$ in the x–y plane is an ellipse and find the direction and lengths of its principal axes. Sketch the ellipse together with its principal axes in the x–y plane.

Solution

The curve may be written in the form $\mathbf{x}^{\mathrm{T}}\mathbf{A}\mathbf{x} = k$, where $\mathbf{A}$ is the symmetric matrix

$$\mathbf{A} = \begin{pmatrix} 5 & 3 \\ 3 & 5 \end{pmatrix}, \quad \mathbf{x} = \begin{pmatrix} x \\ y \end{pmatrix},$$

and $k = 8$. We start by finding the eigenvalues of $\mathbf{A}$. These are given by

$$\begin{vmatrix} 5-\lambda & 3 \\ 3 & 5-\lambda \end{vmatrix} = 0.$$

Thus, $\lambda_1 = 2$ and $\lambda_2 = 8$. The eigenvectors then follow from the equations

$$\begin{pmatrix} 5 & 3 \\ 3 & 5 \end{pmatrix}\begin{pmatrix} x \\ y \end{pmatrix} = \lambda \begin{pmatrix} x \\ y \end{pmatrix}.$$

For $\lambda_1 = 2$ and $\lambda_2 = 8$ this gives

$$\mathbf{u}^{(1)} = \frac{1}{\sqrt{2}}\begin{pmatrix} -1 \\ 1 \end{pmatrix} \quad \text{and} \quad \mathbf{u}^{(2)} = \frac{1}{\sqrt{2}}\begin{pmatrix} 1 \\ 1 \end{pmatrix},$$

respectively. So the principle axes are along the directions $y = -x$ and $y = x$. To find the lengths of the principal axes, we first find the matrix that diagonalises $\mathbf{A}$. This is

$$\mathbf{P} = \frac{1}{\sqrt{2}}\begin{pmatrix} -1 & 1 \\ 1 & 1 \end{pmatrix} \quad \text{with} \quad \mathbf{P}^{-1} = \frac{1}{\sqrt{2}}\begin{pmatrix} -1 & 1 \\ 1 & 1 \end{pmatrix}.$$

So in terms of the transformed variables $\mathbf{x}' = \mathbf{P}^{-1}\mathbf{x}$, the curve becomes

$$(x')^2 + 4(y')^2 = 4,$$

in analogy to (10.35). This is an ellipse and the lengths of the principal axes are

$$2(1/\sqrt{\lambda_1}) = 2(1/\sqrt{2}) = \sqrt{2} \text{ and } 2(1/\sqrt{\lambda_2}) = 2(1/\sqrt{8}) = 1/\sqrt{2}.$$

The resulting ellipse is shown in Figure 10.6.

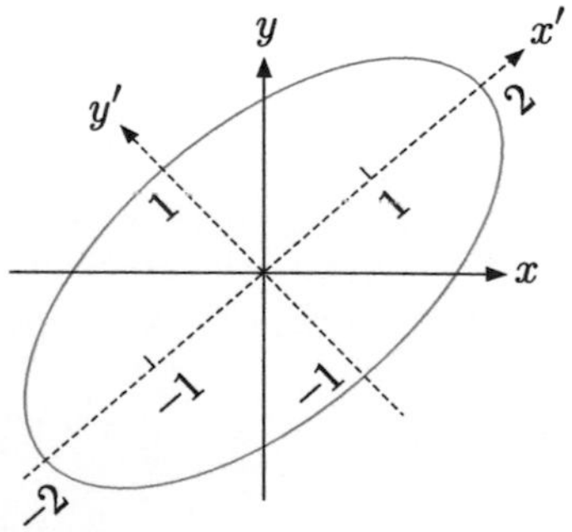

Figure 10.6

Problems 10

10.1 Given that one of the eigenvalues of the matrix

$$\begin{pmatrix} 1 & 2 & -1 \\ 1 & 2 & 0 \\ -1 & 1 & 2 \end{pmatrix}$$

is $\lambda = 3$, find the other two eigenvalues, and hence the associated eigenvectors. Are the eigenvectors orthogonal?

10.2 Verify that the sum of the eigenvalues of the matrix

$$\mathbf{A} = \begin{pmatrix} 1 & 3 & 2 \\ 1 & 2 & 3 \\ 3 & 2 & 1 \end{pmatrix}$$

is equal to its trace and that their product is equal to $\det \mathbf{A}$.

10.3 Verify that the eigenvalues of the matrix

$$\mathbf{A} = \begin{pmatrix} 2 & 1 & 1 \\ -1 & 0 & 1 \\ 1 & 2 & 2 \end{pmatrix}$$

are the inverses of the eigenvalues of $\mathbf{A}^{-1}$.

10.4 If $\mathbf{A}$ is an $n \times n$ matrix with eigenvalues λ_i $(i = 1, 2, \ldots, n)$, show that the transpose matrix $\mathbf{A}^{\mathrm{T}}$ also has eigenvalues λ_i, and that the inverse matrix $\mathbf{A}^{-1}$, if it exists, has eigenvalues λ_i^{-1}.

10.5 (a) Prove that the eigenvalues of a unitary matrix have unit modulus.

(b) Show that an anti-unitary matrix $\mathbf{U}^\dagger = -\mathbf{U}^*$ has no eigenvalues.

10.6 Find the linearly independent eigenvectors of the matrix

$$\mathbf{A} = \begin{pmatrix} i & 0 & -i \\ 0 & 1 & 0 \\ i & 0 & -i \end{pmatrix}.$$

Is the matrix defective?

10.7 Show that the eigenvalues of an anti-Hermitian matrix $\mathbf{A}^\dagger = -\mathbf{A}$ are purely imaginary, and that the eigenvectors corresponding to distinct eigenvectors are orthogonal.

10.8 (a) Find the eigenvalues and eigenvectors of the matrix

$$\mathbf{A} = \begin{pmatrix} 1 & i \\ -2i & 1 \end{pmatrix}.$$

Are the eigenvectors orthogonal?

(b) Verify that the eigenvectors of the Hermitian matrix

$$\mathbf{A} = \begin{pmatrix} 1 & 1+i \\ 1-i & 1 \end{pmatrix}$$

are orthogonal.

10.9 Confirm, by explicit calculation, that the eigenvalues of the real, symmetric matrix

$$\mathbf{A} = \begin{pmatrix} 2 & 1 & 2 \\ 1 & 2 & 2 \\ 2 & 2 & 1 \end{pmatrix}$$

are real, and its eigenvectors are orthogonal.

10.10 Use the Gram–Schmidt orthogonalisation process of Section 10.1.3 to construct the orthonormalised vectors $\hat{\mathbf{x}}^{(i)}$ $(i = 1, 2, 3)$ corresponding to the vectors

$$\mathbf{u}^{(1)} = (1, 0, 1),\ \mathbf{u}^{(2)} = (2, 1, 0),\ \mathbf{u}^{(3)} = (0, 1, 2).$$

10.11 Source a computer matrix-manipulation application on the internet (there are several free ones) and use it to find the determinant, the inverse, the eigenvalues and the eigenvectors of the matrix

$$\mathbf{A} = \begin{pmatrix} 1 & 0 & -2 & 1 & 2 & 1 \\ 0 & 2 & 1 & 1 & 0 & 2 \\ 2 & 1 & -1 & 0 & 0 & 1 \\ 1 & 2 & 3 & 0 & 2 & 1 \\ 0 & -2 & 1 & 3 & -1 & 1 \\ 1 & 1 & -1 & 0 & 2 & 1 \end{pmatrix}.$$

***10.12** Find the matrix that diagonalises the matrix

$$\mathbf{A} = \begin{pmatrix} 1 & 2 & 1 \\ 0 & 2 & 0 \\ 2 & 1 & 1 \end{pmatrix}.$$

Verify this result by finding the form of the resulting diagonal matrix.

***10.13** Consider three masses on the x-axis joined by springs that obey Hooke's law with a common spring constant k, as shown in Figure 10.7. If the three masses remain on the x-axis, find the normal modes, in which they all move with the same frequency. (This type of system provides a simple model of molecules like CO_2 that is, carbon dioxide, where the three atoms are arranged linearly.

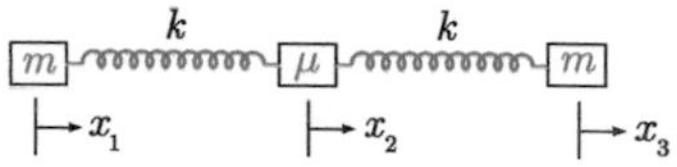

Figure 10.7

***10.14** (a) A mass m, connected to two fixed points by identical stretched strings each of length l and with tension T, is displaced transversely from its equilibrium position by a distance y, as shown in Figure 10.8a. Assuming that for small displacements the change in the tension T can be neglected, show that

$$\frac{\mathrm{d}^2 y}{\mathrm{d}t^2} = -2\omega_0^2 y \quad \text{where} \quad \omega_0^2 = \frac{T}{ml}.$$

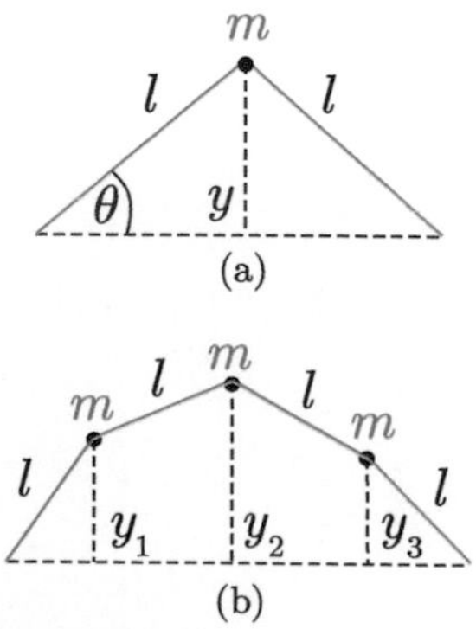

Figure 10.8

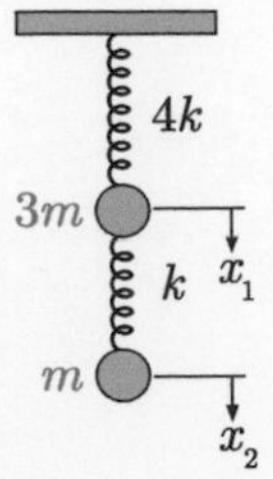

Figure 10.9

(b) Three masses m, connected to two fixed points and to each other by four identical strings of length ℓ and with tension T, undergo small transverse displacements y_1, y_2, y_3, as shown in Figure 10.8b. Deduce the frequencies and normal modes, and sketch the latter.

***10.15** Two masses, m and 3 m, suspended from two springs with force constants 4 k and k, respectively, are displaced downwards from their equilibrium positions by x_1 and x_2, as shown in Figure 10.9. If they are released from rest at $x_1 = 0$, $x_2 = 1$ at time $t = 0$, what will their positions be at time $t = (m/k)^{1/2}$?

***10.16** Consider the surface described by the equation

$$11x^2 + 5y^2 + 2z^2 + 16xy + 20yz - 4xz + 9 = 0.$$

By writing this in the quadratic form $\mathbf{x}^{\mathrm{T}}\mathbf{A}\mathbf{x} = k$, find the principal axes, and show that it is a two-sheet hyperboloid. What is the distance between the two sheets? *Hint*: One of the eigenvalues of $\mathbf{A}$ is $\lambda_1 = 18$.

***10.17** Classify the surfaces described by the quadratic forms $\mathbf{x}^{\mathrm{T}}\mathbf{A}\mathbf{x} = k > 0$, as ellipsoid or spheroid (specify which type in either case), when

$$\text{(a) } \mathbf{A} = \begin{pmatrix} 1 & 0 & 0 \\ 0 & 2 & -1 \\ 0 & -1 & 2 \end{pmatrix} \text{ (b) } \mathbf{A} = \begin{pmatrix} 1 & 0 & -1 \\ 0 & 2 & 1 \\ -1 & 1 & 1 \end{pmatrix}$$

***10.18** Show that the quadratic form

$$Q = \mathbf{x}\mathbf{A}^{\mathrm{T}}\mathbf{x} \geq \lambda_m$$

for any unit vector $\mathbf{x}$, where λ_m is the smallest eigenvalue of $\mathbf{A}$. Hence state the condition for Q to be positive definite ($Q > 0$) for all $\mathbf{x}$, except for the null vector $\mathbf{x} = \mathbf{0}$.

***10.19** Show that the curve described by the equation

$$3x^2 - 3y^2 - 8xy - 5 = 0$$

is a hyperbola. Find the angle between the principal axes and the x and y axes, and sketch the hyperbola in the x–y plane. What are the x and y co-ordinates of the points at which the two branches are closest together?

11

Line and multiple integrals

In Chapter 7 we extended the discussion of differentiation given in Chapter 3 to functions of several variables. In this chapter we will extend the discussion of integration given in Chapter 4 in a similar way. We will begin by discussing functions of two variables, which we will usually take to be the Cartesian co-ordinates x, y, although they could equally well be, for example, a position and a time. The discussion will then be generalised to three or more variables and to other co-ordinate systems, especially polar co-ordinates in three dimensions. This will form the basis for important applications in vector analysis, which is an essential tool in understanding topics such as electromagnetic fields, fluid dynamics and potential theory, and which will be discussed extensively in Chapter 12.

11.1 Line integrals

In this section, we first introduce line integrals and their properties in two dimensions and then briefly indicate their extension to three dimensions, which is relatively straightforward. In both cases we will use Cartesian co-ordinates.

11.1.1 Line integrals in a plane

Suppose $y = f(x)$ is a real single-valued monotonic continuous function of x defined in some interval $x_1 < x < x_2$, as represented by the curve C shown in Figure 11.1a. Then, if $P(x, y)$ is a real single-valued continuous function of x and y for all points on the curve C, the integral

$$\int_C P(x, y)\mathrm{d}x, \tag{11.1}$$

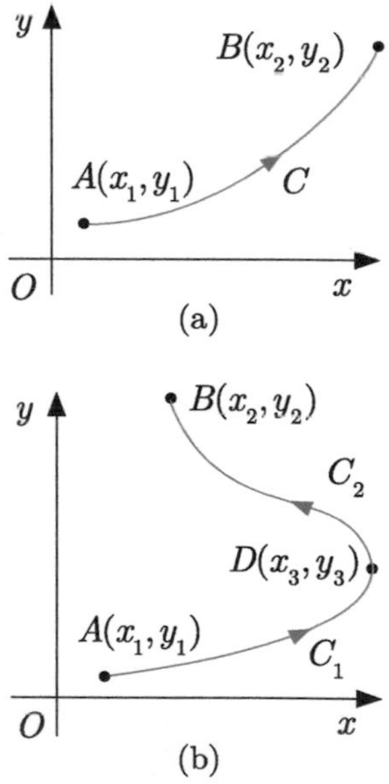

Figure 11.1 Integration path for (a) a single-valued function and (b) a two-valued function.

Mathematics for Physicists, First Edition. B.R. Martin and G. Shaw.

Companion website: www.wiley.com/go/martin/mathsforphysicists

is called a *line integral* and the symbol C on the integration sign indicates that the path, or contour, of integration from the initial point A to the end point B is along the curve C. The formal definition of line integral is closely related to that of ordinary integrals as discussed in Chapter 4. Thus for a function $P(x, y)$,

$$\int_C P(x, y)\mathrm{d}x = \lim_{\delta x_i \to 0} \sum P(x_i, y)\delta x_i, \tag{11.2}$$

where the sum is over all elements δx_i on the curve C. Since $y = f(x)$, the integral (11.1) is equivalent to an ordinary integral with respect to a single variable x. Thus,

$$\int_C P(x, y)\mathrm{d}x = \int_{x=x_1}^{x=x_2} P[x, f(x)]\mathrm{d}x. \tag{11.3}$$

We could also consider the integral along C as being with respect to the variable y by inverting the relation $y = f(x)$ to give x as a function of y along C, for example $x = g(y)$. Then if $Q(x, y)$ is another real single-valued continuous function of x and y for all points on the curve C, a line integral analogous to (11.3) is

$$\int_C Q(x, y)\mathrm{d}y = \int_{y=y_1}^{y=y_2} Q[g(y), y]\mathrm{d}y. \tag{11.4}$$

Alternatively, we can convert line integrals over x and y into line integrals over y and x by writing

$$\int_C P(x, y)\mathrm{d}x = \int_{y=y_1}^{y=y_2} P[g(y), y]g'(y)\mathrm{d}y \tag{11.5}$$

and

$$\int_C Q(x, y)\mathrm{d}y = \int_{x=x_1}^{x=x_2} Q[x, f(x)]\, f'(x)\mathrm{d}x \tag{11.6}$$

where we have used $\mathrm{d}x = g'(y)\mathrm{d}y$ and $\mathrm{d}y = f'(x)\mathrm{d}x$ respectively. In what follows, it is often useful to write line integrals along a given curve C in the general form

$$\int [P(x, y)\, \mathrm{d}x + Q(x, y)\, \mathrm{d}y], \tag{11.7}$$

where P and Q are given functions.

In the above discussion, we have assumed that the contour of integration C can be described by a single-valued function $y = f(x)$. This is not always the case. Consider the curve shown in Figure 11.1b.

For some values of x, two different values of y are obtained, that is, $f(x)$ is not single-valued. In this case, the integral must be divided into two parts (or more if $f(x)$ is multi-valued) in each of which it is single-valued, and by the results of Chapter 4, we may write

$$\int_C P(x,y)\mathrm{d}x = \int_{C_1} P[x, f_1(x)]\mathrm{d}x + \int_{C_2} P[x, f_2(x)]\mathrm{d}x,$$

where C_1 is the path from A to D, which is described by the function $f_1(x)$, and C_2 is the path from D to B, which is described by the function $f_2(x)$.

Finally, the path C may be defined by an implicit relationship between the x and y co-ordinates, and in particular by parametric forms $x = x(t)$ and $y = y(t)$. Here, both x and y are defined by single-valued differentiable functions of a single parameter t, so that as t goes from t_A, the value of t at A, to t_B, the value of t at B, the path between A and B is traced out in the right direction once and once only. Any line integral of the general form (11.7) can then be transformed into a definite integral over t:

$$I_C = \int_{t_A}^{t_B} \left[P(x,y)\frac{\mathrm{d}x}{\mathrm{d}t} + Q(x,y)\frac{\mathrm{d}y}{\mathrm{d}t}\right] \mathrm{d}t, \qquad (11.8)$$

by substituting the given forms $x(t), y(t)$.

Example 11.1

Evaluate the line integral

$$I = \int_C [2xy\,\mathrm{d}x - (x^2 + y^2)\,\mathrm{d}y],$$

between the points $(x, y) = (0, 1)$ and $(1, 2)$, where the contour C is: (a) the curve $y = x^2$ and (b) the line $y = x + 1$.

Solution

(a) Using $y = x^2$, with $\mathrm{d}y = 2x\,\mathrm{d}x$, gives

$$I = \int_0^1 [2x^3 - (x^2 + x^4)2x]\mathrm{d}x = -2\int_0^1 x^5\mathrm{d}x = -1/3.$$

(b) Using $y = x + 1$, with $\mathrm{d}y = \mathrm{d}x$, gives

$$I = \int_0^1 \{2x(x+1) - [x^2 + (x+1)^2]\}\mathrm{d}x = -\int_0^1 \mathrm{d}x = -1.$$

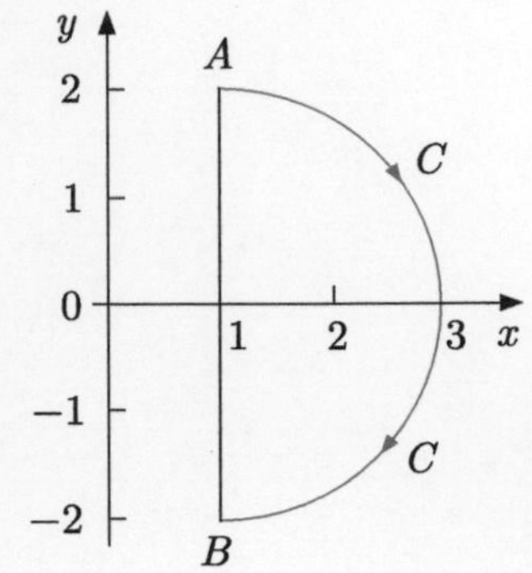

Figure 11.2

Example 11.2

Evaluate the line integral

$$I = \int_C (x - 2y)\mathrm{d}x$$

from point A to point B, around the semi-circular curve C, as shown in Figure 11.2.

Solution

The equation of the circle of which C forms a part is

$$y^2 = 4 - (x-1)^2,$$

and thus is not single-valued. The contour C must therefore be split into two sections, with

$$y(x) = \begin{cases} \sqrt{4-(x-1)^2} & \text{in the upper half} \\ -\sqrt{4-(x-1)^2} & \text{in the lower half} \end{cases}$$

Then

$$I = \int_1^3 [x - 2\sqrt{4-(x-1)^2}]\,\mathrm{d}x + \int_3^1 [x + 2\sqrt{4-(x-1)^2}]\,\mathrm{d}x$$

$$= -4\int_1^3 \sqrt{4-(x-1)^2}\,\mathrm{d}x.$$

This integral may be evaluated by changing variables to t, where $\sin t = \frac{1}{2}(1-x)$ and $-\pi/2 < t < 0$ corresponds to the range $1 < x < 3$. Then

$$I = 8\sin^{-1}\left[\tfrac{1}{2}(1-x)\right] + 2(1-x)(-x^2+2x+3)^{1/2} = -4\pi.$$

Example 11.3

Evaluate the line integral

$$I = \int_C (x^2 + 2y)\mathrm{d}x,$$

where C is the path starting from the point $(x, y) = (1, 0)$ and finishing at $(x, y) = (-1, 0)$ moving along the circle $x^2 + y^2 = 1$.

Solution

Here we can use a parametric representation by writing

$$x = \cos t, \quad y = \sin t,$$

with $\mathrm{d}x = -\sin t\,\mathrm{d}t$, and the integration is from $t = 0$ to $t = \pi$. Then

$$I = -\int_0^{\pi} (\cos^2 t + 2\sin t)\sin t\,\mathrm{d}t = -\int_0^{\pi} \left(\sin t\cos^2 t + 1 - \cos 2t\right)\mathrm{d}t$$

$$= -\left[-\frac{1}{3}\cos^3 t + t - \frac{1}{2}\sin 2t\right]_0^{\pi} = -\left(\frac{2+3\pi}{3}\right).$$

11.1.2 Integrals around closed contours and along arcs

In Figure 11.1a the integral is along the path C from A to B, and it is clear that the value of the line integral depends on the functional form $y = f(x)$ of the curve, and so in general the integral will be different for different paths between the same two points, although later we shall meet examples where this is not true and the line integral only depends on the end points of the integral. We could of course also take the integration from B to A. It follows from the results established in Chapter 4 for ordinary integrals that

$$\int_{A\to B} P(x,y)\mathrm{d}x = -\int_{B\to A} P(x,y)\mathrm{d}x, \qquad (11.9)$$

where it is understood that the integration is still along the path C, but in the reverse direction. It then follows that a line integral from A to B, and returning to A along the *same* path is zero. However, for a closed contour where the return path from B to A is *not* the same as that from A to B, in general the integral is non-zero, although again we will meet examples later where this is not true.

If the integration path is a simple closed plane curve, that is, one that does not cross itself, such as that shown in Figure 11.3, the integral is written

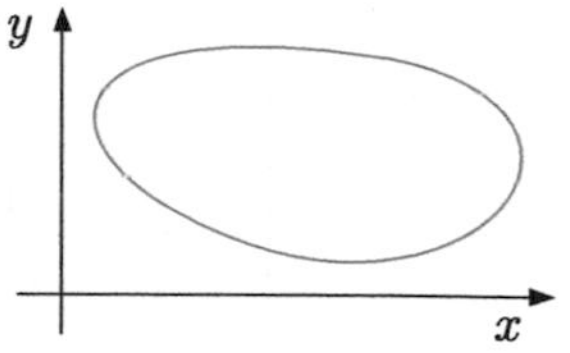

Figure 11.3 A simple closed plane curve.

$$\oint P(x,y)\mathrm{d}x \quad \text{or} \quad \oint_C P(x,y)\mathrm{d}x. \qquad (11.10)$$

It is conventionally assumed that the integration is in the counter-clockwise direction, but to be totally unambiguous, the direction of travel around the closed contour can be indicated by an arrow on the circle, i.e.

$$\ointctrclockwise P(x,y)\mathrm{d}x \quad \text{or} \quad \ointclockwise P(x,y)\mathrm{d}x,$$

where the symbols indicate integration in the counter-clockwise (positive) and clockwise (negative) directions, respectively. A closed curve cannot be represented by a single-valued function, so when evaluating integrals like (11.10), the technique of breaking the contour of integration into sections must be used.

We may also consider line integrals of the form

$$\int_C P(x,y)\mathrm{d}l, \qquad (11.11a)$$

where dl is an infinitesimal arc length of the curve C. For the simple case $P(x,y)=1$, the integral

$$L=\int_A^B \mathrm{d}l \qquad (11.11b)$$

gives the length of the curve $f(x)$ from A to B. The integrals (11.11a) and (11.11b) may be converted to the standard form (11.1), with a modified function P, by using the result

$$\mathrm{d}l=[(\mathrm{d}x)^2+(\mathrm{d}y)^2]^{1/2}=\left[1+\left(\frac{\mathrm{d}y}{\mathrm{d}x}\right)^2\right]^{1/2}\mathrm{d}x, \qquad (11.12a)$$

where $y=f(x)$, or

$$\mathrm{d}l=\left[\left(\frac{\mathrm{d}x}{\mathrm{d}t}\right)^2+\left(\frac{\mathrm{d}y}{\mathrm{d}t}\right)^2\right]^{1/2}\mathrm{d}t \qquad (11.12b)$$

if x and y are given in parametric forms as functions of a parameter t.

Example 11.4

Evaluate the integral

$$\oint(x\mathrm{d}y+y\mathrm{d}x)$$

around the contour shown in Figure 11.4.

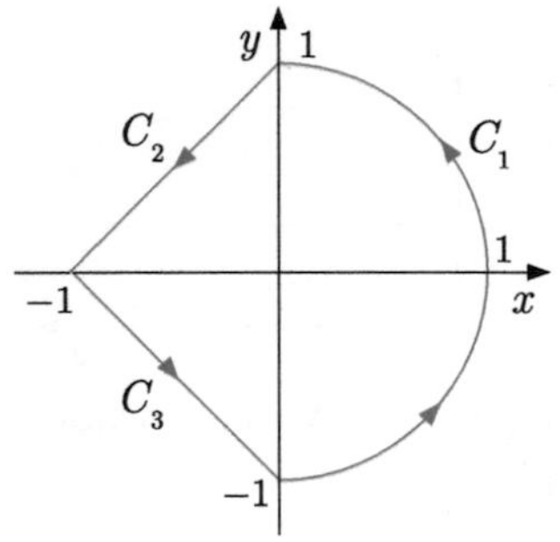

Figure 11.4

Solution

The three parts of the contour may be parametrised as follows:

$$C_1\colon x=\cos t,\quad y=\sin t,\qquad -\pi/2\le t\le\pi/2$$
$$C_2\colon y=x+1,\qquad -1\le x\le 0$$
$$C_3\colon y=-x-1,\qquad -1\le x\le 0$$

Along C_1, using $\mathrm{d}x/\mathrm{d}t=-\sin t$ and $\mathrm{d}y/\mathrm{d}t=\cos t$ gives

$$I_1=\int_{-\pi/2}^{\pi/2}[\cos t\cos t-\sin t(-\sin t)]\mathrm{d}t=\pi.$$

Along C_2,

$$I_2 = \int_0^{-1} (x\mathrm{d}y + y\,\mathrm{d}x) = [x^2 + x]_0^{-1} = 0.$$

and along C_3,

$$I_3 = \int_{-1}^{0} (x\mathrm{d}y + y\,\mathrm{d}x) = \int_{-1}^{0} [-x - (x+1)]\,\mathrm{d}x = -[x^2 + x]_{-1}^{0} = 0.$$

Finally, $I = I_1 + I_2 + I_3 = \pi$.

Example 11.5

Find the length along the curve $y = x^2$ between the points $x = 0$ and $x = 1$. Note the integral

$$\int (1 + nx^2)^{1/2}\mathrm{d}x = \frac{1}{2}x(1 + nx^2)^{1/2} + \frac{1}{2\sqrt{n}}\ln(\sqrt{n}x + \sqrt{1 + nx^2}).$$

Solution

Using (11.12a) gives

$$L = \int_0^1 \left[1 + \left(\frac{\mathrm{d}y}{\mathrm{d}x}\right)^2\right]^{1/2} \mathrm{d}x = \int_0^1 (1 + 4x^2)^{1/2}\mathrm{d}x,$$

which using the integral given is

$$L = \frac{1}{4}\left[2\sqrt{5} + \ln(2 + \sqrt{5})\right].$$

11.1.3 Line integrals in three dimensions

The extension of the above ideas to functions of three real variables x, y, z is straightforward, and will be summarised here very briefly. In an obvious notation, the general line integral (11.7) becomes

$$\int_C (P\mathrm{d}x + Q\mathrm{d}y + R\mathrm{d}z), \qquad (11.13)$$

where P, Q, R are single-valued functions of x, y, z, and if $y = f_1(x)$, $z = f_2(x)$, with $x_1 < x < x_2$ along the path of integration, then

$$\int_C P(x, y, z)\mathrm{d}x = \int_{x_1}^{x_2} P\left[x, f_1(x), f_2(x)\right]\mathrm{d}x \qquad (11.14)$$

in analogy with (11.3), with similar expressions analogous to (11.4) for the other terms in (11.13). Alternatively, if the path is specified by three functions $x(t)$, $y(t)$, $z(t)$ of a single parameter t, with $t_A < t < t_B$, then (11.13) becomes a single integral

$$\int_C \left(P\frac{\mathrm{d}x}{\mathrm{d}t} + Q\frac{\mathrm{d}y}{\mathrm{d}t} + R\frac{\mathrm{d}z}{\mathrm{d}t} \right) \mathrm{d}t. \tag{11.15}$$

Finally, we may again consider integrals of the form

$$\int_C P(x, y, z)\mathrm{d}l \tag{11.16a}$$

in analogy to (11.11), where the element of arc length $\mathrm{d}l$ is now given by

$$\mathrm{d}l = \left[1 + \left(\frac{\mathrm{d}y}{\mathrm{d}x}\right)^2 + \left(\frac{\mathrm{d}z}{\mathrm{d}x}\right)^2 \right]^{1/2} \mathrm{d}x, \tag{11.16b}$$

or, if the path C is specified by a real parameter t,

$$\mathrm{d}l = \left[\left(\frac{\mathrm{d}x}{\mathrm{d}t}\right)^2 + \left(\frac{\mathrm{d}y}{\mathrm{d}t}\right)^2 + \left(\frac{\mathrm{d}z}{\mathrm{d}t}\right)^2 \right] \mathrm{d}t. \tag{11.16c}$$

Example 11.6

The relations

$$x = d\cos t, \quad y = d\sin t, \quad z = pt$$

define a helix with diameter d and pitch p. Evaluate the integral

$$I = \int (x^2\mathrm{d}x + y^2\mathrm{d}y + z^2\mathrm{d}z)$$

from $t = 0$ to $t = 2\pi$ along the path of the helix.

Solution

Using (11.16c), I may be written

$$I = \int_0^{2\pi} \left(d^2\cos^2 t\frac{\mathrm{d}x}{\mathrm{d}t} + d^2\sin^2 t\frac{\mathrm{d}y}{\mathrm{d}t} + p^2t^2\frac{\mathrm{d}z}{\mathrm{d}t} \right) \mathrm{d}t$$

$$= \int_0^{2\pi} \left(-d^3\sin t\cos^2 t + d^3\cos t\sin^2 t + p^3t^2 \right) \mathrm{d}t$$

$$= \left[\frac{d^3}{3}(\sin^3 t + \cos^3 t) + \frac{p^3t^3}{3} \right]_0^{2\pi} = \frac{8}{3}\pi^3p^3.$$

11.2 Double integrals

The ideas discussed in Chapter 4 for defining and evaluating definite integrals may be extended to evaluate integrals over two or more variables. We start by considering double integrals, that is, integrals over two variables, again using Cartesian co-ordinates. These may be written in a number of forms:

$$\iint_S f(x, y)\mathrm{d}x\,\mathrm{d}y \quad \text{or} \quad \int \mathrm{d}x \int f(x, y)\,\mathrm{d}y \quad \text{or} \quad \int \mathrm{d}y \int f(x, y)\,\mathrm{d}x, \tag{11.17a}$$

where S is an area, which we will assume is defined by a simple boundary curve, that is, one that does not cross itself, and where the limits on the x and y integrations will be specified shortly. In addition, we will assume that the function $f(x, y)$ is finite, single-valued and continuous within and on the boundary.

In Chapter 4 we defined a definite integral of a function $f(x)$ of a single variable x by dividing the range of integration into n small intervals of width δx_n, and then taking the limit

$$\lim_{\delta x_n \to 0} \sum_n f(x_n)\,\delta x_n,$$

where $f(x_n)$ is the value of $f(x)$ at the mid-point of the interval. The double integrals (11.17a) are defined in an analogous way by dividing S into small elements by a grid of lines parallel to the x and y axes, as shown in Figure 11.5a. If the grid widths are $j_1, j_2, \ldots, j_r$ in the x direction, and $k_1, k_2, \ldots, k_s$ in the y direction, the area of a rectangle rs is $j_r k_s$. If $f(x_r, y_s)$ is a point within this rectangle, the double integral is defined as a sum of contributions

$$\sum_{r,\,s} f(x_r,\, y_s) j_r k_s. \tag{11.17b}$$

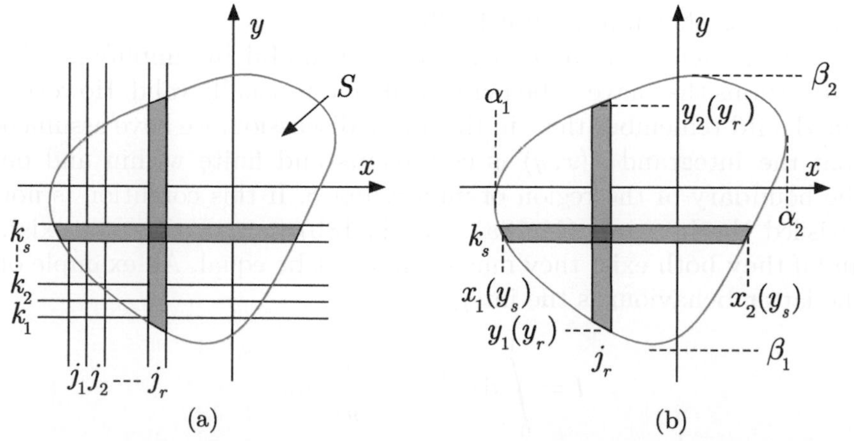

Figure 11.5 Constructions for defining a double integral.

in the limit that all j_r and k_s tend to zero, in which case the number of rectangles tends to infinity. This sum can be conveniently rewritten in the form

$$\sum_{r,\,s} f(x_r,\, y_s) j_r k_s = \sum_s k_s \sum_r f(x_r,\, y_s) j_r,$$

where, for fixed s, the sum over r is the contribution from the horizontal shaded strip in Figure 11.5a and y_s can be assumed to be constant for all terms in the sum. In the limit $j_r \to 0$, this sum is the integral

$$g(y_s) = \int_{x_1(y_s)}^{x_2(y_s)} f(x, y_s)\,\mathrm{d}x,$$

where the limits of integration are shown in Figure 11.5b. The double sum then becomes

$$\lim_{k_s \to 0} \sum_s k_s g(y_s) = \int_{\beta_1}^{\beta_2} g(y)\mathrm{d}y,$$

where β_1 and β_2 are the minimum and maximum values of y in the region S. Thus the double integral is

$$\int_{\beta_1}^{\beta_2} \mathrm{d}y \int_{x_1(y)}^{x_2(y)} f(x, y)\mathrm{d}x. \tag{11.18a}$$

Alternatively, we could have done the sum over s first, followed by the sum over r. In this case the double integral would be

$$\int_{\alpha_1}^{\alpha_2} \mathrm{d}x \int_{y_1(x)}^{y_2(x)} f(x, y)\mathrm{d}y, \tag{11.18b}$$

where α_1 and α_2 are the minimum and maximum values of x in the region S, as shown in Figure 11.4b.

Interchanging the order can often be useful in simplifying the integrations that have to be performed, and is usually valid. However, one should remember that, in the above discussion, we have assumed that the integrand $f(x, y)$ is continuous and finite within and on the boundary of the region of integration S. If this condition is not satisfied the integrals (11.18a) and (11.18b) may or may not exist; and if they both exist they may or may not be equal. An example of the latter behaviour is the integral,

$$I = \int_0^1 \mathrm{d}x \int_0^1 \frac{x - y}{(x + y)^3}\,\mathrm{d}y,$$

where the region of integration S is bounded by the lines $x = 0$, $x = 1, y = 0,\ y = 1$. The integrand has a discontinuity on the boundary of S at the point $(0, 0)$ and thus violates the above condition, so that it is not necessarily safe to invert the two integrations. This is confirmed by setting $y = u - x$, when it is easily shown that $I = 1/2$. However, inverting the order of integration gives

$$I' = \int_0^1 dy \int_0^1 \frac{x - y}{(x + y)^3}\, \mathrm{d}x$$

and using the same substitution gives $I' = -1/2$.

Example 11.7

Evaluate the following double integrals

$$\text{(a) } I_1 = \iint_{S_1} (1 + xy)\mathrm{d}x\, \mathrm{d}y, \quad \text{(b) } I_2 = \iint_{S_2} (2xy)\mathrm{d}x\, \mathrm{d}y$$

where S_1 is the area between the curves $y = x$ and $y = x^2$, between their intersections at $x = 0$ and $x = 1$, and S_2 is the area bounded by the lines $x = 0,\ y = 0,\ x = -2$ and $y = -(9 - x^2)^{1/2}$.

Solution

(a) Integrating with respect to y first and then with respect to x gives

$$I = \int_0^1 [y + xy^2/2]_{x^2}^{x}\, \mathrm{d}x = \int_0^1 (x + x^3/2 - x^2 - x^5/2)\, \mathrm{d}x$$

$$= \left[\frac{x^2}{2} + \frac{x^4}{8} - \frac{x^3}{3} - \frac{x^6}{12}\right]_0^1 = \frac{5}{24}.$$

(b) Integrating with respect to y first and then with respect to x gives

$$I = 2\int_{-2}^{0} x[y^2/2]_{-(9-x^2)^{1/2}}^{0}\, \mathrm{d}x = -\int_{-2}^{0} x(9 - x^2)\mathrm{d}x$$

$$= -\left[\frac{9x^2}{2} - \frac{x^4}{4}\right]_{-2}^{0} = -14.$$

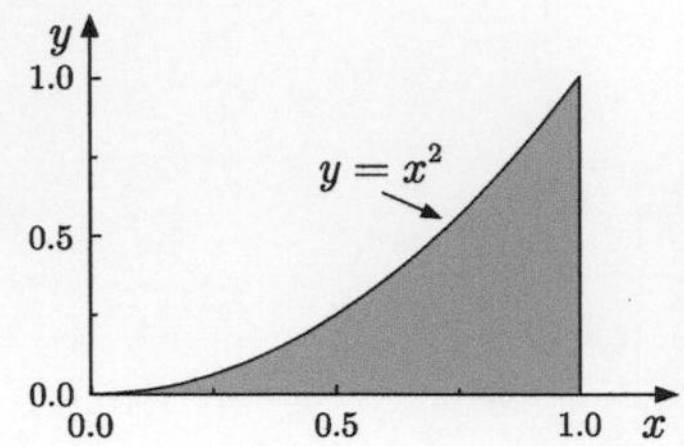

Figure 11.6

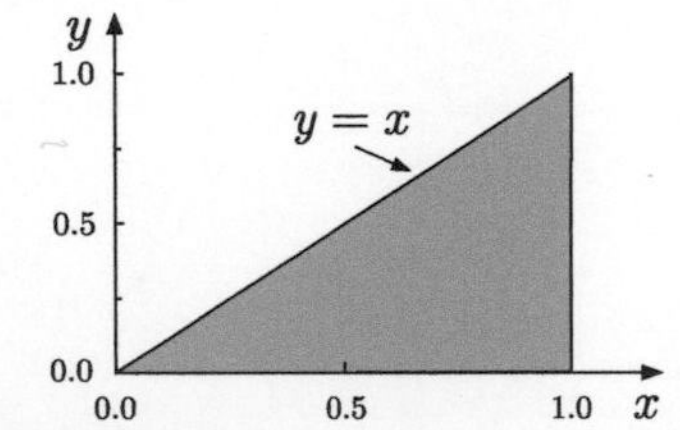

Figure 11.7

Example 11.8

Evaluate the following integrals by reversing the order of integration.

$$\text{(a) } I_1 = \int_0^1 \mathrm{d}y \int_{\sqrt{y}}^1 \sqrt{2+x^3}\,\mathrm{d}x \qquad \text{(b) } I_2 = \int_0^1 \mathrm{d}y \int_y^1 \exp(x^2)\,\mathrm{d}x.$$

Solution

(a) The area of integration is shown in Figure 11.6. If we integrate over y first the integral becomes

$$I = \int_0^1 \mathrm{d}x \int_0^{x^2} (2+x^3)^{1/2}\mathrm{d}y = \int_0^1 x^2(2+x^3)^{1/2}\mathrm{d}x$$

$$= \frac{2}{9}\left[(2+x^3)^{3/2}\right]_0^1 = \frac{2}{9}(3\sqrt{3} - 2\sqrt{2}).$$

(b) The area of integration is shown in Figure 11.7. If we integrate over y first, the integral becomes

$$I = \int_0^1 \mathrm{d}x \int_0^x \exp(x^2)\mathrm{d}y = \int_0^1 x\exp(x^2)\mathrm{d}x$$

$$= \frac{1}{2}[\exp(x^2)]_0^1 = \frac{e-1}{2}.$$

11.2.1 Green's theorem in the plane and perfect differentials

It is quite common for a line integral to be taken around a closed loop and we have seen in Section 11.1.1 how to evaluate such integrals.

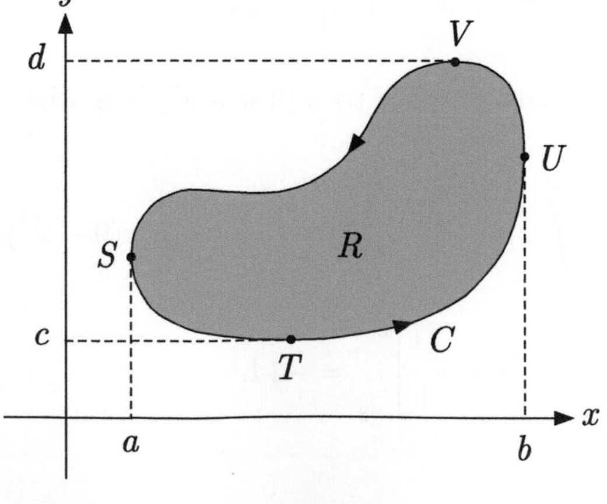

Figure 11.8 Figure used in the derivation of Green's theorem in the plane.

Green's theorem in the plane shows how to relate them to double integrals over the region enclosed by the loop, which is often easier to evaluate.

Let $P(x, y)$ and $Q(x, y)$ be two functions of x and y with continuous, finite partial derivatives in a region R and on the boundary C, as shown in Figure 11.8. Then

$$\iint_R \frac{\partial P}{\partial y} \mathrm{d}x\,\mathrm{d}y = \int_a^b \mathrm{d}x \int_{y_1(x)}^{y_2(x)} \frac{\partial P}{\partial y}\,\mathrm{d}y,$$

where $y_1(x)$ is the curve STU and $y_2(x)$ is the curve SVU. Evaluating the right-hand side gives

$$\begin{aligned}&\int_a^b \{P\,[x, y_2(x)] - P\,[x, y_1(x)]\}\,\mathrm{d}x \\ &\quad = -\int_a^b P\,[x, y_1(x)]\,\mathrm{d}x - \int_b^a P\,[x, y_2(x)]\,\mathrm{d}x = -\oint_C P\,\mathrm{d}x,\end{aligned} \tag{11.19}$$

where the notation in the final integral means the integral is around the closed curve C. In an analogous way, if we start with the integral

$$\iint_R \frac{\partial Q}{\partial y} \mathrm{d}x\,\mathrm{d}y$$

and let $x_1(y)$ be the curve TSV and $x_2(y)$ be the curve TUV, we have

$$\begin{aligned}&\int_a^b \{P\,[x, y_2(x)] - P[x, y_1(x)]\}\,\mathrm{d}x \\ &\quad = \int_d^c Q[x_1(y), y]\,\mathrm{d}y + \int_c^d Q[x_2(y), y]\,\mathrm{d}y = \oint_C Q\,\mathrm{d}y\,.\end{aligned}$$

Subtracting (11.19) from this equation gives

$$\oint_C (P\mathrm{d}x + Q\mathrm{d}y) = \iint_R \left(\frac{\partial Q}{\partial x} - \frac{\partial P}{\partial y}\right) \mathrm{d}x\,\mathrm{d}y, \tag{11.20}$$

which is Green's theorem in the plane.

Green's theorem in the plane shows that a line integral of the general form (11.7), where C is a loop, can be converted to a double integral over the area enclosed by the loop. It also shows that if

$$\frac{\partial P(x, y)}{\partial y} = \frac{\partial Q(x, y)}{\partial x}, \tag{11.21a}$$

then the line integral around the loop vanishes, i.e.

$$\oint (P\mathrm{d}x + Q\mathrm{d}y) = 0. \tag{11.22a}$$

Equation (11.21a) is also the condition that

$$\mathrm{d}I = P(x, y)\mathrm{d}x + Q(x, y)\mathrm{d}y \tag{11.21b}$$

is an exact, or perfect, differential (cf. Section 7.2.2) with

$$\frac{\partial I(x, y)}{\partial x} = P(x, y), \quad \frac{\partial I(x, y)}{\partial y} = Q(x, y). \tag{11.21c}$$

Hence if (11.21a) is satisfied, the line integral from $A \to B$ along any path is given by

$$\int_{A \to B} (P\mathrm{d}x + Q\mathrm{d}y) = \int_{A \to B} \mathrm{d}I = I_B - I_A, \tag{11.22b}$$

where I_A and I_B are the values of I at the points A and B, respectively, independent of the path connecting A to B.

To summarise, the necessary and sufficient condition for any loop integral (11.8c) to vanish for a closed loop and for the integral (11.22d) to be independent of the path for all paths is that (11.21b) is a perfect differential. This result extends to three dimensions, that is, the general line integral in three dimensions (11.13) is also independent of the path if

$$\mathrm{d}I = P(x, y, z)\mathrm{d}x + Q(x, y, z)\mathrm{d}y + R(x, y, z)\mathrm{d}z \tag{11.23a}$$

is a perfect differential, that is, if [cf. (7.19b)]

$$\frac{\partial P}{\partial y} - \frac{\partial Q}{\partial x} = \frac{\partial Q}{\partial z} - \frac{\partial R}{\partial y} = \frac{\partial R}{\partial x} - \frac{\partial P}{\partial z} = 0 \tag{11.23b}$$

is satisfied.

Example 11.9

Use Green's theorem in the plane to evaluate the following line integrals.

(a) $\oint (y\mathrm{d}x - x\mathrm{d}y)$ around a circle of radius 3.

(b) $\oint [(3x + 2y)\mathrm{d}x + (x - y)\mathrm{d}y]$ around an ellipse with semi-minor axis $a = 1$ and semi-major axes $b = 3$.

Solution

(a) In the notation of (11.21), $P = y$ and $Q = -x$, and with A the area of the circle,

$$\oint (y\mathrm{d}x - x\mathrm{d}y) = -2 \iint_A \mathrm{d}x\,\mathrm{d}y = -18\pi.$$

(b) In the notation of (11.21), $P = 3x + 2y$ and $Q = x - y$, and with $A = \pi ab$ the area of the ellipse,

$$\oint [(3x + 2y)\mathrm{d}x + (x - y)\mathrm{d}y] = -\iint_A \mathrm{d}x\mathrm{d}y = -A = -3\pi.$$

Example 11.10

If the integrands below are perfect differentials, find the values of the integral between the given points A and B.

(a) $\int_{A\to B} [(y + z)\mathrm{d}x + (x + z)\mathrm{d}y + (x + y)\mathrm{d}z]$,
$A = (0, 2, 0)$, $B = (1, 2, 3)$

(b) $\int_{A\to B} [(xy^2 + z)\mathrm{d}x + (x^2y + 2)\mathrm{d}y + x\,\mathrm{d}z]$,
$A = (1, 1, 1)$, $B = (0, 1, 2)$

Solution

(a) In the notation of (11.23a), (11.23b) is satisfied, so that the integrand is a perfect differential $\mathrm{d}I$, with

$$\frac{\partial I}{\partial x} = y + z, \quad \frac{\partial I}{\partial y} = x + z, \quad \frac{\partial I}{\partial z} = x + y$$

Hence

$$I = xy + xz + yz + c,$$

where c is an arbitrary constant, and the integral is given by $I_B - I_A = 11$.

(b) In this case,

$$\frac{\partial I}{\partial x} = xy^2 + z, \quad \frac{\partial I}{\partial y} = x^2y + 2, \quad \frac{\partial I}{\partial z} = x$$

so that

$$I = \tfrac{1}{2}x^2y^2 + 2y + xz + c$$

and the integral is given by $I_B - I_A = -3/2$.

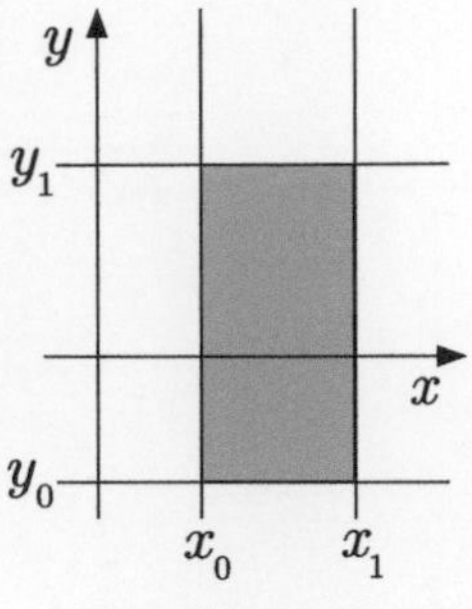

(a)

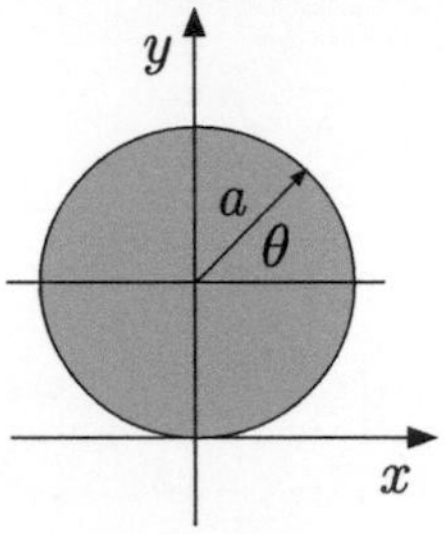

(b)

Figure 11.9 Two co-ordinate systems.

11.2.2 Other co-ordinate systems and change of variables

Up to now we have used mainly the Cartesian system of co-ordinates, but in real applications it is often useful to take advantage of any symmetry the system may have by choosing a different co-ordinate system. Consider the example shown in Figure 11.9a. The shaded area corresponds to the ranges $x_0 \le x \le x_1$, $y_0 \le y \le y_1$; and in Figure 11.9b the shaded area corresponds to either $|x| \le a$, $|y| \le (a^2 - x^2)^{1/2}$, or $|y| \le a$, $|x| \le (a^2 - y^2)^{1/2}$. The latter illustrates that in general the ranges of the two variables are not independent. However, had we used plane polar co-ordinates $(r,\ \theta)$, then the shaded area would correspond to the ranges $0 \le r \le a$, $0 \le \theta < 2\pi$, which are independent. This illustrates the usefulness of choosing co-ordinates to fit the specific problem, and we will see that the evaluation of double integrals like (11.7) can sometimes be considerably simplified if appropriate co-ordinates can be found. However, in order to do this, it is necessary to show how such double integrals can be expressed in variables other than Cartesian co-ordinates.

To do this, let us suppose we are using co-ordinates u_1, u_2 such that the corresponding Cartesian co-ordinates are given by continuous, differentiable functions $x(u_1, u_2)$ and $y(u_1, u_2)$. Such variables are called *curvilinear co-ordinates* because fixing u_1 and allowing u_2 to vary leads to a family of curves in the x–y plane, as shown in Figure 11.10, and fixing u_2 while u_1 varies leads to a different family of curves, also shown in Figure 11.10. The value of a function $f(x, y)$ at any point can be expressed in terms of curvilinear co-ordinates, i.e.

$$f(x, y) = f[x(u_1, u_2),\ y(u_1, u_2)] \equiv F(u_1, u_2),$$

and a double integral of $f(x, y)$ over the area S bounded by the curve in Figure 11.10 is given by

$$\lim_{\delta u_1, \delta u_2 \to 0} \sum_{\delta S_{rs}} F(u_1,\ u_2)\delta S_{rs} \tag{11.24}$$

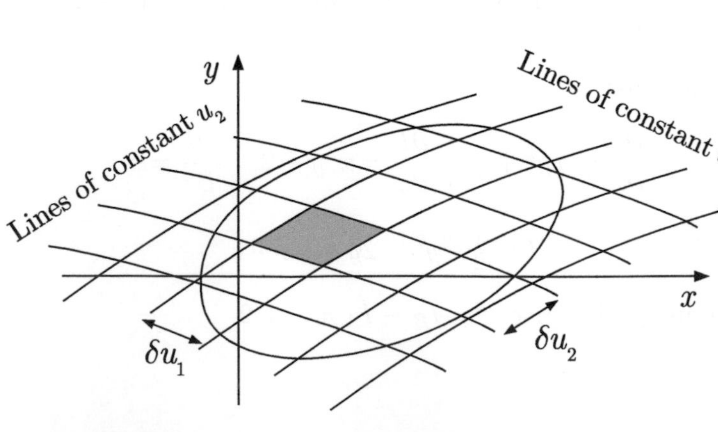

Figure 11.10 Curvilinear co-ordinates in a plane, showing lines of constant u_1 and u_2, spaced by δu_1 and δu_2, respectively. The area S to be integrated over is the interior of the closed loop and the shaded region is one of the areas δS_{rs}.

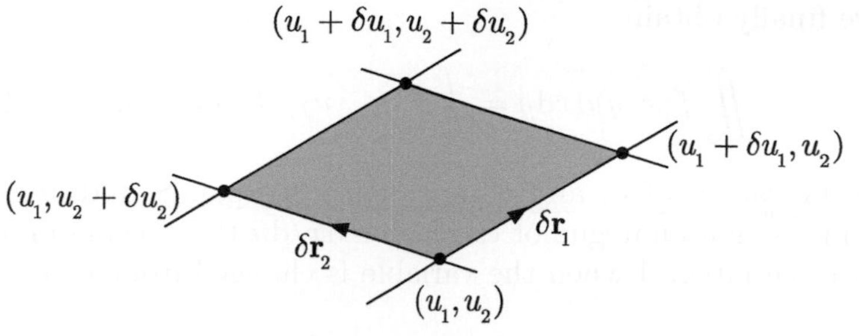

Figure 11.11 Construction to define the area δS_{rs} in the limit that δu_1 and δu_2 become infinitesimally small.

where the δS_{rs} are the small areas bounded by u_i and $u_i + \delta u_i$ where $i = 1, 2$ as shown in Figure 11.10.

In the limit where the separations δu_1 and δu_2 between such curves tend to zero, the shaded area shown in Figure 11.10 becomes a parallelogram, and to evaluate (11.24) we need to find its area. Referring to Figure 11.11, we write

$$\delta\mathbf{r}_i \equiv (\delta x_i, \delta y_i), \quad i = 1, 2.$$

If δx_1 is the displacement in the x direction, then

$$\delta x_1 = x(u_1 + \delta u_1, u_2) - x(u_1, u_2) \approx \frac{\partial x(u_1, u_2)}{\partial u_1}\delta u_1,$$

and similarly for δy_1. So

$$\delta\mathbf{r}_1 \approx \left(\frac{\partial x}{\partial u_1}, \frac{\partial y}{\partial u_1}\right)\delta u_1, \quad \delta\mathbf{r}_2 \approx \left(\frac{\partial x}{\partial u_2}, \frac{\partial y}{\partial u_2}\right)\delta u_2$$

and the area of the parallelogram δS_{rs} is then given by $|\delta\mathbf{r}_1||\delta\mathbf{r}_2|\sin\theta$, where θ is the angle between $\delta\mathbf{r}_1$ and $\delta\mathbf{r}_2$. Hence

$$\delta S_{rs} \approx |\delta\mathbf{r}_1 \times \delta\mathbf{r}_2| = |J|\delta u_1 \delta u_2, \tag{11.25a}$$

where the determinant

$$J \equiv \begin{vmatrix} \partial x/\partial u_1 & \partial y/\partial u_1 \\ \partial x/\partial u_2 & \partial y/\partial u_2 \end{vmatrix} \tag{11.25b}$$

is called the *Jacobian* and is also written in the shorthand form

$$J \equiv \frac{\partial(x, y)}{\partial(u_1, u_2)}. \tag{11.25c}$$

The sum (11.24) now becomes

$$\lim_{\delta u_1, \delta u_2 \to 0} \sum_{\delta S_{rs}} F(u_1, u_2)|J|\,\delta u_1 \delta u_2 = \iint_S F(u_1, u_2)|J|\,\mathrm{d}u_1 \mathrm{d}u_2$$

and we finally obtain

$$\iint_S f(x, y)\mathrm{d}x\mathrm{d}y = \int_S F(u_1,\, u_2)|J\,|\,\mathrm{d}u_1\mathrm{d}u_2, \tag{11.26}$$

where the ranges of u_1 and u_2 are chosen to span S, and $|J\,|$ is the two-dimensional analogue of the factor $\mathrm{d}x/\mathrm{d}u$ that occurs in a one-dimensional integral when the variable is changed from x to u.

Example 11.11

Evaluate the integral

$$\iint_S \exp[-\alpha(x^2 + y^2)]\mathrm{d}x\,\mathrm{d}y, \quad \alpha > 0$$

where S is the interior of a circle of radius R centred at the origin, and use the result to verify the standard integral

$$I_\alpha = \int_{-\infty}^{\infty} e^{-\alpha x^2}\,\mathrm{d}x = \left(\frac{\pi}{\alpha}\right)^{1/2}, \quad \alpha > 0. \tag{11.27}$$

Solution

Choosing plane polar co-ordinates $(u_1,\, u_2) \equiv (r, \theta)$, where $x = r\cos\theta,\ y = r\sin\theta$, the Jacobian

$$J = \begin{vmatrix} \partial x/\partial r & \partial y/\partial r \\ \partial x/\partial \theta & \partial y/\partial \theta \end{vmatrix} = r > 0,$$

so that the integral becomes

$$\int_0^R \mathrm{d}r \int_0^{2\pi} re^{-\alpha r^2}\,\mathrm{d}\theta = 2\pi\int_0^R re^{-\alpha r^2}\,\mathrm{d}r = \frac{\pi}{\alpha}(1 - e^{-\alpha R^2}),$$

where we have used the substitution $z = r^2$ to evaluate the integral over r. In the limit $R \to \infty$ this becomes an integral over the whole x-y plane, so that

$$\frac{\pi}{\alpha} = \int_{-\infty}^{\infty}\int_{-\infty}^{\infty} e^{-\alpha(x^2+y^2)}\mathrm{d}x\,\mathrm{d}y = \int_{-\infty}^{\infty} e^{-\alpha x^2}\,\mathrm{d}x \int_{-\infty}^{\infty} e^{-\alpha y^2}\,\mathrm{d}y = I_\alpha^2.$$

Hence, since I_α is clearly positive, we obtain (11.27) as required.

Example 11.12

Use the change of variables $s = xy$ and $t = xy^2$ to evaluate the integral

$$I = \int_S xy^2\mathrm{d}A,$$

where A is the area bounded by $xy = 1,\ xy = 4,\ xy^2 = 1,\ xy^2 = 4$.

Solution

Solving for x and y gives $x = s^2/t$ and $y = t/s$, and so the Jacobian of the transformation is

$$J = \begin{vmatrix} \partial x/\partial s & \partial y/\partial s \\ \partial x/\partial t & \partial y/\partial t \end{vmatrix} = \begin{vmatrix} 2s/t & -t/s^2 \\ -s^2/t^2 & 1/s \end{vmatrix} = 2/t - 1/t = 1/t.$$

The transformed integral is then

$$I = \iint_S xy^2 J \,\mathrm{d}s\,\mathrm{d}t = \int_1^4 \mathrm{d}s \int_1^4 \mathrm{d}t = 9.$$

11.3 Curvilinear co-ordinates in three dimensions

Before extending the discussion to include triple integrals, it will be convenient to consider co-ordinate systems other than Cartesian co-ordinates in three dimensions. To do this, we suppose that we have three co-ordinates u_1, u_2, u_3, such that the Cartesian co-ordinates are given by single-valued differentiable functions $x(u_1, u_2, u_3)$, $y(u_1, u_2, u_3)$, and $z(u_1, u_2, u_3)$, and each set of values u_1, u_2, u_3 corresponds to a single point in space:

$$\mathbf{r} = x(u_1, u_2, u_3)\mathbf{i} + y(u_1, u_2, u_3)\mathbf{j} + z(u_1, u_2, u_3)\mathbf{k}, \quad (11.28)$$

where $\mathbf{i}, \mathbf{j}, \mathbf{k}$ are as usual unit vectors along the x, y, z axes, respectively. Alternatively, we can define unit vectors

$$\mathbf{e}_i \equiv \frac{1}{|\partial\mathbf{r}/\partial u_i|}\frac{\partial\mathbf{r}}{\partial u_i}, \qquad i = 1, 2, 3 \quad (11.29)$$

so that if

$$\mathrm{d}\mathbf{r} \equiv \mathbf{r}\,(u_1 + \mathrm{d}u_1, u_2 + \mathrm{d}u_2, u_3 + \mathrm{d}u_3) - \mathbf{r}\,(u_1, u_2, u_3),$$

we have

$$\mathrm{d}\mathbf{r} = \sum_{i=1}^{3} \frac{\partial\mathbf{r}}{\partial u_i}\,\mathrm{d}u_i = \sum_{i=1}^{3} h_i \mathbf{e}_i \mathrm{d}u_i, \quad (11.30\text{a})$$

where

$$h_i = |\partial\mathbf{r}/\partial u_i|, \quad i = 1, 2, 3. \quad (11.30\text{b})$$

The unit vectors $\mathbf{e}_i$ in general depend on the position $\mathbf{r}$, as we shall shortly demonstrate by example, and since they act as basis vectors at each $\mathbf{r}$, they are written without 'hats', even though they are unit vectors. Finally, if

$$\mathbf{e}_i \cdot \mathbf{e}_j = \delta_{ij} \quad (11.31)$$

at all $\mathbf{r}$, then u_1, u_2, u_3 are called *orthogonal curvilinear co-ordinates* and it follows from (11.30) and (11.31) that

$$\mathrm{d}\mathbf{r}^2 = h_1^2 \mathrm{d}u_1^2 + h_2^2 \mathrm{d}u_2^2 + h_3^2 \mathrm{d}u_3^2. \quad (11.32\text{a})$$

Similarly, the parallelepiped with adjacent sides given by

$$h_1 \mathrm{d}u_1 \mathbf{e}_1, \quad h_2 \mathrm{d}u_2 \mathbf{e}_2, \quad h_3 \mathrm{d}u_3 \mathbf{e}_3$$

reduces to a cuboid with volume

$$\mathrm{d}v = h_1 h_2 h_2 \mathrm{d}u_2 \mathrm{d}u_2 \mathrm{d}u_3 \tag{11.32b}$$

if the co-ordinates are orthogonal. This is called the element of volume and plays a crucial role in evaluating triple integrals in orthogonal curvilinear co-ordinates, as we shall see in Section 11.4.1.

We shall now illustrate these ideas by introducing the two most important examples of orthogonal curvilinear co-ordinates: cylindrical and spherical polar co-ordinates, which are used for situations with cylindrical or spherical symmetry, respectively.

11.3.1 Cylindrical and spherical polar co-ordinates

Cylindrical polar co-ordinates in three dimensions are denoted by ρ, ϕ and z and are shown in Figure 11.12a. They are related to Cartesian co-ordinates by

$$x = \rho\cos\phi, \quad y = \rho\sin\phi, \quad z = z \tag{11.33a}$$

and lie in the ranges

$$0 \le \rho \le \infty, \quad 0 \le \phi \le 2\pi, \quad -\infty < z < \infty. \tag{11.33b}$$

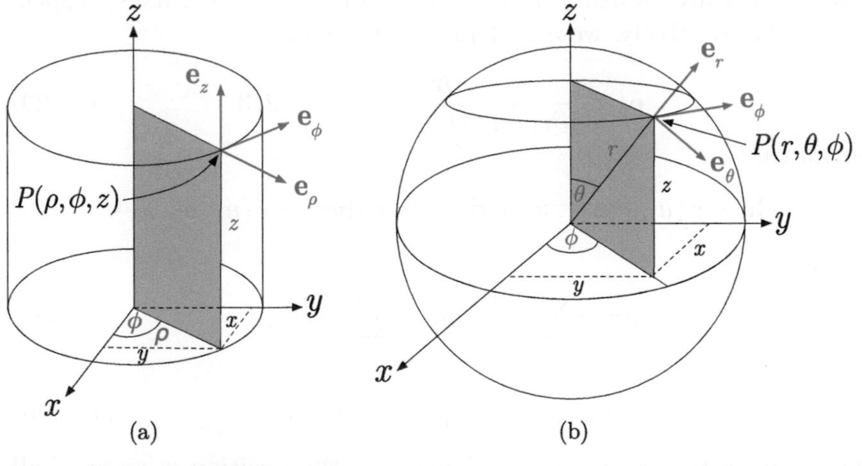

Figure 11.12 (a) Cylindrical polar co-ordinates ρ, ϕ, z, and the associated unit vectors $\mathbf{e}_\rho$, $\mathbf{e}_\phi$, $\mathbf{e}_z$. The vector $\mathbf{e}_\rho$ is in the direction of the radius vector ρ; $\mathbf{e}_\phi$ is in the xy-plane, tangential to the circle through P, and in the direction of increasing ϕ; $\mathbf{e}_z$ is in the z-direction. The three vectors $\mathbf{e}_\rho$, $\mathbf{e}_\phi$, $\mathbf{e}_z$ are mutually orthogonal. (b) Spherical polar co-ordinates r, θ, ϕ, and the associated unit vectors $\mathbf{e}_r$, $\mathbf{e}_\theta$, $\mathbf{e}_\phi$. The vector $\mathbf{e}_r$ is in the direction of the radius vector $\mathbf{r}$; $\mathbf{e}_\phi$ is in the xy-plane, tangential to the circle through P, and in the direction of increasing ϕ; $\mathbf{e}_\theta$ is at right angles to $\mathbf{e}_r$ in the direction of increasing θ. The three vectors $\mathbf{e}_r$, $\mathbf{e}_\theta$, $\mathbf{e}_\phi$ are mutually orthogonal.

The position vector

$$\mathbf{r} = \rho\cos\phi\,\mathbf{i} + \rho\sin\phi\,\mathbf{j} + z\,\mathbf{k}, \tag{11.34}$$

and identifying u_1, u_2, u_3 with ρ, ϕ, z, one finds, in an obvious notation,

$$\mathbf{e}_\rho = \cos\phi\,\mathbf{i} + \sin\phi\,\mathbf{j}, \quad \mathbf{e}_\phi = -\sin\phi\,\mathbf{i} + \cos\phi\,\mathbf{j}, \quad \mathbf{e}_z = \mathbf{k}, \tag{11.35}$$

while (11.30a) and (11.30b) give

$$\mathrm{d}\mathbf{r} = \mathrm{d}\rho\,\mathbf{e}_\rho + \rho\,\mathrm{d}\phi\,\mathbf{e}_\phi + \mathrm{d}z\,\mathbf{e}_z. \tag{11.36}$$

Note the factor ρ in the second term. Thus, unlike the case of Cartesian co-ordinates, if $\phi \to \phi + \mathrm{d}\phi$ for fixed ρ and z, the distance moved is *not* $\mathrm{d}\phi$, but $\rho\,\mathrm{d}\phi$. Another difference from Cartesian co-ordinates is that the basis vectors (11.35), which are also shown in Figure 11.12a, are not constants, but depend on the position $\mathbf{r}$. However, one easily verifies using (11.35) that they are orthogonal,

$$(\mathrm{d}\mathbf{r})^2 = \mathrm{d}\mathbf{r}\cdot\mathrm{d}\mathbf{r} = (\mathrm{d}\rho)^2 + \rho^2(\mathrm{d}\phi)^2 + (\mathrm{d}z)^2. \tag{11.37}$$

Finally, because the basis vectors are orthogonal, the parallelepiped defined by the vectors

$$\mathrm{d}\rho\,\mathbf{e}_\rho, \quad \rho\mathrm{d}\phi\,\mathbf{e}_\phi, \quad \mathrm{d}z\,\mathbf{e}_z$$

is actually a cuboid, as shown in Figure 11.13a, with a volume given by

$$\mathrm{d}v = \rho\,\mathrm{d}\rho\,\mathrm{d}\phi\,\mathrm{d}z. \tag{11.38}$$

This is called the volume element in cylindrical polar co-ordinates.

Spherical polar co-ordinates in three dimensions are $(r,\ \theta,\ \phi)$ and are shown in Figure 11.12b; $r = |\mathbf{r}|$ is called the *radial* co-ordinate,

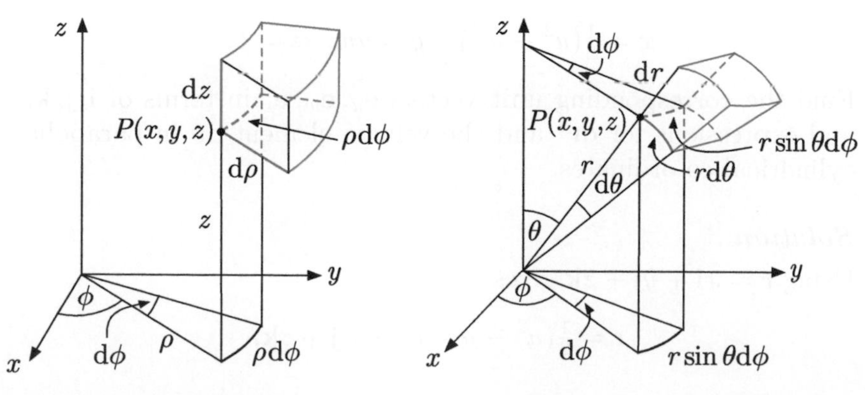

Figure 11.13 The volume element in (a) cylindrical polar co-ordinates; (b) spherical polar co-ordinates.

θ is the *polar angle* between r and the z-axis; and ϕ is the *azimuthal angle.* As can be seen, they are related to Cartesian co-ordinates by

$$x = r\sin\theta\cos\phi, \quad y = r\sin\theta\sin\phi, \quad z = r\cos\theta, \tag{11.39a}$$

and are restricted to the ranges

$$r \geq 0, \quad 0 \leq \theta \leq \pi, \quad 0 \leq \phi \leq 2\pi, \tag{11.39b}$$

in order to cover the space once, except for the origin, which is given by $(r, \theta, \phi) = (0, \theta, \phi)$ for any θ and ϕ. The position vector is now

$$\mathbf{r} = r\sin\theta\cos\phi\,\mathbf{i} + r\sin\theta\sin\phi\,\mathbf{j} + z\cos\theta\,\mathbf{k} \tag{11.40}$$

so that using (11.27), one finds, in an obvious notation,

$$\begin{aligned} \mathbf{e}_r &= \sin\theta\cos\phi\,\mathbf{i} + \sin\theta\sin\phi\,\mathbf{j} + \cos\theta\,\mathbf{k}, \\ \mathbf{e}_\theta &= \cos\theta\cos\phi\,\mathbf{i} + \cos\theta\sin\phi\,\mathbf{j} - \sin\theta\,\mathbf{k}, \\ \mathbf{e}_\phi &= -\sin\phi\,\mathbf{i} + \cos\phi\,\mathbf{j}, \end{aligned} \tag{11.41}$$

while

$$\mathrm{d}\mathbf{r} = \mathrm{d}r\,\mathbf{e}_r + r\mathrm{d}\theta\,\mathbf{e}_\theta + r\sin\theta\mathrm{d}\phi\,\mathbf{e}_\phi. \tag{11.42}$$

The unit vectors are shown in Figure 11.12b. They are again orthogonal, so that

$$(\mathrm{d}\mathbf{r})^2 = \mathrm{d}\mathbf{r}\cdot\mathrm{d}\mathbf{r} = (\mathrm{d}r)^2 + r^2(\mathrm{d}\theta)^2 + r^2\sin^2\theta(\mathrm{d}\phi)^2. \tag{11.43}$$

Similarly, the volume element $\mathrm{d}v$ is the volume of the cuboid defined by the vectors

$$\mathrm{d}r\,\mathbf{e}_r, \quad r\mathrm{d}\theta\,\mathbf{e}_\theta, \quad r\sin\theta\,\mathrm{d}\phi\,\mathbf{e}_\phi,$$

and is given by

$$\mathrm{d}v = r^2\sin\theta\,\mathrm{d}r\,\mathrm{d}\theta\,\mathrm{d}\phi. \tag{11.44}$$

It is shown in Figure 11.13b.

Example 11.13

Parabolic cylindrical co-ordinates u, w, z are related to Cartesian co-ordinates by

$$x = \tfrac{1}{2}(u^2 - w^2), \quad y = uw, \quad z = z.$$

Find the corresponding unit vectors $\mathbf{e}_u, \mathbf{e}_w, \mathbf{e}_z$ in terms of $\mathbf{i}, \mathbf{j}, \mathbf{k}$, and expressions for $\mathrm{d}\mathbf{r}^2$ and the volume element $\mathrm{d}v$ in parabolic cylindrical co-ordinates.

Solution

Using $\mathbf{r} = x\mathbf{i} + y\mathbf{j} + z\mathbf{k}$ gives

$$\mathbf{r} = \tfrac{1}{2}(u^2 - w^2)\,\mathbf{i} + uw\,\mathbf{j} + z\mathbf{k},$$

Hence by (11.39) we have

$$\mathbf{e}_u = \frac{u\mathbf{i} + w\mathbf{j}}{\sqrt{u^2 + w^2}}, \quad \mathbf{e}_w = \frac{-w\mathbf{i} + u\mathbf{j}}{\sqrt{u^2 + w^2}}, \quad \mathbf{e}_z = \mathbf{k},$$

from which we see that they are orthogonal, i.e.

$$\mathbf{e}_u \cdot \mathbf{e}_w = \mathbf{e}_u \cdot \mathbf{e}_z = \mathbf{e}_w \cdot \mathbf{e}_z = 0.$$

From (11.30), we have

$$\mathrm{d}\mathbf{r} = h_u \mathbf{e}_u \mathrm{d}u + h_w \mathbf{e}_w \mathrm{d}w + h_z \mathbf{e}_z \mathrm{d}z,$$

where

$$h_u = h_w = \sqrt{u^2 + w^2}, \quad h_z = 1$$

and hence

$$\mathrm{d}\mathbf{r}^2 = (u^2 + w^2)(\mathrm{d}u^2 + \mathrm{d}w^2) + \mathrm{d}z^2.$$

Similarly, the element of volume defined by the vectors

$$h_u \mathbf{e}_u \mathrm{d}u, \quad h_w \mathbf{e}_w \mathrm{d}w, \quad h_z \mathbf{e}_z \mathrm{d}z$$

is

$$\mathrm{d}v = h_u h_w h_z \mathrm{d}u\, \mathrm{d}w\, \mathrm{d}z = (u^2 + w^2)\mathrm{d}u\, \mathrm{d}w\, \mathrm{d}z$$

since the co-ordinates are orthogonal.

11.4 Triple or volume integrals

We turn next to triple or volume integrals, denoted by

$$\iiint_\Omega f(x, y, z)\mathrm{d}x\, \mathrm{d}y\, \mathrm{d}z, \tag{11.45}$$

where Ω is the region of space to be integrated over and $f(x, y, z)$ is continuous, single-valued and finite within and on the boundary of the region. Since they are a direct generalisation of double integrals, we shall discuss their properties rather briefly.

In Section 11.2.1, we defined double integrals by dividing the region of integration S into small rectangles of side lengths j_r, k_s, as shown in Figure 11.5, and taking the limit of the weighted sum (11.17a) as both j_r and k_s tend to zero. Triple integrals are defined in a similar way by dividing Ω into small cuboids with sides of lengths j_r, k_s and l_t, and taking the limit of

$$\sum_{r,s,t} f(x_r, y_s, z_t) j_r k_s l_t,$$

as j_r, k_s and l_t tend to zero, where x_r, y_s, z_t is any point within the cuboid r, s, t. As in the two-dimensional case, the order of summation determines the order of integration in the final expression. In particular, if we sum over t, then s, then r, we obtain

$$\iiint_{\Omega} f(x,y,z)\mathrm{d}x\,\mathrm{d}y\,\mathrm{d}z = \int_{\alpha_1}^{\alpha_2} \mathrm{d}x \int_{y_1(x)}^{y_2(x)} \mathrm{d}y \int_{z_1(x,y)}^{z_2(x,y)} f(x,y,z)\mathrm{d}z. \qquad (11.46)$$

Here α_2 and α_1 are the maximum and minimum values of x in the region Ω, $y_2(x)$ and $y_1(x)$ are the maximum and minimum values of y at fixed x in the region Ω, and $z_2(x,y)$ and $z_1(x,y)$ are the maximum and minimum values of z at fixed values of x and y in the same region. Other orderings of the summation lead to different orderings of the x, y and z integrations, with appropriate limits, but provided $f(x,y,z)$ is single-valued, finite and continuous, they all yield the same value for the integral. Finally, it follows directly from this definition that

$$V = \iiint_{\Omega} \mathrm{d}x\,\mathrm{d}y\,\mathrm{d}z \qquad (11.47)$$

is the volume of the region Ω.

Example 11.14

A cube of unit side is made of a material having a density $\rho(x,y,z) = 1 + 2xyz$ in some units. If one corner is at the origin and the sides are aligned parallel to the Cartesian axes, calculate the mass of the cube.

Solution

The mass of the cube is given by

$$\begin{aligned} M &= \iiint_V \rho(x,y,z)\mathrm{d}x\,\mathrm{d}y\,\mathrm{d}z = \int_0^1\int_0^1\int_0^1 (1+2xyz)\mathrm{d}x\,\mathrm{d}y\,\mathrm{d}z \\ &= \int_0^1\int_0^1 [x + x^2yz]_{x=0}^{x=1}\mathrm{d}y\,\mathrm{d}z = \int_0^1\int_0^1 (1+yz)\mathrm{d}y\,\mathrm{d}z \\ &= \int_0^1 [y + y^2z/2]_0^1\mathrm{d}z = \int_0^1 (1+z/2)\mathrm{d}z = 5/4. \end{aligned}$$

11.4.1 Change of variables

The discussion of changing variables in double integrals given in Section 11.2.2 extends in a straightforward manner to triple integrals, except that instead of summing over infinitesimal parallelograms as in Figures 11.10 and 11.11, we now have to sum over infinitesimal parallelepipeds in three dimensions. We shall not reproduce the derivation but merely state the result, which is a direct generalisation of (11.25) and (11.26). Specifically, if we consider curvilinear co-ordinates u_1, u_2, u_3 (which need not be orthogonal) then

$$\iiint_\Omega f(x, y, z) \mathrm{d}x \, \mathrm{d}y \, \mathrm{d}z = \iiint_\Omega F(u_1, u_2, u_3) |J| \mathrm{d}u_1 \, \mathrm{d}u_2 \, \mathrm{d}u_3, \qquad (11.48a)$$

where

$$F(u_1, u_2, u_3) = f[x(u_i), y(u_i), z(u_i)], \qquad (11.48b)$$

and the Jacobian

$$J = \frac{\partial(x, y, z)}{\partial(u_1, u_2, u_3)} = \begin{vmatrix} \partial x/\partial u_1 & \partial y/\partial u_1 & \partial z/\partial u_1 \\ \partial x/\partial u_2 & \partial y/\partial u_2 & \partial z/\partial u_2 \\ \partial x/\partial u_3 & \partial y/\partial u_3 & \partial z/\partial u_3 \end{vmatrix}. \qquad (11.48c)$$

Finally, the integrals (11.48a) are often written as

$$\int_\Omega f \, \mathrm{d}v \qquad (11.49a)$$

without specifying any particular co-ordinate system. However, to evaluate them, a particular co-ordinate system must be chosen with the volume element

$$\mathrm{d}v = |J| \mathrm{d}u_1 \mathrm{d}u_2 \mathrm{d}u_3, \qquad (11.49b)$$

which reduces to $\mathrm{d}v = \mathrm{d}x \, \mathrm{d}y \, \mathrm{d}z$ in Cartesian co-ordinates $(u_1, u_2, u_3) = (x, y, z)$. In particular, one easily verifies that (11.49b) is identical to our previous results (11.38) and (11.44) for the volume elements in cylindrical and spherical polar co-ordinates.

Example 11.15

Evaluate the integral

$$I = \iiint_V xyz(a^2 - x^2 - y^2 - z^2)^{1/2} \, \mathrm{d}x \, \mathrm{d}y \, \mathrm{d}z,$$

where V is the volume of the positive octant $(x \geq 0, \ y \geq 0, \ z \geq 0)$ of a sphere of radius a.

Solution

We will use spherical polar co-ordinates (11.39a) and the Jacobian

$$J = \begin{vmatrix} \dfrac{\partial x}{\partial r} & \dfrac{\partial y}{\partial r} & \dfrac{\partial z}{\partial r} \\ \dfrac{\partial x}{\partial \theta} & \dfrac{\partial y}{\partial \theta} & \dfrac{\partial z}{\partial \theta} \\ \dfrac{\partial x}{\partial \phi} & \dfrac{\partial y}{\partial \phi} & \dfrac{\partial z}{\partial \phi} \end{vmatrix} = r^2 \sin\theta$$

with the range of variables

$$0 \le r \le a, \quad 0 \le \theta \le \pi/2, \quad 0 \le \phi \le \pi/2.$$

Then the integral may be written

$$\begin{aligned} I &= \int_0^{\pi/2}\int_0^{\pi/2}\int_0^a [r^3 \sin^2\theta \cos\theta \sin\phi \cos\phi\, (a^2 - r^2)^{1/2} r^2 \sin\theta]\, \mathrm{d}r\, \mathrm{d}\theta\, \mathrm{d}\phi \\ &= \int_0^{\pi/2} \sin\phi \cos\phi\, \mathrm{d}\phi \int_0^{\pi/2} \sin^3\theta \cos\theta \mathrm{d}\theta \int_0^a r^5 (a^2 - r^2)^{1/2} \mathrm{d}r. \end{aligned}$$

The three integrals are

$$I_\phi \equiv \int_0^{\pi/2} \sin\phi \cos\phi\, \mathrm{d}\phi = \left[-\tfrac{1}{4}\cos 2\phi\right]_0^{\pi/2} = \tfrac{1}{2},$$

$$I_\theta \equiv \int_0^{\pi/2} \sin^3\theta \cos\theta \mathrm{d}\theta = \left[\tfrac{1}{4}\sin^4\theta\right]_0^{\pi/2} = \tfrac{1}{4},$$

$$I_r \equiv \int_0^a r^5 (a^2 - r^2)^{1/2} \mathrm{d}r = \tfrac{8}{105} a^7.$$

The integral over r can be done by first substituting $r = \sin\theta$ and then using the methods of Section 4.3.2. Thus,

$$I_r = a^7 \int_0^{\pi/2} (\sin^5\theta - \sin^7\theta)\, \mathrm{d}\theta,$$

and then setting $z = \cos\theta$ gives

$$I_r = a^7 \int_0^1 (z^2 - 2z^4 + z^6) \mathrm{d}z = 8a^7/105.$$

Finally, $I = a^7/105$.

Problems 11

11.1 Evaluate the line integral

$$I = \int_C xy\, \mathrm{d}x$$

for two paths: (a) the straight line joining the points $A(1,1)$ and $B(3,4)$, and (b) the straight line joining $A(1,1)$ to $C(0,3)$, followed by the straight line joining $C(0,3)$ to $B(3,4)$.

11.2 Evaluate the line integral

$$I = \int_C (x^2 + y)\mathrm{d}x$$

from the point $(0,0)$ to the point $(1,1)$ along the curve $y = x^3$: (a) by expressing $I(a)$ as a function of x only and (b) as a function of y only.

11.3 Evaluate the integral

$$\oint (x^2 + xy + y^2)\mathrm{d}l,$$

where the contour is the circle $x^2 + y^2 = 1$.

11.4 Evaluate the line integral

$$\oint [(x + 2y)\mathrm{d}x - 2x\,\mathrm{d}y]$$

round the following closed paths, taken to be counter-clockwise: (a) the circle $x^2 + y^2 = 1$, (b) the square joining the points (1, 1), (−1, 1), (−1, −1) and (1, −1).

11.5 Evaluate the line integral

$$I = \int_C (y^2\mathrm{d}x + 2x\,\mathrm{d}y + \mathrm{d}z),$$

where the path C is (a) the straight line connecting (0, 0, 0) to (1, 1, 1), and (b) the three connecting straight lines $(0,0,0) \to (1,0,0) \to (1,1,0) \to (1,1,1)$.

11.6 Evaluate the integral

$$I = \iint \frac{\mathrm{d}x\,\mathrm{d}y}{(2 - x - y)^2}$$

over the triangle bounded by the axes $x = 0$ and $y = 0$, and the line $x + y = 1$.

11.7 Evaluate the integral

$$\int_{y=0}^{y=2}\int_{x=1}^{x=4} (x + 3y)\mathrm{d}x\,\mathrm{d}y$$

by first integrating with respect x and then with respect to y. Then repeat using the reversed order of integration. Comment on your result.

11.8 Invert the order of integration in the double integral

$$I = \int_0^a \mathrm{d}y \int_0^{(4a^2-4ay)^{1/2}} f(x,y)\,\mathrm{d}x$$

assuming that $f(x, y)$ is well-behaved within the region of integration.

11.9 Reverse the order of integration and hence evaluate the following integrals:

$$(a)\ \int_0^1 x\,\mathrm{d}x \int_x^{2-x} y^{-1}\mathrm{d}y,\quad (b)\ \int_0^1 \mathrm{d}x \int_0^x [y(2-y)]^{1/2}\mathrm{d}y.$$

11.10 Evaluate the integral

$$I = \int_0^6 \int_{x/3}^2 x(y^3+1)^{1/2}\mathrm{d}y\,\mathrm{d}x$$

by reversing the order of integration.

11.11 Evaluate

$$I = \oint [(e^x y + \cos x \sin y)\mathrm{d}x + (e^x + \sin x \cos y)\mathrm{d}y]$$

around the ellipse $x^2/a^2 + y^2/b^2 = 1$.

11.12 Evaluate

$$\oint_C (xy\,\mathrm{d}x + x^2\mathrm{d}y)$$

around the sides of a square with vertices $A(0,0), B(1,0), C(1,1)$ and $D(0,1)$ in an anti-clockwise direction. Then convert the line integral to a double integral and verify Green's theorem in a plane.

11.13 Use Green's theorem in the plane to evaluate the integral

$$\int_C [e^x \cos y\,\mathrm{d}x - e^x \sin y\,\mathrm{d}y]$$

from the point $(\ln 2,\ 0)$ to $(0,\ 1)$ and then to $(-\ln 2,\ 0)$.

11.14 If the integrands below are perfect differentials, find the values of the integrals between the given points A and B.

(a) $\int_{A\to B} [(y+z)\mathrm{d}x + (x+z)\mathrm{d}y + (x+y)\mathrm{d}z]$, $A = (0,2,0)$, $B = (1,2,3)$

(b) $\int_{A\to B} [(xy^2+z)\mathrm{d}x + (x^2y+2)\mathrm{d}y + x\,\mathrm{d}z]$, $A = (1,1,1)$, $B = (0,1,2)$

11.15 The quantity

$$\mathrm{d}f(x,y) = (x^2+y^2)\mathrm{d}x + 2xy\,\mathrm{d}y$$

is an exact differential. Confirm this by integrating it between the points $(0,0)$ and $(2,2)$ along the following paths: (a) $y = x^2/2$, (b) the straight line joining $(0,0)$ to $(2,0)$, followed by the straight line joining $(2,0)$ to $(2,2)$, (c) the curve defined by the parametric forms $x = t^2/2$ and $y = t$.

11.16 Integrate the function

$$z = x^3 y \left(1 - \frac{x^2}{a^2} - \frac{y^2}{b^2}\right)^{-1/2}$$

over the region of the first quadrant inside the ellipse

$$x^2/a^2 + y^2/b^2 = 1,$$

using the substitutions $x = a \sin\theta \cos\phi$, $y = b \sin\theta \sin\phi$.

11.17 Evaluate the integral

$$I = \int_S 2xy(x^2 + y^2) \exp[-(x^2 + y^2)^2] \, dx \, dy$$

over the coloured area shown in Figure 11.14, which extends to infinity in the x and y directions.

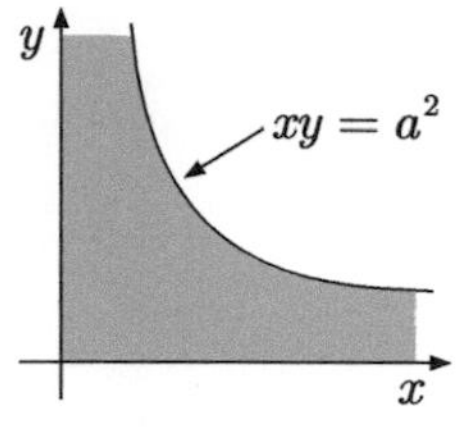

Figure 11.14

11.18 Paraboloidal co-ordinates u, w, ϕ are related to Cartesian co-ordinates by

$$x = uw \cos\phi, \quad y = uw \sin\phi, \quad z = \tfrac{1}{2}(u^2 - w^2).$$

Find the corresponding unit vectors $\mathbf{e}_u$, $\mathbf{e}_w$, $\mathbf{e}_\phi$ in terms of $\mathbf{i}$, $\mathbf{j}$, $\mathbf{k}$, and expressions for $d\mathbf{r}^2$ and the volume element dv in paraboloidal co-ordinates.

11.19 Elliptic co-ordinates in a plane are defined by

$$x = \alpha \cosh u \cos w, \quad y = \alpha \sinh u \sin w,$$

where α is a positive constant, with $0 \le u < \infty$ and $0 \le w < 2\pi$. Show that (u, w) are orthogonal co-ordinates and that the lines $u = \text{constant}, w = \text{constant}$ correspond to an ellipse and a hyperbola, respectively. Take $0 < w < \pi/2$, so that the point of intersection of these lines lies in the positive quadrant. Sketch the lines, and indicate the co-ordinate axes $\mathbf{e}_u$, $\mathbf{e}_w$ at this point.

11.20 If $\mathbf{F} = xz\mathbf{i} + x\mathbf{j} - 2y^2\mathbf{k}$, evaluate the integral

$$I = \iiint_\Omega \mathbf{F}(x, y, z) \, dx \, dy \, dz$$

where Ω is the volume bounded by the surfaces $x = 0$, $y = 0$, $y = 3$, $z = x^2$, $z = 2$.

11.21 Evaluate the integral

$$I = \iiint_\Omega \left[xz^2 \exp\left(\frac{x^2 + y^2 + z^2}{a^2}\right)\right] dx \, dy \, dz$$

over the octant bounded by the co-ordinate planes $x = 0$, $y = 0$, $z = 0$ and the sphere $x^2 + y^2 + z^2 = a^2$.
[Hint: the integral

$$\int t^2 \exp(\alpha t) dt = \frac{e^{\alpha t}(\alpha^2 t - 2\alpha t + 2)}{\alpha^3}$$

may be useful]

11.22 A container in the shape of a hemisphere of radius R is held so that its flat top is horizontal, and filled with liquid to a height $h < R$. What is the volume occupied by the liquid?

11.23 Using the result

$$\int (a-p)^{n-1} \ln(a-p)\mathrm{d}p = -\left[\frac{(a-p)^n}{n}\ln(a-p) - \frac{(a-p)^n}{n^2}\right],$$

evaluate the integral

$$\iiint \ln(1-x-y-z)\mathrm{d}x\,\mathrm{d}y\,\mathrm{d}z$$

over the tetrahedron bounded by the co-ordinate planes and the plane $P: \; x+y+z=1$.

12

Vector calculus

In Chapter 8 we introduced the idea of a vector as a quantity with both magnitude and direction and we discussed vector algebra, particularly as applied to analytical geometry, and the differentiation and integration of vectors with respect to a scalar parameter. In this chapter we extend our discussion to include directional derivatives and integration over variables that are themselves vectors. This topic is called *vector calculus* or *vector analysis*. It plays a central role in many areas of physics, including fluid mechanics, electromagnetism and potential theory.

12.1 Scalar and vector fields

If scalars and vectors can be defined as continuous functions of position throughout a region of space, they are referred to as *fields* and the region of space in which they are defined is called a *domain*. An example of a scalar field would be the distribution of temperature T within a fluid. At each point the temperature is represented by a scalar field $T(\mathbf{r})$ whose value depends on the position $\mathbf{r}$ at which it is measured. A useful concept when discussing scalar fields is that of an *equipotential surface*, that is, a surface joining points of equal value. This is somewhat analogous to the contour lines on a two-dimensional map, which join points of equal height. An example of a vector field is the distribution of velocity $\mathbf{v}(\mathbf{r})$ in a fluid. At every point $\mathbf{r}$, the velocity is represented by a vector of definite magnitude and direction, both of which can change continuously throughout the domain. In this case, we can define *flow lines* such that the tangent to a flow line at any point gives the direction of the vector at that point. Flow lines cannot intersect. This is illustrated in Figure 12.1.

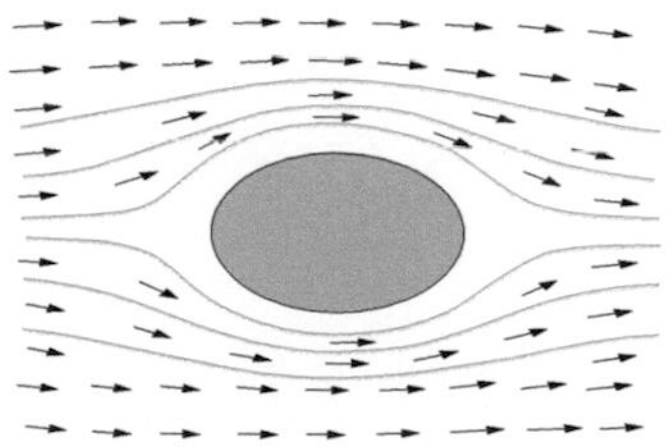

Figure 12.1 The motion of a fluid around a smooth solid. The coloured lines are the flow lines and the arrows show the direction of the vector field, in this case the velocity $\mathbf{v}(\mathbf{r})$.

Mathematics for Physicists, First Edition. B.R. Martin and G. Shaw.

Companion website: www.wiley.com/go/martin/mathsforphysicists

In the rest of this section, we shall extend our discussion of differentiation to embrace scalar and vector fields. Since we are primarily interested in applications where these fields are physical quantities, we shall assume throughout this chapter that they and their first derivatives are single-valued, continuous and differentiable.

12.1.1 Gradient of a scalar field

If we consider the rate of change of a scalar field $\psi(\mathbf{r})$ as $\mathbf{r}$ varies, this leads to a vector field called the *gradient* of $\psi(\mathbf{r})$, written as $\mathrm{grad}\,\psi(\mathbf{r})$. We will define and derive an expression for this by reference to Figure 12.2.

Consider the point P on an equipotential surface $\psi = \psi_P$ of the scalar field $\psi(\mathbf{r})$. Let R be a point on the normal to the surface through P that also lies on an equipotential surface $\psi_R > \psi_P$. If we take the two surfaces to be close, then they will be approximately parallel to each other and $\mathrm{grad}\,\psi(\mathbf{r})$ at P is defined as a *vector* in the direction $\mathbf{PR}$ of magnitude

$$|\mathrm{grad}\,\psi(\mathbf{r})| = \lim_{PR\to 0}\left(\frac{\psi_R - \psi_P}{|\mathbf{PR}|}\right).$$

This definition is similar to the definition of a derivative; hence the name gradient. Now let PQ be the signed distance from P to the surface ψ_R measured in the positive x-direction, and let α be the angle between $\mathbf{PR}$ and the x-direction, as shown in Figure 12.2. Then as $PR \to 0$, $PQ \to |\mathbf{PR}|\sec\alpha$, and thus the component of $\mathrm{grad}\,\psi(\mathbf{r})$ in the x-direction is

$$\lim_{PR\to 0}\left(\frac{\psi_R - \psi_P}{|\mathbf{PR}|}\right)\cos\alpha = \lim_{PQ\to 0}\left(\frac{\psi_Q - \psi_P}{PQ}\right), \tag{12.1}$$

since the point Q is on the surface $\psi = \psi_R$. The right-hand side of (12.1) is $\partial\psi(x,y,z)/\partial x$. Similarly, the components of $\mathrm{grad}\,\psi(\mathbf{r})$ in the y and z directions are $\partial\psi/\partial y$ and $\partial\psi/\partial z$, and hence

$$\mathrm{grad}\,\psi(\mathbf{r}) = \frac{\partial\psi}{\partial x}\mathbf{i} + \frac{\partial\psi}{\partial y}\mathbf{j} + \frac{\partial\psi}{\partial z}\mathbf{k}. \tag{12.2}$$

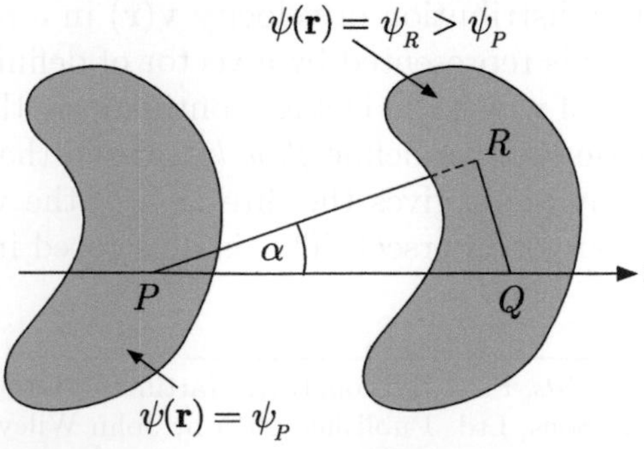

Figure 12.2 Diagram for the derivation of the gradient of a scalar field $\psi(\mathbf{r})$. PR is normal to the equipotential surface ψ_R at the point of intersection P.

Further, if we make a small displacement $\delta\mathbf{r} = (\delta x, \delta y, \delta z)$ from P in any direction, we have

$$\delta\psi \approx \frac{\partial\psi}{\partial x}\delta x + \frac{\partial\psi}{\partial y}\delta y + \frac{\partial\psi}{\partial z}\delta z, \tag{12.3}$$

which using (12.2) is

$$\delta\psi \approx \text{grad}\,\psi \cdot \delta\mathbf{r}. \tag{12.4a}$$

Similarly the corresponding differential [cf. (7.10), (7.11)] is given by

$$\mathrm{d}\psi = \text{grad}\,(\psi) \cdot \mathrm{d}\mathbf{r}. \tag{12.4b}$$

It follows from (12.4a) and (12.4b) that the rate of change of the field with respect to an infinitesimal displacement depends on the direction of travel. For this reason, it is called the *directional derivative.* To find it, consider moving a distance $\mathrm{d}s$ in the direction specified by a unit vector $\hat{\mathbf{n}}$, so that $\mathrm{d}\mathbf{r} = \hat{\mathbf{n}}\,\mathrm{d}s$. Substituting this expression into (12.4b) and dividing by $\mathrm{d}s$ then gives

$$\frac{\mathrm{d}\psi}{\mathrm{d}s} = \text{grad}\ (\psi) \cdot \hat{\mathbf{n}} \tag{12.5}$$

as the directional derivative of ψ in the direction $\hat{\mathbf{n}}$.

Another way of writing (12.2) and (12.5) is in terms of an object $\boldsymbol{\nabla}$, called *del*, and defined in the Cartesian system by

$$\boldsymbol{\nabla} \equiv \mathbf{i}\frac{\partial}{\partial x} + \mathbf{j}\frac{\partial}{\partial y} + \mathbf{k}\frac{\partial}{\partial z}, \tag{12.6}$$

so that (12.2) becomes

$$\text{grad}\,\psi(\mathbf{r}) = \boldsymbol{\nabla}\psi = \frac{\partial\psi}{\partial x}\mathbf{i} + \frac{\partial\psi}{\partial y}\mathbf{j} + \frac{\partial\psi}{\partial z}\mathbf{k}. \tag{12.7}$$

Del is called a *vector operator*, meaning that it acts on (i.e. operates on) a scalar field ψ to give a vector field $\boldsymbol{\nabla}\psi$. It is also an example of a differential operator in that it involves derivatives and, like all operators,[1] it acts only on objects to its right.

The directional derivative (12.5) can also be rewritten using (12.7) and the definition of $\hat{\mathbf{u}}$, when it becomes

$$\frac{\mathrm{d}\psi}{\mathrm{d}s} = \boldsymbol{\nabla}\psi \cdot \hat{\mathbf{u}}. \tag{12.8}$$

Since $\text{grad}\,\psi(\mathbf{r})$ is, by definition, normal to the equipotential surface at $P(\mathbf{r})$, (12.8) satisfies the requirement that $\mathrm{d}\psi/\mathrm{d}s = 0$ if $\hat{\mathbf{u}}$ is

[1] We have already met operators, and differential operators in particular, in Section 9.3.2 [cf. (9.46a)].

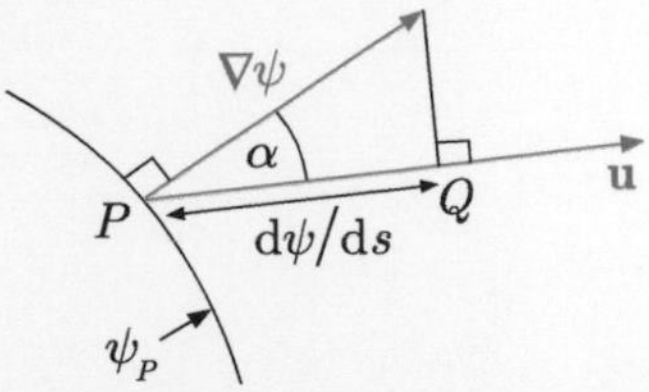

Figure 12.3 Gradient and directional derivative, where the vector **u** indicates the direction of $\hat{\mathbf{u}}$ defined in the text.

along the direction of a tangent to the equipotential surface at P and it attains its maximum value of $|\boldsymbol{\nabla}\psi|$ when $\hat{\mathbf{u}}$ and $\boldsymbol{\nabla}\psi$ are in the same direction. The relation between these quantities is shown in Figure 12.3.

Example 12.1

If $u = xy$ and $w = yz$, find expressions for grad (uw), u grad w and w grad u.

Solution

$$\operatorname{grad}(uw) = \boldsymbol{\nabla}(uw) = \frac{\partial}{\partial x}(xy^2z)\mathbf{i} + \frac{\partial}{\partial y}(xy^2z)\mathbf{j} + \frac{\partial}{\partial z}(xy^2z)\mathbf{k}$$
$$= y^2z\,\mathbf{i} + 2xyz\,\mathbf{j} + xy^2\,\mathbf{k},$$

$$u\operatorname{grad} w = u\boldsymbol{\nabla}w = u\left[\frac{\partial}{\partial x}(yz)\mathbf{i} + \frac{\partial}{\partial y}(yz)\mathbf{j} + \frac{\partial}{\partial z}(yz)\mathbf{k}\right]$$
$$= uz\,\mathbf{j} + uy\,\mathbf{k} = xyz\,\mathbf{j} + xy^2\,\mathbf{k},$$

and

$$w\operatorname{grad} u = w\boldsymbol{\nabla}u = w\left[\frac{\partial}{\partial x}(xy)\mathbf{i} + \frac{\partial}{\partial y}(xy)\mathbf{j} + \frac{\partial}{\partial z}(xy)\mathbf{k}\right]$$
$$= wy\,\mathbf{i} + wx\,\mathbf{j} = y^2z\,\mathbf{i} + xyz\,\mathbf{j}.$$

Example 12.2

A scalar field is given by $\psi = x^2 - y^2z$. Find $\boldsymbol{\nabla}\psi$ at the point $(1,1,1)$; the directional derivative there in the direction $\mathbf{i} - 2\mathbf{j} + \mathbf{k}$; the equation of the line passing through and normal to the surface $\psi = 0$ at the point $(1,1,1)$.

Solution

The gradient of ψ is

$$\boldsymbol{\nabla}\psi = 2x\,\mathbf{i} - 2yz\,\mathbf{j} - y^2\mathbf{k},$$

and at (1, 1, 1) this is $\boldsymbol{\nabla}\psi = 2\mathbf{i} - 2\mathbf{j} - \mathbf{k}$. The directional derivative is, from (12.8),

$$\mathrm{d}\psi/\mathrm{d}s = \boldsymbol{\nabla}\psi \cdot \hat{\mathbf{u}},$$

where $\hat{\mathbf{u}} = \frac{1}{\sqrt{6}}(\mathbf{i} - 2\mathbf{j} + \mathbf{k})$, so that $\mathrm{d}\psi/\mathrm{d}s = 5/\sqrt{6}$. Since the direction of the normal is

$$\boldsymbol{\nabla}\psi = 2\mathbf{i} - 2\mathbf{j} - \mathbf{k},$$

the equation of the line normal to the surface at (1,1,1) is [cf. (8.37a)]

$$\mathbf{r} = (x\mathbf{i} + y\mathbf{j} + z\mathbf{k}) + t(2\mathbf{i} - 2\mathbf{j} - \mathbf{k}),$$

where t is a variable parameter.

12.1.2 Div, grad and curl

Because del is a vector operator, it can form both scalar and vector products with vector fields. Thus if $\mathbf{V}$ is the vector field

$$\mathbf{V} = V_x\mathbf{i} + V_y\mathbf{j} + V_z\mathbf{k},$$

then the scalar product with del is

$$\boldsymbol{\nabla} \cdot \mathbf{V} = \frac{\partial V_x}{\partial x} + \frac{\partial V_y}{\partial y} + \frac{\partial V_z}{\partial z}. \tag{12.9}$$

This is called the *divergence* of $\mathbf{V}$ and is written div$\mathbf{V}$. Therefore div$\mathbf{V} \equiv \boldsymbol{\nabla} \cdot \mathbf{V}$. The vector product of del with $\mathbf{V}$ is

$$\begin{aligned}\boldsymbol{\nabla} \times \mathbf{V} &= \begin{vmatrix} \mathbf{i} & \mathbf{j} & \mathbf{k} \\ \partial/\partial x & \partial/\partial y & \partial/\partial z \\ V_x & V_y & V_z \end{vmatrix} \\ &= \left(\frac{\partial V_z}{\partial y} - \frac{\partial V_y}{\partial z}\right)\mathbf{i} - \left(\frac{\partial V_z}{\partial x} - \frac{\partial V_x}{\partial z}\right)\mathbf{j} + \left(\frac{\partial V_y}{\partial x} - \frac{\partial V_x}{\partial y}\right)\mathbf{k}\end{aligned} \tag{12.10}$$

and is called the *curl* of $\mathbf{V}$. Thus curl $\mathbf{V} \equiv \boldsymbol{\nabla} \times \mathbf{V}$. The origin of these names will emerge later in the chapter. Note that in both cases del is to the left of $\mathbf{V}$ because it is an operator. Thus, for example, $\boldsymbol{\nabla} \cdot \mathbf{V}$ and $\mathbf{V} \cdot \boldsymbol{\nabla}$ do *not* have the same meaning, even though they are both scalar products of the same quantities. The former is the simple scalar given in (12.9), the latter is the scalar differential operator

$$\mathbf{V} \cdot \boldsymbol{\nabla} = V_x\frac{\partial}{\partial x} + V_y\frac{\partial}{\partial y} + V_z\frac{\partial}{\partial z}.$$

Various combinations of div, grad and curl can also be formed. For example, if ψ is a scalar field, then $\boldsymbol{\nabla}\psi$ is a vector field. Hence, if we choose $\mathbf{V} = \boldsymbol{\nabla}\psi$, we can take its divergence to give

$$\text{div grad } \psi = \boldsymbol{\nabla} \cdot \boldsymbol{\nabla}\psi = \boldsymbol{\nabla}^2\psi, \tag{12.11a}$$

where the scalar operator

$$\boldsymbol{\nabla}^2 \equiv \boldsymbol{\nabla} \cdot \boldsymbol{\nabla} = \frac{\partial^2}{\partial x^2} + \frac{\partial^2}{\partial y^2} + \frac{\partial^2}{\partial z^2}, \tag{12.12}$$

is called the *Laplacian operator* and $\boldsymbol{\nabla}^2\psi$ is called the *Laplacian* of ψ. The Laplacian is an important operator in physical science and occurs very frequently, for example, in the wave equation

$$\boldsymbol{\nabla}^2\psi = \frac{1}{v^2}\frac{\partial^2\psi}{\partial t^2},$$

where v is the wave velocity. Similarly we note that, since the divergence of a vector field $\mathbf{V}$ is itself a scalar field, we can take its gradient to give

$$\text{grad div }\mathbf{V} = \boldsymbol{\nabla}(\boldsymbol{\nabla}\cdot\mathbf{V}). \tag{12.11b}$$

From this we see that grad div acts on a vector field $\mathbf{V}$ to give a vector field (12.11b) and is quite different from div grad, which acts on a scalar field ψ to give a scalar field (12.11a). This illustrates again that care is required with the order of factors when operators are involved. The Laplacian can also operate on a vector field $\mathbf{V}$ to give another vector field $\boldsymbol{\nabla}^2\mathbf{V}$ defined by

$$\boldsymbol{\nabla}^2\mathbf{V} = \boldsymbol{\nabla}^2 V_x\,\mathbf{i} + \boldsymbol{\nabla}^2 V_y\,\mathbf{j} + \boldsymbol{\nabla}^2 V_y\,\mathbf{k}. \tag{12.13}$$

The combination div grad and grad div are two of only five valid combinations of pairs of div, grad and curl. The other three, together with important identities which they satisfy, are

$$\text{curl grad}\,\psi = \boldsymbol{\nabla}\times\boldsymbol{\nabla}\psi = 0, \tag{12.14a}$$

$$\text{div curl}\,\mathbf{V} = \boldsymbol{\nabla}\cdot(\boldsymbol{\nabla}\times\mathbf{V}) = 0 \tag{12.14b}$$

$$\text{curl curl }\mathbf{V} = \boldsymbol{\nabla}\times(\boldsymbol{\nabla}\times\mathbf{V}) = \boldsymbol{\nabla}(\boldsymbol{\nabla}\cdot\mathbf{V}) - \boldsymbol{\nabla}^2\mathbf{V}. \tag{12.14c}$$

For example, from (12.2) and (12.10) we have

$$\text{curl grad }\psi = \begin{vmatrix} \mathbf{i} & \mathbf{j} & \mathbf{k} \\ \partial/\partial x & \partial/\partial y & \partial/\partial z \\ \partial\psi/\partial x & \partial\psi/\partial y & \partial\psi/\partial z \end{vmatrix} = 0,$$

and the other two identities also follow from the definitions of div, grad and curl.

In addition, there are many other identities involving del and two or more scalar or vector fields. They can all be verified by using the previous formulas, taking $\boldsymbol{\nabla}$ to be a differential vector operator. Some useful identities involving two fields are given in Table 12.1, where $\mathbf{a}$, $\mathbf{b}$ are arbitrary vector fields, and ψ and ϕ are arbitrary scalar fields.

Table 12.1 Some useful identities involving del

$\boldsymbol{\nabla}\times(\psi\mathbf{a}) = \psi(\boldsymbol{\nabla}\times\mathbf{a}) - \mathbf{a}\times(\boldsymbol{\nabla}\psi)$
$\boldsymbol{\nabla}\cdot(\mathbf{a}\times\mathbf{b}) = \mathbf{b}\cdot(\boldsymbol{\nabla}\times\mathbf{a}) - \mathbf{a}\cdot(\boldsymbol{\nabla}\times\mathbf{b})$
$\boldsymbol{\nabla}\times(\mathbf{a}\times\mathbf{b}) = -(\mathbf{a}\cdot\boldsymbol{\nabla})\mathbf{b} + (\mathbf{b}\cdot\boldsymbol{\nabla})\mathbf{a} + \mathbf{a}(\boldsymbol{\nabla}\cdot\mathbf{b}) - \mathbf{b}(\boldsymbol{\nabla}\cdot\mathbf{a})$
$\boldsymbol{\nabla}(\mathbf{a}\cdot\mathbf{b}) = \mathbf{b}\times(\boldsymbol{\nabla}\times\mathbf{a}) + (\mathbf{b}\cdot\boldsymbol{\nabla})\mathbf{a} + \mathbf{a}\times(\boldsymbol{\nabla}\times\mathbf{b}) + (\mathbf{a}\cdot\boldsymbol{\nabla})\mathbf{b}$
$\boldsymbol{\nabla}\cdot(\boldsymbol{\nabla}\phi\times\boldsymbol{\nabla}\psi) = 0$
$\boldsymbol{\nabla}\cdot(\psi\mathbf{a}) = \psi(\boldsymbol{\nabla}\cdot\mathbf{a}) + \mathbf{a}\cdot(\boldsymbol{\nabla}\psi)$

Example 12.3

Given that

$$\mathbf{A} = xy\,\mathbf{i} - y^2z\,\mathbf{j} + xz^2\mathbf{k}, \quad \mathbf{B} = xy^2\,\mathbf{i} + xz\,\mathbf{j} - 3xy\,\mathbf{k}, \quad \phi = 2xyz^2,$$

find: (a) $\boldsymbol{\nabla}\phi$, (b) $\operatorname{div}\mathbf{A}$, (c) $\operatorname{curl}\mathbf{B}$, (d) $(\boldsymbol{\nabla}\cdot\mathbf{B})\mathbf{A}$, (e) $(\mathbf{B}\cdot\boldsymbol{\nabla})\mathbf{A}$, (f) $\boldsymbol{\nabla}^2\phi$, (g) $\boldsymbol{\nabla}^2\mathbf{A}$.

Solution

(a) $$\boldsymbol{\nabla}\phi = \frac{\partial\phi}{\partial x}\mathbf{i} + \frac{\partial\phi}{\partial y}\mathbf{j} + \frac{\partial\phi}{\partial z}\mathbf{k} = 2yz^2\mathbf{i} + 2xz^2\mathbf{j} + 4xyz\,\mathbf{k},$$

(b) $$\boldsymbol{\nabla}\cdot\mathbf{A} = \frac{\partial}{\partial x}(xy) + \frac{\partial}{\partial y}(-y^2) + \frac{\partial z}{\partial z}(xz^2) = y - 2yz + 2xz,$$

(c) $$\boldsymbol{\nabla}\times\mathbf{B} = \begin{vmatrix} \mathbf{i} & \mathbf{j} & \mathbf{k} \\ \dfrac{\partial}{\partial x} & \dfrac{\partial}{\partial y} & \dfrac{\partial}{\partial z} \\ xy^2 & xz & -3xy \end{vmatrix} = (-3x - x)\mathbf{i} - (-3y)\mathbf{j}$$
$$+(z - 2xy)\mathbf{k} = -4x\,\mathbf{i} + 3y\,\mathbf{j} + (z - 2xy)\mathbf{k}.$$

(d) $$(\boldsymbol{\nabla}\cdot\mathbf{B})\mathbf{A} = (y^2)\mathbf{A} = xy^3\mathbf{i} - y^4z\,\mathbf{j} + xy^2z^2\mathbf{k}$$

(e) $$\begin{aligned}(\mathbf{B}\cdot\boldsymbol{\nabla})\mathbf{A} &= \left(xy^2\frac{\partial}{\partial x} + xz\frac{\partial}{\partial y} - 3xy\frac{\partial}{\partial z}\right)\mathbf{A} \\ &= xy^2(y\,\mathbf{i} + z^2\mathbf{k}) + xz(x\,\mathbf{i} - 2yz\,\mathbf{j}) \\ &\quad - 3xy(-y^2\mathbf{j} + 2xz\,\mathbf{k}) \\ &= (xy^3 + x^2z)\mathbf{i} + (-2xyz^2 + 3xy^3)\mathbf{j} \\ &\quad + (xy^2z^2 - 6x^2yz)\mathbf{k},\end{aligned}$$

(f) $$\boldsymbol{\nabla}^2\phi = \left(\frac{\partial^2}{\partial x^2} + \frac{\partial^2}{\partial y^2} + \frac{\partial^2}{\partial z^2}\right)(2xyz^2) = 4xy$$

(g) $$\begin{aligned}\boldsymbol{\nabla}^2\mathbf{A} &= \left(\mathbf{i}\frac{\partial^2}{\partial x^2} + \mathbf{j}\frac{\partial^2}{\partial y^2} + \mathbf{k}\frac{\partial^2}{\partial z^2}\right)(xy\,\mathbf{i} - y^2z\,\mathbf{j} \ + xz^2)\,\mathbf{k} \\ &= -2z + 2x\end{aligned}$$

Example 12.4

If $\mathbf{E}$ and $\mathbf{B}$ are electric and magnetic fields, *Maxwell's equations* in free space in the absence of charges and currents are

$$\operatorname{curl}\mathbf{E} = -\frac{\partial\mathbf{B}}{\partial t}; \quad \operatorname{curl}\mathbf{B} = \frac{1}{c^2}\frac{\partial\mathbf{E}}{\partial t},$$

with

$$\text{div } \mathbf{E} = \text{div } \mathbf{B} = 0,$$

where c is the speed of light in a vacuum. Show that $\mathbf{E}$ and $\mathbf{B}$ satisfy

$$\boldsymbol{\nabla}^2\mathbf{U} = \frac{1}{c^2}\frac{\partial^2 \mathbf{U}}{\partial t^2}, \qquad (\mathbf{U} = \mathbf{E}, \mathbf{B}\,).$$

Solution

Taking the curl of the first equation and using the second, we have:

$$\begin{aligned}\text{curl curl } \mathbf{E} &= \boldsymbol{\nabla} \times (\boldsymbol{\nabla} \times \mathbf{E}) = \boldsymbol{\nabla} \times \left(-\frac{\partial \mathbf{B}}{\partial t}\right) \\ &= -\frac{\partial}{\partial t}(\boldsymbol{\nabla} \times \mathbf{B}\,) = -\frac{1}{c^2}\frac{\partial^2 \mathbf{E}}{\partial t^2}.\end{aligned}$$

But from (12.14c)

$$\boldsymbol{\nabla} \times (\boldsymbol{\nabla} \times \mathbf{E}) = -\boldsymbol{\nabla}^2\mathbf{E} + \boldsymbol{\nabla}(\boldsymbol{\nabla} \cdot \mathbf{E}) = -\boldsymbol{\nabla}^2\mathbf{E}$$

since div $\mathbf{E} = 0$. Thus,

$$\boldsymbol{\nabla}^2\mathbf{E} = \frac{1}{c^2}\frac{\partial^2 \mathbf{E}}{\partial t^2}.$$

An analogous procedure for $\mathbf{B}$ shows that $\boldsymbol{\nabla}^2\mathbf{B} = \dfrac{1}{c^2}\dfrac{\partial^2 \mathbf{B}}{\partial t^2}$.

12.1.3 Orthogonal curvilinear co-ordinates

So far, we have defined div, grad and curl in Cartesian co-ordinates. However, in problems with spherical or cylindrical symmetry, it is much easier to work in spherical or cylindrical polar co-ordinates, which reflect the symmetry of the problem. As we saw in Section 11.3, these two co-ordinate systems are examples of orthogonal curvilinear co-ordinates $u_i (i = 1, 2, 3)$, which are such that distances $\mathbf{dr}$ are obtained from formulas of the type

$$\mathbf{dr} = \sum_{1}^{3} h_i \, du_i \, \mathbf{e}_i, \tag{12.15}$$

where the unit vectors $\mathbf{e}_i$ are orthogonal. The scale factors h_i are given by (11.30b), which for the special case of polar co-ordinates are [cf. (11.36) and (11.42)]

$$\begin{aligned}\text{cylindrical} \quad & h_r = 1 \quad h_\phi = \rho \quad h_z = 1, \\ \text{spherical} \quad & h_r = 1 \quad h_\theta = r \quad h_\phi = r\sin\theta.\end{aligned} \tag{12.16}$$

Here we shall first give the forms of $\boldsymbol{\nabla}$, $\boldsymbol{\nabla}^2$ etc. in orthogonal curvilinear co-ordinates, and then obtain the corresponding expressions from them for the cases of spherical and cylindrical polar co-ordinates.

Consider firstly the gradient of a scalar ψ, that is, $\mathrm{grad}\,\psi = \boldsymbol{\nabla}\psi$. Returning to Figure 12.2, we let PQ be in the direction of u_1, rather than x as before. Then the component of $\boldsymbol{\nabla}\psi$ in the direction of u_1 (with u_2 and u_3 held fixed) is the directional derivative $\mathrm{d}\psi/\mathrm{d}s$, where $\mathrm{d}s = h_1\,\mathrm{d}u_1$, that is, the component of $\boldsymbol{\nabla}\psi$ in the direction $\mathbf{e}_1$, is

$$\frac{1}{h_1}\frac{\partial\psi}{\partial u_1},$$

and similarly for the other directions. Thus,

$$\boldsymbol{\nabla}\psi = \sum_{i=1}^{3}\frac{\mathbf{e}_i}{h_i}\frac{\partial\psi}{\partial u_i}. \tag{12.17}$$

For example, for spherical polar co-ordinates

$$\mathbf{e}_1 = \mathbf{e}_r; \quad \mathbf{e}_2 = \mathbf{e}_\theta; \quad \mathbf{e}_3 = \mathbf{e}_\phi$$

and using (12.16), gives

$$\boldsymbol{\nabla}\psi = \frac{\partial\psi}{\partial r}\mathbf{e}_r + \frac{1}{r}\frac{\partial\psi}{\partial\theta}\mathbf{e}_\theta + \frac{1}{r\sin\theta}\frac{\partial\psi}{\partial\phi}\mathbf{e}_\phi.$$

The derivations of the corresponding results for div $\mathbf{V}$, curl $\mathbf{V}$ and $\boldsymbol{\nabla}^2\mathbf{V}$ using the technique above are more difficult. The derivations are much easier using results we will obtain in Sections 12.3 and 12.4, and will be given there. For the present we will just quote the results

$$\boldsymbol{\nabla}\cdot\mathbf{V} = \frac{1}{h_1h_2h_3}\left[\frac{\partial}{\partial u_1}(h_2h_3V_1) + \frac{\partial}{\partial u_2}(h_1h_3V_2) + \frac{\partial}{\partial u_3}(h_1h_2V_3)\right], \tag{12.18}$$

$$\boldsymbol{\nabla}^2\psi = \frac{1}{h_1h_2h_3}\left[\frac{\partial}{\partial u_1}\left(\frac{h_2h_3}{h_1}\frac{\partial\psi}{\partial u_1}\right) + \frac{\partial}{\partial u_2}\left(\frac{h_1h_3}{h_2}\frac{\partial\psi}{\partial u_2}\right) + \frac{\partial}{\partial u_3}\left(\frac{h_1h_2}{h_3}\frac{\partial\psi}{\partial u_3}\right)\right], \tag{12.19}$$

and

$$\boldsymbol{\nabla}\times\mathbf{V} = \frac{1}{h_1h_2h_3}\begin{vmatrix} h_1\mathbf{e}_1 & h_2\mathbf{e}_2 & h_3\mathbf{e}_3 \\ \dfrac{\partial}{\partial u_1} & \dfrac{\partial}{\partial u_2} & \dfrac{\partial}{\partial u_3} \\ h_1V_1 & h_2V_2 & h_3V_3 \end{vmatrix}. \tag{12.20}$$

The expressions for $\boldsymbol{\nabla}\psi$, $\boldsymbol{\nabla}\cdot\mathbf{V}$, $\boldsymbol{\nabla}\times\mathbf{V}$ and $\boldsymbol{\nabla}^2\mathbf{V}$ given earlier for the special case of Cartesian co-ordinates are easily regained

Table 12.2 Grad, div, curl and the Laplacian in polar co-ordinates

	Spherical polar	Cylindrical polar
$\nabla\psi$	$\dfrac{\partial\psi}{\partial r}\mathbf{e}_r + \dfrac{1}{r}\dfrac{\partial\psi}{\partial\theta}\mathbf{e}_\theta + \dfrac{1}{r\sin\theta}\dfrac{\partial\psi}{\partial\phi}\mathbf{e}_\phi$	$\dfrac{\partial\psi}{\partial\rho}\mathbf{e}_\rho + \dfrac{1}{\rho}\dfrac{\partial\psi}{\partial\phi}\mathbf{e}_\phi + \dfrac{\partial\psi}{\partial z}\mathbf{e}_z$
$\nabla\cdot\mathbf{V}$	$\dfrac{1}{r^2}\dfrac{\partial}{\partial r}\left(r^2V_r\right) + \dfrac{1}{r\sin\theta}\dfrac{\partial}{\partial\theta}\left(\sin\theta V_\theta\right) + \dfrac{1}{r\sin\theta}\dfrac{\partial V_\phi}{\partial\phi}$	$\dfrac{1}{\rho}\dfrac{\partial}{\partial\rho}(\rho V_\rho) + \dfrac{1}{\rho}\dfrac{\partial V_\phi}{\partial\phi} + \dfrac{\partial V_z}{\partial z}$
$\nabla\times\mathbf{V}$	$\dfrac{1}{r^2\sin\theta}\begin{vmatrix}\mathbf{e}_r & r\,\mathbf{e}_\theta & r\sin\theta\,\mathbf{e}_\phi \\ \partial/\partial r & \partial/\partial\theta & \partial/\partial\phi \\ V_r & rV_\theta & r\sin\theta\,V_\phi\end{vmatrix}$	$\dfrac{1}{\rho}\begin{vmatrix}\mathbf{e}_\rho & \rho\,\mathbf{e}_\phi & \mathbf{e}_z \\ \partial/\partial\rho & \partial/\partial\phi & \partial/\partial z \\ V_\rho & \rho V_\phi & V_z\end{vmatrix}$
$\nabla^2\psi$	$\dfrac{1}{r^2}\dfrac{\partial}{\partial r}\left(r^2\dfrac{\partial\psi}{\partial r}\right) + \dfrac{1}{r^2\sin\theta}\dfrac{\partial}{\partial\theta}\left(\sin\theta\dfrac{\partial\psi}{\partial\theta}\right) + \dfrac{1}{r^2\sin^2\theta}\dfrac{\partial^2\psi}{\partial\phi^2}$	$\dfrac{1}{\rho}\dfrac{\partial}{\partial\rho}\left(\rho\dfrac{\partial\psi}{\partial\rho}\right) + \dfrac{1}{\rho^2}\dfrac{\partial^2}{\partial\phi^2} + \dfrac{\partial^2\psi}{\partial z^2}$

by setting $(u_1, u_2, u_3) = (x, y, z)$ and $h_x = h_y = h_z = 1$ in (12.17) to (12.20), respectively. The corresponding results for spherical polar and cylindrical polar co-ordinates are similarly obtained using (12.16) for the scale factors and are shown in Table 12.2.

Example 12.5

If $\psi = 2xyz$ and $\mathbf{V} = y\,\mathbf{i} + z\,\mathbf{j}$, express (a) ψ, (b) $\nabla\psi$, (c) $\mathbf{V}$ and (d) $\nabla\times\mathbf{V}$ in cylindrical co-ordinates.

Solution

From (11.33a) and (11.35) we have

$$x = \rho\cos\phi, \quad y = \rho\sin\phi, \quad z = z$$

and

$$\mathbf{e}_\rho = \cos\phi\,\mathbf{i} + \sin\phi\,\mathbf{j}, \quad \mathbf{e}_\phi = -\sin\phi\,\mathbf{i} + \cos\phi\,\mathbf{j}, \quad \mathbf{e}_z = \mathbf{k}.$$

Hence,

(a) $\psi = 2\rho^2 z\sin\phi\cos\phi = \rho^2 z\sin(2\phi)$,

(b) From table 12.1,

$$\begin{aligned}\nabla\psi &= \frac{\partial\psi}{\partial\rho}\mathbf{e}_\rho + \frac{1}{\rho}\frac{\partial\psi}{\partial\phi}\mathbf{e}_\phi + \frac{\partial\psi}{\partial z}\mathbf{e}_z \\ &= (2\rho\, z\sin 2\phi)\mathbf{e}_\rho + (2\rho^2 z\cos 2\phi)\mathbf{e}_\phi + (\rho^2\sin 2\phi)\mathbf{e}_z.\end{aligned}$$

(c) If $\mathbf{V} = V_\rho \mathbf{e}_\rho + V_\phi \mathbf{e}_\phi + V_z \mathbf{e}_z$, then since $\mathbf{e}_\rho$, $\mathbf{e}_\phi$, $\mathbf{e}_z$ are orthogonal,

$$V_\rho = \mathbf{e}_\rho \cdot \mathbf{V} = y\cos\phi + z\sin\phi = \rho\sin\phi\cos\phi + z\sin\phi,$$
$$V_\phi = \mathbf{e}_\phi \cdot \mathbf{V} = -y\sin\phi + z\cos\phi = z\cos\phi - \rho\sin^2\phi.$$
$$V_z = \mathbf{e}_z \cdot \mathbf{V} = 0,$$

so

$$\mathbf{V} = (\tfrac{1}{2}\rho\sin 2\phi + z\sin\phi)\mathbf{e}_\rho + (z\cos\phi - \rho\sin^2\phi)\mathbf{e}_\phi.$$

(d) Again from Table 12.1,

$$\rho\boldsymbol{\nabla}\times\mathbf{V} = \mathbf{e}_\rho\left[\frac{\partial V_z}{\partial\phi} - \frac{\partial}{\partial z}(\rho V_\phi)\right] - \rho\,\mathbf{e}_\phi\left[\frac{\partial V_z}{\partial\rho} - \frac{\partial V_\rho}{\partial z}\right]$$
$$+\mathbf{e}_z\left[\frac{\partial}{\partial\rho}(\rho V_\phi) - \frac{\partial V_\rho}{\partial\phi}\right],$$

so

$$\boldsymbol{\nabla}\times\mathbf{V} = -\cos\phi\,\mathbf{e}_\rho + \sin\phi\,\mathbf{e}_\phi - (2\sin^2\phi + \cos 2\phi)\,\mathbf{e}_z.$$

12.2 Line, surface, and volume integrals

In this section, we shall extend the discussion of line integrals given in Chapter 11 to embrace integrals of a vector field and use vector methods to define integrals of a vector field over a curved surface.

12.2.1 Line integrals

In Section 8.3.1, we saw that the running vector

$$\mathbf{r} = \mathbf{a} + s\,\mathbf{b} \tag{12.21}$$

where $\mathbf{a}$ and $\mathbf{b}$ are fixed vectors, described a straight line passing through the point $\mathbf{r} = \mathbf{a}$ in the direction $\hat{\mathbf{b}}$ as the scalar parameter s varied in the range $-\infty < s < \infty$. More generally, any running vector $\mathbf{r}(s)$, where $\mathbf{r}$ is a differentiable function of s, will describe a curve in space and for any given s the differential

$$\mathrm{d}\mathbf{r} = \frac{\mathrm{d}\mathbf{r}}{\mathrm{d}s}\mathrm{d}s \tag{12.22}$$

is an infinitesimal vector directed along the tangent to the curve, as shown in Figure 12.4 for the point P corresponding to $s = s_0$. For example, in Cartesian co-ordinates

$$\mathbf{r}(s) = s\,\mathbf{i} + (4as)^{1/2}\mathbf{j}$$

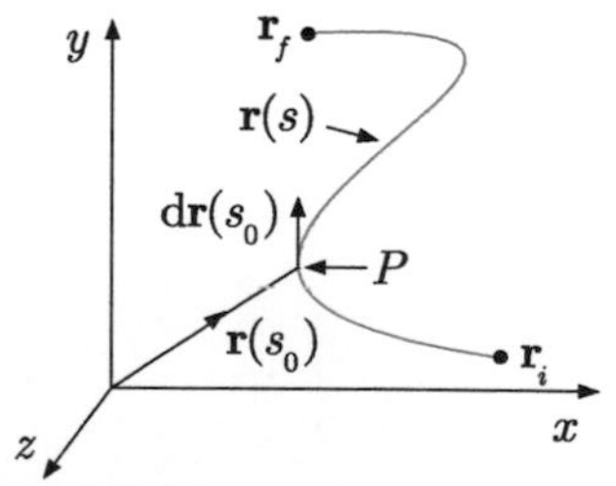

Figure 12.4 Definition of a space curve for the line integral of a scalar product. The path is defined in terms of a parameter s, so that $\mathbf{r} = \mathbf{r}(s)$.

is the vector equation of the parabola $y^2 = 4ax$ lying in the plane $z = 0$, and the corresponding differential

$$\mathbf{dr} = \mathrm{d}s\,\mathbf{i} + \left(\frac{a}{s}\right)^{1/2} \mathrm{d}s\,\mathbf{j}$$

is an infinitesimal vector directed along the tangent to the curve, as shown in Figure 12.4 for the point P corresponding to $s = s_0$.

Now suppose we have a vector field $\mathbf{V}(\mathbf{r})$. Then we can define two line integrals

$$\int_C \mathbf{V} \cdot \mathbf{dr} \quad \text{and} \quad \int_C \mathbf{V} \times \mathbf{dr}, \tag{12.23}$$

where C as usual denotes the path, or contour, of integration. The first of these is by far the most important in physics, and is the only one we shall discuss. One important example is the work done by a force field $\mathbf{F}(\mathbf{r})$. If $\mathbf{F}(\mathbf{r})$ is the force acting at the position $\mathbf{r}$, then $\mathbf{F}(\mathbf{r}) \cdot \mathbf{dr}$ is the work done by the force in moving from $\mathbf{r}$ to $\mathbf{r} + \mathbf{dr}$, and the integral

$$W = \int_C \mathbf{F}(\mathbf{r}) \cdot \mathbf{dr} \tag{12.24}$$

is the work done in moving from the initial position $\mathbf{r}_i$ to the final position $\mathbf{r}_f$ along the path C. If the force returns to $\mathbf{r}_i$, then $\mathbf{r}_i = \mathbf{r}_f$ and the integral is denoted

$$\oint_C \mathbf{F}(\mathbf{r}) \cdot \mathbf{dr},$$

where the circle on the integral sign emphasises that the path is a closed loop.

So far, we have not used any co-ordinates to define the integrals. In general, if we use a set of co-ordinates u_1, u_2, u_3, such that

$$\mathbf{dr} = h_1\mathbf{e}_1\mathrm{d}u_1 + h_2\mathbf{e}_2\mathrm{d}u_2 + h_3\mathbf{e}_3\mathrm{d}u_3$$

and

$$\mathbf{V}(\mathbf{r}) = V_1\mathbf{e}_1 + V_2\mathbf{e}_2 + V_3\mathbf{e}_3,$$

then,

$$\int_C \mathbf{V} \cdot \mathbf{dr} = \int_C (V_1 h_1 \mathrm{d}u_1 + V_2 h_2 \mathrm{d}u_2 + V_3 h_3 \mathrm{d}u_3). \tag{12.25}$$

In Cartesian co-ordinates

$$\mathbf{dr} = \mathrm{d}x\,\mathbf{i} + \mathrm{d}y\,\mathbf{j} + \mathrm{d}z\,\mathbf{k},$$

so that (12.25) becomes

$$\int_C \mathbf{V} \cdot \mathbf{dr} = \int_C (V_x \mathrm{d}x + V_y \mathrm{d}y + V_z \mathrm{d}z). \tag{12.26a}$$

Here x, y, z (and in general u_1, u_2, u_3) are not independent variables along the path C, but are specified by a single parameter s, so that $C = \mathbf{r}(s)$ and (12.26a) becomes

$$\int_C \mathbf{V} \cdot \mathrm{d}\mathbf{r} = \int \left(V_x \frac{\mathrm{d}x}{\mathrm{d}s} + V_y \frac{\mathrm{d}y}{\mathrm{d}s} + V_z \frac{\mathrm{d}z}{\mathrm{d}s} \right) \mathrm{d}s, \qquad (12.26\text{b})$$

using (12.22). In particular, s may be chosen to be one of the coordinates themselves, for example x, when (12.26a) may be used directly together with the relations $y = y(x)$, $z = z(x)$ along the contour C.

At this point we note that (12.26) is identical with the line integral (11.13) discussed in Section 11.1.3, if the functions Q, R, P are replaced by functions V_x, V_y, V_z. Hence the methods and results discussed in Section 11.1.3 can be carried over, with a trivial relabeling, to the line integrals (12.26) as we shall illustrate in the next section.

Example 12.6

The vector $\mathbf{V}$ is given by

$$\mathbf{V} = (3x + 6y^2)\mathbf{i} - 4x^2yz\,\mathbf{j} + 2xz^3\mathbf{k}.$$

Evaluate the line integral

$$\int \mathbf{V} \cdot \mathrm{d}\mathbf{r},$$

from the point (0,0,0) to the point (1,1,1) to four significant figures, where the curve is given (a) by the parametric form

$$x = t^2, \quad y = t, \quad z = t^3$$

and (b) by a straight line between the two end points.

Solution

The integral is

$$\begin{aligned}\int \mathbf{V} \cdot \mathrm{d}\mathbf{r} &= \int [(3x + 6y^2)\mathbf{i} - 4x^2yz\mathbf{j} + 2xz^3\mathbf{k}] \cdot (\mathrm{d}x\,\mathbf{i} + \mathrm{d}y\,\mathbf{j} + \mathrm{d}z\,\mathbf{k}) \\ &= \int (3x + 6y^2)\mathrm{d}x - \int (4x^2yz)\mathrm{d}y + \int (2xz^3)\mathrm{d}z\end{aligned}$$

(a) Using $x = t^2$, $y = t$, $z = t^3$ gives

$$\int \mathbf{V} \cdot \mathrm{d}\mathbf{r} = \int_0^1 (18t^3 - 4t^8 + 16t^{13})\mathrm{d}t = 4.484,$$

since (0,0,0) and (1,1,1) correspond to $t = 0$ and $t = 1$, respectively.

(b) In parametric form, $x = y = z = t$ and so

$$\int \mathrm{V} \cdot \mathrm{d}\mathbf{r} = \int_0^1 (3t + 6t^2 - 2t^4)\mathrm{d}t = 3.100$$

Example 12.7

Evaluate the integral

$$\int_C \mathbf{V} \cdot \mathrm{d}\mathbf{r} = \int_C [y\mathrm{d}x - y(x-1)\mathrm{d}y + y^2 z\,\mathrm{d}z],$$

where C is the curve given by the intersection of the sphere

$$x^2 + y^2 + z^2 = 4$$

and the cylinder

$$(x-1)^2 + y^2 = 1,$$

in the positive octant $x, y, z > 0$, between the points $A(2, 0, 0)$ and $B(0, 0, 2)$.

Solution

The solid black lines in Figure 12.5 show an octant of the sphere; the dashed lines show the cylinder. Where they intersect defines

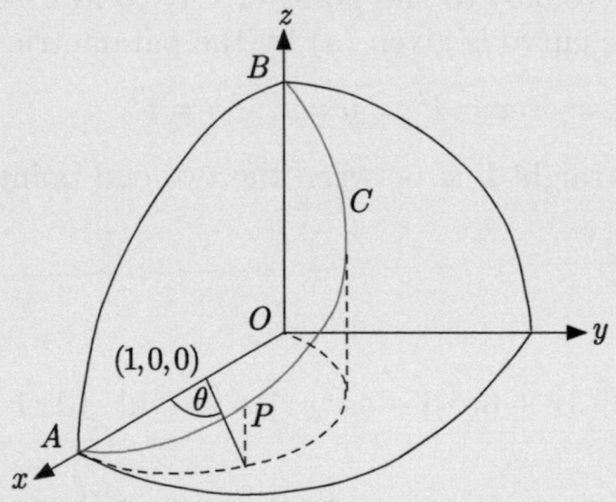

Figure 12.5

the path C of the integral, which is shown in blue. The point P is one such intersection point. We need to express C in a parametric form, by using the fact that points on C satisfy

$$x^2 + y^2 + z^2 = 4 \quad \text{and} \quad (x-1)^2 + y^2 = 1.$$

Since the point P lies on the cylinder, we can write

$$x - 1 = \cos\theta, \quad y = \sin\theta,$$

where θ is the angle shown in Figure 12.5 and $0 \leq \theta \leq \pi$ in the positive octant. Substituting into the equation for the sphere gives

$$1 + 2\cos\theta + \cos^2\theta + \sin^2\theta + z^2 = 4 \Rightarrow z = 2\sin(\theta/2) \quad (z > 0).$$

Then the integral becomes

$$\int_0^\pi [\sin\theta\, \mathrm{d}(1+\cos\theta) - \sin\theta\cos\theta\, \mathrm{d}(\sin\theta)$$
$$+2\sin^2\theta\sin(\theta/2)\,\mathrm{d}(2\sin\theta/2)]$$
$$= \int_0^\pi (-\sin^2\theta - \sin\theta\cos^2\theta + \sin^3\theta)\mathrm{d}\theta = 2/3 - \pi/2,$$

where the final integral is evaluated using the methods of Section 4.3.2.

12.2.2 Conservative fields and potentials

The result of a line integral of a vector between any two points will in general depend on the path taken between them. If, however, the line integral is independent of the path for any choice of end points within the field, the vector field $\mathbf{V}$ is said to be *conservative*. Conservative fields play an important role in physics, as we shall now see.

Suppose that

$$\operatorname{curl}\mathbf{V} = \boldsymbol{\nabla}\wedge\mathbf{V} = 0. \tag{12.27}$$

Then, using the expression (12.10) for the curl in Cartesian coordinates, we see that (12.27) implies that

$$\frac{\partial V_x}{\partial y} - \frac{\partial V_y}{\partial x} = \frac{\partial V_y}{\partial z} - \frac{\partial V_z}{\partial y} = \frac{\partial V_z}{\partial x} - \frac{\partial V_x}{\partial z} = 0,$$

which is precisely the condition [cf. (11.23b) and (7.19b)] that

$$\mathrm{d}\psi = V_x\mathrm{d}x + V_y\mathrm{d}y + V_z\mathrm{d}z \tag{12.28}$$

is an exact, or perfect, differential. From (12.28) we immediately see that

$$V_x = \frac{\partial\psi}{\partial x}, \quad V_y = \frac{\partial\psi}{\partial y}, \quad V_z = \frac{\partial\psi}{\partial z},$$

i.e.

$$\mathbf{V} = \operatorname{grad}\psi, \tag{12.29}$$

and that

$$\int_{A\to B}\mathbf{V}\cdot\mathrm{d}\mathbf{r} = \int_{A\to B}\mathrm{d}\psi = \psi_B - \psi_A, \tag{12.30}$$

where ψ_A and ψ_B are the values of ψ at the points A and B. Hence $\mathbf{V}$ is a conservative field and can be derived from a scalar field ψ, called a *potential field,* or just a *potential.* We note that $\psi(\mathbf{r})$ is only defined up to a constant by (12.29) and (12.30). This is usually chosen by requiring that ψ has a given value ψ_0 at a reference point $\mathbf{r}_0$, or sometimes that $\psi(\mathbf{r}) \to 0$ as $|\mathbf{r}| \to \infty$.

The above argument shows that (12.27) is a sufficient condition for $\mathbf{V}$ to be a conservative field. That it is also a necessary condition is seen by reversing the argument. If $V(\mathbf{r})$ is a conservative field, then we can *define* a potential by

$$\psi(\mathbf{r}) \equiv \int_{\mathbf{r}_0}^{\mathbf{r}} \mathbf{V} \cdot \mathrm{d}\mathbf{r} + \psi_0,$$

since the integral is independent of the chosen path between the reference point $\mathbf{r}_0$ and the point $\mathbf{r}$. This implies that

$$\mathrm{d}\psi(\mathbf{r}) = \psi(\mathbf{r} + \mathrm{d}\mathbf{r}) - \psi(\mathbf{r}) = \mathbf{V} \cdot \mathrm{d}\mathbf{r},$$

so that $\mathbf{V} = \operatorname{grad} \psi$ and $\operatorname{curl} \mathbf{V} = 0$ by (12.14a). Hence $\operatorname{curl} \mathbf{V} = 0$ is not only a sufficient condition, but also a necessary condition for $\mathbf{V}$ to be a conservative field.

An important example of a conservative field is the gravitational field. In general, if $\mathbf{F}(\mathbf{r})$ is the force acting on a particle at a position $\mathbf{r}$, it is usual to introduce the *potential energy* due to gravity such that

$$\mathbf{F} = -\boldsymbol{\nabla}\phi, \tag{12.31}$$

that is, the force acts in the direction of maximally decreasing potential energy. The work W done when $\mathbf{F}$ moves a particle from A to B is

$$W = \int_A^B \mathbf{F} \cdot \mathrm{d}\mathbf{r} = \phi_A - \phi_B, \tag{12.32}$$

so that the work done by the force equals the loss of potential energy.

Of course not all forces are conservative. If dissipative forces such as friction are involved, then energy will be lost in moving from A to B in a way that depends on the path and a potential cannot be defined.

Example 12.8

In spherical co-ordinates, the gravitational force on a body of mass m at position $\mathbf{r}$ relative to the centre of the earth is given by

$$\mathbf{F} = -\frac{GMm}{r^2}\mathbf{e}_r,$$

where G is Newton's gravitational constant, M is the mass of the earth and we assume $|\mathbf{r}| > R$, the earth's radius. Show that $\mathbf{F}$ is a conservative force field, and find the potential $\phi(\mathbf{r})$ satisfying $\mathbf{F} = -\boldsymbol{\nabla}\phi$, where we take $\phi \to 0$ as $|\mathbf{r}| \to \infty$.

Solution

Using the expression for curl $\mathbf{V}$ in spherical co-ordinates given in Table 12.1, one easily see that $\boldsymbol{\nabla} \times \mathbf{F} = \mathbf{0}$, that is, the force is conservative. The potential is obviously spherically symmetric, so we can write $\phi = \phi(r)$ and hence (12.31) becomes

$$\mathbf{F} = -\frac{GMm}{r^2}\mathbf{e}_r = -\boldsymbol{\nabla}\phi(r) = -\frac{\partial\phi(r)}{\partial r}\mathbf{e}_r,$$

using the expression for $\boldsymbol{\nabla}$ in spherical co-ordinates given in Table 12.1. Hence

$$\frac{\partial\phi(r)}{\partial r} = \frac{GMm}{r^2},$$

with the solution $\phi(r) = -GMm/r$, where we have imposed thc boundary condition $\phi(r) \to 0$ as $r \to \infty$.

Example 12.9

Show that the vector

$$\mathbf{V} = (2xy - z^3)\mathbf{i} + x^2\mathbf{j} - (3xz^2 + 1)\mathbf{k}$$

is conservative and find a scalar potential ϕ such that $\mathbf{V} = -\boldsymbol{\nabla}\phi$.

Solution

To show that $\mathbf{V}$ is a conservative field we calculate

$$\begin{aligned}\text{curl } \mathbf{V} &= \begin{vmatrix} \mathbf{i} & \mathbf{j} & \mathbf{k} \\ \partial/\partial x & \partial/\partial y & \partial/\partial z \\ 2xy - z^3 & x^2 & -(3xz^2 + 1) \end{vmatrix} \\ &= (0)\mathbf{i} - (-3z^2 + 3z^2)\mathbf{j} + (2x - 2x)\mathbf{k} = \mathbf{0}.\end{aligned}$$

Thus $\mathbf{V}$ is a conservative field. Then

$$W = \int_A^B \mathbf{V} \cdot \mathrm{d}\mathbf{r} = \int_A^B [(2xy - z^3)\mathrm{d}x + x^2\mathrm{d}y - (3xz^2 + 1)\mathrm{d}z]$$

is independent of the path of integration. We choose A to be the origin and integrate to an arbitrary point B, with co-ordinates (x, y, z), along the path consisting of the three segments

$$(0, 0, 0) \to (x, 0, 0) \to (x, y, 0) \to (x, y, z).$$

Then,

for segment 1: $y = z = 0 \Rightarrow \mathrm{d}y = \mathrm{d}z = 0$,
for segment 2: $x = \text{constant},\ z = 0 \Rightarrow \mathrm{d}x = \mathrm{d}z = 0$,
for segment 3: $x = \text{constant},\ y = \text{constant} \Rightarrow \mathrm{d}x = \mathrm{d}y = 0$.

So we have (the first integral is zero),

$$W = \int_0^y x^2 \mathrm{d}y - \int_0^z (3xz^2 + 1)\mathrm{d}z = x^2 y - xz^3 - z.$$

Finally, using (12.32) and setting $\phi = 0$ at the origin, we have

$$\phi = -W = -x^2 y + xz^3 + z.$$

12.2.3 Surface integrals

In three dimensions, a surface is defined by an equation of the form

$$f(x, y, z) = d, \tag{12.33}$$

where $f(x, y, z)$ is a given function and d is a constant.[2] Simple examples are the equation of a plane [cf (8.45a)],

$$ax + by + cz = d, \tag{12.34}$$

which is an example of an open surface; and the equation of a sphere

$$(x - a)^2 + (y - b)^2 + (z - c)^2 = r^2, \tag{12.35}$$

which is an example of a closed surface. In addition, since (12.33) defines an equipotential surface for the scalar field f, it follows from the discussion of $\boldsymbol{\nabla} f$ in Section 12.1.1 (cf. Figure 12.2) that at any point on the surface

$$\hat{\mathbf{n}} = \pm \frac{1}{|\boldsymbol{\nabla} f|} \boldsymbol{\nabla} f \tag{12.36}$$

are the two vectors normal to the surface.

We now introduce integrals of a vector field $\mathbf{V}(\mathbf{r})$ over a surface S as follows. Given a small surface element $\mathrm{d}s$, we form a vector surface element

$$\mathrm{d}\mathbf{s} = \mathrm{d}s\, \hat{\mathbf{n}}, \tag{12.37}$$

[2] The constant d is often taken to the left-hand side and (12.33) written in the form $g(x, y, z) = f(x, y, z) - d = 0$.

where $\hat{\mathbf{n}}$ is a unit vector normal to the surface at the position of ds, so that the direction of $\hat{\mathbf{n}}$ varies continuously over S. Surface integrals can now be defined of the form

$$\iint_s \mathbf{V} \cdot \mathrm{d}\mathbf{s} \quad \text{and} \quad \iint_s \mathbf{V} \times \mathrm{d}\mathbf{s}. \tag{12.38}$$

In each case, the integral is a double integral over a surface S, which may be open or closed. If the surface is closed, $\hat{\mathbf{n}}$ is chosen to point outwards from the closed region. If the surface is open it must be two-sided, that is, it is only possible to get from one side to the other by crossing the curve bounding the surface. Figure 12.6a shows a two-sided surface, whereas Figure 12.6b shows a so-called Mobius strip, which is one-sided.

For open surfaces, one must choose the direction for $\hat{\mathbf{n}}$. However, if a direction is associated with a boundary curve that surrounds the surface, as it is in some very important applications, then $\hat{\mathbf{n}}$ is chosen to be 'right-handed'. To see what this means, let us suppose that the surface and its boundary curve were to be projected onto a plane. Then, as shown in Figure 12.6c, $\hat{\mathbf{n}}$ is chosen so that the direction of integration around the contour of integration corresponds to that of a right-hand screw pointing in the direction of $\hat{\mathbf{n}}$.

Of the two integrals (12.37), the scalar integrals are by far the more important. Their evaluation is often facilitated by choosing an appropriate co-ordinate system. For example, if one is integrating over a planar surface that lies in the x–y plane, then $\mathrm{d}\mathbf{s} = \mathrm{d}x\,\mathrm{d}y\,\mathbf{k}$ and

$$\iint_s \mathbf{V} \cdot \mathrm{d}\mathbf{s} = \iint_s V_z \mathrm{d}x\,\mathrm{d}y.$$

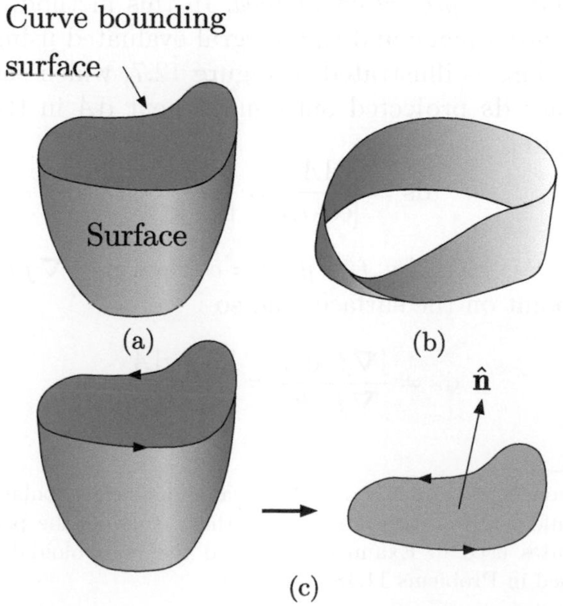

Figure 12.6 Examples of open surfaces that are (a) two-sided (b) one-sided and (c) the use of a projection of a two-sided surface onto a plane to define the direction of $\hat{\mathbf{n}}$.

On the other hand, suppose S lies on surface of a sphere of radius a. Then if we take the origin to be at the centre, the equation of the sphere in spherical co-ordinates is $r = a$, and one sees from Figure 11.14 that

$$\mathrm{d}\mathbf{s} = a^2 \sin\theta \, \mathrm{d}\theta \, \mathrm{d}\phi \, \mathbf{e}_r. \tag{12.39a}$$

Similarly, $\theta = \theta_0$ is the equation of a cone with its axis along the z-direction and, from Figure 11.14, one sees that in this case

$$\mathrm{d}\mathbf{s} = r \sin\theta_0 \, \mathrm{d}r \, \mathrm{d}\phi \, \mathbf{e}_\theta. \tag{12.39b}$$

More generally, given any set of orthogonal curvilinear co-ordinates (u_1, u_2, u_3), keeping any one of them constant defines a surface with, for example,

$$\mathrm{d}\mathbf{s} = h_2 h_3 \mathrm{d}u_2 \mathrm{d}u_3 \mathbf{e}_1 \tag{12.40}$$

if u_1 is constant. Hence if

$$V(\mathbf{r}) = V_1 \mathbf{e}_1 + V_2 \mathbf{e}_2 + V_3 \mathbf{e}_3,$$

an integral over a surface on which u_1 is constant reduces to

$$\iint_S \mathbf{V} \cdot \mathrm{d}\mathbf{s} = \int V_1 h_2 h_3 \, \mathrm{d}u_2 \, \mathrm{d}u_3,$$

with similar expressions if either u_2 or u_3 is constant.[3] These are straightforward double integrals, which can be evaluated using the methods discussed in Section 11.2.

If the surface does not correspond to a constant value of a suitably chosen orthogonal curvilinear co-ordinate, the integral can be evaluated using the *projection method.* In this method, the surface is projected onto a plane and the integral evaluated using Cartesian co-ordinates. This is illustrated in Figure 12.7, which shows an element of surface ds projected onto an element dA in the xy plane. From this figure,

$$\mathrm{d}s = \frac{\mathrm{d}A}{|\cos\alpha|} = \frac{\mathrm{d}A}{|\hat{\mathbf{n}} \cdot \mathbf{k}|}.$$

If the surface S is given by $f(x, y, z) = d$, then $\hat{\mathbf{n}} = \boldsymbol{\nabla} f / |\boldsymbol{\nabla} f|$ evaluated at the point on the surface, and so

$$\mathrm{d}s = \frac{|\boldsymbol{\nabla} f| \, \mathrm{d}A}{\boldsymbol{\nabla} f \cdot \mathbf{k}} = \frac{|\boldsymbol{\nabla} f| \, \mathrm{d}A}{\partial f / \partial z}. \tag{12.41}$$

[3] Although we concentrate mainly on cylindrical and spherical polar co-ordinates, the reader should be aware that there are others, such as the parabolic cylindrical co-ordinates used in Example 11.13 and the paraboloidal and elliptical co-ordinates used in Problems 11.18 and 11.19.

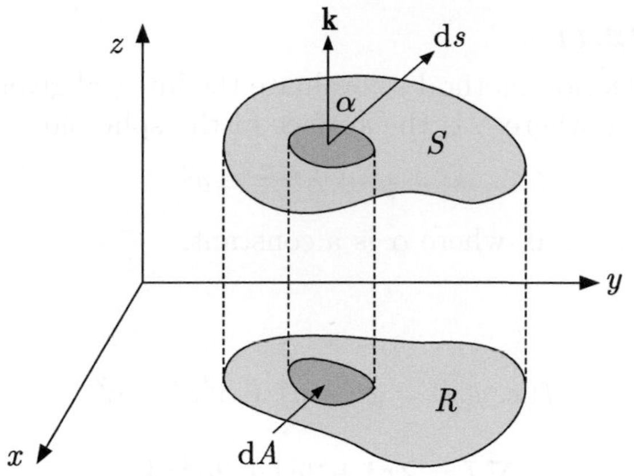

Figure 12.7 Diagram used to illustrate the evaluation of surface integrals by the projection method.

This general formula can be used to convert an integral over a curved surface S to an integral over A in the xy plane, as illustrated in Example 12.11 below.

Example 12.10

Evaluate the integral

$$I = \iint_s \mathbf{V} \cdot \mathrm{d}\mathbf{s},$$

where $\mathbf{V} = x\mathbf{i}$ and S is the surface of the hemisphere

$$x^2 + y^2 + z^2 = a^2, \quad z \geq 0$$

Solution

In spherical polar co-ordinates, the surface is $r = a$ with $0 \leq \theta \leq \pi/2$ and $0 \leq \phi \leq 2\pi$, while the surface element $\mathrm{d}\mathbf{s} = r^2 \sin\theta \, \mathrm{d}\theta \, \mathrm{d}\phi \, \mathbf{e}_r$. Hence the integral is

$$I = a^2 \int_0^{\pi/2} \mathrm{d}\theta \int_0^{2\pi} (\mathbf{V} \cdot \mathbf{e}_r) \sin\theta \, \mathrm{d}\phi,$$

where, for $r = a$,

$$\mathbf{V} \cdot \mathbf{e}_r = a \sin\theta \cos\phi (\mathbf{i} \cdot \mathbf{e}_r) = a \sin^2\theta \cos^2\phi$$

using (11.41) for $\mathbf{e}_r$. Therefore,

$$I = -a^3 \int_0^{\pi/2} \sin^2\theta \, \mathrm{d}\cos\theta \int_0^{2\pi} \cos^2\phi \, \mathrm{d}\phi = \frac{2\pi a^3}{3}.$$

Example 12.11

Use the projection method to evaluate the integral given in Example 12.10, but where S is the surface of the spheroid

$$x^2 + y^2 + \alpha^2 z^2 = a^2$$

in the region $z > 0$, where α is a constant.

Solution

Writing

$$f(x, y, z) = x^2 + y^2 + \alpha^2 z^2 = a^2$$

we have

$$\boldsymbol{\nabla} f = 2x\,\mathbf{i} + 2y\,\mathbf{j} + 2\alpha^2 z\,\mathbf{k}$$

so that

$$|\boldsymbol{\nabla} f\,| = 2(x^2 + y^2 + \alpha^2 z^2)^{1/2},$$

and

$$\hat{\mathbf{n}} = \frac{x\,\mathbf{i} + y\,\mathbf{j} + \alpha z\,\mathbf{k}}{(x^2 + y^2 + \alpha^2 z^2)^{1/2}}.$$

Hence on projecting onto the x–y plane and using (12.41) we have

$$I = \iint_A (\mathbf{V} \cdot \hat{\mathbf{n}}) \frac{|\boldsymbol{\nabla} f| \mathrm{d}A}{\partial f/\partial z},$$

where $\mathrm{d}A = \mathrm{d}x\,\mathrm{d}y$. The other terms are

$$\mathbf{V} \cdot \hat{\mathbf{n}} = \frac{x^2}{(x^2 + y^2 + \alpha^2 z^2)^{1/2}} = \frac{x^2}{a},$$

$$|\boldsymbol{\nabla} f\,| = 2(x^2 + y^2 + \alpha^2 z^2)^{1/2} = 2a$$

and

$$\partial f/\partial z = 2\alpha^2 z = 2\alpha(a^2 - x^2 - y^2)^{1/2}.$$

Putting these in I gives

$$I = \frac{1}{\alpha} \iint_A \frac{x^2}{(a^2 - x^2 - y^2)^{1/2}} \mathrm{d}x\,\mathrm{d}y,$$

where A is the interior of the circle $x^2 + y^2 = a^2$. This may be evaluated using plane polar co-ordinates, so that

$$I = \frac{1}{\alpha} \int_0^{2\pi} \mathrm{d}\theta \int_0^{a} \mathrm{d}r \frac{r^3 \cos^2\theta}{(a^2 - r^2)^{1/2}}.$$

Finally, setting $r = a \sin\phi$ gives

$$I = \frac{1}{\alpha} \int_0^{2\pi} \cos^2\theta\,\mathrm{d}\theta \int_0^{\pi/2} a^3 \sin^3\phi\,\mathrm{d}\phi = \frac{2\pi a^3}{3\alpha}.$$

12.2.4 Volume integrals: moments of inertia

In Section 11.4, we considered volume integrals of the form

$$\int_{\Omega} f\,\mathrm{d}v = \iiint_{\Omega} f(x, y, z)\mathrm{d}x\,\mathrm{d}y\,\mathrm{d}z \tag{12.42}$$

in Cartesian co-ordinates, where the integral extends over the region Ω, and the abbreviated notation on the left-hand side is sometimes used for convenience in what follows. We can now also define similar integrals over a vector field, i.e.

$$\int_{\Omega} \mathbf{V}\mathrm{d}v = \mathbf{i}\int_{\Omega} V_x\mathrm{d}v + \mathbf{j}\int_{\Omega} V_y\mathrm{d}v + \mathbf{k}\int_{\Omega} V_z\mathrm{d}v, \tag{12.43}$$

whose evaluation essentially involves evaluating three integrals of the form (12.42).

To illustrate this, let us consider a solid body with variable density ρ occupying a region of space Ω. Then since the mass of a volume element is $\rho\,\mathrm{d}v$, the total mass of the body is given by

$$M = \int_{\Omega} \rho\,\mathrm{d}v, \tag{12.44}$$

while the formula

$$M\,\bar{\mathbf{r}} = \sum_i m_i\mathbf{r}_i$$

for the centre-of-mass $\bar{\mathbf{r}}$ of a system of point particles of masses m_i at positions $\mathbf{r}$, becomes

$$M\,\bar{\mathbf{r}} = \int_{\Omega} \rho\,\mathbf{r}\,\mathrm{d}v \tag{12.45}$$

for an arbitrary solid body. Similarly, the formula

$$I = \sum_i m_i r_{oi}^2$$

for the moment of inertia I, where r_{oi} is the perpendicular distance from the mass m_i to the axis of rotation, becomes

$$I = \int_{\Omega} \rho\, r_0^2\,\mathrm{d}v. \tag{12.46}$$

Example 12.12

Prove the *theorem of parallel axes*

$$I = I_{\mathrm{CM}} + Md^2 \tag{12.47}$$

for an arbitrary solid body of mass M, where I_{CM} is the moment of inertia about an axis through the centre of mass, and I is the moment of inertia about a parallel axis at a perpendicular distance d from the centre of mass.

Solution

We will use Cartesian co-ordinates with the origin at the centre of mass of the body and the z-axis along the direction of rotation, so that the parallel axis must lie along a line $-\infty < z < \infty$ at fixed co-ordinates x, y with $x^2 + y^2 = d^2$. Then if $\rho\, \mathrm{d}v_1$ is an element of mass lying within the body at position (x_1, y_1, z_1), (12.45) becomes

$$\mathbf{0} = \int_\Omega \rho\, \mathbf{r}_1 \mathrm{d}v_1, \tag{12.48}$$

since the centre of mass is at the origin, and (12.46) becomes

$$I_{\mathrm{CM}} = \int_\Omega \rho(x_1^2 + y_1^2)\mathrm{d}v_1.$$

Similarly, the moment of inertia about the parallel axis is given by

$$I = \int_\Omega \rho \left[(x - x_1)^2 + (y - y_1)^2\right] \mathrm{d}v_1,$$

which, on expanding the brackets, becomes

$$I = (x^2 + y^2)\int_\Omega \rho \mathrm{d}v_1 - 2x\int_\Omega \rho x_1 \mathrm{d}v_1 - 2y\int_\Omega \rho y_1 \mathrm{d}v_1$$
$$+ \int_\Omega \rho(x_1^2 + y_1^2)\mathrm{d}v_1 = Md^2 + I_{\mathrm{CM}},$$

since $d^2 = x^2 + y^2$ and the second and third integrals vanish by (12.48).

12.3 The divergence theorem

The *divergence theorem*[4] states that, for any vector field $\mathbf{V}$,

$$\iiint_\Omega \boldsymbol{\nabla} \cdot \mathbf{V}\, \mathrm{d}v = \oint\!\!\oint_S \mathbf{V} \cdot \mathrm{d}\mathbf{s}, \tag{12.49}$$

for any surface S enclosing a region Ω. The quantity $\mathbf{V} \cdot \mathrm{d}\mathbf{s}$ is called the *flux* of $\mathbf{V}$ through $\mathrm{d}\mathbf{s}$ and the circle on the double integral is to emphasise that S is a closed surface, by analogy with closed paths

[4] This theorem is also sometimes called *Green's theorem*, or *Gauss' divergence theorem*.

in line integrals. This circle is sometimes omitted and (12.49) is also sometimes written in the abbreviated form

$$\int_{\Omega} \boldsymbol{\nabla} \cdot \mathbf{V} \mathrm{d}v = \int_{S} \mathbf{V} \cdot \mathbf{ds}$$

already used in (11.49) for volume integrals. However, in whatever form it is written, the theorem states that the volume integral of the divergence of $\mathbf{V}$ is equal to the total flux out of the bounding surface S, since $\mathbf{ds} = \mathrm{d}s\,\hat{\mathbf{n}}$ points out of a closed surface.

We shall derive the divergence theorem and two well-known identities resulting from it in Section 12.3.1 below. Before that, we point out that the divergence theorem is central to the physical interpretation of divergence. To see this, we apply (12.49) to the case when S encloses a small volume element that shrinks to a point as $\mathrm{d}v \to 0$. In this limit, the variation of $\mathbf{V}$ in $\mathrm{d}v$ can be neglected, so the left-hand side of (12.49) becomes div $\mathbf{V}\mathrm{d}v$, implying

$$\text{div}\ \mathbf{V} = \lim_{\mathrm{d}v \to 0} \left(\frac{1}{\mathrm{d}v} \iint_{S} \mathbf{V} \cdot \mathbf{ds} \right). \tag{12.50}$$

In other words, div $\mathbf{V}$ at a point $\mathbf{r}$ is the flux per unit volume out of an infinitesimal volume $\mathrm{d}v$ surrounding $\mathbf{r}$. For example, if $\mathbf{V} = \rho\,\mathbf{v}$, where ρ is the density and $\mathbf{v}$ is the velocity field of a fluid, the flux $\mathbf{V} \cdot \mathbf{ds}$ is the rate of flow of mass through the surface. Hence if div $\mathbf{V}(\mathbf{r})$ is greater than zero, there is a net flow of mass away from $\mathbf{r}$, so that either the density is decreasing at the point, or a *source* (i.e. a point where fluid is entering the system) is present. On the other hand, if there is no source or *sink* (i.e. a point where fluid is leaving the system) at $\mathbf{r}$, and the density is constant, which is normally a good approximation for a liquid, then div $\mathbf{V} = 0$. In this latter case $\mathbf{V}$ is called a *solenoidal field*. Although we have chosen the example of a fluid, the same ideas may be applied to other situations, including the flow of electric current.

12.3.1 Proof of the divergence theorem and Green's identities

To derive the divergence theorem, we consider a segment through the region Ω lying parallel to the x-axis and with constant infinitesimal cross section $\mathrm{d}y\,\mathrm{d}z$, as shown in Figure 12.8[5]. Further, let the unit vectors $\hat{\mathbf{n}}_1$ and $\hat{\mathbf{n}}_2$ be the outward normals on the surface elements 1

[5] This theorem is often 'derived' by approximating the region by a sum of little cuboids and taking the limit when they become infinitesimally small. However, while the volume integral is well-defined in this limit and approaches the exact integral over Ω, the limit of the corresponding surface integral is not well defined.

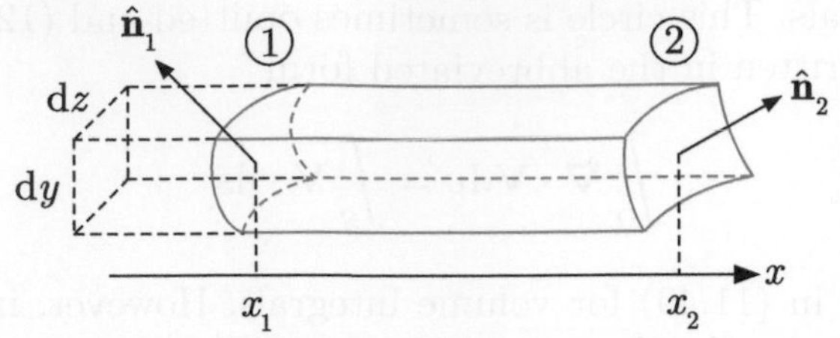

Figure 12.8 A segment through the region Ω lying parallel to the x-axis and with constant infinitesimal cross section $\mathrm{d}y\,\mathrm{d}z$, used in the derivation of the divergence theorem.

and 2, respectively, where the segment intersects the surface of the region Ω, so that $\mathbf{ds}_1 = \hat{\mathbf{n}}_1 \mathrm{d}s_1$, $\mathbf{ds}_2 = \hat{\mathbf{n}}_2 \mathrm{d}s_2$ and

$$\mathrm{d}y\,\mathrm{d}z = -(\hat{\mathbf{n}}_1 \cdot \mathbf{i})\,\mathrm{d}s_1 = (\hat{\mathbf{n}}_2 \cdot \mathbf{i})\,\mathrm{d}s_2.$$

Then, since

$$\int_{1\to 2} \frac{\partial V_x}{\partial x}\mathrm{d}x = \int_{1\to 2} \mathrm{d}V_x = V_x(2) - V_x(1)$$

at fixed y, z, we have

$$\left[\int_{1\to 2} \frac{\partial V_x}{\partial x}\mathrm{d}x\right]\mathrm{d}y\,\mathrm{d}z = V_x(2)(\mathbf{i}\cdot\mathbf{ds}_2) + V_x(1)(\mathbf{i}\cdot\mathbf{ds}_1), \qquad (12.51)$$

where the right-hand side is the net flux through the surface elements 1 and 2 from the x-component $V_x\mathbf{i}$.

All that remains now is to add together the contributions from enough segments to cover the whole region Ω, so that (12.51) becomes

$$\iiint_\Omega \frac{\partial V_x}{\partial x}\mathrm{d}x\,\mathrm{d}y\,\mathrm{d}z = \iint_S V_x(\mathbf{i}\cdot\mathbf{ds}_1).$$

The contributions from the y and z components of $\mathbf{V}$ can be calculated in a similar way, and adding all three components we obtain

$$\iiint_\Omega \left(\frac{\partial V_x}{\partial x} + \frac{\partial V_y}{\partial y} + \frac{\partial V_z}{\partial z}\right)\mathrm{d}x\,\mathrm{d}y\,\mathrm{d}z = \iint_S (V_x\mathbf{i} + V_y\mathbf{j} + V_z(\mathbf{k})\cdot\mathbf{ds},$$

which is the divergence theorem.[6]

Finally, we use the divergence theorem to derive two other useful results as follows. Let ϕ and ψ be two scalar fields continuous

[6]For simplicity, we have assumed simple regions Ω, such that a segment like that shown in Figure 12.8 only crosses the surface in two places. However, the result can easily be extended to more complicated regions by dividing Ω into subregions, each of which does satisfy this requirement.

and differentiable in some region Ω bounded by a closed surface S. Applying the divergence theorem to $(\phi\boldsymbol{\nabla}\psi)$ gives

$$\begin{aligned}\iint_S (\phi\,\boldsymbol{\nabla}\psi)\cdot\mathrm{d}\mathbf{s} &= \iiint_\Omega \boldsymbol{\nabla}\cdot(\phi\boldsymbol{\nabla}\psi)\mathrm{d}v\\ &= \iiint_\Omega [\phi\boldsymbol{\nabla}^2\psi + (\boldsymbol{\nabla}\phi)\cdot(\boldsymbol{\nabla}\psi)]\,\mathrm{d}v.\end{aligned} \tag{12.52a}$$

This is known as *Green's first identity.* Similarly, interchanging ϕ and ψ gives

$$\begin{aligned}\iint_S (\psi\,\boldsymbol{\nabla}\phi)\cdot\mathrm{d}\mathbf{s} &= \iiint_\Omega \boldsymbol{\nabla}\cdot(\psi\boldsymbol{\nabla}\phi)\mathrm{d}v\\ &= \iiint_\Omega [\psi\boldsymbol{\nabla}^2\phi + (\boldsymbol{\nabla}\psi)\cdot(\boldsymbol{\nabla}\phi)]\,\mathrm{d}v.\end{aligned}$$

Subtracting these two equations gives

$$\iint_S (\phi\,\boldsymbol{\nabla}\psi - \psi\,\boldsymbol{\nabla}\phi)\cdot\mathrm{d}\mathbf{s} = \iiint_\Omega (\phi\boldsymbol{\nabla}^2\psi - \psi\boldsymbol{\nabla}^2\phi)\,\mathrm{d}v, \tag{12.52b}$$

which is *Green's second identity.*

Example 12.13

Derive the general relation

$$V = \frac{1}{3}\iint_S \mathbf{r}\cdot\mathrm{d}\mathbf{s},$$

where V is the volume of the region enclosed by the surface S, and hence evaluate the integral

$$\iint_S (\mathbf{r}\cdot\hat{\mathbf{n}})\,\mathrm{d}s$$

over the closed surface of a cylinder of height h and radius a.

Solution

Since $\mathbf{r} = x\,\mathbf{i} + y\,\mathbf{j} + z\,\mathbf{k}$, div $\mathbf{r} = 3$ and the divergence theorem gives,

$$\iint_S \mathbf{r}\cdot\mathrm{d}\mathbf{s} = \iiint_\Omega \mathrm{div}\,\mathbf{r}\,\mathrm{d}v = 3\iiint_\Omega \mathrm{d}v = 3V,$$

independent of the shape of Ω, giving the desired result. In the case of the cylinder, $V = \pi a^2 h$, so that

$$\iint_S \mathbf{r}\cdot\mathrm{d}\mathbf{s} = 3\pi a^2 h.$$

This result could have been obtained directly by evaluating the surface integral, but the calculation is much longer. In general, surface integrals are more difficult to evaluate than volume integrals, so the divergence theorem is often used to evaluate flux integrals over closed surfaces more easily.

*12.3.2 Divergence in orthogonal curvilinear co-ordinates

Having derived the divergence theorem (12.49) using Cartesian co-ordinates, then the corollary (12.50) follows, and can be regarded as an alternative definition of the divergence, independent of the co-ordinate system. In particular, it can be used to find the general expression (12.18) for the divergence in an arbitrary set of orthogonal curvilinear co-ordinates (u_1, u_2, u_3).

To do this, we consider the region bounded by surfaces of constant u_i and constant $u_i + \delta u_i$ as shown in Figure 12.9. The edges AB, AD and AA' are along the orthogonal co-ordinate axes, and so are of approximate length $h_1\delta u_1$, $h_2\delta u_2$ and $h_3\delta u_3$, where h_i are the coefficients defined in (11.30b). We first calculate the contribution to the integral

$$\iint_S (\mathbf{V} \cdot \hat{\mathbf{n}})\,\mathrm{d}s$$

from the faces $ABCD$ and $A'B'C'D'$. If V_1, V_2, V_3 are the components of $\mathbf{V}$ along u_1, u_2, u_3, then the contribution from the face $ABCD$ is approximately

$$h_1\delta u_1 h_2\delta u_2 \mathbf{V} \cdot \mathbf{n} = -h_1\delta u_1 h_2\delta u_2 V_3$$

evaluated at u_3, while the contribution from $A'B'C'D'$ is approximately

$$h_1\delta u_1 h_2\delta u_2 \mathbf{V} \cdot \mathbf{n} = h_1\delta u_1 h_2\delta u_2 V_3$$

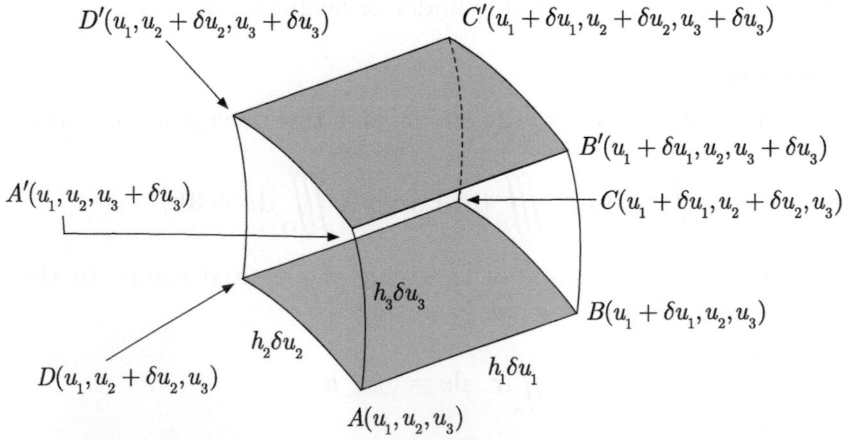

Figure 12.9 Construction to derive the divergence in orthogonal curvilinear co-ordinates.

evaluated at $u_3 + \delta u_3$, where terms of third order in δu_i have been neglected. Applying the Taylor series to $h_1 h_2 V_3$ at fixed u_3 and neglecting terms of order $(\delta u_3)^2$ gives

$$h_1 h_2 V_3(u_3 + \delta u_3) = h_1 h_2 V_3(u_3) + \delta u_3 \frac{\partial}{\partial u_3}\left[h_1 h_2 V_3(u_3)\right],$$

so that the net contribution from these two faces is

$$\delta u_1 \delta u_2 \delta u_3 \frac{\partial}{\partial u_3}(V_3 h_1 h_2).$$

Now the volume element is

$$\delta v \approx h_1 h_2 h_3 \delta u_1 \delta u_2 \delta u_3$$

to the same order and so from (12.50) the contribution to the divergence is

$$\frac{1}{h_1 h_2 h_3}\frac{\partial}{\partial u_3}(V_3 h_1 h_2)$$

on taking the limit $\delta v \to 0$. Contributions from other pairs of faces may be found in a similar way and putting these together yields

$$\operatorname{div}\mathbf{V} = \frac{1}{h_1 h_2 h_3}\left[\frac{\partial}{\partial u_1}(V_1 h_2 h_3) + \frac{\partial}{\partial u_2}(V_2 h_3 h_1) + \frac{\partial}{\partial u_3}(V_3 h_1 h_2)\right],$$

which is the required result (12.18). We leave it as an exercise for the reader to show that the corresponding result (12.19) for the Laplacian follows from combining this result with (12.17) for the gradient, and that the corresponding results for cylindrical and spherical spherical co-ordinates given in Table 12.2 follow on substituting the appropriate values for h_1, h_2 and h_3.

*12.3.3 Poisson's equation and Gauss' theorem

The electrostatic field $\mathbf{E}$ obeys the fundamental equation

$$\operatorname{div}\mathbf{E}(\mathbf{r}) = \varepsilon_0^{-1}\rho(\mathbf{r}), \qquad (12.53)$$

where ρ is the electric charge density and the constant ε_0 is the electric permittivity of free space. This equation is called Poisson's equation. Since div $\mathbf{E}$ is the flux of E per unit volume away from the point at which it is evaluated, the interpretation of Poisson's equation is that the electric charge is the source of the electrostatic field.

If we now apply the divergence theorem (12.49) to the field $\mathbf{E}$, and use (12.53), we immediately obtain

$$\iint_S \mathbf{E}\cdot d\mathbf{s} = \varepsilon_0^{-1}\iiint_\Omega \rho(\mathbf{r})dv. \qquad (12.54)$$

This relation is called *Gauss' theorem*. It says that the electric flux through a closed surface S is equal to ε_0^{-1} times the total charge enclosed by the surface.

Gauss' theorem is useful in that it allows the field due to a given charge distribution $\rho(\mathbf{r})$ to be evaluated relatively easily in cases where there is a high degree of symmetry. For example, let us suppose that we have a charged sphere centred at the origin with radius R and total charge Q, and that the charge density within the sphere is also spherically symmetric, that is, $\rho(\mathbf{r}) = \rho(r)$. Then the resulting field must also be spherically symmetric, that is, it must be of the form

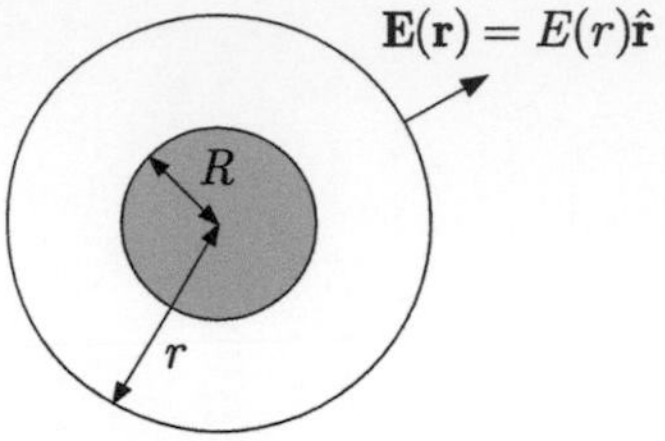

Figure 12.10 The spherical surface S used to calculate the electric field due to a spherical charge distribution of radius R for $r > R$.

$$E(\mathbf{r}) = E(r)\hat{\mathbf{r}}, \tag{12.55}$$

so that $E(\mathbf{r})$ points away from (or towards) the origin and its magnitude is the same in all directions. Hence if we choose the surface S to be a sphere of radius $r > R$, as shown in Figure 12.10, then $E(\mathbf{r})$ is perpendicular to S and

$$\iint_S \mathbf{E} \cdot d\mathbf{s} = \iint_S E(r)ds = 4\pi r^2 E(r) = \varepsilon_0^{-1} Q$$

by Gauss' theorem. Consequently,

$$\mathbf{E}(\mathbf{r}) = \frac{Q}{4\pi\varepsilon_0}\frac{1}{r^2}\hat{\mathbf{r}}, \quad r > R \tag{12.56}$$

which reduces to Coulomb's law

$$\mathbf{E}(\mathbf{r}) = \frac{Q}{4\pi\varepsilon_0}\frac{1}{r^2}\hat{\mathbf{r}}, \quad r > 0 \tag{12.57}$$

for a point charge at the origin if we allow $R \to 0$ at fixed Q.

This analysis is easily generalised to other inverse square law forces. In particular, if $\mathbf{g}$ is the gravitational field, so that the force on a point particle of mass is $\mathbf{F} = m\mathbf{g}$, then $\mathbf{g}$ obeys the Poisson equation

$$\text{div}\mathbf{g} = -4\pi G\rho, \tag{12.58}$$

where ρ is the mass density and G is the gravitational constant. The result corresponding to (12.56) for a spherically symmetric sphere of total mass M is

$$\mathbf{g} = -\frac{GM}{r^2}\hat{\mathbf{r}}, \quad r > R,$$

which reduces to

$$\mathbf{g} = -\frac{GM}{r^2}\hat{\mathbf{r}}, \quad r > 0,$$

when $R \to 0$ at fixed M. This is basis of the approximation that the Earth may be treated as if all its mass were concentrated at its centre when calculating its gravitational field for $r > R$. However, the approximation is not exact, because the earth is flattened at the poles.

Finally, we note that both the electrostatic and gravitational fields are conservative, satisfying

$$\text{curl}\,\mathbf{E} = \mathbf{0}, \quad \text{curl}\,\mathbf{g} = \mathbf{0}$$

so that we can introduce scalar potentials ϕ and ψ by

$$\mathbf{E} = -\boldsymbol{\nabla}\phi, \quad \mathbf{g} = -\boldsymbol{\nabla}\psi \tag{12.59}$$

in accordance with the discussion of Section 12.2.2. For the electrostatic case, substituting (12.59) into (12.53) gives

$$\boldsymbol{\nabla}^2\phi = -\varepsilon_0^{-1}\rho, \tag{12.60}$$

which is Poisson's equation for the electrostatic potential; and if one requires $\phi \to 0$ as $r \to \infty$, one easily shows that the potential corresponding to (12.57) is the familiar Coulomb potential

$$\phi = \frac{Q}{4\pi\varepsilon_0}\frac{1}{r}. \tag{12.61}$$

Example 12.14

Calculate the electric field $\mathbf{E}$ close to the surface of a conductor if the equilibrium surface charge density is σ.

Solution

Since the charges are in equilibrium, the components of $\mathbf{E}$ parallel to the surface must be zero, otherwise the charges would move. Hence, close to the surface, $\mathbf{E}$ must be perpendicular to the surface. Similarly, $\mathbf{E}$ must be zero inside the conductor because, if it were not, current would flow until equilibrium was reached. By Gauss' theorem

$$\iint_S \mathbf{E}\cdot \mathrm{d}\mathbf{s} = 0$$

for an arbitrary closed surface inside the conductor, since all the charge is on the surface.

Next, consider an infinitesimal cylinder drawn perpendicular to the surface, as shown in Figure 12.11, so that the variation of $\mathbf{E}$ above the surface may be neglected within the cylinder and the surface intersected by the cylinder can be approximated by a flat disc. Then by Gauss' theorem,

$$\iint_S \mathbf{E}\cdot \mathrm{d}\mathbf{s} = \varepsilon_0^{-1}\sigma \mathrm{d}A$$

and the flux through the top of the cylinder is $\mathbf{E}\,\mathrm{d}A$. There is no flux through the sides or bottom of the cylinder, so

$$\mathbf{E}\,\mathrm{d}A = \varepsilon_0^{-1}\sigma \mathrm{d}A \quad \text{and} \quad \mathbf{E} = \varepsilon_0\sigma\hat{\mathbf{n}},$$

where $\hat{\mathbf{n}}$ is perpendicular to the surface

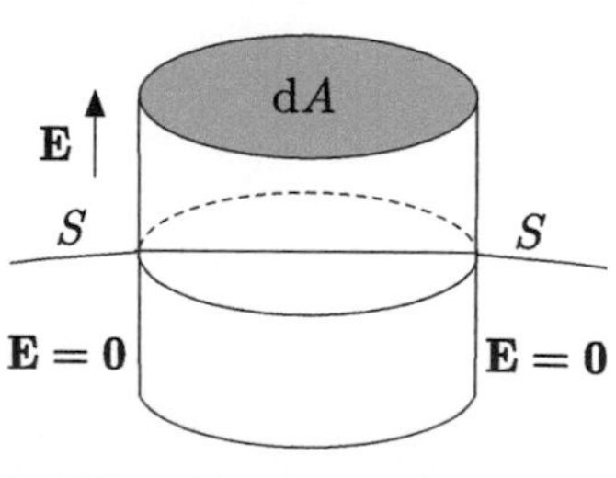

Figure 12.11 The infinitesimal cylinder used in Example 12.14, where S is the surface of the conductor and $\mathbf{E} = \mathbf{0}$ within the conductor.

*12.3.4 The continuity equation

Let us consider a fluid of density $\rho(\mathbf{r}, t)$ with a velocity field $\mathbf{v}(\mathbf{r}, t)$ at time t. Then if we consider a surface element $\mathrm{d}\mathbf{s}$, which may lie within the body of the fluid, the mass of liquid passing through $\mathrm{d}\mathbf{s}$ in unit time is $\mathbf{j} \cdot \mathrm{d}\mathbf{s}$, where $\mathbf{j} = \rho\mathbf{v}$ is the current vector. If mass is conserved, the rate of change of the mass

$$M = \iiint_\Omega \rho(\mathbf{r}, t)\mathrm{d}v$$

contained in a given region Ω must be balanced by the rate at which mass flows out through the surface S bounding Ω, i.e.

$$\frac{\partial}{\partial t}\iiint_\Omega \rho(\mathbf{r}, t)\mathrm{d}v + \iint_S \mathbf{j} \cdot \mathrm{d}\mathbf{s} = 0. \qquad (12.62)$$

Equation (12.62) is the statement of mass conservation in integral form. However, it is often more convenient to express it in differential, or local, form, that is, one that refers only to quantities at a single point in space. This can be achieved by using the divergence theorem on the right-hand side of (12.62) and taking the derivative inside the integral on the left-hand side to give

$$\iiint_\Omega \left(\frac{\partial \rho}{\partial t} + \boldsymbol{\nabla} \cdot \mathbf{j}\right) \mathrm{d}\Omega = 0.$$

Since this must hold for any region Ω, we must have

$$\boxed{\frac{\partial \rho}{\partial t} + \boldsymbol{\nabla} \cdot \mathbf{j} = 0} \qquad (12.63)$$

at any point in space.

Equation (12.63) is called the *equation of continuity* and is the statement of mass conservation in differential, or local, form. Furthermore, any $\rho(\mathbf{r}, t)$ that satisfies a relation of the form (12.63), whatever the relation between the density ρ and the current $\mathbf{j}$, is the density of a conserved quantity. This is because the argument can be reversed, that is, (12.62) follows from (12.63) using the divergence theorem. Then, if we let the surface S recede to infinity, we obtain

$$\frac{\partial}{\partial t}\iiint \rho\, \mathrm{d}v = 0, \qquad (12.64)$$

where the integral extends over all space, provided ρ, $\mathbf{j} \to 0$ sufficiently rapidly at infinity, as they usually do. Many examples of conserved quantities occur in physics, including electric charge and energy. However, the relation between the density ρ and the current $\mathbf{j}$ is not always as simple as $\mathbf{j} = \rho\mathbf{v}$, as shown in Example 12.15.

Example 12.15

In non-relativistic quantum mechanics, the equation of motion of a point particle of mass m in a potential $V(\mathbf{r})$ is

$$-\frac{\hbar^2}{2m}\boldsymbol{\nabla}^2\psi + V(\mathbf{r})\psi = i\hbar\frac{\partial\psi}{\partial t}, \qquad (12.65)$$

where $\hbar \equiv h/2\pi$, h is Planck's constant, and $\psi(\mathbf{r}, t)$ is the Schrödinger wave function. Show that

$$\rho = \psi^*(\mathbf{r}, t)\psi(\mathbf{r}, t) \qquad (12.66)$$

satisfies the continuity equation (12.62) and find the form of the corresponding current density $\mathbf{j}$.

Solution

On multiplying (12.65) by ψ^* on the left we obtain

$$i\hbar\psi^*\frac{\partial\psi}{\partial t} = -\frac{\hbar^2}{2m}\psi^*\boldsymbol{\nabla}^2\psi + V(\mathbf{r})\psi^*\psi. \qquad (12.67a)$$

Taking the complex conjugate then gives

$$-i\hbar\psi\frac{\partial\psi^*}{\partial t} = -\frac{\hbar^2}{2m}\psi\,\boldsymbol{\nabla}^2\psi^* + V(\mathbf{r})\psi^*\psi$$

and subtracting this from (12.67a), we obtain

$$i\hbar\frac{\partial}{\partial t}(\psi^*\psi) = -\frac{\hbar^2}{2m}\left(\psi^*\boldsymbol{\nabla}^2\psi - \psi\,\boldsymbol{\nabla}^2\psi^*\right). \qquad (12.67b)$$

Using the identity (cf. Table 12.1)

$$\boldsymbol{\nabla}\cdot(\phi\mathbf{a}) = \phi\boldsymbol{\nabla}a + \mathbf{a}\cdot\boldsymbol{\nabla}\phi$$

on the right-hand side of (12.67b) gives

$$i\hbar\frac{\partial}{\partial t}(\psi^*\psi) = -\frac{\hbar^2}{2m}\boldsymbol{\nabla}\cdot(\psi^*\boldsymbol{\nabla}\psi - \psi\,\boldsymbol{\nabla}\psi^*).$$

On multiplying by $-i/\hbar$ and comparing with (12.66), we obtain the continuity equation (12.63) where the current

$$\mathbf{j} = -\frac{i\hbar}{2m}(\psi^*\boldsymbol{\nabla}\psi - \psi\,\boldsymbol{\nabla}\psi^*).$$

12.4 Stokes' theorem

Given a closed contour C, spanned by a surface S, and a vector field $\mathbf{V}$ defined on S, then *Stokes' theorem* states that

$$\iint_S \boldsymbol{\nabla}\times\mathbf{V}\cdot d\mathbf{s} = \oint_C \mathbf{V}\cdot d\mathbf{r}, \qquad (12.68)$$

where the sense of the vector element $\mathbf{ds}$ is given by a right-handed screw rule with respect to the direction of integration around C.

The line integral on the right-hand side of (12.68) is called the *circulation* of $\mathbf{V}$ around the loop C. Thus the theorem states that the surface integral of curl $\mathbf{V}$ is equal to the circulation of $\mathbf{V}$ around the bounding curve C. This is closely related to the interpretation of curl. To see this, we apply (12.68) to a loop C that encloses a small surface element $\mathbf{ds} = \hat{\mathbf{n}}\,\mathrm{d}s$, which shrinks to a point when $\mathrm{d}s \to 0$. In this limit, the variation of V and $\hat{\mathbf{n}}$ can be neglected on $\mathbf{d}s$, so that the left-hand side of (12.68) becomes $\mathrm{curl}\,\mathbf{V}\cdot\hat{\mathbf{n}}\,\mathrm{d}s$, implying

$$\mathrm{curl}\,\mathbf{V}\cdot\hat{\mathbf{n}} = \lim_{\mathrm{d}s\to 0}\left(\frac{1}{\mathrm{d}s}\oint \mathbf{V}\cdot\mathrm{d}\mathbf{r}\right). \tag{12.69}$$

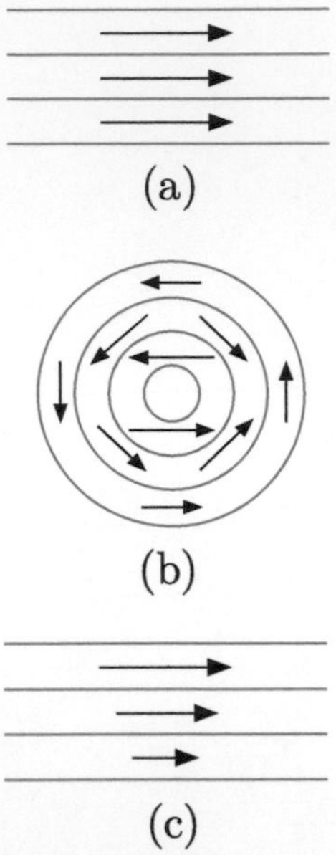

Figure 12.12 Flow of a fluid. The coloured lines are the flow lines; the arrows show the direction of the vector field $\mathbf{V}$. Their lengths show the relative magnitudes of $\mathbf{V}$.

In other words, curl $\mathbf{V}$ at a point $\mathbf{r}$ is the circulation per unit area around the boundary of an infinitesimal surface $\mathrm{d}s$ containing the point $\mathbf{r}$. For example, let us again consider a vector field $\mathbf{V} = \rho\mathbf{v}$, where ρ is the density and $\mathbf{v}$ is the velocity field of a fluid. Then for a uniform flow pattern, such as that shown in Figure 12.12a, $\mathrm{curl}\,\mathbf{V} = \mathbf{0}$ and $\mathbf{V}$ is said to be *irrotational*. On the other hand, at the centre of a vortex, like that shown in Figure 12.12b, clearly $\mathrm{curl}\,\mathbf{V} \neq \mathbf{0}$. It is also non-zero in a non-uniform parallel motion, as shown in Figure 12.12c, since the velocities on either side of a point are different. Essentially, $\mathrm{curl}\,\mathbf{V} \neq \mathbf{0}$ when there is rotational motion in addition to, or opposed to, translational motion. A practical viewpoint is to consider what would happen if one inserted a small 'paddle wheel', which is free to rotate about its axis. In the flow pattern of Figure 12.12a, where $\mathrm{curl}\,\mathbf{V} = \mathbf{0}$, it would not rotate: the motion is irrotational. In Figures 12.12b and 12.12c, where $\mathrm{curl}\,\mathbf{V} \neq \mathbf{0}$, it would rotate.

In the rest of this section we will first derive Stokes' theorem, and then consider some applications.

12.4.1 Proof of Stokes' theorem

We start by considering a closed curve C surrounding a plane surface S parallel to the x–y plane, so that z is constant. Then

$$\mathrm{d}\mathbf{r} = \mathrm{d}x\,\mathbf{i} + \mathrm{d}y\,\mathbf{j}$$

and

$$\oint_C \mathbf{V}\cdot\mathrm{d}\mathbf{r} = \int_C (V_x\mathrm{d}x + V_y\mathrm{d}y).$$

But by Green's theorem in the plane (11.20), we have

$$\int_C (V_x\mathrm{d}x + V_y\mathrm{d}y) = \iint_S \left(\frac{\partial V_y}{\partial x} - \frac{\partial V_x}{\partial y}\right)\mathrm{d}x\,\mathrm{d}y,$$

so that

$$\iint_S \mathrm{curl}\,\mathbf{V}\,\mathrm{d}s = \oint_C \mathbf{V}\cdot\mathrm{d}\mathbf{r}, \tag{12.70}$$

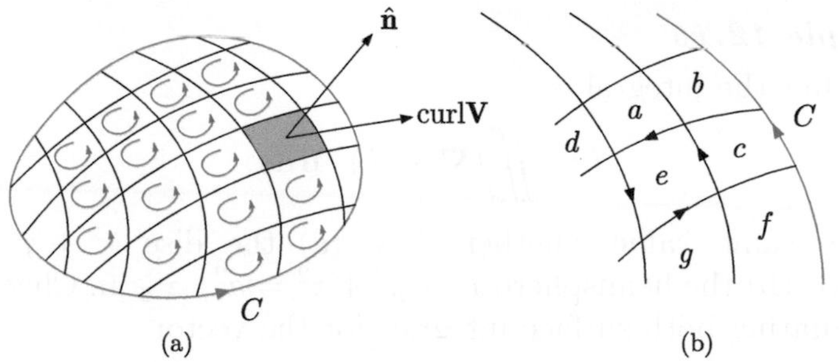

Figure 12.13 Construction to derive Stokes' theorem.

where $\hat{\mathbf{n}} = \mathbf{k}$, a unit vector in the z-direction. Furthermore, in the limit $\mathrm{d}s \to 0$, where the variation of $\nabla \times \mathbf{V}$ over the surface can be neglected, we obtain

$$(\nabla \times \mathbf{V}) \cdot \hat{\mathbf{n}} = \lim_{\mathrm{d}s \to 0} \frac{1}{\mathrm{d}s} \oint \mathbf{V} \cdot \mathrm{d}\mathbf{r}. \tag{12.71}$$

As there is nothing special about the z-direction – we may choose it in any direction we like – it follows that (12.70) and (12.71) hold for any finite or infinitesimal planar surface, respectively, where $\hat{\mathbf{n}}$ is the normal defined in the usual sense. We will now use this result to derive Stokes' theorem.

Consider an open surface, which must be two-sided, divided into small regions ds, as shown in Figure 12.13a. As $\mathbf{ds} \to \mathbf{0}$, each element, irrespective of its shape, approaches ever more closely to an element of the plane tangential to the surface at the centre of the surface ds. Therefore, (12.71) implies (12.69), and (12.70) becomes

$$\int_C \mathbf{V} \cdot \mathrm{d}\mathbf{r} = \iint_{\mathrm{d}s} (\nabla \times \mathbf{V}) \cdot \hat{\mathbf{n}}\, \mathrm{d}s,$$

as $\mathrm{d}s \to 0$, where the circulation is around the boundary of $\mathrm{d}s$. If we sum over all $\mathrm{d}s$,

$$\sum_{\mathrm{d}s} \oint \mathbf{V} \cdot \mathrm{d}\mathbf{r} = \iint_S (\nabla \times \mathbf{V}) \cdot \hat{\mathbf{n}}\, \mathrm{d}s,$$

and from the enlarged section shown in Figure 12.13b it is clear that all interior contributions to the circulation will vanish, resulting in

$$\oint_C \mathbf{V} \cdot \mathrm{d}\mathbf{r} = \iint_S (\nabla \times \mathbf{V}) \cdot \mathrm{d}\mathbf{s}.$$

This is Stokes' theorem as required. It is worth emphasising that the right-hand side is an integral over *any* surface that is bounded by the curve C. Note also the direction of the circulation, which is 'right-handed' relative to the directions ds, as discussed in Section 12.2.3 [cf. Figure 12.6]. In the following subsections we will consider some applications of this theorem.

Example 12.16

Show that the integral

$$I = \iint_S (\boldsymbol{\nabla} \times \mathbf{V}) \cdot \hat{\mathbf{n}}\, ds,$$

has the same value whether S is: (a) the disc $x^2 + y^2 < a^2$, $z = 0$, or (b) the hemisphere $x^2 + y^2 + z^2 = a^2$, $z \geq 0$. Check this by evaluating both surface integrals for the vector

$$\mathbf{V} = 3y\,\mathbf{i} + x\,\mathbf{j} + 2z\,\mathbf{k}.$$

Solution

By Stokes' theorem, in both cases

$$I = \iint_S (\boldsymbol{\nabla} \times \mathbf{V}) \cdot \mathrm{d}\mathbf{s} = \oint_C \mathbf{V} \cdot \mathrm{d}\mathbf{r},$$

where C is the circle $x^2 + y^2 = a^2$.

(a) In this case, $\hat{\mathbf{n}} = \mathbf{k}$. But direct evaluation of the curl gives $\boldsymbol{\nabla} \times \mathbf{V} = -2\mathbf{k}$, and hence $I = -2\pi a^2$.

(b) In spherical polar co-ordinates the hemisphere corresponds to $r = a$, with

$$\mathrm{d}\mathbf{s} = a^2 \mathrm{d}\cos\theta\, \mathrm{d}\phi\, \mathbf{e}_r$$

by (12.59) and $\mathbf{e}_r \cdot \mathbf{k} = \cos\theta$ by (11.41), so that

$$I = \int_0^{2\pi} \mathrm{d}\phi \int_0^1 \mathrm{d}(\cos\theta)\,(-2\cos\theta) = -2\pi a^2.$$

*12.4.2 Curl in curvilinear co-ordinates

Having derived Stokes' theorem (12.68) and its corollary (12.69), the latter can be regarded as an alternative definition of curl, independent of the co-ordinate system. Here we shall use it to obtain the expression for curl in an arbitrary system of orthogonal linear co-ordinates.[7]

To do this, we consider the infinitesimal surface element $\mathrm{d}\mathbf{s} = \mathrm{d}s\,\mathbf{e}_3$ swept out when $u_1 \to u_1 + \mathrm{d}u_1$ and $u_2 \to u_2 + \mathrm{d}u_2$ at constant u_3, as shown in Figure 12.14. Then from (12.69), we have

$$\mathbf{e}_3 \cdot \operatorname{curl} \mathbf{V} = \lim_{\mathrm{d}s \to 0} \left(\frac{1}{\mathrm{d}s} \oint_C \mathbf{V} \cdot \mathrm{d}\mathbf{r} \right), \qquad (12.72)$$

[7] The argument is similar to that given for divergence in Section 12.3.2, and will therefore be summarised rather briefly.

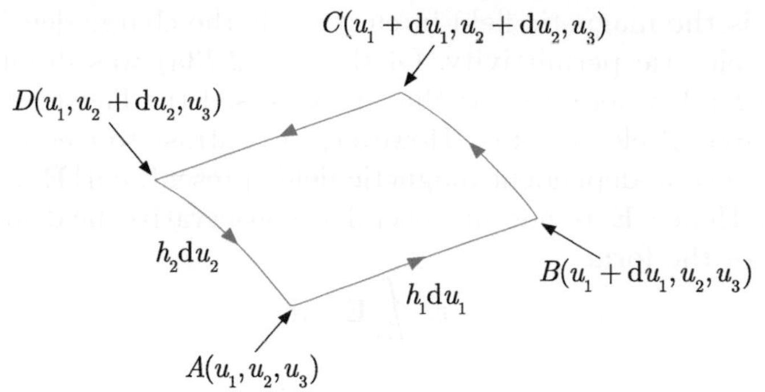

Figure 12.14 Construction to derive Stokes' theorem in curvilinear co-ordinates.

where C is the contour $ABCD$ shown in Figure 12.4. We now write

$$\mathbf{V} = V_1\mathbf{e}_1 + V_2\mathbf{e}_2 + V_3\mathbf{e}_3,$$

where $\mathbf{e}_1$ and $\mathbf{e}_2$ are unit vectors along the directions $\mathbf{AB}$ and $\mathbf{AD}$, respectively, and use the fact that in the limit that $\mathrm{d}u_1$, $\mathrm{d}u_2$ tend to zero the corresponding lines may be approximated by straight lines, and $ABCD$ may be approximated by a rectangle, since $\mathbf{e}_1$ and $\mathbf{e}_2$ are orthogonal. Hence the contribution of V_1 to the line integral arises solely from the arcs AB and DC and is

$$\begin{aligned} &V_1(\text{arc } AB) - V_1(\text{arc } DC) \\ &= V_1 h_1 \mathrm{d}u_1 - \left[V_1 h_1 \mathrm{d}u_1 + \mathrm{d}u_2 \frac{\partial}{\partial u_1}(V_1 h_1 \mathrm{d}u_1)\right] \\ &= -\mathrm{d}u_1 \mathrm{d}u_2 \frac{\partial}{\partial u_2}(V_1 h_1)\,. \end{aligned}$$

Similarly, the contribution from V_2 is

$$\mathrm{d}u_1 \mathrm{d}u_2 \frac{\partial}{\partial u_1}(V_2 h_2)\,,$$

and $\mathrm{d}s = h_1 h_2 u_1 u_2$, so that on substituting into (12.72) we obtain

$$\mathbf{e}_3 \cdot \operatorname{curl} \mathbf{V} = \frac{1}{h_1 h_2}\left[\frac{\partial}{\partial u_1}(h_2 V_2) - \frac{\partial}{\partial u_2}(h_1 V_1)\right].$$

This identical to the $\mathbf{e}_3$ component given in (12.20) and analogous results follows for the other components.

*12.4.3 Applications to electromagnetic fields

Finally we illustrate the use of Stokes' theorem by applying it to the behaviour of electric and magnetic fields, starting with the electric field $\mathbf{E}$. In free space, this is determined by the fundamental equations

$$\text{(a)}\ \ \boldsymbol{\nabla} \cdot \mathbf{E} = \varepsilon_0^{-1}\rho \ \ \text{and}\ \ \text{(b)}\ \ \boldsymbol{\nabla} \times \mathbf{E} = -\frac{\partial \mathbf{B}}{\partial t}, \qquad (12.73)$$

where $\mathbf{B}$ is the magnetic field intensity, ρ is the charge density, and ε_0 is the electric permittivity. Of these, (12.73a) was discussed in Section 12.3.3, where we saw that it expressed the fact that charge is the source of electric flux. However, in contrast to electrostatics, if there are time-dependent magnetic fields present, curl $\mathbf{E}$ no longer vanishes. Hence $\mathbf{E}$ is not in general a conservative field and loop integrals of the form

$$\varepsilon_C \equiv \oint_C \mathbf{E} \cdot \mathbf{dr}$$

no longer vanish. Rather, by Stokes' theorem and (12.73), we have

$$\varepsilon_C \equiv \oint_C \mathbf{E} \cdot \mathbf{dr} = -\frac{\partial}{\partial t} \iint_S \mathbf{B} \cdot \mathbf{ds}, \tag{12.74}$$

where S is any open surface spanning the loop C. This is *Faraday's law of induction* that states that the 'emf' ε_C induced around a loop C is equal to minus the rate of change of the magnetic flux through the loop. We also note that the argument can be reversed: if (12.74) holds, then Stokes' theorem gives

$$\iint_S \boldsymbol{\nabla} \times \mathbf{E} \cdot \mathbf{ds} = -\iint_S \frac{\partial \mathbf{B}}{\partial t} \cdot \mathbf{ds}$$

which can only hold for an arbitrary open surface S if (12.74) is satisfied. Equation (12.73b) and (12.74) are the differential and integral forms of Faraday's law.

Equations (12.73a) and (12.73b) are the first two *Maxwell's equations* in free space. The remaining two are

$$\text{(a)}\ \boldsymbol{\nabla} \cdot \mathbf{B} = 0, \quad \text{(b)}\ \boldsymbol{\nabla} \times \mathbf{B} = \mu_0 \mathbf{j} + \frac{1}{c^2}\frac{\partial \mathbf{B}}{\partial t}, \tag{12.75}$$

where $\mathbf{j}$ is the electric current density, μ_0 is the magnetic permeability of free space, and the speed of light $c = (\mu_0 \varepsilon_0)^{-1/2}$. On comparing with (12.73a), we see that (12.75a) reflects the experimental observation that there are no free magnetic charges. The second equation (12.75b) indicates that non-zero magnetic fields can be generated by currents or time-dependent electric fields. In the absence of the latter, it becomes

$$\boldsymbol{\nabla} \times \mathbf{B} = \mu_0 \, \mathbf{j}. \tag{12.76}$$

By Stokes' theorem

$$\oint_C \mathbf{B} \cdot \mathbf{dr} = \iint_S (\boldsymbol{\nabla} \times \mathbf{B}) \cdot \mathbf{ds},$$

giving

$$\oint_C \mathbf{B}\cdot d\mathbf{r} = \mu_0 \iint_S \mathbf{j}\cdot d\mathbf{s} \equiv \mu_0 I_{\text{encl}}, \tag{12.77}$$

where S is any surface spanning the loop C. This is called *Ampère's law* and it states that the line integral of $\mathbf{B}$ around a closed loop is equal to μ_0 times the total current I_{encl} flowing through the loop. It enables the magnetic field to be calculated quickly in symmetrical situations, as we shall illustrate.

Example 12.17

An infinitely long thin wire is aligned along the z-axis and carries a current I. Find the form of the generated magnetic field $\mathbf{B}$, assuming that $\mathbf{B} \to 0$ infinitely far from the wire.

Solution

We have cylindrical symmetry about the z-axis, so that if we use cylindrical polar co-ordinates, $\mathbf{B}$ must be independent of z and ϕ, so that

$$\mathbf{B} = B_\rho(\rho)\mathbf{e}_\rho + B_\phi(\rho)\mathbf{e}_\phi + B_z(\rho)\mathbf{e}_z, \tag{12.78}$$

where $\mathbf{e}_\rho$, $\mathbf{e}_\phi$, $\mathbf{e}_z$ are the unit vectors shown in Figure 11.13.

We next impose (12.75a) and (12.76) on (12.78) at $\mathbf{r} \neq \mathbf{0}$, where $\mathbf{j} = \mathbf{0}$. Using the forms of div and curl in cylindrical polar co-ordinates given in Table 12.1, this gives

$$\boldsymbol{\nabla}\cdot\mathbf{B} = \frac{\partial B_\rho}{\partial\rho} = 0$$

and

$$\boldsymbol{\nabla}\times\mathbf{B} = -\mathbf{e}_\phi\frac{\partial B_z}{\partial\rho} + \frac{1}{\rho}\mathbf{e}_z\frac{\partial}{\partial\rho}(\rho B_\phi) = \mathbf{0},$$

so that

$$\frac{\partial B_z}{\partial\rho} = 0, \quad \frac{\partial B_\rho}{\partial\rho} = 0, \quad \frac{\partial}{\partial\rho}(\rho B_\phi) = 0.$$

This can only be satisfied, subject to the boundary condition $\mathbf{B} \to \mathbf{0}$ as $\rho \to \infty$, by

$$B_z = 0, \quad B_\rho = 0, \quad B_\phi = k/\rho,$$

where k is a constant. This can now be determined by applying Ampère's law (12.77) to a circle in the x–y plane shown in Figure 12.15, giving

$$\oint_C \mathbf{B}\cdot d\mathbf{r} = 2\pi\rho B_\phi(\rho) = \mu_0 I,$$

so that, finally,

$$\mathbf{B}(\mathbf{r}) = \frac{\mu_0 I}{2\pi\rho}\mathbf{e}_\phi. \tag{12.79}$$

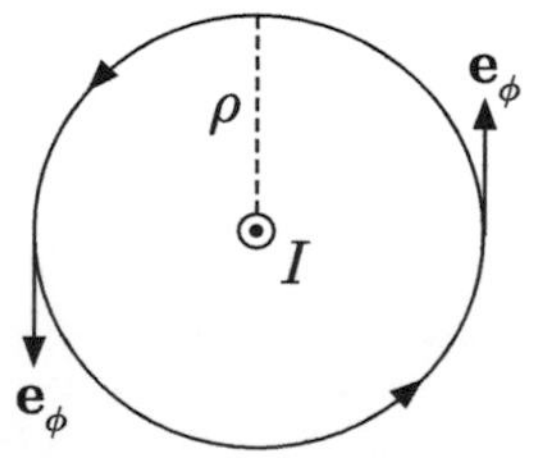

Figure 12.15 The circuit C used to derive (12.79), where the current I at the centre is directed out of the page.

Problems 12

12.1 A scalar field electrostatic potential is given by $\phi = x^2 - y^2$ and the associated electric field $\mathbf{E}$ is given by $\mathbf{E} = -\boldsymbol{\nabla}\phi$. What is the magnitude and direction of $\mathbf{E}$ at (2, 1)? In what direction does ϕ increase most rapidly at the point (−3, 2) and what is the rate of change of ϕ at the point (1, 2) in the direction $3\mathbf{i} - \mathbf{j}$?

12.2 Given the scalar function $\psi = x^2 - y^2 z$, find (a) $\boldsymbol{\nabla}\psi$ at (1, 1, 1); (b) the derivative of ψ at (1, 1, 1) in the direction $\mathbf{i} - 2\mathbf{j} + \mathbf{k}$; (c) the equation of the normal to the surface $\psi = x^2 - y^2 z = 0$ at (1, 1, 1).

12.3 If $\mathbf{A} = 2xz^2\mathbf{i} - yz\,\mathbf{j} + 3xz^3\mathbf{k}$ and $S = x^2yz$, find in Cartesian co-ordinates (a) curl $\mathbf{A}$, (b) curl $(S\mathbf{A})$, (c) curl curl $\mathbf{A}$, (d) grad $(\mathbf{A} \cdot \text{curl}\,\mathbf{A})$ and (e) curl grad S.

12.4 Given a scalar field ψ and a vector field $\mathbf{V}$, show (a) that

$$\text{curl}\,(\psi\mathbf{V}) = \psi\,\text{curl}\,\mathbf{V} - \mathbf{V} \times \boldsymbol{\nabla}\psi.$$

Hence show (b) that if $\alpha\mathbf{V} = \text{grad}\,\psi$, where α is a scalar field, then

$$\mathbf{V} \cdot \text{curl}\,\mathbf{V} = 0.$$

12.5 Show, without explicitly writing out the components, that

$$\boldsymbol{\nabla} \times \mathbf{a}(\boldsymbol{\nabla} \cdot \mathbf{a}) + \mathbf{a} \times [\boldsymbol{\nabla} \times (\boldsymbol{\nabla} \times \mathbf{a})] + \mathbf{a} \times \boldsymbol{\nabla}^2\mathbf{a} = (\boldsymbol{\nabla} \cdot \mathbf{a})(\boldsymbol{\nabla} \times \mathbf{a}).$$

12.6 If $\psi = 2yz$ and $\mathbf{V} = x\mathbf{j} - y\mathbf{k}$, express

$$\text{(a) } \psi,\ \text{(b) } \mathbf{V},\ \text{(c) } \boldsymbol{\nabla}\psi \text{ and (d) } \boldsymbol{\nabla} \times \mathbf{V}$$

in spherical polar co-ordinates.

12.7 Directly evaluate the line integral

$$\oint_C \mathbf{V} \cdot d\mathbf{r},$$

where

$$\mathbf{V} = (x^2 + y^2)y\,\mathbf{i} - (x^2 + y^2)x\,\mathbf{j} + (a^3 + z^3)\,\mathbf{k},$$

around the circle $(x^2 + y^2) = a^2$ in the x–y plane. Verify your result using Green's theorem in the plane.

12.8 A force field is

$$\mathbf{F} = (y + z)\mathbf{i} - (x + z)\mathbf{j} + (x + y)\mathbf{k}.$$

Find the work done in moving a particle round a closed curve from the origin to the point $(x, y, z) = (0, 0, 2\pi)$ along the path

$$x = 1 - \cos t, \quad y = \sin t, \quad z = t$$

and then back to the origin along the z-axis.

12.9 Find the work done by a force **F** given by

$$\mathbf{F} = (x - 2y^2)\mathbf{i} + (3x + 2y)\mathbf{j} + (3x^2 - 2y)\mathbf{k}$$

when moving a particle clockwise along a semicircle of unit radius in the x–y plane from $x = -1$ to $x = 1$ with $y \geq 0$.

12.10 Find the work done by a force $\mathbf{F} = (x^2 + y^2)\mathbf{j}$ when moving between the points $A(x = a,\ y = 0,\ z = 0)$ and $B(x = 0,\ y = a,\ z = 0)$ along a path C, where C is (a) along the x-axis to the origin, then along the y-axis to B, and (b) along the arc of the circle $x^2 + y^2 = a^2,\ z = 0$, in the positive quadrant.

12.11 A force $\mathbf{F} = xy\mathbf{i} - y^2\mathbf{j}$ moves around a closed loop starting at the origin along the curve $x = 2\sqrt{y}$ to (2, 1), then parallel to the x axis to (0, 1) and finally returning to the origin along the y axis. Use Green's theorem in the plane to calculate the work done by the force.

12.12 Show that

$$\mathbf{V} = y^2 z \sinh(2xz)\mathbf{i} + 2y\cosh^2(xz)\mathbf{j} + xy^2 \sinh(2xz)\mathbf{k}$$

is a conservative field and find a scalar potential ϕ, such that $\mathbf{V} = -\boldsymbol{\nabla}\phi$.

12.13 A force field

$$\mathbf{F} = (x + 2y + az)\mathbf{i} + (bx - 3y - z)\mathbf{j} + (4x + cy + 2z)\mathbf{k},$$

where a, b, c, are constants. For what values of a, b, c, is **F** a conservative field? Find the scalar potential in this case.

12.14 Let $\bar{S}$ be that part of the surface of the cylinder

$$x^2 + y^2 = 4, \quad 0 < z < 3$$

for which $x > 0,\ y > 0$. What is the value of the surface integral

$$I = \iint_S \mathbf{A} \cdot \mathrm{d}\mathbf{s}$$

if $\mathbf{A} = 6y\mathbf{i} + (2x + z)\mathbf{j} - x\mathbf{k}$ and S is that part of $\bar{S}$ that lies on the curved surface of the cylinder?

12.15 Evaluate the integral

$$\oint_S \mathbf{V} \cdot \mathrm{d}\mathbf{s}.$$

where $\mathbf{V} = x^2\mathbf{i} + \frac{1}{2}y^2\mathbf{j} + \frac{1}{2}z^2\mathbf{k}$ and S is the surface of a unit cube $0 \leq x \leq 1,\ 0 \leq y \leq 1,\ 0 \leq z \leq 1$, without using the divergence theorem. The vector **s** is defined in the outward direction from each face of the cube.

12.16 Evaluate the integral

$$I = \iint_S \sigma(y^2 + z^2)\mathrm{d}s$$

over the curved surface of a hemisphere of radius a with its centre at the origin and base in the x–y plane, where $ds = |\mathbf{ds}|$ and σ is a constant.

12.17 A sphere of uniform density ρ has mass M and radius a. Calculate its moment of inertia about (a) a tangent to the sphere and (b) an axis through the centre of the sphere.

12.18 A cylinder of uniform density ρ_0 has mass M, radius a and length d. Calculate its moment of inertia about an axis that lies in the plane of the base of the cylinder and passes through the centre of the base.

12.19 Use the divergence theorem to evaluate

$$\iint_S (\mathbf{F}\cdot\mathbf{n})\mathrm{d}s,$$

where $\mathbf{F} = 4xz\,\mathbf{i} - y^2\,\mathbf{j} + yx\,\mathbf{k}$ and S is the surface of the cube bounded by

$$x = 0,\ x = 1;\quad y = 0,\ y = 1;\quad z = 0,\ z = 1.$$

12.20 Scalar fields $\phi_i (i = 1, 2, \ldots)$ are solutions of the equations

$$\boldsymbol{\nabla}^2\phi_i = \gamma_i\phi_i,$$

where the γ_i are constants, within a region Ω, subject to the boundary conditions $\phi_i = 0$ on the closed surface S enclosing Ω. Show that

$$\iiint_\Omega \phi_i\phi_j \mathrm{d}v = 0,\quad i = 1, 2, \ldots;\ j = 1, 2, \ldots$$

if $i \neq j\ \gamma_i \neq \gamma_j$.

12.21 Prove the identity

$$\boldsymbol{\nabla}\cdot(\psi\,\boldsymbol{\nabla}\psi) = (\boldsymbol{\nabla}\psi)^2 + \psi\,\boldsymbol{\nabla}^2\psi$$

for any scalar field ψ. A scalar field ψ satisfies the conditions $\psi = 0$ on S and $\boldsymbol{\nabla}^2\psi = 0$ in Ω, where S is the closed surface surrounding the region Ω. Show that $\psi = 0$ in Ω.

***12.22** State Gauss' theorem for the gravitational field. A homogeneous spherical shell has mass M, inner radius a and outer radius $b > a$. Find an expression for the gravitational field due to the shell for (a) $r > b$, (b) $r < a$ and (c) $a < r < b$, where $r = |\mathbf{r}|$. Finally, calculate the potential at any point with $r < a$ assuming the potential goes to zero as $r \to \infty$.

***12.23** (a) Prove the relation

$$\mathrm{div}(\phi\mathbf{E}) = \phi\,\mathrm{div}\mathbf{E} + \mathbf{E}\cdot\mathrm{grad}\,\phi,$$

where ϕ is a scalar field and $\mathbf{E}$ is a vector field.

(b) Let $\rho(\mathbf{r})$ be an electric charge density, which vanishes outside a finite region Ω_1 enclosing the origin, with total charge

$$Q = \iiint_{\Omega_i} \rho(\mathbf{r})\mathrm{d}v.$$

Write down an approximate value for the electrostatic field $\mathbf{E}$ and potential ϕ on a sphere centred at the origin with radius R, assuming that R is very large compared to the dimensions of Ω_1.
(c) Show that, in the same approximation,

$$\iiint_{\Omega_2} \rho\phi \mathrm{d}v = \varepsilon_0 \iiint_{\Omega_2} \mathbf{E}^2 \mathrm{d}v + \frac{c}{R},$$

where Ω_2 is the interior of the sphere of radius R, and find the value of the constant c.
[Hint: use Poisson's equation $\boldsymbol{\nabla} \cdot \mathbf{E} = \varepsilon_0^{-1}\rho$.]

***12.24** In a homogeneous continuous medium, Maxwell's equations take the form

$$\boldsymbol{\nabla} \cdot \mathbf{D} = \rho, \quad \boldsymbol{\nabla} \times \mathbf{E} = -\frac{\partial \mathbf{B}}{\partial t},$$

$$\boldsymbol{\nabla} \cdot \mathbf{B} = 0, \quad \boldsymbol{\nabla} \times \mathbf{H} = \mathbf{j} + \frac{\partial \mathbf{D}}{\partial t},$$

with $\mathbf{D} = \varepsilon\mathbf{E}$, $\mathbf{H} = \mathbf{B}/\mu$, and the constants ε and μ are the permittivity and permeability of the medium.
(a) Show that Maxwell's equations imply that ρ is the density of a conserved charge.
(b) In a conductor, the current obeys Ohm's law $\mathbf{j} = \sigma\mathbf{E}$, where σ is the conductivity. Find the charge density ρ as a function of time if $\rho(\mathbf{r}, t = 0) = \rho_0(\mathbf{r})$.

12.25 Verify Stokes' theorem for the vector

$$\mathbf{A} = (2x - y)\mathbf{i} - yz^2\mathbf{j} - y^2 z\mathbf{k},$$

where S is the surface of the hemisphere

$$x^2 + y^2 + z^2 = 1, \quad z > 0,$$

and C is the boundary of S.

12.26 Use Stokes' theorem (12.68) to prove the relation

$$\iint_S \mathrm{d}\mathbf{s} \times \boldsymbol{\nabla}\phi = \oint_C \phi \, \mathrm{d}\mathbf{s},$$

where ϕ is a scalar fields and S is an open surface bounded by a closed curve C. [Hint: apply Stokes' theorem to the vector field $\mathbf{V} = \phi\,\mathbf{c}$, where $\mathbf{c}$ is a constant vector.]

12.27 A force field $\mathbf{F} = y^2\mathbf{i} + x^2\mathbf{j}$ acts on a particle. Write down a line integral corresponding to the work done by the field when the particle moves once round the circle $x^2 + y^2 = a^2$, $z = 0$ in the anticlockwise direction. Evaluate this integral (a) directly and (b) by converting it to a surface integral. (c) Is the force field $\mathbf{F}$ conservative?

13

Fourier analysis

In Chapter 5 we discussed how functions could be represented as power series using the expansions of Taylor and Maclaurin. Those expansions are valid only within certain radii of convergence, where the functions must be continuous and infinitely differentiable. However, this is not the only way that functions can be expressed as a series. In this chapter we will consider another expansion, which may also be used for functions that are neither continuous nor differentiable at certain points. To start with, the discussion will be centred on functions $f(x)$ that are *periodic*, that is, they obey the relation $f(x) = f(x + np)$, where p is the *period*, and $n = 1, 2, \ldots$. Many functions that occur in physical science are of this type. For example, solutions of the equations for problems concerning wave motion involve sinusoidal functions. The form of $f(x)$ can be arbitrarily complicated, such as the continuous function shown in Figure 13.1. We shall also see that the method can be applied to functions that are only defined in a finite range of x. This leads naturally to an important extension in which non-periodic functions$f(x)$, defined for all x, can be expressed in terms of simple sinusoidal functions, provided $f(x) \to 0$ rapidly enough as $x \to \infty$. Such expressions, called *Fourier transforms*, are extremely useful and will be discussed in Section 13.3.

13.1 Fourier series

Initially, we will assume for convenience that the function to be expanded is periodic with a period $p = 2\pi$, so that

$$f(x + 2n\pi) = f(x) \tag{13.1}$$

for any integer n. Then, in certain circumstances, $f(x)$ may be written as a sum of trigonometric functions of the form

$$f_N(x) \equiv \frac{a_0}{2} + \sum_{n=1}^{N} a_n \cos nx + \sum_{n=1}^{N} b_n \sin nx, \tag{13.2}$$

Mathematics for Physicists, First Edition. B.R. Martin and G. Shaw.

Companion website: www.wiley.com/go/martin/mathsforphysicists

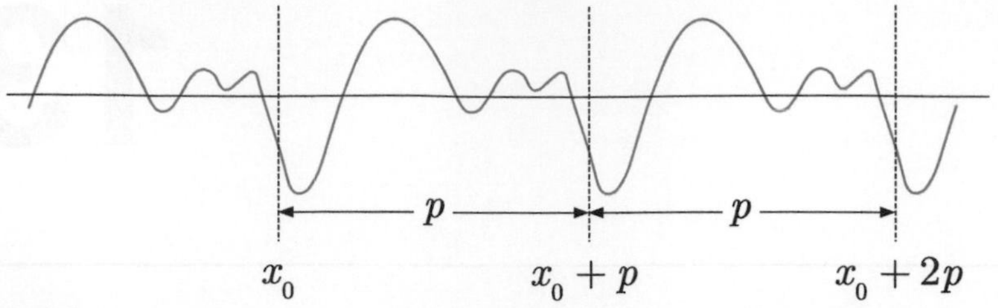

Figure 13.1 An arbitrary periodic function.

where the coefficients a_n and b_n are chosen so that $f_N(x)$ is the best representation of $f(x)$. (The reason for the factor $\frac{1}{2}$ in the first term will be made clear presently.) Since $f_N(x)$ also satisfies (13.1), it is sufficient to consider only the range $-\pi \leq x \leq \pi$, because outside this range both $f(x)$ and $f_N(x)$ repeat themselves. Two questions now have to be considered: firstly, what do we mean by 'best', and secondly, does the convergence of the series in (13.2) ensure that $f_N(x) \to f(x)$ as $N \to \infty$?

13.1.1 Fourier coefficients

There is no unique way of defining 'best', but the one that is most convenient is to choose the coefficients to minimise the integral

$$I_N \equiv \int_{-\pi}^{\pi} [f(x) - f_N(x)]^2 \mathrm{d}x. \tag{13.3}$$

In this case, $f_N(x)$ is said to be the best *approximation in the mean* to $f(x)$. To find the coefficients a_n and b_n, we start by substituting (13.2) into (13.3), giving

$$I_N = \int_{-\pi}^{\pi} \left\{ f(x) - \left[a_0/2 + \sum_{n=1}^{N} a_n \cos nx + \sum_{n=1}^{N} b_n \sin nx \right] \right\}^2. \tag{13.4a}$$

Then expanding the integrand on the right-hand side of (13.4a) gives

$$\begin{aligned} I_N = & \int_{-\pi}^{\pi} f^2(x)\,\mathrm{d}x - a_0 \int_{-\pi}^{\pi} f(x)\,\mathrm{d}x - 2\sum_{n=1}^{N} a_n \int_{-\pi}^{\pi} f(x) \cos nx\,\mathrm{d}x \\ & - 2\sum_{n=1}^{N} b_n \int_{-\pi}^{\pi} f(x) \sin nx\,\mathrm{d}x \\ & + \int_{-\pi}^{\pi} \left[a_0/2 + \sum_{n=1}^{N} a_n \cos nx + \sum_{n=1}^{N} b_n \sin nx\,\mathrm{d}x \right]^2. \end{aligned} \tag{13.4b}$$

This may be simplified by using the general integrals

$$\int_{-p/2}^{p/2} \cos\left(\frac{2n\pi x}{p}\right) \sin\left(\frac{2m\pi x}{p}\right) \mathrm{d}x = 0, \quad \text{for all } m, n, \tag{13.5a}$$

$$\int_{-p/2}^{p/2} \cos\left(\frac{2n\pi x}{p}\right) \cos\left(\frac{2m\pi x}{p}\right) \mathrm{d}x = \begin{cases} 0 & m \neq n \\ p/2 & m = n \neq 0 \\ p & m = n = 0 \end{cases}, \tag{13.5b}$$

$$\int_{-p/2}^{p/2} \sin\left(\frac{2n\pi x}{p}\right) \sin\left(\frac{2m\pi x}{p}\right) \mathrm{d}x = \begin{cases} 0 & m \neq n \\ p/2 & m = n \neq 0 \\ 0 & m = n = 0 \end{cases}, \tag{13.5c}$$

which for $p = 2\pi$ reduce to

$$\int_{-\pi}^{\pi} \cos(nx) \sin(mx)\, \mathrm{d}x = 0, \quad \text{for all } m, n, \tag{13.6a}$$

$$\int_{-\pi}^{\pi} \cos(nx) \cos(mx)\, \mathrm{d}x = \begin{cases} 0 & m \neq n \\ \pi & m = n \neq 0 \\ 2\pi & m = n = 0 \end{cases}, \tag{13.6b}$$

$$\int_{-\pi}^{\pi} \sin(nx) \sin(mx)\, \mathrm{d}x = \begin{cases} 0 & m \neq n \\ \pi & m = n \neq 0 \\ 0 & m = n = 0 \end{cases}. \tag{13.6c}$$

Finally, we define

$$A_n \equiv \frac{1}{\pi} \int_{-\pi}^{\pi} f(x) \cos(nx)\, \mathrm{d}x \quad \text{and} \quad B_n \equiv \frac{1}{\pi} \int_{-\pi}^{\pi} f(x) \sin(nx)\, \mathrm{d}x \tag{13.7}$$

for $n = 0, 1, 2, \ldots$. Then, using (13.6) and (13.7) in (13.4b), we have, after simplification,

$$I_N = \int_{-\pi}^{\pi} f^2(x)\, \mathrm{d}x + \frac{\pi}{2}(a_0 - A_0)^2 + \pi \sum_{n=1}^{N} (a_n - A_n)^2 + \pi \sum_{n=1}^{N} (b_n - B_n)^2 - \frac{\pi}{2} A_0^2 - \pi \sum_{n=1}^{N} (A_n^2 + B_n^2),$$

which is a minimum when $a_n = A_n$ and $b_n = B_n$ for all $n = 0, 1, 2, \ldots, N$.

Setting $a_n = A_n$ and $b_n = B_n$ for all $n = 0, 1, 2, \ldots, N$ in the equation for I_N, and using the fact that from its definition $I_N \geq 0$, gives *Bessel's inequality*

$$\frac{1}{2}a_0^2 + \sum_{n=1}^{N}\left(a_n^2 + b_n^2\right) \leq \frac{1}{\pi}\int_{-\pi}^{\pi} f^2(x)\,\mathrm{d}x, \tag{13.8}$$

which becomes an equality if $f_N(x)$ is an exact representation of $f(x)$ in the mean, that is, if $I_N = 0$. It can be shown that this occurs for all reasonably well-behaved functions in the limit $N \to \infty$. In this case the expansion (13.2) becomes the *Fourier series*

$$f(x) \equiv \frac{a_0}{2} + \sum_{n=1}^{\infty} a_n \cos nx + \sum_{n=1}^{\infty} b_n \sin nx, \tag{13.9}$$

where the *Fourier coefficients* a_n and b_n are given by

$$a_n \equiv \frac{1}{\pi}\int_{-\pi}^{\pi} f(x)\cos(nx)\,\mathrm{d}x, \tag{13.10a}$$

$$b_n \equiv \frac{1}{\pi}\int_{-\pi}^{\pi} f(x)\sin(nx)\,\mathrm{d}x. \tag{13.10b}$$

The Fourier series (13.9) is simplified if the function $f(x)$ has a definite symmetry. If$f(x)$ is an even function, that is, $f(-x) = f(x)$, then the integral (13.10b) vanishes for all n [cf. (4.32c)] so that $b_n = 0$ and the Fourier expansion (13.9) reduces to a *cosine series*

$$f(x) = \frac{a_0}{2} + \sum_{n=1}^{\infty} a_n \cos nx, \quad f(x) = f(-x) \tag{13.11a}$$

Likewise, if $f(x)$ is an odd function, so that $f(-x) = -f(x)$, then the coefficients $a_n = 0$ and the expansion is a *sine series*

$$f(x) = \sum_{n=1}^{\infty} b_n \sin nx, \quad f(x) = -f(-x) \tag{13.11b}$$

For example, if $f(x) = x$ for $-\pi < x < \pi$, and has period 2π, then $a_n = 0$ by symmetry and the coefficients b_n are given by

$$b_n = \frac{1}{\pi}\int_{-\pi}^{\pi} x\sin(nx)\mathrm{d}x = \frac{2}{\pi}\int_{0}^{\pi} x\sin(nx)\mathrm{d}x$$

$$= -\frac{2\cos n\pi}{n} = \frac{2(-1)^{n+1}}{n}, \tag{13.12a}$$

on integrating by parts, and the Fourier series becomes

$$f(x) = 2\sum_{n=1}^{\infty} \frac{(-1)^{n+1}}{n} \sin nx = 2\left[\sin x - \frac{\sin 2x}{2} + \frac{\sin 3x}{3} - \cdots\right]. \tag{13.12b}$$

Example 13.1

Draw a diagram of the function $f(x)$ with period 2π given that

$$f(x) = \begin{cases} 0 & -\pi < x \le 0 \\ x & 0 < x < \pi \end{cases} \tag{1}$$

and find its Fourier decomposition.

Solution

The function is shown in Figure 13.2. If the Fourier coefficients are a_n and b_n, then

$$a_n = \frac{1}{\pi}\int_{-\pi}^{\pi} f(x)\cos nx\,\mathrm{d}x = \frac{1}{\pi}\int_{0}^{\pi} x\cos nx\,\mathrm{d}x.$$

For $n > 1$, this integral can be integrated by parts to give

$$a_n = \frac{1}{n\pi}\int_0^{\pi} x\frac{\mathrm{d}\sin nx}{\mathrm{d}x}\,\mathrm{d}x = \frac{1}{\pi}\left[\frac{x}{n}\sin nx\right]_0^{\pi} - \frac{1}{\pi n}\int_0^{\pi}\sin nx\,\mathrm{d}x$$

$$= \frac{1}{\pi}\left[\frac{1}{n^2}\cos nx\right]_0^{\pi} = \frac{1}{\pi n^2}[(-1)^n - 1].$$

Thus, $a_1 = -2/\pi$, $a_2 = 0$, $a_3 = -2/9\pi$, etc. Similarly,

$$a_0 = \frac{1}{\pi}\int_{-\pi}^{\pi} f(x)\,\mathrm{d}x = \frac{1}{\pi}\int_0^{\pi} x\,\mathrm{d}x = \frac{\pi}{2},$$

and

$$b_n = \frac{1}{\pi}\int_{-\pi}^{\pi} f(x)\sin nx\,\mathrm{d}x = \frac{1}{\pi}\int_0^{\pi} x\sin nx\,\mathrm{d}x = \frac{(-1)^{n+1}}{\pi n}.$$

So, finally

$$f(x) = \frac{\pi}{4} - \frac{2}{\pi}\left(\cos x + \frac{1}{9}\cos 3x + \frac{1}{25}\cos 5x + \cdots\right)$$
$$+ \frac{1}{\pi}\left(\sin x - \frac{1}{2}\sin 2x + \frac{1}{3}\sin 3x - \frac{1}{5}\sin 5x + \cdots\right).$$

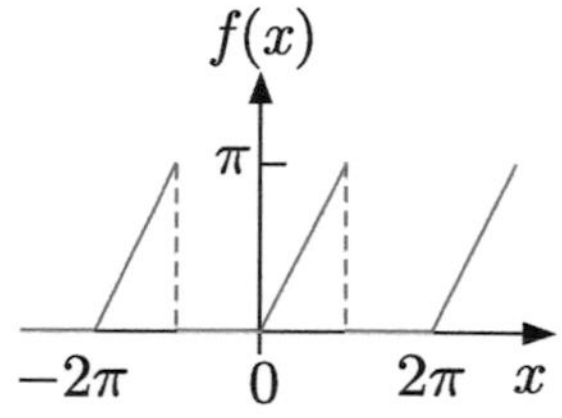

Figure 13.2

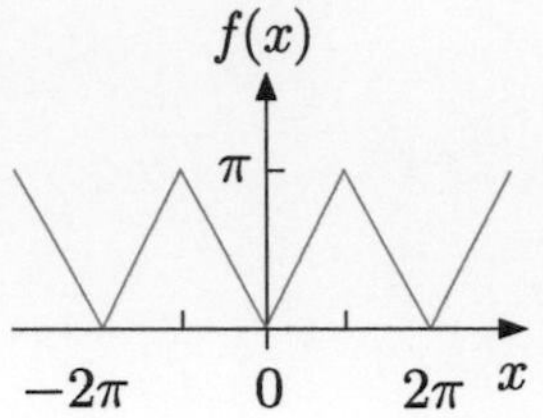

Figure 13.3

Example 13.2

Draw a diagram of the function $f(x)$ of period 2π, such that

$$f(x) = |x|, \quad -\pi < x < \pi$$

and find its Fourier decomposition.

Solution

The function is shown in Figure 13.3. The series is even, and so contains only cosines, with coefficients

$$a_n = \frac{1}{\pi}\int_{-\pi}^{\pi} |x| \cos nx \, dx = \frac{2}{\pi}\int_{0}^{\pi} x \cos nx \, dx.$$

Hence $a_0 = \pi$, and for $n \neq 0$, integrating by parts,

$$a_n = \frac{2}{\pi}\int_{0}^{\pi} x \cos nx \, dx = \frac{2}{\pi n^2}\left[(-1)^n - 1\right].$$

So the series is

$$f(x) = \frac{\pi}{2} - \frac{4}{\pi}\left[\cos x + \frac{1}{9}\cos 3x + \frac{1}{25}\cos 5x + \cdots\right].$$

This result can also be obtained from the result of Example 13.1 by noting that $f(x) = g(x) + g(-x)$, where $g(x)$ is the function (1) defined in that example.

13.1.2 Convergence

We now return to the second question posed above: under what conditions does the Fourier series converge to $f(x)$? This is important because even if the Fourier series converges in the mean to $f(x)$, it does not follow that it converges to $f(x)$ for all values of x. The proof of this convergence property of Fourier series is lengthy and we will not reproduce it here,[1] but the conditions under which convergence holds, called the *Dirichlet conditions*, may be summarised by the following theorem.

Dirichlet theorem *If $f(x)$ is periodic with interval 2π, and in the interval $-\pi$ to π is a single-valued continuous function, except for a*

[1] It is given, for example, in H.S.W. Massey and H. Kestleman (1963) *Ancillary Mathematics*, Isaac Pitman & Sons Ltd., London, p. 790.

possible finite number of jump discontinuities, with a finite number of maxima and minima, and in addition the integral

$$\int_{-\pi}^{\pi} |f(x)|\,\mathrm{d}x \tag{13.13}$$

is finite, then the Fourier series converges to $f(x)$ at all points where $f(x)$ is continuous. At discontinuous points, the series converges to the mid-point of the jump discontinuity.

An example of a discontinuous function is shown in Figure 13.4, and in accordance with the Dirichlet theorem, as $x \to x_0$ the series converges to $\pi/2$.

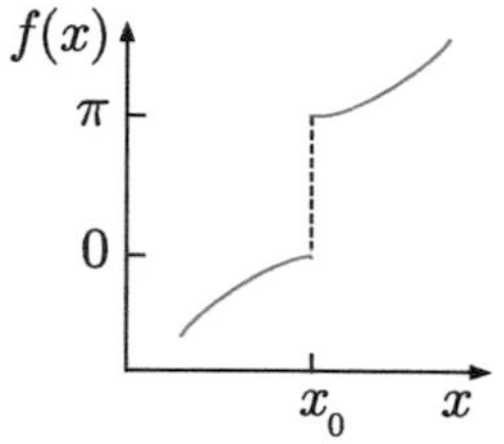

Figure 13.4 A discontinuous function.

To establish convergence in principle requires the integral (13.13) to be evaluated. However, this is usually not necessary in practice, because if the value of $f(x)$ is bounded, that is, it lies between $\pm B$ for some positive value B, then

$$\int_{\pi}^{\pi} |f(x)|\,\mathrm{d}x \le \int_{-\pi}^{\pi} B\,\mathrm{d}x = 2\pi B,$$

and so is finite. It is worth noting that the converse of Dirichlet's theorem is *not* true. Thus, if a function fails to satisfy the conditions of the theorem, it *may* still be possible to expand it as a Fourier series. However, functions of this type are seldom encountered in physical situations.

In practice, only a finite numbers of terms can be evaluated in a Fourier series, so it is of interest to see how well such a series represents a given function when a limited number of terms is used. For simple functions, the convergence of the series is usually quite rapid. For example, consider the 'rectified half-wave' function. This is a continuous function periodic in x with period 2π and satisfies the Dirichlet conditions. For the interval $-\pi \le x \le \pi$ it is given by

$$f(x) = \begin{cases} 0 & \text{for} \quad -\pi \le x < 0 \\ \sin x & \text{for} \quad 0 < x \le \pi \end{cases}, \tag{13.14}$$

and is shown in Figure 13.5. The coefficients of its Fourier series may be found using (13.10). Thus, from (13.10a),

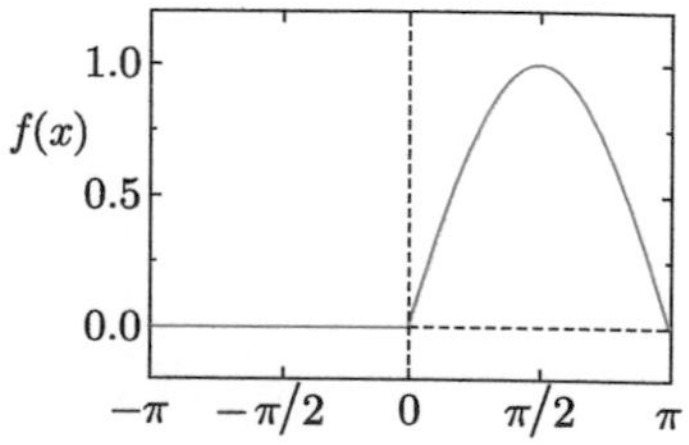

Figure 13.5 The function $f(x)$ given by Eq. (13.14).

$$a_n = \frac{1}{\pi}\int_{-\pi}^{\pi} f(x)\cos(nx)\mathrm{d}x = \frac{1}{\pi}\int_{0}^{\pi} \sin x\cos(nx)\mathrm{d}x.$$

For $n = 0$,

$$a_0 = \frac{1}{\pi}\int_0^{\pi} \sin x \, dx = -\frac{1}{\pi}\left[\cos x\right]_0^{\pi} = \frac{2}{\pi}$$

and for $n = 1$,

$$a_1 = \frac{1}{\pi}\int_0^{\pi} \sin x \cos x \, dx = -\frac{1}{4\pi}\left[\cos 2x\right]_0^{\pi} = 0,$$

while for $n \neq 0, 1$, using (2.36c), we have

$$a_n = \frac{1}{\pi}\int_0^{\pi} \sin x \cos(nx) dx = \frac{1}{2\pi}\int_0^{\pi}\left[\sin(n+1)x - \sin(n-1)x\right] dx$$

$$= \frac{1}{2\pi}\left[\frac{-\cos(n+1)x}{n+1} + \frac{\cos(n-1)x}{n-1}\right]_0^{\pi} = \frac{\left[1+(-1)^n\right]}{\pi(n^2-1)},$$

using $\cos(n+1)\pi = \cos(n-1)\pi = (-1)^{n+1}$. Similarly, from (13.10b),

$$b_n = \frac{1}{\pi}\int_{-\pi}^{\pi} f(x)\sin(nx) dx = \frac{1}{\pi}\int_0^{\pi} \sin x \sin(nx) dx.$$

So, for $n = 1$,

$$b_1 = \frac{1}{\pi}\int_0^{\pi} \sin^2 x \, dx = \frac{1}{2\pi}\int_0^{\pi}(1 - \cos 2x) dx = \frac{1}{2},$$

and for $n \neq 1$, using (2.36e), we have

$$b_n = \frac{1}{\pi}\int_0^{\pi} \sin x \sin(nx) dx = \frac{1}{2\pi}\int_0^{\pi}\left[\cos(n-1)x - \cos(n+1)x\right] dx = 0.$$

Finally, collecting terms, the Fourier series is

$$f(x) = \frac{1}{\pi} + \frac{1}{2}\sin x - \frac{2}{\pi}\sum_{n=1}^{\infty}\frac{\cos 2nx}{(2n)^2 - 1}$$

$$= \frac{1}{\pi} + \frac{1}{2}\sin x - \frac{2}{\pi}\left[\frac{\cos 2x}{3} + \frac{\cos 4x}{15} + \frac{\cos 6x}{35} + \cdots\right]. \quad (13.15)$$

The goodness of this representation is shown in Figure 13.6, where it can be seen that convergence is very rapid.

A second example is the function shown in Figure 13.7. This is discontinuous and coincides with the $[0, 1]$ step function in the range $-\pi \leq x \leq \pi$, i.e.

$$f(x) = \begin{cases} 0 & -\pi \leq x \leq 0 \\ 1 & 0 < x \leq \pi \end{cases}. \quad (13.16)$$

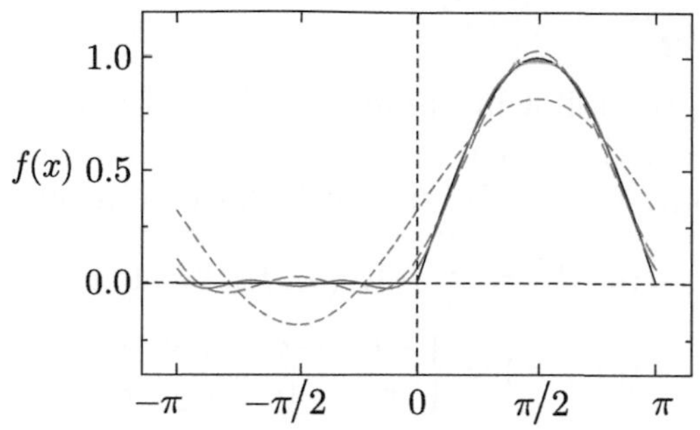

Figure 13.6 The function (13.14) (solid black line), together with its Fourier representations derived from (13.15). The short-dashed blue line is the constant plus the sine term; the long-dashed blue line also includes the first cosine term; and the solid blue line includes the first two cosine terms.

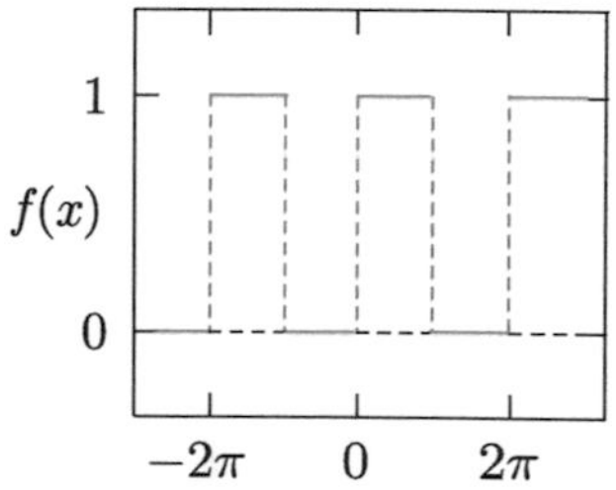

Figure 13.7 The function $f(x)$ given by (13.16).

The coefficients of its Fourier expansion are found in an analogous way to those for the previous example, and without further detail, the Fourier series is

$$f(x) = \frac{1}{2} + \frac{2}{\pi}\sum_{n=1}^{\infty}\frac{\sin(2n-1)x}{2n-1}$$
$$= \frac{1}{2} + \frac{2}{\pi}\left[\sin x + \frac{1}{3}\sin 3x + \frac{1}{5}\sin 5x + \cdots\right]. \quad (13.17)$$

Figure 13.8 shows the comparison of the Fourier approximation (13.17) with the exact function (13.16), for 4 and 11 terms in the series. Again the convergence is rapid and tends towards the mid-point of the jump discontinuity at $x = 0$ in accordance with the Dirichlet theorem.

An interesting point about Fourier series for discontinuous functions is that near to a discontinuity, the series will overshoot the true value of the function being represented, the size of the overshoot being proportional to the magnitude of the discontinuity. This behaviour is known as the *Gibbs phenomenon*, and is clearly seen in Figure 13.8. Although the overshoot moves closer to the discontinuity as the number of terms is increased, it does not vanish even with an infinite number of terms.[2]

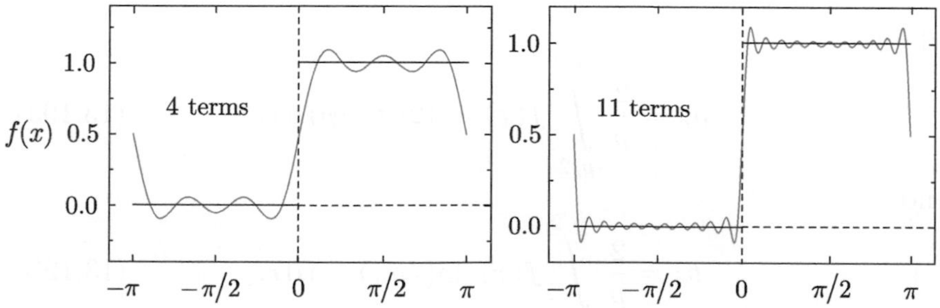

Figure 13.8 The function (13.16) (black lines), together with its Fourier representations (coloured lines) derived from (13.17) with 4 and 11 terms (including the constant), respectively.

[2] See, for example, G.B. Arfken and H.J. Weber (2005) *Mathematical Methods for Physicists*, 6th edn., Academic Press, San Diego, California, Section 14.5

Example 13.3

State whether the following functions $f(x)$ of periodicity 2π, and defined below in the range $-\pi < x < \pi$, are guaranteed to have convergent Fourier series by the Dirichlet conditions.

$$(a)\ f^2 = x^2 + 1,\ (b)\ f(x) = 1/x,$$

$$(c)\ f(x) = \begin{cases} -x & -\pi \le x < 0 \\ x & 0 < x \le \pi \end{cases}.$$

Solution

(a) Does not satisfy the Dirichlet conditions because $f(x)$ is not single-valued.

(b) Does not satisfy the Dirichlet conditions because the integral of $1/x$ between $-\pi$ to π is infinite.

(c) The function is shown in Figure 13.3. It is single-valued and continuous, with a finite number of maxima in the specified range, and in addition the integral (13.13) is finite. It therefore satisfies the Dirichlet conditions.

13.1.3 Change of period

In the above discussion we have assumed that $f(x)$ is periodic with a period $p = 2\pi$, but this can easily be generalised to an arbitrary period p, by using the trigonometric functions

$$\sin(2n\pi x/p) \quad \text{and} \quad \cos(2n\pi x/p),$$

which also have period p. The Fourier expansion then becomes

$$f(x) = \frac{a_0}{2} + \sum_{n=1}^{\infty} a_n \cos(2n\pi x/p) + \sum_{n=1}^{\infty} b_n \sin(2n\pi x/p), \tag{13.18}$$

where

$$a_n = \frac{2}{p} \int_{-p/2}^{p/2} f(x) \cos(2n\pi x/p)\, \mathrm{d}x, \tag{13.19a}$$

and

$$b_n = \frac{2}{p} \int_{-p/2}^{p/2} f(x) \sin(2n\pi x/p)\, \mathrm{d}x. \tag{13.19b}$$

In (13.19) the integration range is symmetric, over an interval $-p/2 < x < p/2$. However, for any periodic function $f(x)$, the integral over a single period $x_0 < x < x_0 + p$ is independent of x_0, since

$$\int_{x_0}^{x_0+p} f(x)\mathrm{d}x = \int_{-p/2}^{p/2} f(x)\mathrm{d}x + \int_{p/2}^{x_0+p} f(x)\mathrm{d}x - \int_{-p/2}^{x_0} f(x)\mathrm{d}x \qquad (13.20)$$

and the last two integrals cancel because $f(x+p) = f(x)$. Hence (13.19) can be replaced by the more general forms

$$a_n = \frac{2}{p}\int_{x_0}^{x_0+p} f(x)\cos(2n\pi x/p)\,dx, \qquad (13.21a)$$

$$b_n = \frac{2}{p}\int_{x_0}^{x_0+p} f(x)\sin(2n\pi x/p)\,dx, \qquad (13.21b)$$

where x_0 can be any convenient value.

13.1.4 Non-periodic functions

Up to now we have assumed that the function is periodic. If we want to find the Fourier decomposition of a non-periodic function only within a fixed range, then we may continue the function outside this range to make it periodic. The Fourier series of the resulting periodic function will represent the non-periodic function in the desired range. Since the function can often be continued in a number of different ways, it is convenient to make the extension so that the resulting function has a definite symmetry.

For example, consider the non-periodic function shown in Figure 13.9 and defined for $0 < x < L$. Three ways of continuing this outside the interval $0 < x < L$ are shown in Figure 13.9. In (b) there is no definite symmetry. The coefficients are given by (13.21) with $x_0 = 0$ and $p = L$, and both a_n and b_n are in general non-zero in the Fourier expansion (13.18). In (c) there is odd symmetry about $x = 0$. The coefficients are now given by (13.19) with period $p = 2L$ and the a_n will vanish by symmetry leaving a sine series of the form (13.11b); whereas in (d) there is even symmetry about $x = 0$ leading to a cosine series of the form (13.11a). Using either of (c) or (d) therefore reduces the number of integrals to be calculated and leads to a simpler series. Other than this, provided the series is only used to represent the function in the range $0 < x < L$, the choice of which continuation to use is unimportant.

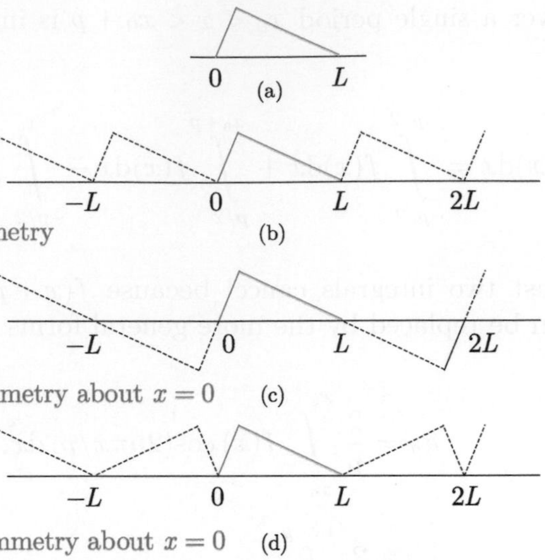

Figure 13.9 Possible continuations of the non-periodic function shown in (a). These have the following symmetries about $x = 0$: (b) none, (c) odd and (d) even.

Example 13.4

Consider the function $f(x) = x^2$ for $0 \le x \le 2$. Continue $f(x)$ outside this interval and hence find a Fourier cosine series that represents $f(x)$ in the range $0 \le x \le 2$.

Solution

If the resulting series is to contain only cosines, the continued $f(x)$ must be symmetric about $x = 0$. The given function is shown in Figure 13.10, together with an extension to all x values such that $f(x) = f(-x)$, giving an even function of period 4. The Fourier series of the continued function will then correctly represent the original function in the interval $0 < x < 2$ in which it is defined.

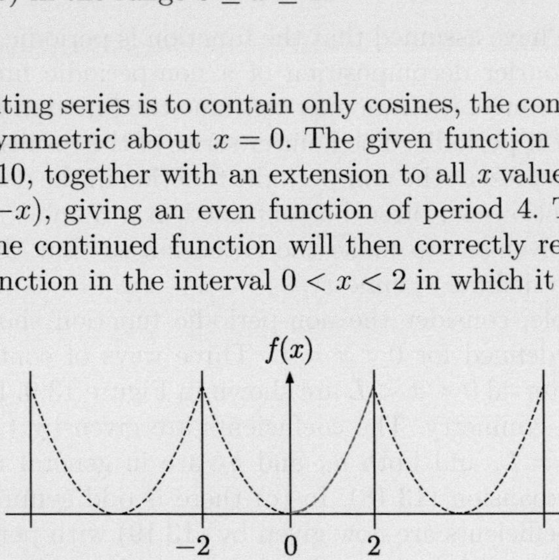

Figure 13.10 The function $f(x) = x^2$ for $0 \le x \le 2$, together with an extension to the interval $-2 \le x \le 2$, so that the resulting function is even about $x = 0$.

It remains to calculate the coefficients. Because the extended function is even, $b_n = 0$ and (13.19) gives

$$a_n = \frac{2}{4}\int_{-2}^{2} x^2 \cos\left(\frac{2\pi n x}{4}\right) dx = \int_{0}^{2} x^2 \cos\left(\frac{\pi n x}{2}\right) dx.$$

Integrating by parts for $n \neq 0$ gives

$$a_n = \frac{2}{\pi n} \int_0^2 x^2 \frac{\mathrm{d}}{\mathrm{d}x} \sin\left(\frac{\pi n x}{2}\right) \mathrm{d}x$$

$$= \left[\frac{2}{\pi n} x^2 \sin\left(\frac{\pi n x}{2}\right)\right]_0^2 - \frac{4}{\pi n} \int_0^2 x \sin\left(\frac{\pi n x}{2}\right) \mathrm{d}x.$$

The first term vanishes, and integrating by parts a second time gives

$$a_n = \frac{8}{\pi^2 n^2} \left[x \cos\left(\frac{\pi n x}{2}\right)\right]_0^2 - \frac{8}{\pi^2 n^2} \int_0^2 \cos\left(\frac{\pi n x}{2}\right) \mathrm{d}x$$

$$= \frac{16}{\pi^2 n^2} \cos(n\pi) = \frac{16}{\pi^2 n^2} (-1)^n.$$

For $n = 0$,

$$a_0 = \int_0^2 x^2 \, \mathrm{d}x = \frac{8}{3},$$

so the final expression is

$$x^2 = \frac{4}{3} + 16 \sum_{n=1}^{\infty} \frac{(-1)^n}{\pi^2 n^2} \cos\left(\frac{\pi n x}{2}\right), \quad 0 \leq x \leq 2.$$

13.1.5 Integration and differentiation of Fourier series

In Section 5.4.2, we saw that the Taylor expansion of a given function $f(x)$ can be integrated or differentiated term by term to give a new series that converges to the integral or derivative of the original function. Thus, for example, the power series expansion for $\cos x$ given in Table 5.1 can be obtained by differentiating the corresponding series for $\sin x$, assuming that this series has already been obtained.

Similarly, integrating a well-defined Fourier expansion for a function $f(x)$ yields a series that converges to the integral of $f(x)$. For simplicity, consider a function of period 2π that has a given functional form $f(x)$ in the range $-\pi < x < \pi$. Then integrating (13.9), one obtains

$$\int f(x)\mathrm{d}x = c + \frac{a_0 x}{2} + \sum_{n=1}^{\infty} \frac{1}{n}[a_n \sin(nx) - b_n \cos(nx)], \qquad (13.22)$$

for the indefinite integral in the range $-\pi < x < \pi$, where c is an integration constant that can be evaluated by equating the right-hand

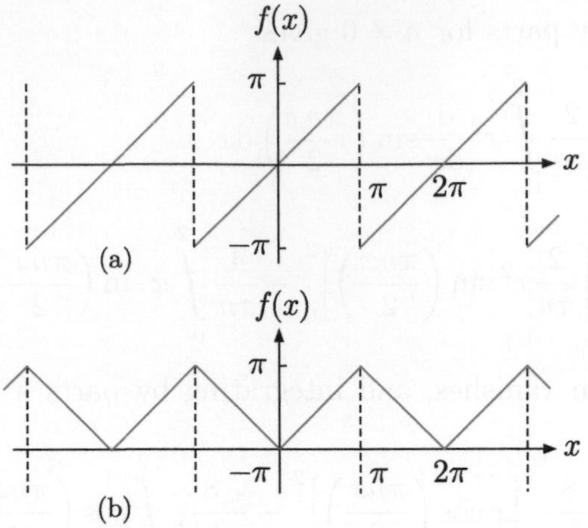

Figure 13.11 The functions defined by (a) Eq. (13.24) and (b) Eq. (13.25).

and left-hand sides of (13.22) at any value in the range $-\pi < x < \pi$. However, because of the second term in (13.22), this convergent series only reduces to a Fourier series if the mean value vanishes, i.e.

$$\frac{a_0}{2} = \frac{1}{2\pi} \int_{-\pi}^{\pi} f(x) \, dx = 0. \tag{13.23}$$

The differentiation of Fourier series is more problematic. Consider, for example, the function

$$f(x) = x, \; -\pi < x < \pi; \; f(x + 2\pi) = f(x), \tag{13.24}$$

shown in Figure 13.11a. Since $f(x) = -f(-x)$, it has a sine series, with coefficients

$$b_n = \frac{1}{\pi} \int_{-\pi}^{\pi} x \sin(nx) \, dx = -\frac{2(-1)^n}{n},$$

on integrating by parts. The Fourier series is therefore

$$x = -2 \sum_{n=1}^{\infty} \frac{(-1)^n}{n} \sin nx = 2 \left[\sin x - \frac{\sin 2x}{2} + \frac{\sin 3x}{3} - \cdots \right],$$

where we have assumed $-\pi < x < \pi$ on the left-hand side. Differentiating this gives

$$1 = -2 \sum_{n=1}^{\infty} \cos nx = -2(\cos x - \cos 2x + \cos 3x - \cdots),$$

which is a meaningless result because the series on the right-hand side diverges by (5.15), since $\cos nx \nrightarrow 0$ as $n \to \infty$. On the other hand, if we consider the function

$$f(x) = |x|, \quad -\pi < x < \pi; \quad f(x + 2\pi) = f(x) \qquad (13.25)$$

shown in Figure 13.11b. The corresponding Fourier series can be differentiated without difficulty, as shown in Example 13.5 below.

Clearly, a criterion is needed that tells us whether the expansion of a given function can be differentiated or not. It is provided by the following theorem:

If $f(x)$ is periodic and continuous for all values of x, then at each point for which $f(x)$ is continuous, the Fourier series for $f(x)$ may be differentiated term by term and the resultant Fourier series will converge to $f'(x)$, provided the latter satisfies the Dirichlet conditions.

(13.26)

From Figure 13.11, we see that the function defined in (13.24) fails these conditions because $f(x)$ is discontinuous at $x = n\pi$, whereas the function defined in (13.25) is continuous at these points.

It remains to prove this theorem. In doing so, we shall initially assume for simplicity that the function has period 2π. Then, since $f'(x)$ satisfies the Dirichlet conditions, it has a Fourier expansion

$$f'(x) = \tfrac{1}{2}A_0 + \sum_{n=0}^{\infty} A_n \cos nx + \sum_{n=1}^{\infty} B_n \sin nx, \qquad (13.27)$$

where

$$A_n = \frac{1}{\pi}\int_{-\pi}^{\pi} f'(x)\cos nx\,\mathrm{d}x \quad \text{and} \quad B_n = \frac{1}{\pi}\int_{-\pi}^{\pi} f'(x)\sin nx\,\mathrm{d}x.$$

(13.28)

Furthermore, by Dirichlet's theorem, this converges to $f'(x)$ at all points where $f'(x)$ is continuous. Hence, to establish the theorem, we just have to show that (13.27) is identical to the series

$$f'(x) = \sum_{n=1}^{\infty}(-na_n \sin nx) + \sum_{n-1}^{\infty}(nb_n \cos nx), \qquad (13.29)$$

obtained by differentiating the Fourier series (13.9) for $f(x)$ itself. To do this, we integrate (13.28) by parts to obtain

$$A_n = \frac{1}{\pi}[f(x)\cos nx]_{-\pi}^{\pi} + \frac{n}{\pi}\int_{-\pi}^{\pi} f(x)\sin(x)\mathrm{d}x$$

$$= \frac{1}{\pi}[f(\pi) - f(-\pi)]\cos n\pi + nb_n = nb_n, \qquad (13.30a)$$

using (13.10) together with $f(\pi) = f(-\pi)$ by periodicity. Similarly,

$$B_n = \frac{1}{\pi}[f(\pi) - f(-\pi)] \sin n\pi - na_n = -na_n. \quad (13.30b)$$

Then substituting (13.30) into (13.27) gives (13.29) as required.

This proof is easily extended to an arbitrary period p. The essential point is that, provided the conditions (13.26) on $f(x)$ and $f'(x)$ are satisfied, the Fourier series can be safely differentiated away from discontinuities, where $f(x)$ is ill-defined.

Example 13.5

The Fourier series for (13.25) is given in Example 13.2. (a) Verify that differentiating it leads to the correct Fourier series for

$$F(x) = \begin{cases} -1 & -\pi < x < 0 \\ 1 & 0 < x < \pi \end{cases}$$

with $F(x + 2\pi) = F(x)$ elsewhere. (b) What value does the series assign to $F(x)$ at $x = n\pi$?

Solution

(a) The function $F(x) = f'(x)$, where $f(x)$ is defined in (13.25). Differentiating the Fourier series for $f(x)$ given in Example 13.2 gives

$$F(x) = \frac{4}{\pi}\left(\sin x + \frac{\sin 3x}{3} + \frac{\sin 5x}{5} + \cdots\right).$$

Evaluating the Fourier series directly gives a sine series, since $F(x) = -F(-x)$, with

$$b_n = \frac{1}{\pi}\int_{-\pi}^{\pi} F(x) \sin nx \,\mathrm{d}x = \frac{2}{\pi}\int_{0}^{\pi} \sin nx \,\mathrm{d}x$$

$$= -\frac{2}{n\pi}[\cos nx]_0^\pi = -\frac{2}{n\pi}[(-1)^n - 1].$$

Hence

$$F(x) = \frac{4}{\pi}\left(\sin x + \frac{\sin 3x}{3} + \frac{\sin 5x}{5} + \cdots\right),$$

which is the same result as obtained above by differentiating the series for $f(x)$.

(b) At $x = n\pi$, $F(x)$ is discontinuous, jumping from –1 to +1 for even n and from +1 to –1 for odd n. Dirichlet's theorem therefore says that the series gives $F(x) = 0$, which is the case since $\sin k\pi = 0$ for any integer k.

Example 13.6

In $-\pi < x < \pi$, the function $f(x) = x^2$ may be expanded as a Fourier series

$$x^2 = \frac{\pi^2}{3} - 4\left[\cos x - \frac{1}{4}\cos 2x + \cdots + \frac{(-1)^{n+1}}{n^2}\cos nx + \cdots\right].$$

Show, by integrating this series, that

$$\pi^3 = 32\sum_{k=0}^{\infty}\frac{(-1)^k}{(2k+1)^3}.$$

Solution

Integrating the series for $f(x)$ term by term gives

$$\frac{x^3}{3} = c + \frac{\pi^2 x}{3} - 4\left[\sin x - \frac{1}{8}\sin 2x + \cdots + \frac{(-1)^{n+1}\sin nx}{n^3} + \cdots\right],$$

where c is an integration constant. Setting $x = 0$ gives $c = 0$ and then setting $x = \pi/2$ gives

$$\frac{\pi^3}{32} = \sum_{n=1}^{\infty}(-1)^{n+1}\frac{\sin n\pi/2}{n^3}.$$

For even $n = 2k$, $\sin n\pi/2 = 0$, and for odd $n = 2k+1$, $\sin n\pi/2 = (-1)^k$, so that

$$\pi^3 = 32\sum_{k=0}^{\infty}\frac{(-1)^k}{(2k+1)^3}$$

as required.

13.1.6 Mean values and Parseval's theorem

If $f(x)$ is a function periodic in some interval, then *Parseval's theorem* relates the average, or mean, value of $[f(x)]^2$ over the interval, to the coefficients of the Fourier expansion of $f(x)$ in the interval. From (13.19a), we see that the mean value $\bar{f}$ of $f(x)$ over a single period is related to the first Fourier coefficient a_0. Explicitly,

$$\bar{f} \equiv \frac{1}{p}\int_{-p/2}^{p/2} f(x)\mathrm{d}x = \frac{a_0}{2}. \tag{13.31}$$

In addition, the average value of the squared function $f^2(x)$ can be related to the Fourier coefficients that occur in the expansion of $f(x)$. If

$$f(x) = \frac{a_0}{2} + \sum_{n=1}^{\infty} a_n\cos(2n\pi x/p) + \sum_{n=1}^{\infty} b_n\sin(2n\pi x/p),$$

then

$$\int [f(x)]^2 \mathrm{d}x = \frac{1}{2}a_0 \int f(x)\,\mathrm{d}x + \int \sum_{n=1}^{\infty} a_n \cos(2\pi nx/p) f(x)\,\mathrm{d}x$$
$$+ \int \sum_{n=1}^{\infty} b_n \sin(2\pi nx/p) f(x)\,\mathrm{d}x$$
$$= \frac{a_0^2 p}{4} + \sum_{n=1}^{\infty} a_n(pa_n/2) + \sum_{n=1}^{\infty} b_n(pb_n/2),$$

where the integrals extend from $-p/2$ to $p/2$ and we have used (13.19a) and (13.19b). Hence

$$\overline{f^2} \equiv \frac{1}{p}\int_{-p/2}^{p/2} [f(x)]^2\,\mathrm{d}x = \left(\frac{a_0}{2}\right)^2 + \frac{1}{2}\sum_{n=1}^{\infty}\left(a_n^2 + b_n^2\right). \quad (13.32)$$

This is Parseval's theorem.

Example 13.7

Use Parseval's theorem and the Fourier series for $f(x) = x^2$ in the interval $-2 \le x \le 2$ deduced in Example 13.4, to calculate the sum

$$S = \sum_{n=1}^{\infty} \frac{1}{n^4}.$$

Solution

Firstly, we find the average value of $f^2(x)$ over the interval $-2 \le x \le 2$. This is

$$\frac{1}{4}\int_{-2}^{2} x^4\,\mathrm{d}x = \left[\frac{x^5}{20}\right]_{-2}^{2} = \frac{16}{5}.$$

By Parseval's theorem this is equal to

$$\frac{1}{2}\left[\frac{a_0^2}{2} + \sum_{n=1}^{\infty}\left(a_n^2 + b_n^2\right)\right],$$

where we have already deduced from Example 13.4 that for $-2 \le x \le 2$,

$$a_0 = \frac{8}{3}, \quad a_n(n \neq 0) = \frac{16(-1)^n}{n^2\pi^2}, \quad b_n = 0 \text{ (all } n).$$

Substituting these values and rearranging then gives

$$\sum_{n=1}^{\infty} \frac{1}{n^4} = \frac{\pi^4}{90}.$$

13.2 Complex Fourier series

In this section, we show how to rewrite Fourier series in a complex form. This is simpler in some ways than the original series, and serves as the foundation of the discussion of Fourier transforms that follows. We also use it to point out the strong analogy between Fourier expansions and the expansion (9.18) of a vector in terms of a set of orthogonal basis vectors.

Using Euler's formula (6.18),

$$e^{i\theta} = \cos\theta + i\sin\theta, \tag{13.33}$$

one easily shows that the *complex Fourier series*

$$f(x) = \sum_{-\infty}^{\infty} c_n \exp(i2n\pi x/p) = c_0 + \sum_{n=1}^{\infty} [c_n e^{i2n\pi x/p} + c_{-n} e^{-i2n\pi x/p}] \tag{13.34}$$

is identical to the Fourier series (13.18), provided that

$$\begin{aligned} c_0 &= a_0/2, \\ c_n &= (a_n - ib_n)/2, \quad n > 0, \\ c_{-n} &= (a_n + ib_n)/2, \quad n > 0. \end{aligned} \tag{13.35}$$

Together with (13.19a) and (13.19b), these equations imply

$$c_n = \frac{1}{p} \int_{-p/2}^{p/2} f(x)\, e^{-i2\pi nx/p}\, \mathrm{d}x, \tag{13.36}$$

for all integer n, irrespective of sign. Equations (13.35) also allow Parseval's theorem (13.32) to be expressed in terms of the coefficients c_n, giving

$$\overline{f^2} = \frac{1}{p} \int_{-p/2}^{p/2} f^2(x)\, \mathrm{d}x = \sum_{n=-\infty}^{\infty} |c_n|^2. \tag{13.37}$$

In the above discussion, we have stressed the equivalence of the complex Fourier series (13.34) to the real Fourier series (13.18) and obtained key results, such as (13.36) and (13.37), by relating the complex coefficients c_n to the real coefficients a_n, b_n. However, it is important to note that they can be obtained directly from (13.34), without reference to the original series. In doing this, it will be instructive to introduce basis functions

$$f_n(x) = \frac{1}{\sqrt{p}} \exp\left(\frac{i2n\pi x}{p}\right), \qquad (13.38a)$$

which are easily shown to satisfy the *orthogonality relations*

$$\int_{-p/2}^{p/2} f_n(x) f_m^*(x)\, \mathrm{d}x = \delta_{nm}, \qquad (13.38b)$$

where δ_{nm} is the Kronecker delta symbol (9.24b). The complex Fourier series (13.34) then becomes

$$f(x) = \sum_{-\infty}^{\infty} c_n \sqrt{p}\, f_n(x) \qquad (13.39a)$$

and on multiplying by $f_m^*(x)$ and integrating using (13.38b), one obtains

$$c_m \sqrt{p} = \int_{-p/2}^{p/2} f(x)\, f_m^*(x)\, \mathrm{d}x \qquad (13.39b)$$

for any m, which is identical to (13.36) as required. Parseval's theorem in the form (13.37) can also be derived directly from the complex Fourier series (13.34) by using (13.38b) in a similar way.

Example 13.8

Find a complex Fourier series for $f(x) = x$ in the interval $-2 \le x \le 2$.

Solution

From (13.36), the coefficients of the expansion are

$$c_n = \frac{1}{4} \int_{-2}^{2} x \exp(-i\pi n x/2)\, \mathrm{d}x.$$

Integrating by parts gives

$$c_n = \frac{i}{2\pi n} \int_{-2}^{2} x \frac{\mathrm{d}}{\mathrm{d}x}[\exp(-i\pi n x/2)]\, \mathrm{d}x,$$

which is

$$c_n = \left[-\frac{x}{2i\pi n}\exp(-i\pi nx/2)\right]_{-2}^{2} + \frac{1}{2i\pi n}\int_{-2}^{2}\exp(-i\pi nx/2)\,dx$$

$$= \frac{2i}{n\pi}\cos(n\pi) + \frac{2i}{n^2\pi^2}\sin(n\pi) = \frac{2i}{n\pi}(-1)^n.$$

So,

$$x = \sum_{n=-\infty}^{\infty}\frac{2i(-1)^n}{n\pi}\exp\left(\frac{i\pi nx}{2}\right)$$

*13.2.1 Fourier expansions and vector spaces

The discussion of complex Fourier series also highlights the similarities between the Fourier expansion (13.39a) and the expansion (9.24) of an n-dimensional vector $\mathbf{a}$ in terms of orthonormal basis vectors $\mathbf{e}_i$, discussed in Section 9.2. On comparing (13.39a) and (9.24), one sees that $f(x)$ is the analogue of the vector $\mathbf{a}$, the $f_n(x)$ are the analogues of the basis vectors $\mathbf{e}_i$, while the integral on the left-hand side of (13.38b) plays the same role in the discussion as the scalar product $(\mathbf{e}_i, \mathbf{e}_j)$. To emphasise this, we define the scalar product of any two periodic functions $f(x)$ and $g(x)$ to be

$$(f, g) \equiv \int_{-p/2}^{p/2} f^*(x)g(x)\,dx. \tag{13.40}$$

With this notation, (9.38b) and (9.39b) take the form

$$(f_m,\ f_n) = \delta_{mn} \tag{13.41a}$$

and

$$c_m\sqrt{p} = (f_m,\ f), \tag{13.41b}$$

corresponding to [cf. (9.24) and (9.25)]

$$(\mathbf{e}_m,\ \mathbf{e}_n) = \delta_{mn}$$

and

$$a_m = (\mathbf{e}_m,\ \mathbf{a}),$$

where we have relabelled the indices i, j to m, n in the vector case, Furthermore, if we substitute (13.39a) into (13.40) with $g = f$, and use the orthogonality relation (13.41a), we obtain

$$(f, f) = \sum_m\sum_n c_m^* c_m\, p(f_n, f_n) = \sum_n p|c_n|^2, \tag{13.41c}$$

compared to

$$(\mathbf{a},\ \mathbf{a}) = \sum_n |a_n|^2$$

in the vector case. At this point, we note that there is no need to restrict $f(x)$ to real functions in (13.39), but if we do $c_{-n} = c_n^*$ and (13.41c) reduces to Parseval's theorem (13.37) for real $f(x)$. On the other hand, if we do not restrict ourselves to real $f(x)$, the coefficients c_n and c_{-n} are in general independent, and Parseval's theorem becomes

$$\overline{f^*f} = \frac{1}{p}\int_{-p/2}^{p/2} |f(x)|^2\,\mathrm{d}x = \sum_{n=-\infty}^{\infty} |c_n|^2.$$

This discussion shows that the analogy between the expansion of a vector in terms of a complete set of basis vectors and the expansion of a periodic function in terms of its Fourier components is very close, and one can introduce a more general formalism that includes both.[3] We shall not pursue this further here, except to note that the expansion of functions in terms of sets of basis functions is not confined to periodic functions. Two further examples will occur later, in Sections 15.3.1 and 15.4.2.

13.3 Fourier transforms

Fourier series are widely used in the study of waves because they allow complicated waveforms to be decomposed into a sum of *harmonic waves*, that is, simple sines or cosines. However, many situations occur in practice where waves are localised in space and time. For example, at a given time, the waveform could be described by the function

$$f(x) = \exp(-x^2/2a^2)e^{iqx}, \quad a \gg q^{-1} \tag{13.42}$$

whose real part is shown in Figure 13.12 for the case $a = 10$, $q = 1$. When they move through space, such localised waveforms are called *wave packets* and fail to satisfy the periodicity condition

$$f(x+p) = f(x), \tag{13.43}$$

required to formulate a Fourier series for any finite period p. This is one reason, among others, why it is useful to take the limit $p \to \infty$ in the complex Fourier series (13.34), so that a single period covers the whole x-axis $-\infty < x < \infty$.

[3] It is called the theory of linear vector spaces. See, for example, P. Dennery and A. Krzywicki (1966) *Mathematics for Physicists*, Dover Publications, New York, Chapters 2 and 3.

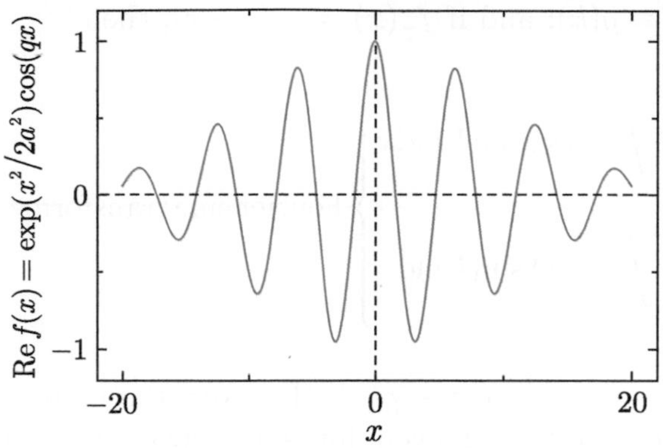

Figure 13.12 The real part of the wave form described by the function (13.42).

We will show below that taking this limit leads to the expansion

$$f(x) = \int_{-\infty}^{\infty} g(k)e^{ikx}\mathrm{d}k, \tag{13.44a}$$

where $g(k)$ is given by

$$g(k) = \frac{1}{2\pi}\int_{-\infty}^{\infty} f(x)e^{-ikx}\mathrm{d}x. \tag{13.44b}$$

The relation (13.44b) is called a *Fourier transform* of $f(x)$ and (13.44a) is called the *inverse Fourier transform* of $g(k)$. The latter is a generalisation of the complex Fourier series (13.34) and allows a non-periodic function $f(x)$ to be decomposed into a continuous sum of harmonic waves of definite wave number $k \equiv 2\pi/\lambda$, where λ is the wavelength. This distribution, which is characterised by $g(k)$, is called the spectrum of $f(x)$. It converges provided that the integral

$$\int_{-\infty}^{\infty} |f(x)|\,\mathrm{d}x \tag{13.45}$$

converges and that $f(x)$ is continuous except for finite (or jump) discontinuities, where, as for the Fourier series, the expansion (13.44) converges to the value of $f(x)$ at the mid-point of the discontinuity.

In the case of functions with definite symmetry, either even or odd, the Fourier transforms can be written as *Fourier cosine* and *Fourier sine transforms*, respectively. Explicitly, if $f_+(x) = f_+(-x)$, then

$$\left.\begin{aligned} f_+(x) &= 2\int_0^{\infty} g_+(k)\cos(kx)\mathrm{d}k \\ g_+(k) &= \frac{1}{\pi}\int_0^{\infty} f_+(x)\cos(kx)\,\mathrm{d}x \end{aligned}\right\}\ \text{Fourier cosine transforms,} \tag{13.46a}$$

where $g_+(k) \equiv g(k)$; and if $f_-(x) = -f_-(-x)$, then

$$\left.\begin{aligned} f_-(x) &= 2\int_0^\infty g_-(k)\sin(kx)\mathrm{d}k \\ g_-(k) &= \frac{1}{\pi}\int_0^\infty f_-(x)\sin(kx)\mathrm{d}x \end{aligned}\right\}\text{Fourier sine transforms,} \tag{13.46b}$$

where in this case $g_-(k) \equiv -ig(k)$. However, the more general form (13.44) is often retained even for symmetric and antisymmetric functions.

It remains to obtain the key relations (13.44) by taking the limit $p \to \infty$ in the complex Fourier series (13.34). To do this, we define variables

$$k_n \equiv 2\pi n/p,$$

with spacing

$$\Delta k_n \equiv k_{n+1} - k_n = 2\pi/p,$$

which tends to zero as $p \to \infty$. In terms of these, the Fourier expansion (13.34) becomes

$$f(x) = \sum_{n=-\infty}^{\infty} c_n e^{ik_n x}, \tag{13.47a}$$

where

$$c_n = \frac{\Delta k_n}{2\pi}\int_{-p/2}^{p/2} f(z)e^{-ik_n z}\mathrm{d}z, \tag{13.47b}$$

and the dummy variable x in (13.47b) has been relabelled z for later convenience. Substituting (13.47b) into (13.47a) then gives

$$f(x) = \frac{1}{2\pi}\sum_{-\infty}^{\infty} F(k_n)\Delta k_n, \tag{13.48a}$$

where

$$F(k_n) = \int_{-p/2}^{p/2} f(z)e^{ik_n(x-z)}\mathrm{d}z. \tag{13.48b}$$

Comparing (13.48a) with the Riemann definition of an integral (4.15), and recalling that $\Delta k_n \to 0$ as $p \to \infty$, then gives

$$\lim_{\Delta k_n \to 0}\left[\frac{1}{2\pi}\sum_n F(k_n)\Delta k_n\right] = \frac{1}{2\pi}\int_{-\infty}^{\infty} F(k)\mathrm{d}k,$$

where k is now a continuous variable, and since $p \to \infty$,

$$F(k) = \int_{-\infty}^{\infty} f(z) e^{ik(x-z)} \mathrm{d}z.$$

Hence (13.48a) becomes

$$f(x) = \frac{1}{2\pi} \int_{-\infty}^{\infty} F(k) \mathrm{d}k$$

$$= \frac{1}{2\pi} \int_{-\infty}^{\infty} \mathrm{d}k \int_{-\infty}^{\infty} f(z) e^{ik(x-z)} \mathrm{d}z = \int_{-\infty}^{\infty} g(k) e^{ikx}\, \mathrm{d}k,$$

where

$$g(k) = \frac{1}{2\pi} \int_{-\infty}^{\infty} f(z) e^{\ ikz} \mathrm{d}z = \frac{1}{2\pi} \int_{-\infty}^{\infty} f(x) e^{-ikx} \mathrm{d}x.$$

These are identical to the relations (13.44a) and (13.44b).

Finally, it is worth noting that some authors use different definitions to those above. Thus the Fourier transform is sometimes defined as

$$\tilde{g}(k) = (2\pi)^{1/2} g(k) \equiv \frac{1}{\sqrt{2\pi}} \int_{-\infty}^{\infty} f(x) e^{-ikx} \mathrm{d}x,$$

so that

$$f(x) = \frac{1}{\sqrt{2\pi}} \int_{-\infty}^{\infty} \tilde{g}(k)\, e^{ikx} \mathrm{d}k,$$

and the external factors in the Fourier and inverse Fourier transforms are then the same. The only requirement is that the product of both external factors must $1/2\pi$. We will use convention (13.44) throughout.

The rest of this section will be devoted to a discussion of the properties of Fourier transforms (13.44b) and the corresponding Fourier expansion (13.44a).[4]

Example 13.9

Find the Fourier transforms of the functions

$$\text{(a) } f(x) = \exp(-|x|) \text{ and (b) } f(x) = \begin{cases} \pi & \text{for } |x| < 4 \\ 0 & \text{for } |x| \geq 4 \end{cases}$$

[4]Extensive tables of Fourier transforms may be found in books of mathematical formulas, for example, Alan Jeffreys and Hui-Hui Dai (2008) *Handbook of Mathematical Formulas and Integrals*, 4th edn., Academic Press, New York, pp. 356–362.

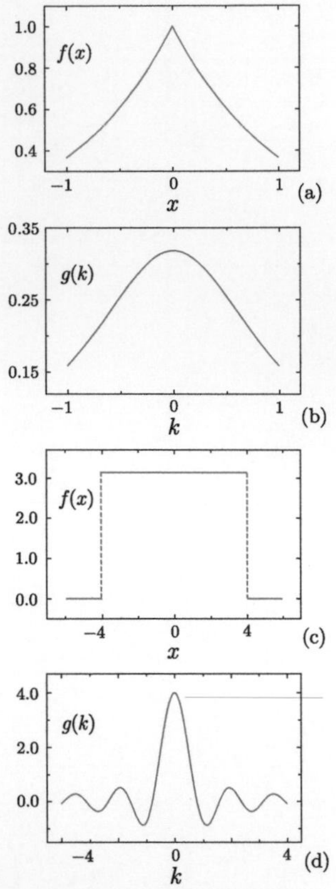

Figure 13.13 The functions: (a) $f(x) = \exp(-|x|)$ and (b) its Fourier transform $1/[\pi(1+k^2)]$; and (c) the function $f(x) = \pi$ ($|x| < 4$), 0 ($|x| \geq 4$) and (d) its Fourier transform $\sin(4k)/k$.

Solution

(a) The Fourier transform is

$$g(k) = \frac{1}{2\pi}\int_{-\infty}^{\infty} e^{-|x|}e^{-ixk}\mathrm{d}x$$
$$= \frac{1}{2\pi}\int_{-\infty}^{0} e^{-x(ik-1)}\mathrm{d}x + \frac{1}{2\pi}\int_{0}^{\infty} e^{-x(ik+1)}\mathrm{d}x$$
$$= \frac{1}{2\pi}\left(\frac{1}{1-ik} + \frac{1}{1+ik}\right) = \frac{1}{\pi}\frac{1}{(1+k^2)}.$$

The functions $f(x)$ and $g(k)$ are shown in Figure 13.13a and 13.13b, respectively.

(b) The Fourier transform is

$$g(k) = \frac{1}{2\pi}\int_{-4}^{4} \pi e^{-ikx}\mathrm{d}x = \frac{i}{2k}[e^{-ikx}]_{-4}^{4}$$
$$= \frac{i}{2k}(e^{-4ik} - e^{4ik}) = 4\mathrm{sinc}(4k),$$

where $\mathrm{sinc}\, x \equiv (\sin x)/x$, as defined in Eq. (5.63). The functions $f(x)$ and $g(k)$ are shown in Figure 13.13c and 13.13d, respectively.

13.3.1 Properties of Fourier transforms

In this section we will introduce some general properties of Fourier transforms. In doing so, we shall use the abbreviation $F[f(x)]$ to mean the Fourier transform of the function $f(x)$, so that (13.44b) becomes

$$F[f(x)] = g(k) \tag{13.49a}$$

and the abbreviation $F^{-1}[g(k)]$ to mean the inverse Fourier transform, so that (13.44a) becomes

$$F^{-1}[g(k)] = f(x). \tag{13.49b}$$

The operators F and F^{-1} are linear, so for example

$$F[a_1 f_1(x) + a_2 f_2(x)] = a_1 F[f_1(x)] + a_2 F[f_2(x)] \tag{13.49c}$$

and $FF^{-1} = 1$. Similarly, if in the inverse Fourier transform (13.44a) we re-label the variables k, x *as* x, $-k$, it becomes

$$f(-k) = \int_{-\infty}^{\infty} g(x) e^{-ikx} \mathrm{d}x,$$

which, on dividing by 2π, becomes

$$F\,[g(x)] = \frac{1}{2\pi} f(-k). \tag{13.50}$$

Equation (13.50) is just one of a number of relations which, given the Fourier transform of a function $f(x)$, enable the Fourier transforms of related functions – in this case $g(x)$, where $g(k)$ is the Fourier transform of $f(x)$ – to be written down immediately, without further calculation. For example, since

$$\int_{-\infty}^{\infty} f^*(x)\, e^{-ikx} \mathrm{d}x = \left[\int_{-\infty}^{\infty} f(x)\, e^{ikx} \mathrm{d}x \right]^*,$$

it follows from (13.44a) that

$$F[f^*(x)] = [g(-k)]^* \tag{13.51a}$$

and similarly,

$$F[f(-x)] = g(-k), \tag{13.51b}$$

where $g(k)$ is defined by (13.44b) as usual. From (13.51b) we see that the Fourier transform of a symmetric or antisymmetric function is itself symmetric or antisymmetric, i.e.

$$f(x) = \pm f(x) \Rightarrow g(-k) = \pm g(k) \tag{13.52}$$

However, from (13.49a) and (13.51b) one sees that the Fourier transform of a real function $f(x) = f^*(x)$ is *not* a real function unless $g(k) = g(-k)$. Hence the Fourier transform of a real symmetric function is also real and symmetric, as shown in Figure 13.13 above.

Further useful results are as follows.

(i) *Differentiation*

$$\mathrm{F}[f'(x)] = ikg(k), \tag{13.53a}$$

$$\mathrm{F}[f''(x)] = (ik)^2 g(k) = -k^2 g(k), \tag{13.53b}$$

and so on for higher derivatives of $f(x)$.

(ii) *Multiplication by a constant*

$$\mathrm{F}[f(ax)] = \frac{1}{a} g\left(\frac{k}{a}\right), \tag{13.54}$$

where a is a constant.

(iii) *Translation*

$$F[f(x+a)] = \exp(ika)g(k), \tag{13.55a}$$

where a is a constant. Inverting (13.55a), we also have

$$F[e^{iax}g(x)] = \frac{1}{2\pi}f(a-k). \tag{13.55b}$$

These results are easily proved, as shown in Example 13.10. Here we illustrate a very important property of Fourier transforms arising from (13.54) by considering the Fourier transform

$$F\left[\exp(-|x|)\right] = \frac{1}{\pi}\frac{1}{(1+k^2)}$$

evaluated in Example 13.9a. Then from (13.54) we see that the Fourier transform of

$$f(x) = \exp(-a|x|) \tag{13.56a}$$

is

$$g(k) = \frac{1}{\pi a}\frac{a^2}{a^2+k^2}. \tag{13.56b}$$

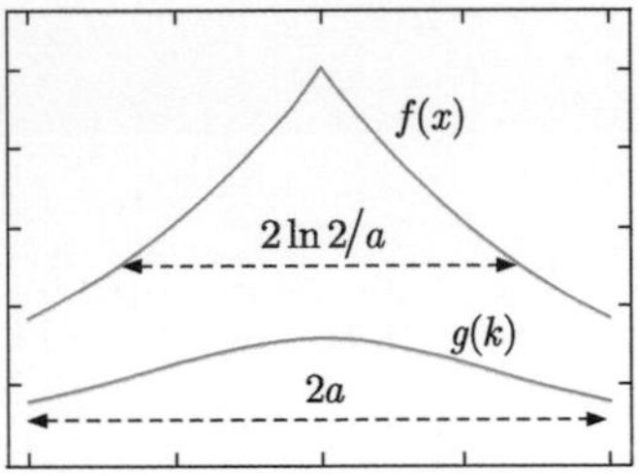

Figure 13.14 The functions $f(x)$ and $g(k)$ of Equations (13.56a) and (13.56b).

The functions $f(x)$ and $g(k)$ are shown in Figure 13.14. They are peaked, with widths at half maximum height of $2\ln 2/a = 1.4/a$ and $2a$, respectively. Hence, if the peak in $f(x)$ is broadened by reducing a, the corresponding peak in $g(k)$ is narrowed; and vice versa if a is increased. This behaviour is a general characteristic of Fourier transforms and is a direct consequence of (13.54). For a Fourier transform to exist, we must have $f(x) \to 0$ as $x \to \pm\infty$ by (13.45); and if the intervening distribution is broadened without changing shape by a transformation $x \to ax$, with $a < 1$, then the Fourier transform $g(k)$ is narrowed by a corresponding transformation $k \to k/a$; and vice versa if $a > 1$. Another example is given in Example 13.12 below.

In physical applications, one is often more interested in the widths of the distributions $|f(x)|^2$ and $|g(k)|^2$ rather than those of $|f(x)|$ and $|g(k)|$ themselves. The former are conveniently characterised by the root mean square deviations Δx and Δk from the means $\bar{x}$ and $\bar{k}$, respectively. These are defined by

$$(\Delta x)^2 \equiv (1/N_x)\int_{-\infty}^{\infty}(x-\bar{x})^2|f(x)|^2\mathrm{d}x,$$

where

$$N_x = \int_{-\infty}^{\infty}|f(x)|\mathrm{d}x$$

and the mean is defined by

$$\bar{x} \equiv (1/N_x) \int_{-\infty}^{\infty} |f(x)|^2 \mathrm{d}x,$$

with analogous expressions for (Δk) and $\bar{k}$ in terms of $g(k)$. For example, if one uses the distribution (13.56), one finds that $\bar{x} = \bar{k}$ because the relevant integrands are odd, and using these results in the integrals for the mean square deviations give $(\Delta x)^2 = 1/(2a^2)$ and $(\Delta k)^2 = a^2$, so that $\Delta x\, \Delta k = 1/\sqrt{2}$. More generally, it can be shown that

$$\Delta x\, \Delta k \geq 1/2, \tag{13.57}$$

where the minimum value is obtained for the Gaussian distribution, whose Fourier transform is given in (13.71) below. This result is called the *bandwidth theorem* and is closely related to an important result in quantum mechanics. In this case, if $f(x)$ is the wavefunction of a particle, then $|f(x)|^2$ is proportional to the probability of finding the particle at position x, and $|g(k)|^2$ is proportional to the probability that the particle has an associated wave number $k = 2\pi/\lambda$. Since in quantum mechanics the momentum p of the particle is given by the de Broglie relation $p = h/\lambda$, where h is the Planck constant, using $k = 2\pi/\lambda$ we may write the bandwidth theorem as

$$\Delta x\, \Delta p \geq \hbar/2,$$

where $\hbar \equiv h/2\pi$ is the 'reduced' Planck constant. This result is called the *Heisenberg uncertainty principle* and sets a fundamental limit on the precision with which the position and momentum of a particle can be simultaneously known.

Example 13.10

Prove the relation (13.53a).

Solution

From the definitions (13.44a) and (13.49a),

$$\mathrm{F}\left[f'(x)\right] = \frac{1}{2\pi} \int_{-\infty}^{\infty} f'(x) e^{-ikx} \mathrm{d}x.$$

Integrating by parts gives

$$\frac{1}{2\pi} \int_{-\infty}^{\infty} f'(x) e^{-ikx} \mathrm{d}x = \frac{1}{2\pi} [e^{-ikx} f(x)]_{-\infty}^{\infty} + \frac{1}{2\pi}(ik) \int_{-\infty}^{\infty} e^{-ikx} f(x)\, \mathrm{d}x. \tag{1}$$

But

$$\int_{-\infty}^{\infty} |\, f(x)\, |\, \mathrm{d}x$$

is finite, and so $f(x) \to 0$ as $x \to \pm\infty$. Thus the first term in (1) is zero and

$$\mathrm{F}\left[f'(x)\right] = \frac{ik}{2\pi} \int_{-\infty}^{\infty} e^{-ikx} f(x)\, \mathrm{d}x = ikg(k).$$

Example 13.11

The time dependence of an oscillating electric field $E(t)$ is given by

$$E(t) = \begin{cases} E_0 e^{i\omega_0 t} e^{-\gamma t/2} & t > 0 \\ 0 & t < 0 \end{cases},$$

where ω_0, E_0 and γ are constants. The spectrum of frequencies is given by the Fourier transform of $E(t)$. Show that the effect of the term in γ is to produce a spread in frequencies about ω_0 in the distribution.

Solution

If $g(\omega)$ is the Fourier transform of $E(t)$, then

$$g(\omega) = \frac{1}{2\pi} \int_{-\infty}^{\infty} E(t) e^{-i\omega t} \mathrm{d}t.$$

Substituting for $E(t)$ gives

$$\begin{aligned} g(\omega) &= \frac{E_0}{2\pi} \int_0^{\infty} \exp\left[i(\omega_0 - \omega)t - \frac{\gamma t}{2}\right] \mathrm{d}t \\ &= \frac{E_0}{2\pi} \left[\frac{e^{-\gamma t/2} e^{i(\omega_0 - \omega)t}}{i(\omega_0 - \omega) - \gamma/2}\right]_0^{\infty} = \frac{E_0}{2\pi} \left[\frac{1}{\gamma/2 - i(\omega_0 - \omega)}\right]. \end{aligned}$$

The spread in frequencies is best seen by plotting the intensity spectrum

$$|g(\omega)|^2 = \frac{E_0^2}{4\pi^2} \left[\frac{1}{\gamma^2/4 + (\omega - \omega_0)^2}\right]$$

as in Figure 13.15. Note that as the exponential fall-off gets faster, that is, γ gets larger, the spread in frequencies increases.

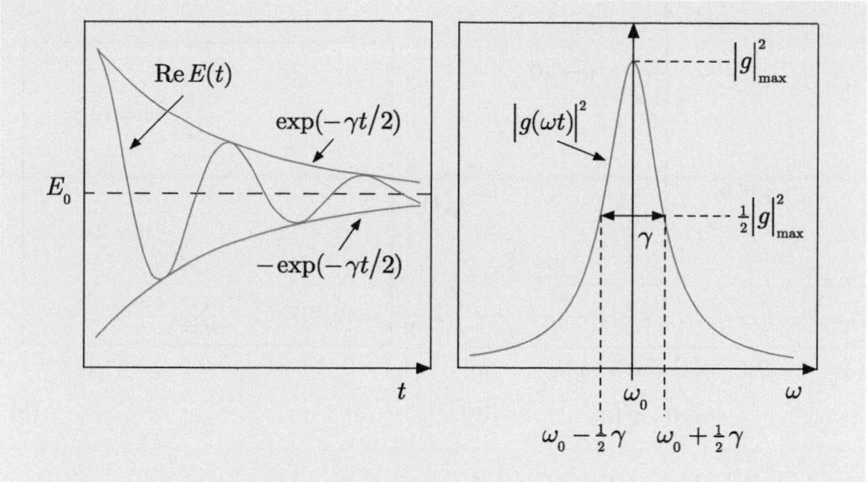

Figure 13.15 The real part of $E(t)$ and the envelope formed by the exponential decay terms (left figure) and the intensity spectrum $|g(\omega)|^2$ (right figure).

*13.3.2 The Dirac delta function

In applications of Fourier transforms, it is often convenient to extend their definition to include $F[e^{iqx}]$, even though $e^{iqx} \not\to 0$ as $x \to \infty$. This achieved by introducing the *Dirac delta function* $\delta(x)$, defined by the relation

$$\int f(x)\delta(x-a)\,\mathrm{d}x \equiv f(a), \tag{13.58a}$$

where a is a real constant that lies within the range of integration, so that

$$\delta(x-a) = 0, \quad x \neq a,$$

and

$$\int_b^c \delta(x-a)\mathrm{d}x = 1, \quad b < a < c.$$

If a lies outside the range of integration, the integral is defined to be zero. In particular,

$$\int_{-\infty}^{\infty} \delta(x)\,\mathrm{d}x = 1 \tag{13.58b}$$

and $\delta(x) = 0$ except at the single point $x = 0$. This property of the δ function is unlike that of any function we have met previously, and for this reason it was viewed with some scepticism by mathematicians when Dirac, a theoretical physicist, first introduced it in his classic formulation of quantum mechanics. It is now recognised as the first of a new class of functions called *generalised functions*,[5] to which a

[5] See, for example, M.J. Lighthill (1958) *An Introduction to Fourier Analysis and Generalised Functions*, Cambridge University Press, Cambridge.

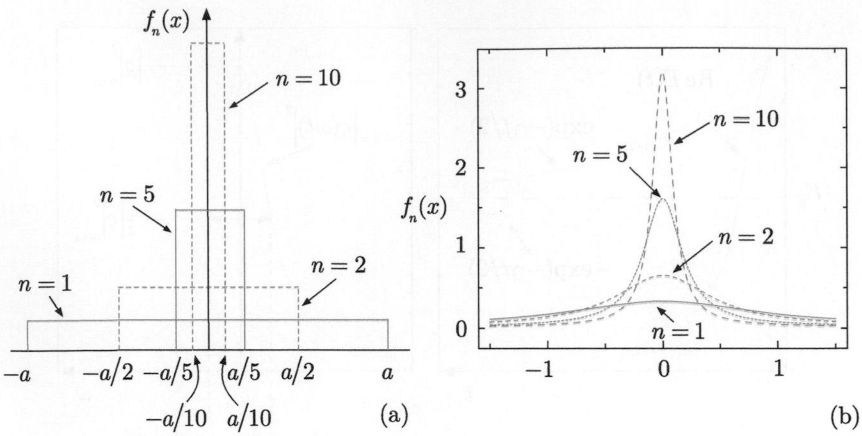

Figure 13.16 Possible forms of $f_n(x)$, which approximate more closely to $\delta(x)$ as $n \to \infty$: (a) from (13.59); (b) from (13.60).

set of well-behaved functions $f_n(x)$ approximates ever more closely as the integer $n \to \infty$. The choice of these functions is not unique and for the δ function, possible sequences include

$$f_n = \begin{cases} n/2a, & |x| < a/n \\ 0, & |x| > a/n \end{cases}, \tag{13.59}$$

and

$$f_n = \frac{n}{\pi} \frac{1}{1 + n^2 x^2}, \tag{13.60}$$

as shown in Figures 13.16a and 13.16b, respectively.

In both cases, $f_n(x)$ satisfies the normalisation condition (13.58b) for all n and $f_n(x) \to 0$ as $n \to \infty$ provided $x \neq 0$. Other useful properties of the δ function that follow from (13.58) are

$$\delta(x) = \delta(-x), \quad \delta(ax) = \frac{1}{|a|}\delta(x), \quad x\delta(x) = 0. \tag{13.61}$$

Finally, let us consider $\delta[f(x)]$, which is clearly non-zero only at points $x = a_i$, where $f(a_i) = 0$. If these are all 'simple zeros', so that

$$f(x) \approx (x - a_i) f'(a_i), \quad f'(a_i) \neq 0$$

near $x = a_i$, then using

$$\delta[f'(a_i)(x - a_i)] = \frac{1}{|f'(a_i)|}\delta(x - a_i) \tag{13.62a}$$

and adding the contributions from all the zeros gives the useful result

$$\delta[f(x)] = \sum_i \frac{1}{|f'(a_i)|}\delta(x - a_i) \tag{13.62b}$$

for any functions $f(x)$ with only simple zeros on the real axis.

The delta function is very useful in physical applications and many situations may be analysed using it even if they are not described exactly by a delta function, but by a very narrow distribution like (13.59) and (13.60) for large n. It also enables us to define the Fourier transform $F\left[e^{iqx}\right]$, even though the integral (13.45) is manifestly divergent in this case. From the Fourier integral formulas (13.44) and the definition (13.58), we have

$$f(a) = \frac{1}{2\pi}\int_{-\infty}^{\infty} dk \int_{-\infty}^{\infty} f(x)e^{ik(x-a)}dx = \int_{-\infty}^{\infty} f(x)\delta(x-a)dx \quad (13.63)$$

for any $f(x)$, so that

$$\delta(x-a) = \frac{1}{2\pi}\int_{-\infty}^{\infty} e^{ik(x-a)}dk \quad (13.64a)$$

or

$$\delta(x) = \frac{1}{2\pi}\int_{-\infty}^{\infty} e^{ikx}dk, \quad (13.64b)$$

implying that the Fourier transform

$$F[e^{iqx}] = \frac{1}{2\pi}\int e^{i(q-k)x}dx = \delta(k-q). \quad (13.65)$$

In any conventional definition the integrals in (13.64) do not exist. Nonetheless, these expressions may be safely used inside integrals,[6] as in (13.63). To illustrate their use, suppose

$$F\left[f_i(x)\right] = g_i(k), \quad i = 1, 2$$

and consider

$$\begin{aligned}\int f_1^*(x)f_2(x)dx &= \int \left[\int g_1^*(k)e^{-ikx}dk\right]\left[\int g_2(q)e^{iqx}dq\right]dx \\ &= 2\pi \int g_1^*(k)\int g_2(q)\delta(q-k)dq\,dk,\end{aligned}$$

where we have integrated over x and used (13.64). Integrating over q using (13.65) then gives *Parseval's relation*

$$\int_{-\infty}^{\infty} f_1^*(x)f_2(x)\,dx = 2\pi \int_{-\infty}^{\infty} g_1^*(k)g_2(k)\,dk. \quad (13.66)$$

[6] For a fuller and more rigorous account, see M.J. Lighthill (1958) *An Introduction to Fourier Analysis and Generalised Functions*, Cambridge University Press, Cambridge.

Finally, setting $f_1 = f_2 = f$ and hence $g_1 = g_2 = g$, we obtain

$$\int_{-\infty}^{\infty} f^*(x) f(x)\,\mathrm{d}x = 2\pi \int_{-\infty}^{\infty} g^*(k) g(k)\,\mathrm{d}k, \qquad (13.67)$$

which is the generalisation of Parseval's theorem for Fourier transforms.

Example 13.12

The Heaviside 'step function' $H(x)$ is defined by

$$H(x) = \begin{cases} 1 & x > 0 \\ 0 & x < 0 \end{cases}.$$

Show that $H'(x) = \delta(x)$.

Solution

Consider the integral

$$\int_{-\infty}^{\infty} f(x) H'(x)\,\mathrm{d}x$$

for any function $f(x)$. Evaluating this by parts gives

$$\int_{-\infty}^{\infty} f(x) H'(x)\,\mathrm{d}x = [f(x)H(x)]_{-\infty}^{\infty} - \int_{-\infty}^{\infty} f'(x) H(x)\,\mathrm{d}x,$$

$$= f(\infty) - [f(x)]_0^{\infty} = f(0),$$

i.e.

$$\int_{-\infty}^{\infty} f(x) H'(x)\,\mathrm{d}x = f(0).$$

Comparing this with the definition (13.58a) of $\delta(x-a)$, yields $H'(x) = \delta(x)$ as required.

Example 13.13

Find the Fourier transform of $f(x) = \sin(qx)$.

Solution

Using the expression

$$\sin(qx) = \frac{1}{2i}[e^{iqx} - e^{-iqx}] = \frac{1}{2i}\int e^{ikx}[\delta(k-q) - \delta(k+q)]\,\mathrm{d}k,$$

gives the Fourier transform

$$F[\sin(qx)] = \frac{1}{2i}[\delta(k-q) - \delta(k+q)].$$

Alternatively, the same result is obtained from

$$\begin{aligned} F[\sin(qx)] &= \frac{1}{2\pi}\int_{-\infty}^{\infty} \sin(qx)e^{-ikx}\,\mathrm{d}x \\ &= \frac{1}{2\pi}\frac{1}{2i}\int_{-\infty}^{\infty} [e^{i(q-k)x} - e^{-i(q+k)x}]\,\mathrm{d}k \\ &= \frac{1}{2i}[\delta(k-q) - \delta(k+q)], \end{aligned}$$

using (13.65) and (13.61).

*13.3.3 The convolution theorem

Given two functions $f_1(x)$ and $f_2(x)$, their *convolution* is defined by

$$f_1 * f_2 \equiv \int_{-\infty}^{\infty} f_1(x-y)f_2(y)\,\mathrm{d}y, \qquad (13.68)$$

assuming that the integral converges. Substituting $z = x - y$ for fixed x gives

$$f_1 * f_2 \equiv \int_{-\infty}^{\infty} f_1(z)f_2(x-z)\,\mathrm{d}z \equiv f_2 * f_1,$$

so that the apparent asymmetry in (13.68) is illusory and $f_1 * f_2 = f_2 * f_1$.

Now suppose we have two functions $f_1(x)$ and $f_2(x)$, which have Fourier transforms

$$F[f_1(x)] = g_1(k) \quad \text{and} \quad F[f_2(x)] = g_2(k)$$

and are such that

$$\int_{-\infty}^{\infty} |\, f_1(x)f_2(x)\,|\,\mathrm{d}x \qquad (13.69)$$

is convergent. Then the *convolution theorem* states that the Fourier transform of the product $f_1(x)f_2(x)$ is the convolution of the Fourier transform $g_1(k)$ and $g_2(k)$, i.e.

$$F[f_1\, f_2] = g_1 * g_2 \tag{13.70a}$$

and conversely, the Fourier transform of the convolution $f_1 * f_2$ is given by

$$F[f_1 * f_2] = 2\pi g_1 g_2. \tag{13.70b}$$

In what follows, we will first derive (13.70b) and then give some examples of its application. The Fourier transform of the convolution is

$$\begin{aligned} F[f_1 * f_2] &= \frac{1}{2\pi}\int_{-\infty}^{\infty} e^{-ikx}\mathrm{d}x \int_{-\infty}^{\infty} f_1(x-y)f_2(y)\mathrm{d}y \\ &= \frac{1}{2\pi}\int_{-\infty}^{\infty} f_2(y)\mathrm{d}y \int_{-\infty}^{\infty} f_1(x-y)e^{-ikx}\mathrm{d}x. \end{aligned}$$

Substituting $z = x - y$ then gives

$$\begin{aligned} F[f_1 * f_2] &= \frac{1}{2\pi}\int_{-\infty}^{\infty} f_2(y)\mathrm{d}y \int_{-\infty}^{\infty} f_1(z)e^{-ik(y+z)}\mathrm{d}z \\ &= \frac{1}{2\pi}\int_{-\infty}^{\infty} f_2(y)e^{-iky}\mathrm{d}y \int_{-\infty}^{\infty} f_1(z)e^{-ikz}\mathrm{d}z = g_1(k)g_2(k), \end{aligned}$$

which is the required result. The proof of (13.70a) proceeds in a similar way and is left as an exercise for the reader.

In the introduction to Fourier transforms, we were motivated by the need to Fourier analyse wave-packets like

$$\phi(x) = \exp(-x^2/2a^2)e^{iqx}.$$

From (13.65) the Fourier transform of e^{iqx} is $\delta(k-q)$, while the Fourier transform of a Gaussian is[7]

$$F[\exp(-x^2/2a^2)] = \frac{a}{\sqrt{2\pi}}\exp(-k^2a^2/2). \tag{13.71}$$

[7]This important result – that the Fourier transform of a Gaussian is itself a Gaussian – is derived indirectly in Problem 14.3. To evaluate it directly from the definition (13.44b) requires the use of the method of contour integration, which is not discussed in this book. See, for example, J. Mathews and R.L. Walker (1965) *Methods of Mathematical Physics*, W.A. Benjamin Inc., New York, Section 3.3.

Hence from the convolution theorem (13.70a), we have

$$F\left[\exp\left(-\frac{x^2}{2a^2}\right)e^{iqx}\right] = \left[\frac{a}{2\pi}\exp\left(-\frac{k^2}{2a^2}\right)\right] * \delta(k-q)$$
$$= \frac{a}{\sqrt{2\pi}}\int_{-\infty}^{\infty}\left[\delta(k'-q)\exp(-(k-k')^2a^2/2)\right]\mathrm{d}k'$$
$$= \frac{a}{\sqrt{2\pi}}\exp(-(k-q)^2a^2/2). \qquad (13.72)$$

This is illustrated in Figure 13.17, where we show the Fourier transforms (13.65) and (13.71) and their convolution, which is the Fourier transform of their product (13.72). From this we can see that the effect of multiplying e^{iqx} by the Gaussian is to smear out its Fourier transform $\delta(k-q)$ into a Gaussian; and the narrower the Gaussian in x, the broader it becomes in k.

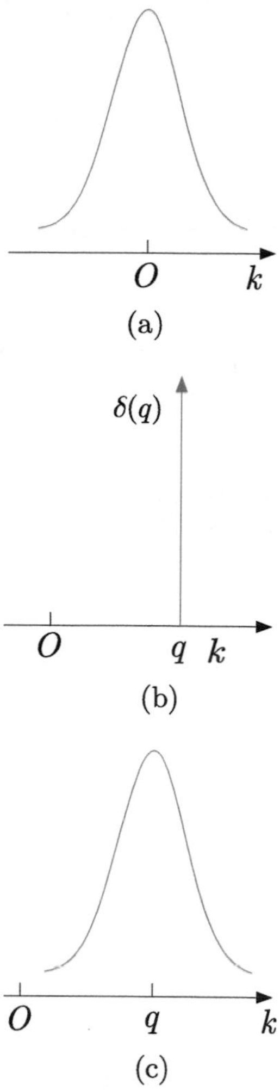

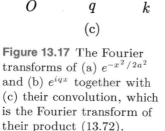

Figure 13.17 The Fourier transforms of (a) $e^{-x^2/2a^2}$ and (b) e^{iqx} together with (c) their convolution, which is the Fourier transform of their product (13.72).

Example 13.14

The 'top hat' function shown in Figure 13.18(a) is defined by

$$T(x) = \begin{cases} 1 & |x| < a \\ 0 & |x| > a \end{cases}$$

Sketch the convolution

$$f(x) = T(x) * [\delta(x-b) + \delta(x+b)],$$

where $b > a$. Show that its Fourier transform has zeros at $k = n\pi/a$ $(n \neq 0)$ and $[(2n+1)\pi]/2b$ (all).

Solution

The convolution

$$f(x) = \int [\delta(x'-b) + \delta(x'+b)]T(x-x')\,\mathrm{d}x' = T(x-b) + T(x+b),$$

is shown in Figure 13.18(b).

By the convolution theorem (13.70b), the Fourier transform is[8]

$$F[f(x)] = F[T(x)]F[\delta(x-b) + \delta(x+b)],$$

[8] This calculation occurs in the theory of 'Young's slits' in optics, when the finite width of the slits is taken into account.

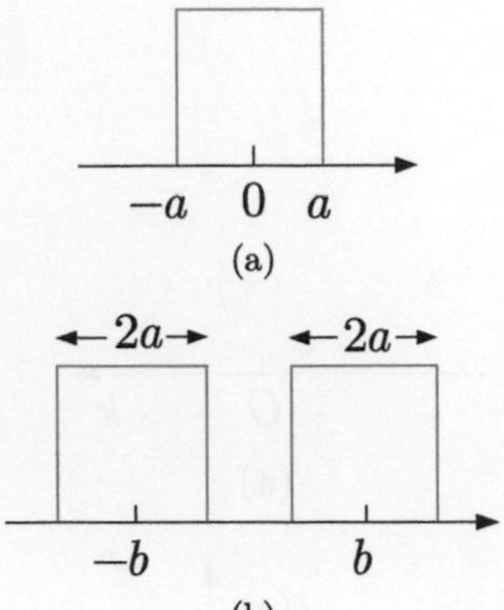

Figure 13.18 (a) The 'top hat' function $T(x)$; (b) the convolution $f(x)$.

where

$$F[T(x)] = \frac{1}{2\pi}\int_{-\infty}^{\infty} T(x)e^{-ikx}\mathrm{d}x = \frac{1}{2\pi}\int_{-a}^{a} e^{-ikx}\mathrm{d}x = \frac{1}{k\pi}\sin(ka),$$

and

$$F[\delta(x-b)+\delta(x+b)] = \frac{1}{2\pi}\int e^{-ikx}[\delta(x-b)+\delta(x+b)]\,\mathrm{d}x$$
$$= \frac{1}{2\pi}[e^{-ikb}+e^{ikb}] = \frac{1}{\pi}\cos(kb),$$

so that

$$F\,[f(x)] = \frac{1}{\pi^2 k}\sin(ka)\cos(kb),$$

with zeros at $k = n\pi/a$ $(n \neq 0)$ and $k = (2n+1)\pi/2b$ (all n). There is no zero at $k = 0$ because $\sin(ka)/k \to a$ as $k \to 0$.

Problems 13

13.1 Find the Fourier expansion of the function

$$f(x) = \begin{cases} |x| & 0 < |x| < 2\pi/3 \\ 2\pi/3 & 2\pi/3 < |x| < \pi \end{cases}$$

with a period of 2π.

13.2 Find the Fourier series for the function

$$f(x) = \begin{cases} -1/2 & -\pi < x < -\pi/2 \\ 0 & -\pi/2 < x < \pi/2 \\ 1/2 & \pi/2 < x < \pi \end{cases}$$

with a period of 2π, and hence deduce the sum of the series

$$1 - \frac{1}{3} + \frac{1}{5} - \frac{1}{7} + \cdots.$$

13.3 Find the Fourier series of the function

$$f(x) = x^2(2\pi^2 - x^2), \quad -\pi \leq x \leq \pi$$

with periodicity 2π.

13.4 Show that the Fourier series with period 2π for the function $f(x) = \cos(\mu x)$, where $-\pi \leq x \leq \pi$ and μ is non-integer, is

$$\cos(\mu x) = \frac{2\mu}{\pi}\sin(\mu\pi)\left[\frac{1}{2\mu^2} + \sum_{n=1}^{\infty}\frac{(-1)^{n+1}}{n^2-\mu^2}\cos(n\pi)\right].$$

Hence deduce an expansion for $\cot(\mu\pi)$ and show that

$$\sum_{n=1}^{\infty} \frac{1}{9n^2 - 1} = \frac{1}{2} - \frac{\pi\sqrt{3}}{18}.$$

13.5 The following functions $f(x)$ have period 2π and are given by (a) $1/(\pi - x)$ (b) $\arcsin x$ (c) $\sinh x$ (d) $x^2 e^{-1/x}$ in the range $-\pi < x < \pi$. Which of these functions are guaranteed to have convergent Fourier series by the Dirichlet conditions? If so, find the series.

13.6 Find the Fourier series of period 2π for the function

$$f(x) = \begin{cases} x^2 & 0 < x < \pi \\ \pi^2 & \pi < x < 2\pi \end{cases}$$

and hence deduce the value of $\sum_{n=1}^{\infty} 1/n^2$.

13.7 Find the Fourier series of the function $f(x)$ of period $2L$ if $f(x) = x$ in the interval $(0, 2L)$.

13.8 Find the Fourier cosine series for the function

$$f(x) = \begin{cases} 1 & 0 < x < \frac{1}{2} \\ 0 & \frac{1}{2} < x < 1 \end{cases}.$$

13.9 The function defined by $f(x) = f(x + 2\pi)$ and

$$f(x) = x^2, \quad -\pi < x < \pi$$

has a Fourier expansion

$$x^2 = \frac{\pi^2}{3} + 4\sum_{n=1}^{\infty} \frac{(-1)^n}{n^2} \cos nx.$$

(a) Integrate this series twice and hence deduce a Fourier series for $f(x) = x^4$ in the interval $[-\pi, \pi]$ and with period 2π. Hence show that

$$\sum_{n=1}^{\infty} \left(\frac{\pi^2}{6n^2} - \frac{1}{n^4} \right) = \frac{\pi^4}{60}.$$

(b) Differentiate the series obtained in (a) twice to obtain candidate series for $f(x) = x^3, x^2$ in the interval $[-\pi, \pi]$ and with period 2π. Which of the series, if any, are valid?

13.10 Use the expansion of x^2 given in Example 13.4 to show that

$$\pi = 4\sum_{n=0}^{\infty} \frac{(-1)^n}{2n + 1}.$$

13.11 Find the Fourier series of period 2 for the function $f(x) = (x - 1)^2$ in the interval $[0, 2]$ and hence evaluate the sum

$$\sum_{n=1}^{\infty} \frac{1}{n^4}.$$

13.12 Use the Fourier series for $\cos(\mu x)$ found in Problem 13.4 to show that

$$\sum_{n=1}^{\infty} \frac{1}{(n^2 - 1/4)^2} = \pi^2 - 8.$$

13.13 Expand the function $f(x) = x^2 + x$ in a complex Fourier series in the interval $-\pi < x < \pi$. A useful integral is

$$\int x^k\, e^{ax}\, \mathrm{d}x = \frac{\mathrm{d}^k}{\mathrm{d}a^k} \int e^{ax}\, \mathrm{d}x,$$

for integer $k \geq 0$.

13.14 Find a complex Fourier series for the function

$$f(x) = \begin{cases} +1 & 0 < x < \pi \\ -1 & -\pi < x < 0. \end{cases}$$

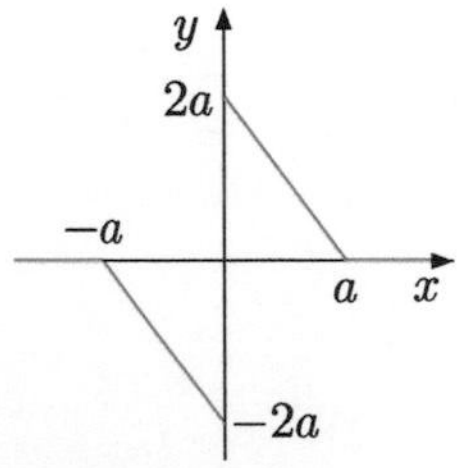

Figure 13.19

13.15 Find the Fourier transform of the function shown in Figure 13.19.

13.16 Derive the Fourier sine transform (13.46b) for antisymmetric functions $f_-(x) = -f(-x)$.

13.17 Find the Fourier sine transform of the function

$$f(x) = \begin{cases} e^{-x} & x > 0 \\ -e^{x} & x < 0 \end{cases}$$

and hence show that

$$\int_0^{\infty} \left[\frac{x \sin(mx)}{x^2 + 1}\right] = \frac{\pi}{2} e^{-m}, \quad m > 0.$$

13.18 Find the Fourier cosine transform of the function $f(x)$ given by

$$f(x) = \begin{cases} 1 - x^2 & 0 \leq |x| < 3 \\ 0 & |x| > 3 \end{cases}$$

and hence deduce the value of

$$\int_0^{\infty} \frac{1}{t^3} [3t \cos(3t) + (4t^2 - 1) \sin(3t)] \cos t \,\mathrm{d}t.$$

13.19 Given the Fourier transform

$$g(k) = F[\exp(-\alpha x^2)] = \frac{1}{2\sqrt{\pi \alpha}} \exp(-k^2/4\alpha),$$

where $\alpha > 0$, deduce the Fourier transforms of the functions

$$\text{(a) } \exp[-\alpha(x-a)^2], \quad \text{(b) } x \exp[-\alpha x^2], \quad \text{(c) } x^2 \exp[-\alpha x^2].$$

13.20 Show that the Fourier transform of a real, antisymmetric function is purely imaginary. What are the Fourier transforms of (a) $f(x) = \cos(2x) \sin(3x)$, (b) $\delta(x^2 - a^2)$?

***13.21** Show that the function

$$\Lambda(x) = \begin{cases} 1+x & -1 < x < 0 \\ 1-x & 0 < x < 1 \\ 0 & |x| > 1 \end{cases}$$

is the convolution of the 'top hat' function

$$T(x) = \begin{cases} 1 & |x| < 1/2 \\ 0 & |x| > 1/2 \end{cases}$$

with itself. Check the convolution theorem in this case by calculating the Fourier transform $F[\Lambda]$ from (a) the convolution theorem and (b) directly from the definition (13.44b).

***13.22** A quantity $f(x)$ is measured using an instrument with known resolution

$$r(x) = \frac{1}{\sqrt{2\pi}\sigma_r} \exp\left[-\frac{1}{2}\left(\frac{x}{\sigma_r}\right)^2\right],$$

which is a Gaussian form centred at zero and with a 'standard deviation' σ_r. The experimentally observed distribution $e(x)$ is also of Gaussian form with standard deviation $\sigma_e > \sigma_r$. Use the result $e = f * r$ and the convolution theorem, to shown that $f(x)$ also has a Gaussian from and deduce its standard deviation.
Note the integral

$$\int_{-\infty}^{\infty} \exp(-ax^2 + ibx)\,\mathrm{d}x = (\pi/a)^{1/2} \exp(-b^2/4a).$$

***13.23** (a) Show that the Fourier transform of the function

$$f(x) = \begin{cases} \pi & |x| < 1 \\ 0 & |x| > 0 \end{cases}$$

is $F[f(x)] = \operatorname{sinc} k$, where [cf. (5.63)], $\operatorname{sinc} x \equiv \sin x/x$. Hence evaluate the integrals

$$I_1 = \int_{-\infty}^{\infty} \operatorname{sinc} x\,\mathrm{d}x \quad \text{and} \quad I_2 = \int_{-\infty}^{\infty} (\operatorname{sinc} x)^2\,\mathrm{d}x.$$

(b) Find the Fourier transform of $\operatorname{sinc} x$, and show that the convolution

$$\operatorname{sinc}(x-a) * \operatorname{sinc}(x+a) = \pi \operatorname{sinc} x.$$

14

Ordinary differential equations

Any equation that expresses the functional dependence of a variable y on its arguments $x_i (i = 1, 2, \ldots)$ and the derivatives of y with respect to those arguments, is called a differential equation. Physical systems are almost always described by such equations. For example, a wave $f(x, t)$ travelling with velocity v in the x-direction obeys the wave equation

$$\frac{\partial^2 f(x,t)}{\partial x^2} - \frac{1}{v^2}\frac{\partial^2 f(x,t)}{\partial t^2} = 0, \tag{14.1a}$$

while the motion of a simple pendulum of length l performing small oscillations θ satisfies

$$\frac{\mathrm{d}^2\theta(t)}{\mathrm{d}t^2} + \frac{g}{l}\sin\theta(t) = 0, \tag{14.1b}$$

where g is a constant. Equation (14.1a) contains partial derivatives because f is a function of more than one variable, and the equation is therefore called a *partial differential equation (PDE)*. These will be discussed in Chapter 16. On the other hand, in Equation (14.1b) the quantity θ depends on the single variable t, and the equation is called an *ordinary differential equation (ODE)*. It is these equations that are the subject of this chapter and the next.

In what follows, we will usually refer to the independent variable as x and the corresponding dependent variable as $y(x)$. Thus, in general, an ordinary differential equation is of the form

$$f\left(x, y, \frac{\mathrm{d}y}{\mathrm{d}x}, \frac{\mathrm{d}^2 y}{\mathrm{d}x^2}, \cdots\right) = 0. \tag{14.2}$$

Mathematics for Physicists, First Edition. B.R. Martin and G. Shaw.

Companion website: www.wiley.com/go/martin/mathsforphysicists

Examples of ODEs are:

$$2xy\frac{\mathrm{d}y}{\mathrm{d}x} = x^2 + y^2, \tag{14.3a}$$

$$\frac{\mathrm{d}^2y}{\mathrm{d}x^2} + 3\frac{\mathrm{d}y}{\mathrm{d}x} + y = \cos x, \tag{14.3b}$$

and

$$\frac{\mathrm{d}^3y}{\mathrm{d}x^3} - \left[\left(\frac{\mathrm{d}y}{\mathrm{d}x}\right)^2 + y\right]^{1/2} = 0 \tag{14.3c}$$

It is convenient to classify ordinary differential equations by their order, degree and linearity. The *order* is defined as the order of the highest derivative in the equation. Thus, Equations (14.3a), (14.3b) and (14.3c) are first, second and third order, respectively. The *degree* of the equation is defined as the power to which the highest order derivative is raised after the equation is rationalised, that is, only integer powers remain. Thus, (14.3a) and (14.3b) are both first degree equations. Equation (14.3c) is of second degree because when the equation is rationalised, it becomes

$$\left(\frac{\mathrm{d}^3y}{\mathrm{d}x^3}\right)^2 - \left[\left(\frac{\mathrm{d}y}{\mathrm{d}x}\right)^2 + y\right] = 0,$$

so that the highest derivative is $\mathrm{d}^3y/\mathrm{d}x^3$ and it is raised to power 2. Finally, any ODE of order n is said to be *linear* if it is linear in the dependent variable y and its first n derivatives. If this is not true, the equation is said to be *non-linear*. Equations (14.3a) and (14.3b) are therefore linear, whereas (14.1b) and (14.3c) are non-linear. The discussion in this chapter will be predominantly about linear ordinary differential equations. This is because, except in a few simple cases, non-linear equations are difficult to solve and one must usually resort to numerical methods. Nonetheless, they are important, especially in describing complex systems, such as the atmosphere, and they play a central role in the so-called chaotic systems.[1]

The solution of an ODE in the variables x and y is defined as a relation between the two variables, such that when substituted into the ODE gives an identity. It may be of the explicit form $y = g(x)$, or the implicit form $\phi(x, y) = 0$, and is obtained, in principle, by repeated integration. Since each such operation will introduce a constant of integration, we can deduce that a solution of an nth order ordinary differential equation cannot be a general solution unless it

[1] A popular introduction to this subject is I.S. Stewart (1989) *Does God Play Dice? The Mathematics of Chaos*, Blackwell, Oxford.

contains n arbitrary constants. In physical situations, these constants are fixed by specifying *boundary conditions*, that is, by requiring that y and/or its derivative has specific values at given points. For example, the linear second-order equation

$$\frac{\mathrm{d}^2y}{\mathrm{d}x^2} + 3\frac{\mathrm{d}y}{\mathrm{d}x} + 2y = 0,$$

has the general solution

$$y = Ae^{-2x} + Be^{-x},$$

which may be verified by direct substitution. A and B are two arbitrary constants, as expected for a second-order equation. To determine these, we could specify the values of y at two values of x. Thus, if we were to require that $y = 0$ when $x = 0$ and $y = 1$ when $x = 1$, then

$$0 = A + B \quad \text{and} \quad 1 = Ae^{-2} + Be^{-1},$$

which gives

$$A = -B = e^2/(1 - e)$$

and hence the specific solution is

$$y(x) = \frac{e^2}{(1-e)}\left(e^{-2x} - e^{-x}\right).$$

Different boundary conditions lead of course to different solutions.

In the following sections we will discuss a number of methods to solve various types of ordinary differential equations, starting with first-order equations of the first degree. First-order equations of higher than first degree rarely occur in physical science and will not be discussed here.

14.1 First-order equations

The general form for equations of this type is

$$\frac{\mathrm{d}y}{\mathrm{d}x} = F(x, y). \tag{14.4}$$

Solutions of (14.4) are found relatively easily if $F(x, y)$ has specific simple forms and we will discuss some of these below.

14.1.1 Direct integration

If $F(x, y) = f(x)$ is a function of x only, then (14.4) takes the form

$$\mathrm{d}y/\mathrm{d}x = f(x) \tag{14.5a}$$

and may be directly integrated to give

$$y(x) = \int f(x)\mathrm{d}x + c,$$

where c is an arbitrary constant. If $F(x, y) = f(y)$ is a function of y only, then we have

$$\mathrm{d}y/\mathrm{d}x = f(y), \tag{14.5b}$$

so that the solution

$$x = \int \frac{\mathrm{d}y}{f(y)} + c$$

is again obtained by direct integration.

Example 14.1

The number of particles N in a radioactive sample decreases at a rate $\mathrm{d}N/\mathrm{d}t = -\alpha N$, where $\alpha > 0$. If $N = N_0$ at time $t = 0$, how many particles remain at time t?

Solution

With a trivial relabelling, the equation for the decay rate is of the form (14.5b) so that

$$t = \int \frac{\mathrm{d}N}{-\alpha N} = -\frac{1}{\alpha}(\ln N + \ln c)$$

and hence

$$N = c^{-1}\exp(-\alpha t),$$

where c is a constant. Since $N(0) = N_0$, we have $c^{-1} = N_0$ and

$$N(t) = N_0 e^{-\alpha t},$$

which is the law of radioactive decay.

14.1.2 Separation of variables

One frequently meets applications in which the right-hand side of (14.4) is a product $F(x, y) = f(x)g(y)$, so that it becomes

$$\frac{\mathrm{d}y}{\mathrm{d}x} = f(x)g(y). \tag{14.6a}$$

If either f or g is a constant, then (14.6a) can be evaluated by direct integration, as in Section 14.1.1. Otherwise, rearranging (14.6a) gives

$$\frac{\mathrm{d}y}{g(y)} = f(x)\,\mathrm{d}x,$$

that is, the variables have been separated. Integrating then gives

$$\int [1/g(y)]\,\mathrm{d}y = \int f(x)\,\mathrm{d}x, \tag{14.6b}$$

which expresses y implicitly in terms of x. In doing the integration one must of course remember to include the constant of integration.

Example 14.2

Find the solutions of

(a) $(xy)\mathrm{d}y/\mathrm{d}x = (x^2+2)/(y-1)$ and

(b) $\mathrm{d}y/\mathrm{d}x + x\sin(x+y) + 1 = 0$.

Solution

(a) Separating variables, we have $y(y-1)\mathrm{d}y = (x+2/x)\mathrm{d}x$, and integrating gives

$$\int (y^2-y)\,\mathrm{d}y = \int (x+2/x)\,\mathrm{d}x,$$

i.e.

$$2y^3 - 3y^2 = 3x^2 + 12\ln x + c,$$

where c is an arbitrary constant. This is the implicit solution for y.

(b) Setting $z = x+y$ so that $\mathrm{d}y/\mathrm{d}x = \mathrm{d}z/dx - 1$ gives

$$\frac{\mathrm{d}z}{\mathrm{d}x} + x\sin z = 0.$$

Separating variables and integrating, we have

$$\int \frac{1}{\sin z}\mathrm{d}z = \ln[\tan(z/2)] = -\frac{x^2}{2} + a,$$

where a is a constant. The implicit solution is therefore $\tan[(x+y)/2] = c\exp(-x^2/2)$, where c is also a constant.

14.1.3 Homogeneous equations

In Section 7.2.4, a function $f(x_1, x_2, \ldots, x_n)$ was defined to be *homogeneous of degree k* if

$$f(\lambda x_1, \lambda x_2, \ldots, \lambda x_n) = \lambda^k f(x_1, x_2, \ldots, x_n)$$

where λ is an arbitrary parameter. Thus, for example, the function $f(x,y) = x^3y + x^2y^2$ is a homogeneous function of degree 4, whereas

$f(x, y) = xy + x/y$ is *inhomogeneous* because it does not satisfy this requirement. Homogeneous equations are of the form

$$\frac{\mathrm{d}y}{\mathrm{d}x} = \frac{g(x, y)}{h(x, y)}, \tag{14.7}$$

where g and h are both homogeneous functions of the same degree. The key property is that the right-hand side of (14.7) can be written as a function of the ratio $z \equiv y/x$, i.e.

$$\frac{g(x, y)}{h(x, y)} = \phi(z)$$

and

$$\frac{\mathrm{d}y}{\mathrm{d}x} = z + x\frac{\mathrm{d}z}{\mathrm{d}x}.$$

Substituting these equations into (14.7) gives

$$\frac{\mathrm{d}z}{\mathrm{d}x} = \frac{\phi(z) - z}{x}. \tag{14.8a}$$

This is a separable equation, which can be integrated to give

$$\int \frac{\mathrm{d}z}{\phi(z) - z} = \int \frac{\mathrm{d}x}{x}, \tag{14.8b}$$

from which z and hence y can be found.

Some equations, although not obviously homogeneous at first sight, may be reduced to this form by suitable transformation of variables. One type that commonly occurs is given in Example 14.3(b).

Example 14.3

Solve the equations:

$$\text{(a) } 2xy\frac{\mathrm{d}y}{\mathrm{d}x} = 3y^2 - x^2 \quad \text{and} \quad \text{(b) } \frac{\mathrm{d}y}{\mathrm{d}x} = \frac{y + x - 2}{y - x + 4}.$$

Solution

(a) The equation may be written

$$\frac{\mathrm{d}y}{\mathrm{d}x} = \frac{3y^2 - x^2}{2xy},$$

which is homogeneous. Introducing $z = y/x$, as above, we have

$$\frac{3y^2 - x^2}{2xy} = \frac{3z^2 - 1}{2z} \equiv \phi(z),$$

so that (14.8b) becomes

$$\int \frac{2z}{z^2-1}\,\mathrm{d}z = \int \frac{1}{x}\,\mathrm{d}x,$$

and integrating gives

$$\ln(z^2-1) = \ln x + \ln c,$$

where c is a constant. Finally, substituting $z = y/x$ and rearranging gives the solution

$$y^2 = x^2(cx+1).$$

(b) This equation is inhomogeneous because of the presence of the constants −2 and 4. However, it can be converted to a homogeneous equation by a simple change of variables. Introducing new variables g and h by $x = x_0 + g$ and $y = y_0 + h$, where x_0 and y_0 are constants, the equation becomes

$$\frac{\mathrm{d}h}{\mathrm{d}g} = \frac{h+g+y_0+x_0-2}{h-g+y_0-x_0+4}.$$

Then choosing x_0 and y_0 so that

$$y_0 + x_0 - 2 = 0 \quad \text{and} \quad y_0 - x_0 + 4 = 0,$$

i.e. $y_0 = -1, \quad x_0 = 3$ gives

$$\frac{\mathrm{d}h}{\mathrm{d}g} = \frac{h+g}{h-g},$$

which is a function of h/g and is therefore homogeneous. Now set $h = zg$, so that

$$g\frac{\mathrm{d}z}{\mathrm{d}g} = \frac{z+1}{z-1} - z = \frac{-z^2+2z+1}{z-1}$$

and hence

$$-\int \frac{z-1}{z^2-2z-1}\,\mathrm{d}z = \int \frac{1}{g}\,\mathrm{d}g.$$

Integrating gives

$$-2\ln g + a = \ln(z^2-2z-1),$$

where a is a constant, and reverting to the original variables gives the implicit solution

$$y^2 - x^2 - 2xy + 8y + 4x = c,$$

where c is also a constant.

14.1.4 Exact equations

If

$$F(x,y) = -\frac{A(x,y)}{B(x,y)},$$

then the first-order equation (14.4) may be written

$$A(x,y) + B(x,y)\frac{\mathrm{d}y}{\mathrm{d}x} = 0. \tag{14.9}$$

If it is possible to find a function $f(x,y)$ such that

$$\frac{\partial f}{\partial x} = A(x,y) \quad \text{and} \quad \frac{\partial f}{\partial y} = B(x,y). \tag{14.10}$$

then, by Equation (7.17a),

$$\mathrm{d}f = A(x,y)\mathrm{d}x + B(x,y)\mathrm{d}y$$

is an exact differential and (14.9) is called an *exact equation.* We then have

$$\frac{\mathrm{d}f}{\mathrm{d}x} = A(x,y) + B(x,y)\frac{\mathrm{d}y}{\mathrm{d}x},$$

and comparing this with (14.9), we see that the latter has the implicit solution

$$f(x,y) = c, \tag{14.11}$$

where c is an integration constant. To see whether a function $f(x,y)$ that satisfies (14.10) exists, we note that (14.10) implies

$$\frac{\partial^2 f}{\partial x \partial y} = \frac{\partial A(x,y)}{\partial y}, \quad \frac{\partial^2 f}{\partial x \partial y} = \frac{\partial B(x,y)}{\partial x},$$

and hence

$$\frac{\partial A(x,y)}{\partial y} = \frac{\partial B(x,y)}{\partial x} \tag{14.12}$$

so that (14.12) is a necessary condition for a relation of the form (14.10) to exist. It can also be shown to be a sufficient condition. If it is satisfied, then integrating A with respect to x at fixed y, and B with respect to y at fixed x gives the results

$$f(x,y) = \int A(x,y)\, dx + f_1(y), \tag{14.13a}$$

and

$$f(x,y) = \int B(x,y)\, dy + f_2(x), \tag{14.13b}$$

where $f_1(y)$ and $f_2(x)$ are arbitrary functions, which may be identified up to a constant by comparing (14.13a) and (14.13b).

Alternatively, (14.13a) may be differentiated with respect to y at fixed x and compared to $\partial f/\partial y = B$ to determine $c_1(y)$. The solution is then given by (14.11), as we shall illustrate by an example.

Example 14.4

Solve the equations

$$\text{(a) } 2xy\frac{\mathrm{d}y}{\mathrm{d}x} + 3x + y^2 = 0, \quad \text{(b) } \frac{\mathrm{d}y}{\mathrm{d}x} = \frac{x^2 + y + 2}{1 - x}.$$

Solution

(a) This is of the form (14.9) with $A = 3x + y^2$ and $B = 2xy$, and the condition (14.12) is satisfied. Equation (14.13a) gives

$$f(x, y) = \int (3x + y^2)\mathrm{d}x + f_1(y) = \tfrac{3}{2}x^2 + xy^2 + f_1(y).$$

Differentiating with respect to y at fixed x and comparing with (14.10) gives

$$\frac{\partial f(x, y)}{\partial y} = 2xy + \frac{\partial f_1(y)}{\partial y} = 2xy$$

so $\partial f_1(y)/\partial y = 0$ and $f(x, y) = \frac{3}{2}x^2 + xy^2 + a$, where a is a constant. The implicit solution (14.11) is therefore

$$3x^2 + 2xy^2 = c,$$

where c is also a constant.

(b) This is of the form (14.9) with $A = x^2 + y + 2$ and $B = x - 1$, and satisfies (14.12). Integrating A with respect to x with y fixed gives

$$f(x, y) = \int (x^2 + y + 2)\,\mathrm{d}x + f_1(y) = \tfrac{1}{3}x^3 + xy + 2x + f_1(y).$$

Differentiating this result with respect to y at fixed x and comparing with B gives

$$x - 1 = x + \frac{\mathrm{d}f_1(y)}{\mathrm{d}y} \quad \Rightarrow \quad f_1(y) = -y + a,$$

where a is a constant. Thus the implicit solution of the equation is

$$x^3 + 3xy + 6x - 3y = c,$$

where c is also a constant.

14.1.5 First-order linear equations

First-order linear ODEs are equations of the form

$$\frac{\mathrm{d}y}{\mathrm{d}x} + p(x)y = q(x), \tag{14.14}$$

where $p(x)$ and $q(x)$ are given functions of x, or constants. If the equation is exact, it may be solved by the method of Section 14.1.4. If it is not of this form, that is, it is *inexact*, it may in principle be solved by multiplying by a function $I(x)$, to be determined below, called an *integrating factor*. We then obtain

$$I(x)\frac{\mathrm{d}y}{\mathrm{d}x} + I(x)p(x)y = I(x)q(x), \tag{14.15a}$$

and $I(x)$ is chosen so that the left-hand side of (14.15a) is equal to $\mathrm{d}[I(x)y]/\mathrm{d}x$, i.e.,

$$\frac{\mathrm{d}[I(x)y]}{\mathrm{d}x} = I(x)\frac{\mathrm{d}y}{\mathrm{d}x} + y\frac{\mathrm{d}I(x)}{\mathrm{d}x} = I(x)\frac{\mathrm{d}y}{\mathrm{d}x} + I(x)p(x)y. \tag{14.15b}$$

Equating the linear terms in y in this equation gives

$$y\frac{\mathrm{d}I(x)}{\mathrm{d}x} = I(x)p(x)y,$$

which on integrating gives (providing $y \neq 0$) the result

$$I(x) = \exp\left[\int p(x)\mathrm{d}x\right] \tag{14.16}$$

for the integrating factor. Finally, from (14.15a) and (14.15b),

$$\frac{\mathrm{d}\left[I(x)y\right]}{\mathrm{d}x} = q(x)I(x),$$

and hence the general solution for y is given by

$$I(x)y = \int q(x)I(x)\mathrm{d}x, \tag{14.17}$$

with $I(x)$ given by (14.16).

Example 14.5

Find the solution of the equation

$$x^2\frac{\mathrm{d}y}{\mathrm{d}x} - xy = \frac{2}{x}$$

if $y = 2$ when $x = 1$.

Solution

Writing the equation in the form (14.13) gives

$$\frac{\mathrm{d}y}{\mathrm{d}x} - \frac{y}{x} = \frac{2}{x^3},$$

so $p(x) = -1/x$ and $q(x) = 2/x^3$. Then from (14.16), the integrating factor is given by

$$I(x) = \exp\left[\int p(x)dx\right] = \exp[-\ln x] = 1/x.$$

Thus, from (14.17), the solution is given by

$$\frac{1}{x}y = \int \frac{2}{x^3}\frac{1}{x}\mathrm{d}x = -\frac{2}{3x^3} + c,$$

i.e.

$$y = -\frac{2}{3x^2} + cx.$$

Finally, using $y = 2$ when $x = 1$ gives $c = 8/3$, and hence the solution is

$$y = \frac{8x}{3} - \frac{2}{3x^2}.$$

14.2 Linear ODEs with constant coefficients

Linear ODEs are of the form

$$\sum_{i=0}^{n} a_i(x)\frac{\mathrm{d}^{n-i}y}{\mathrm{d}x^{n-i}} = f(x), \tag{14.18}$$

where the $a_i(x)$ $(i = 0, 1, 2, \ldots, n)$ and $f(x)$ are given functions of x. For $n = 1$, this reduces to (14.14), and the method of solution has been discussed in Section 14.1.5. In this section we will consider the case of arbitrary n where the $a_i(x)$ are *constants*, that is, $a_i(x) = a_i$ so that

$$\sum_{i=0}^{n} a_i\frac{\mathrm{d}^{n-i}y}{\mathrm{d}x^{n-i}} = f(x). \tag{14.19}$$

Other types of linear ODE will be discussed in Section 14.2.4 and Chapter 15.

The solution of equations of the type (14.19) is found in three steps. Firstly, one finds the general solution y_0 to the *reduced equation* obtained by setting $f(x) = 0$ in (14.19), i.e.

$$\sum_{i=0}^{n} a_i\frac{\mathrm{d}^{n-i}y_0(x; c_1, c_2, \cdots, c_n)}{dx^{n-i}} = 0. \tag{14.20a}$$

The function y_0 contains n free parameters $c_1, c_2, \cdots, c_n$, since it is the general solution to an equation of order n. It is called the *complementary function.* The second step is to find a particular solution $Y(x)$ of (14.19), so that

$$\sum_{i=0}^{n} a_i \frac{\mathrm{d}^{n-i}Y}{\mathrm{d}x^{n-i}} = f(x). \tag{14.20b}$$

The function $Y(x)$ is called the *particular integral.* Finally, one adds the complementary function to the particular integral to obtain

$$y(x) = y_0(x; c_1, c_2, \cdots, c_n) + Y(x). \tag{14.21}$$

On substituting (14.21) into (14.19) and using (14.20a) and (14.20b), one easily shows that it is a solution; and since it contains n arbitrary parameters, it is the general solution.

It is relatively easy to find complementary functions for a specific equation, as we shall show in Section 14.2.1, but there is no general method for finding particular integrals for a given $f(x)$. In Sections 14.2.2 and 14.2.3 we shall discuss two methods that work in a wide variety of cases.

14.2.1 Complementary functions

As defined above, complementary functions are the general solutions of reduced equations of the form

$$\sum_{i=0}^{n} a_i \frac{\mathrm{d}^{n-i}y}{\mathrm{d}x^{n-i}} = 0. \tag{14.22}$$

Equations like (14.22) are often called *homogeneous.*[2] These equations have the important property that if $y_1(x)$ and $y_2(x)$ are solutions, then any linear combination

$$Y(x) = A_1 y_1(x) + A_2 y_2(x) \tag{14.23}$$

is also a solution. This result follows directly on substituting (14.23) into the left-hand side of (14.22). In what follows, we shall discuss second-order equations in some detail, because they are by far the most important in physics applications; then we briefly address the extension to higher-orders.

The second-order homogeneous equation is

$$a\frac{\mathrm{d}^2y}{\mathrm{d}x^2} + b\frac{\mathrm{d}y}{\mathrm{d}x} + cy = 0, \tag{14.24}$$

[2] 'Homogeneous' is an over-used word in mathematics. The usage here is different from that in Section 14.1.3.

where we have relabelled the constants for later convenience. As a trial solution we will take

$$y = e^{mx}, \tag{14.25}$$

since with this form, differentiating y just multiplies it by a constant m. Substituting (14.25) in (14.24) gives

$$(am^2 + bm + c)e^{mx} = 0,$$

and (14.25) is a solution of (14.24) when the bracket vanishes, that is, when m is a root of the *auxiliary equation*

$$am^2 + bm + c = 0. \tag{14.26}$$

This has roots

$$m_1 = \frac{-b + \sqrt{b^2 - 4ac}}{2a} \quad \text{and} \quad m_2 = \frac{-b - \sqrt{b^2 - 4ac}}{2a}, \tag{14.27}$$

and three cases must be distinguished.

(a) If $b^2 > 4ac$, there are two real roots m_1 and m_2, with $m_1 \neq m_2$. The general solution of (14.24) is then, by the superposition principle (14.23),

$$y = A_1 e^{m_1 x} + A_2 e^{m_2 x}, \tag{14.28a}$$

where A_1 and A_2 are arbitrary constants.

(b) If $b^2 = 4ac$, then

$$m_1 = m_2 \equiv m = -b/2a, \tag{14.29}$$

so that (14.28a) would contain only one arbitrary constant $A \equiv A_1 + A_2$ and so cannot be the general solution. In this case, a second solution is obtained by writing

$$y = u(x)e^{mx},$$

where $u(x)$ is a function to be determined. Substituting into (14.24) then gives

$$a\frac{\mathrm{d}^2u}{\mathrm{d}x^2} + (2ma + b)\frac{\mathrm{d}u}{\mathrm{d}x} = 0. \tag{14.30}$$

However, using (14.29), we see that

$$2ma + b = 0,$$

so that the second term in (14.30) vanishes and we are left with the equation $\mathrm{d}^2u/\mathrm{d}x^2 = 0$, with solution

$$u = A_1 + A_2 x,$$

where again A_1 and A_2 are arbitrary constants. The general solution of (14.24) when the auxiliary equation has two equal roots is therefore

$$y = (A_1 + A_2 x)e^{mx}. \tag{14.28b}$$

(c) Finally, if $b^2 < 4ac$, there are two complex solutions

$$m = \alpha + i\beta, \quad m^* = \alpha - i\beta, \tag{14.31}$$

where $\alpha = -b/2a$ and $\beta = \sqrt{4ac - \beta^2}/2a$. Nonetheless, a real solution of (14.24) is obtained by writing

$$y = Ae^{mx} + A^* e^{m^* x},$$

where A is a complex constant. Using (14.31) and $e^z = \cos z + i \sin z$ for any z, this can be rewritten in the form

$$y = e^{\alpha x}(C \cos \beta x + D \sin \beta x), \tag{14.28c}$$

where $C = 2\text{Re}A$ and $D = -2\text{Im}A$ are two arbitrary real constants.

This exhausts the types of solution for homogeneous second-order linear equations with constant coefficients.

In the cases of higher order equations (14.20a), the substitution $y = e^{mx}$ leads to the auxiliary equation

$$a_n m^n + a_{n-1} m^{n-1} + \cdots + a_0 = 0. \tag{14.32}$$

This gives n roots, where degenerate and complex roots are treated in the same way as in the second-order case. In particular, if k roots $m_1, m_2, \cdots, m_k = m$ coincide, then the corresponding term in the solution, analogous to (14.28b), becomes

$$y = (A_1 + A_2 x + \cdots + A_{k-1} x^{k-1})e^{mx}. \tag{14.33}$$

Example 14.6

A *damped harmonic oscillator*[3] is described by an equation of the form

$$\frac{\mathrm{d}^2 x}{\mathrm{d}t^2} + \gamma \frac{\mathrm{d}x}{\mathrm{d}t} + \omega_0^2 x = 0,$$

where ω_0 is a frequency and $\gamma > 0$ is a real damping parameter. If $x(t = 0) = A$, solve the equations for (a) 'light damping', $\gamma < 2\omega_0$, (b) 'critical damping', $\gamma = 2\omega_0$, and (c) 'heavy damping', $\gamma > 2\omega_0$. Sketch the solution in each case

[3] As opposed to a *simple harmonic oscillator* with no damping ($\gamma = 0$) for which the solution is just the sum of sines and cosines.

Solution

(a) On substituting $x = e^{mt}$, the auxiliary equation is

$$m^2 + \gamma m + \omega_0^2 = 0, \qquad (14.34)$$

with solutions

$$m = -\gamma/2 \pm i\left(\omega_0^2 - \gamma^2/4\right)^{1/2}$$

for $\gamma < 2\omega_0$. The general solution is therefore of the form (14.28c),

$$x(t) = e^{-\gamma t/2}(A_1 \cos\omega t + A_2 \sin\omega t), \qquad (14.35a)$$

where $\omega = (\omega_0^2 - \gamma^2/4)^{1/2}$. Applying the boundary conditions $x = A, \mathrm{d}x/\mathrm{d}t = 0$ at $t = 0$ to determine A_1 and A_2 gives

$$x(t) = Ae^{-\gamma t/2}[\cos\omega t - (\gamma/2\omega)\sin\omega t] \approx Ae^{-\gamma t/2}\cos\omega t$$

for $\gamma \ll \omega$ with $\omega \approx \omega_0$. The resulting behaviour is shown in Figure 14.1a, where the distance between maxima is approximately $2\pi/\omega$.

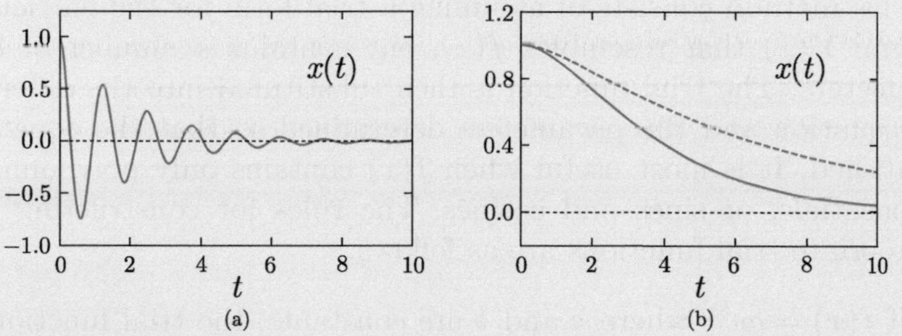

Figure 14.1 Motion $x(t)$ of a damped harmonic oscillator: (a) light damping $\gamma \ll 2\omega_0$; (b) critical damping $\gamma = 2\omega_0$ (solid line), and heavy damping $\gamma > 2\omega_0$ (dashed line). The units of t and $x(t)$ are arbitrary.

(b) For $\gamma = 2\omega_0$, the auxiliary equation (14.34) reduces to $(m + \gamma/2)^2 = 0$, so that we have degenerate roots $m_1 = m_2 = m = -\gamma/2$. The general solution is therefore

$$x(t) = (A_1 + A_2 t)e^{-\gamma t/2}$$

by (14.28b). On imposing the boundary conditions $x = A$, $\mathrm{d}x/\mathrm{d}t = 0$ at $t = 0$, this becomes

$$x(t) = A(1 + \gamma t/2)e^{-\gamma t/2}, \qquad (14.35b)$$

as sketched in Figure 14.1b.

(c) For $\gamma > 2\omega_0$, the solutions of (14.34) are $m = -\alpha_1, -\alpha_2$, where

$$\alpha_1 = \gamma/2 + (\gamma^2/4 - \omega_0^2)^{1/2}, \quad \alpha_2 = \gamma/2 - (\gamma^2/4 - \omega_0^2)^{1/2},$$

so that the general solution is

$$x(t) = A_1 e^{-\alpha_1 t} + A_2 e^{-\alpha_2 t}.$$

Since $\alpha_1, \alpha_2 > 0$ and $\alpha_2 < \gamma/2$, the solution tends to zero as $t \to \infty$ but at a slower rate than for critical damping. An example is shown as the dashed line in Figure 14.1b.

14.2.2 Particular integrals: method of undetermined coefficients

We now turn to the case where $f(x) \neq 0$ in (14.19), again restricting the discussion to the case of constant coefficients. We have already stated that the general solution of such an equation is the sum of the complementary function and a particular integral. The standard method for finding the complementary function has already been discussed, so it only remains to find the particular integral. Unfortunately, there is no general method for doing this, but there is a variety of methods, each of which is appropriate for a range of functions $f(x)$. We will discuss two, starting with that known as the *method of undetermined coefficients.*

This method consists of assuming a trial form for the particular integral $Y(x)$ that resembles $f(x)$, but contains a number of free parameters. The trial function is then substituted into the differential equation and the parameters determined so that the equation is satisfied. It is most useful when $f(x)$ contains only polynomials, exponentials, or sines and cosines. The rules for constructing the appropriate trial functions are as follows.

(i) If $f(x) = ae^{bx}$ where a and b are constants, the trial function is $Y(x) = Ae^{bx}$, where A is a constant.

(ii) If $f(x) = a\sin px + b\cos px$, where a, b and p are constants (a or b may be zero), the trial function is $Y(x) = A\sin px + B\cos px$, where A and B are constants.

(iii) If $f(x) = \sum_{n=0}^{N} a_n x^n$, where some of the coefficients may be zero, the trial function is $Y(x) = \sum_{n=0}^{N} b_n x^n$, where the b_n are constants.

(iv) If any term in these trial functions is contained within the complementary function, then the trial function must be multiplied by the smallest integer power of x such that it then contains no term that is in the complementary function.

(v) Finally, if $f(x)$ is the sum or product of any of these forms, the trial function must be taken to be the sum or product of the appropriate individual trial functions.

Example 14.7

Find the complete solution of the equation

$$\frac{d^2y}{dx^2} - 4\frac{dy}{dx} + 3y = e^{2x} + 3x^2.$$

Solution

The auxiliary equation is

$$(m^2 - 4m + 3) = (m - 1)(m - 3) = 0,$$

with solutions $m = 1, 3$, so the complementary solution is of the form

$$y(x) = Ae^x + Be^{3x},$$

where A and B are constants. The particular integral is found by using the trial function

$$ae^{2x} + (b + cx + dx^2).$$

Substituting this into to the ODE and equating coefficients of e^{2x} and $x^n (n = 0, 1, 2)$ gives

$$a = -1, b = 26/9, c = 8/3, d = 1.$$

Therefore the complete solution is

$$y(x) = Ae^x + Be^{3x} - e^x + \left(\frac{26}{9} + \frac{8}{3}x + x^2\right).$$

Example 14.8

The equation of a forced, damped harmonic oscillator is

$$\frac{d^2x}{dt^2} + \gamma\frac{dx}{dt} + \omega_0^2 x = f_0 \cos\omega t, \tag{1}$$

where γ is a damping coefficient, ω_0 and ω are the natural and forcing frequencies, respectively, and f_0 is a constant that determines the magnitude of the forcing term. Show that at large times $t \to \infty$, the solution can be written in the form

$$x(t) = C\cos(\omega t + \alpha),$$

and find expressions for C and α in terms of γ, ω, ω_0 and f_0.

Solution

The reduced equation is

$$\frac{d^2x}{dt^2} + \gamma\frac{dx}{dt} + \omega_0^2 x = 0.$$

This was solved in Example 14.6 and in all cases the solution goes exponentially to zero as $t \to \infty$. Hence we need only consider the particular integral as $t \to \infty$. Using the method of undetermined coefficients, the trial function is

$$x = A\cos\omega t + B\sin\omega t.$$

Substituting this into (1) gives

$$(\omega_0^2 - \omega^2)(A\cos\omega t + B\sin\omega t) - \gamma\omega(A\sin\omega t - B\cos\omega t) = f_0\cos\omega t.$$

Equating terms in $\sin\omega t$ and $\cos\omega t$ gives

$$B(\omega_0^2 - \omega^2) - A\gamma\omega = 0,$$

and

$$A(\omega_0^2 - \omega^2) + B\gamma\omega = f_0,$$

respectively, so that

$$A = \frac{f_0(\omega_0^2 - \omega^2)}{[(\omega_0^2 - \omega^2)^2 + \gamma^2\omega^2]}, \quad B = \frac{f_0\gamma\omega}{[(\omega_0^2 - \omega^2)^2 + \gamma^2\omega^2]}.$$

Now, writing

$$\begin{aligned} x = A\cos\omega t + B\sin\omega t &= C\cos(\omega t + \alpha) \\ &= C\cos\alpha\cos\omega t - C\sin\alpha\sin\omega t, \end{aligned}$$

we have

$$\tan\alpha = -B/A = \gamma\omega/(\omega^2 - \omega_0^2)$$

and

$$C^2 = A^2 + B^2 = \frac{f_0^2}{[(\omega_0^2 - \omega^2)^2 + \gamma^2\omega^2]},$$

so that, finally,

$$x(t) = \frac{f_0}{[(\omega_0^2 - \omega^2)^2 + \gamma^2\omega^2]^{1/2}}\cos(\omega t + \alpha) \qquad (2)$$

where

$$\alpha = \tan^{-1}\left(\frac{\gamma\omega}{\omega^2 - \omega_0^2}\right)$$

and the complementary function may be neglected in the limit as $t \to \infty$.

We see from (2) that the amplitude of the oscillations peaks when the forcing frequency equals the natural frequency. This is the well-known phenomenon of *resonance*.

*14.2.3 Particular integrals: the *D*-operator method

There are other methods for finding particular integrals, and by way of contrast we will discuss here the so-called *D-operator method*. This method has the advantage that it is not necessary to guess a trial function and the numerical coefficients multiplying the functional form of the particular integral are obtained automatically. However, it does require some experience in identifying which manipulations to use to obtain the solution.

The quantity D, defined by $D \equiv \mathrm{d}/\mathrm{d}x$, was introduced in Chapter 9, Section 9.3.2. It is a differential operator. That is, it only has a meaning when it acts on a function, which is always written to the right of D. Higher differential operators may be constructed from D. For example,

$$D^2 y = D(Dy) = \frac{\mathrm{d}^2 y}{\mathrm{d}x^2} \quad \Rightarrow \quad D^2 = \frac{\mathrm{d}^2}{\mathrm{d}x^2},$$

$$D^3 y = D(D^2 y) = \frac{\mathrm{d}^3 y}{\mathrm{d}x^3} \quad \Rightarrow \quad D^3 = \frac{\mathrm{d}^3}{\mathrm{d}x^3},$$

etc. We see that D satisfies the usual rules of algebra and so may be formally treated as an algebraic quantity, despite the fact that it cannot be evaluated as such to yield a numerical value. From the algebraic rules of differentiation it follows that if $f(x)$ and $g(x)$ are differentiable functions,

$$D[f(x) + g(x)] = Df(x) + Dg(x) \tag{14.36a}$$

and

$$D[cf(x)] = cD[f(x)], \tag{14.36b}$$

where c is a constant, and hence D is a linear operator (see Section 9.3.2).

In terms of D we may now write (14.19) as

$$F(D)y = f(x), \tag{14.37a}$$

where

$$F(D) \equiv a_0 D^n + a_1 D^{n-1} + \cdots + a_{n-1} D + a_n \tag{14.37b}$$

is a polynomial operator in D of order n. Just as D may be formally treated as an algebraic quantity, so may a polynomial function of D, such as $F(D)$, and in particular it may, in suitable cases, be factorised. Moreover the order of the factors is irrelevant provided the coefficients in (14.37) are constants. Thus, for example,

$$(D+1)(D+2)y = (D+2)(D+1)y.$$

However, the order is relevant if the coefficients are functions of x. For example,

$$(D+1)(D+2x)y = D^2 y + Dy + D\,(2xy) + 2xy,$$

whereas

$$(D+2x)(D+1)y = D^2 y + Dy + 2xD\,y + 2xy,$$

so

$$(D+1)(D+2x)y \neq (D+2x)(D+1)y.$$

A number of useful results may be derived when the function $f(x) = e^{kx}$, where k is a constant. For example, using the fact that $D^n e^{kx} = k^n e^{kx}$, it follows that

$$F(D)e^{kx} = F(k)e^{kx}. \tag{14.38a}$$

Similarly, since $D^2 \sin kx = -k^2 \sin kx$, it follows that

$$F(D^2)\sin kx = F(-k^2)\sin kx, \tag{14.38b}$$

and

$$F(D^2)\cos kx = F(-k^2)\cos kx. \tag{14.38c}$$

If the exponential is multiplied by an arbitrary function $V(x)$, then by considering the terms $D\left[e^{kx}V(x)\right]$, $D^2\left[e^{kx}V(x)\right]$, etc. in succession, the result (14.38a) may be generalised in a straightforward way to

$$F(D)\left[e^{kx}V(x)\right] = e^{kx}F(D+k)V(x) \tag{14.38d}$$

Finally, in Chapter 3 we defined indefinite integration as the inverse operation of differentiation. In an analogous way we now defined the inverse operator D^{-1} by

$$D^{-1} \equiv \frac{1}{D} \equiv \int, \tag{14.39}$$

with

$$D\left(D^{-1}y\right) = y. \tag{14.40}$$

Also by analogy with the notation D^n for successive differentiations, we will use $D^{-n} \equiv 1/D^n$ for the operation of n successive integrations.

We now return to the solution of equations of the type (14.19). If these are rewritten in the form (14.37a) and (14.37b), then a particular integral can be obtained by writing

$$y = \frac{1}{F(D)}f(x), \tag{14.41}$$

provided we can interpret and evaluate the right-hand side, using the techniques introduced above. To illustrate this, consider the equation

$$\frac{\mathrm{d}^2y}{\mathrm{d}x^2} + 2\frac{\mathrm{d}y}{\mathrm{d}x} - 3y = \cos x. \tag{14.42}$$

In the D-operator formalism, this is written

$$(D^2 + 2D - 3)y = \cos x,$$

giving the particular integral

$$y(x) = \frac{1}{D^2 + 2D - 3} \cos x$$

We now have to decide how to handle the polynomial in the denominators, using the relations (14.38). If we use (14.38c) with $k = 1$, we have

$$y(x) = \left[\frac{1}{-1 + 2D - 3}\right] \cos x = \frac{1}{2}\left[\frac{D+2}{D^2 - 4}\right] \cos x.$$

Now we can again use (14.38c) to give

$$y(x) = -\frac{1}{10}(D+2)\cos x = \frac{1}{10}(\sin x - 2\cos x)$$

as the desired form for the particular integral.

A different technique is illustrated by considering the equation obtained by replacing $\cos x$ in (14.42) by xe^{2x}. The particular integral is then

$$y(x) = \frac{1}{D^2 + 2D - 3} xe^{2x}$$

and using (14.38d) with $k = 2$ gives

$$y(x) = e^{2x}\left[\frac{1}{(D+2)^2 + 2(D+2) - 3}\right] x = e^{2x}\left[\frac{1}{D^2 + 6D + 5}\right] x.$$

The denominator can be expressed as partial fractions, to give

$$y(x) = e^{2x}\left[\frac{1}{D^2 + 6D + 5}\right] x = \frac{e^{2x}}{4}\left[\frac{1}{1+D} - \frac{1}{5+D}\right] x. \quad (14.43)$$

The point of this decomposition is that each fraction can be expanded as a binomial series

$$(1+D)^{-1} = 1 - D + \cdots \quad \text{and} \quad (5+D)^{-1} = \tfrac{1}{5} - \tfrac{1}{25}D + \cdots,$$

where higher derivatives are not required because when acting on x they will be zero. Using these expansions, (14.43) becomes

$$\frac{e^{2x}}{100}(20 - 24D)x = \frac{e^{2x}}{25}(5x - 6),$$

which is the desired result.

Example 14.9

Evaluate the following expressions:

(a) $(3D^2 - 2D + 5)e^{4x}$, (b) $(2D^2 - 3D - 1)[x^4 e^{2x}]$
(c) $(D^4 - D^2 + 3)\cos 2x$.

Solution

(a) Using (14.38a),

$$(3D^2 - 2D + 5)e^{4x} = (3 \times 4^2 - 2 \times 4 + 5)e^{4x} = 45e^{4x}.$$

(b) Using (14.38d) with $V(x) = x^4$ gives

$$\begin{aligned}(2D^2 - 3D - 1)[x^4 e^{2x}] &= e^{2x}[2(D+2)^2 - 3(D+2) - 1]x^4 \\ &= e^{2x}(2D^2 + 5D + 1)x^4 \\ &= (x^4 + 20x^3 + 24x^2)e^{2x}.\end{aligned}$$

(c) Using (14.38c),

$$\begin{aligned}(D^4 - D^2 + 3)\cos 2x &= \left[(D^2)^2 - D^2 + 3\right]\cos 2x \\ &= \left[(-2^2)^2 - (-2^2) + 3\right]\cos 2x \\ &= 23\cos 2x.\end{aligned}$$

Example 14.10

Use the D-operator method to find the complete solution of the equation

$$\frac{\mathrm{d}^2 y}{\mathrm{d}x^2} - 2\frac{\mathrm{d}y}{\mathrm{d}x} = 3x.$$

Solution

We first find the complementary function as usual by setting $y = e^{mx}$ in the homogeneous equation

$$\frac{\mathrm{d}^2 y}{\mathrm{d}x^2} - 2\frac{\mathrm{d}y}{\mathrm{d}x} = 0.$$

This gives $m^2 - 2m = 0$ and hence $m = 0, 2$. The complementary function is therefore

$$y(x) = A + Be^{2x},$$

where A and B are constants. The particular integral is found from

$$y = \frac{3}{D^2 - 2D}x = \frac{3}{D(D-2)}x,$$

which may be written as partial fractions

$$y = -\frac{3}{2}\left(\frac{1}{D} + \frac{1}{2-D}\right)x.$$

The first term is an integral, by (14.39), and the second may be written as a binomial retaining only terms up to D, so that

$$y = -\frac{3}{2}\left(\frac{1}{D} + \frac{1}{2} + \frac{1}{4}D\right)x = -\frac{3}{8}\left(2x^2 + 2x + 1\right).$$

Finally, adding the complementary function and the particular integral gives the complete solution

$$y(x) = A + Be^{2x} + 3x^2 + \frac{3}{4}x - \frac{3}{8}.$$

*14.2.4 Laplace Transforms

The D-operator method converts a differential equation to an algebraic equation for D. Another method that does something very similar is based on the use of Laplace transforms. The *Laplace transform* $F(p)$ of a function $f(x)$ is defined by

$$L[f(x)] \equiv \int_0^\infty f(x)e^{-px}\,\mathrm{d}x = F(p), \tag{14.44}$$

where the parameter p may in principle be complex, but in the discussion that follows we will take it to be real.

The Laplace transforms of many simple functions may be found by direct evaluation of the defining integral (14.44). Others may then be found by using easily proved general properties of the Laplace transform. These include: *linearity*,

$$L[af(x) + bg(x)] = aL[f(x)] + bL[g(x)],$$

where a and b are constants; and the *shift theorem*

If $L[f(x)] = F(p)$, then $L[e^{ax}f(x)] = F(p-a)$.

A related version of this, called the *translation property*, is

$$L\left[H(x-a)f(x-a)\right] = e^{-ap}F(p),$$

where $H(x)$ is the unit step function

$$H(x) = \begin{cases} 0 & x < 0 \\ 1 & 0 < x \end{cases}$$

For example, the translation property follows by considering the Laplace transform of

$$g(x) = H(x-a)f(x-a),$$

Table 14.1 Laplace transforms of some simple functions[4]

$f(x)$	$L[f(x)] = F(p)$	$f(x)$	$L[f(x)] = F(p)$
1	$\frac{1}{p}, p > 0$	$x^n\ (n = 1, 2, \ldots)$	$\frac{n!}{p^{n+1}}, p > 0, n > -1$
$\sqrt{x}$	$\frac{1}{2p}\sqrt{\frac{\pi}{p}}, p > 0$	$\frac{1}{\sqrt{x}}$	$\sqrt{\frac{\pi}{p}}, p > 0$
$e^{ax}\,(a \neq 0)$	$\frac{1}{p-a}, p > a$	$x^n e^{-ax}$	$\frac{n!}{(p+a)^{n+1}}, p > -a$
$\sin(ax)$	$\frac{a}{p^2+a^2}, p > 0$	$\cos(ax)$	$\frac{p}{p^2+a^2}, p > 0$
$e^{-ax}\sin(bx)$	$\frac{b}{(p+a)^2+b^2}, p > -a$	$e^{-ax}\cos(bx)$	$\frac{p+a}{(p+a)^2+b^2}, p > -a$
$x\sin(ax)$	$\frac{2ap}{(p^2+a^2)^2}, p > 0$	$x\cos(ax)$	$\frac{p^2-a^2}{(p^2+a^2)^2}, p > 0$
$\sinh(ax)$	$\frac{a}{p^2-a^2}, p > 0$	$\cosh(ax)$	$\frac{p}{p^2-a^2}, p > 0$

which has the same form as $f(x)$ but moved a distance a along the x axis. If we let $z = x - a$, then

$$L\left[g(x)\right] = \int_0^\infty e^{-px} g(x)\mathrm{d}x = e^{-ap}\int_0^\infty e^{-pz} f(z)\mathrm{d}z = e^{-ap}F(p).$$

Using a combination of these properties, several useful examples of Laplace transforms may be obtained and are shown in Table 14.1. As an example, we shall derive the result for $e^{-ax}\cos(bx)$. Firstly,

$$\begin{aligned} L[\cos(bx)] &= \int_0^\infty \cos(bx)e^{-px}\mathrm{d}x = \mathrm{Re}\left(\int_0^\infty e^{ibx}e^{-px}\mathrm{d}x\right) \\ &= \mathrm{Re}\left[-\frac{e^{-x(p-ib)}}{p-ib}\right]_0^\infty = \frac{p}{p^2+b^2}. \end{aligned}$$

Then, using the shift theorem,

$$L[e^{-ax}\cos(bx)] = \frac{p+a}{(p+a)^2+b^2}, \quad p > -a.$$

We can also define an *inverse Laplace transform*, denoted L^{-1}, such that

$$f(x) = L^{-1}[F(p)].$$

[4]Extensive tables of Laplace transform pairs may be found in books of mathematical formulas, for example: Alan Jeffreys and Hui-Hui Dai (2008) *Handbook of Mathematical Formulas and Integrals*, 4th edn., Academic Press., New York, pp. 342–352.

It follows that $LL^{-1} = 1$ and because L is linear, so is L^{-1}. Inverse transforms for some simple functions follow from the results of Table 14.1. Thus, since $L[1] = 1/p$, it follows that $L^{-1}[1/p] = 1$, but to find inverse transforms in general requires the techniques of complex variable theory and we will not discuss them here.

In order to use Laplace transforms to solve differential equations, we will also need the transforms of the differentials of a function $y(x)$. For example, consider the Laplace transform of $\mathrm{d}y(x)/\mathrm{d}x$. From the definition (14.44), this is

$$L\left[\frac{\mathrm{d}y(x)}{\mathrm{d}x}\right] = \int_0^\infty e^{-px}\frac{\mathrm{d}y(x)}{\mathrm{d}x}\mathrm{d}x.$$

Integrating by parts, this is

$$\begin{aligned} L\left[\frac{\mathrm{d}y(x)}{\mathrm{d}x}\right] &= \left[ye^{-px}\right]_0^\infty + p\int_0^\infty e^{-px}y(x)\,\mathrm{d}x \\ &= -y(0) + pL[y(x)], \end{aligned}$$

providing $ye^{-px} \to 0$ as $x \to \infty$. So,

$$L\left[\frac{\mathrm{d}y(x)}{\mathrm{d}x}\right] = -y(0) + pF(p). \tag{14.45a}$$

Likewise,

$$\begin{aligned} L\left[\frac{\mathrm{d}^2y(x)}{\mathrm{d}x^2}\right] &= \int_0^\infty e^{-px}\frac{\mathrm{d}^2y(x)}{\mathrm{d}x^2}\,dx \\ &= \left[e^{-px}\frac{\mathrm{d}y(x)}{\mathrm{d}x}\right]_0^\infty + p\int_0^\infty e^{-px}\frac{\mathrm{d}y(x)}{\mathrm{d}x}\,\mathrm{d}x, \end{aligned}$$

so that

$$L\left[\frac{\mathrm{d}^2y(x)}{\mathrm{d}x^2}\right] = p^2F(p) - py(0) - y'(0),$$

where

$$y'(0) \equiv \left[\frac{\mathrm{d}y(x)}{\mathrm{d}x}\right]_{x=0}.$$

This procedure may be repeated to obtain expressions for derivatives of any order. These can be used to convert a differential equation into an algebraic equation in $F(p)$, and if the inverse transform L^{-1} can be found, the solution of the equation for $F(p)$ can be inverted, and a solution of the original equation ODE results. An advantage of the method is that it can easily incorporate boundary conditions on the function and its first derivative, as can be seen from (14.45). Its use is illustrated in Example 14.10.

Fourier transforms can similarly be used to convert a differential equation for $y(x)$ into an algebraic equation for its Fourier transform. However, in contrast to Laplace transforms, their use is restricted to functions that tend to zero as $|x|$ tends to infinity sufficiently rapidly for (13.45) to be satisfied. In these cases, it can be a useful technique, as illustrated in Problem 14.19, but we will not pursue this here.

It frequently happens that part of the expression for which the inverse transform L^{-1} is to be found contains the product of two Laplace transforms. In this the case we can use the method of convolutions that was discussed in Chapter 13, Section 13.3.4 for Fourier transforms. Thus if $g_i(p)$ is the Laplace transform of $f_i(p)$, that is, if

$$g_i(p) \equiv L[f_i(x)], \tag{14.46}$$

where $i = 1,2$, then

$$L[f_1 * f_2] = g_1 g_2 \tag{14.47a}$$

and hence

$$L^{-1}[g_1 g_2] = f_1 * f_2, \tag{14.47b}$$

where the convolution integral is

$$f_1 * f_2 = \int_0^x f_1(x-u) f_2(u)\, \mathrm{d}u.$$

To prove (14.47a), we start from the definition (14.46) and form the product

$$g_1(p)g_2(p) = \int_0^\infty f_1(u)e^{-pu}\mathrm{d}u \int_0^\infty f_2(v)e^{-pv}\mathrm{d}v$$

$$= \int_0^\infty du \int_0^\infty \mathrm{d}v\, f_1(u) f_2(v) e^{-(u+v)}, \tag{14.47c}$$

where u and v are dummy variables. Now letting $x = u + v$, and rewriting in terms of the variables t and u, changes the limits on the integrals, giving

$$g_1(p)g_2(p) = \int_0^\infty f_1(u)\,\mathrm{d}u \int_u^\infty f_2(x-u)e^{-px}\mathrm{d}x.$$

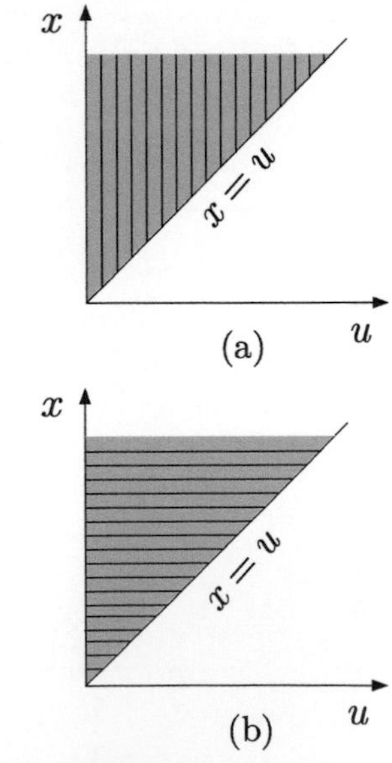

Figure 14.2 Order of integration in (14.47c).

This corresponds to summing the vertical strips on Figure 14.2a. But from the work of Chapter 11, we know that we can equally sum the horizontal strips, as shown in Figure 14.2b, that is, we can reverse the order of integrations. Therefore the double integral may be written

$$g_1 g_2 = \int_0^\infty e^{-px} \mathrm{d}x \int_0^x f_1(u) f_2(x-u)\,\mathrm{d}u = L\left[\int_0^x f_1(u)(x-u)\,\mathrm{d}u\right]$$
$$= L\left[f_1 * f_2\right],$$

and so (14.47a) follows, which completes the proof.

Example 14.11

Use the Laplace transform method to solve the differential equation

$$y''(x) - y(x) = x,$$

subject to the boundary conditions $y(x) = 1$ and $y'(x) = -2$ at $x = 0$.

Solution

Taking the Laplace transform of the ODE we have

$$L[y''] - L[y] = L[x],$$

and using Table 14.1 and Equation (14.45b) gives

$$p^2 F(p) - py(0) - y'(0) - F(p) = 1/p^2.$$

Rearranging and using the boundary conditions, we have the solution

$$F(p) = \frac{p-2}{(p^2-1)} + \frac{1}{p^2(p^2-1)} = \frac{p}{p^2-1} - \frac{1}{p^2-1} - \frac{1}{p^2}.$$

We now find $y(x)$ from the inverse Laplace transform, that is, $y(x) = L^{-1}[F(p)]$. Using Table 13.1 again,

$$y(x) = \cosh(x) - \sinh(x) - x = e^{-x} - x.$$

Note that, in this case, one could alternatively construct the complementary function and then find a particular integral using the method of undetermined coefficients. This leads to the general solution

$$y(x) = ae^{x} + Be^{-x} + x,$$

where the arbitrary constants A and B are determined by the boundary conditions to give the same final answer.

Example 14.12

Solve the equation

$$y''(x) - y(x) = g(x),$$

with the boundary conditions $y(0) = a$ and $y'(0) = b$, where a and b are constants and $g(x)$ is an unknown function of x. Confirm that the solution is consistent with that in Example 14.10.

Solution

Taking Laplace transforms of both sides of the equation and using (14.45a) and (14.45b), we have

$$p^2 F(p) - py(0) - y'(0) - F(p) = G(p),$$

where $g(p)$ is the Laplace transform of $g(x)$. Imposing the boundary conditions gives

$$F(p) = \frac{pa + b}{p^2 - 1} + \frac{G(p)}{p^2 - 1}$$

and hence

$$y(x) = aL^{-1}\left[\frac{p}{p^2 - 1}\right] + bL^{-1}\left[\frac{1}{p^2 - 1}\right] + L^{-1}\left[\frac{G(p)}{p^2 - 1}\right].$$

The first two terms may be evaluated using the results given in Table 14.1 and the third by the use of those results together with (14.47b). This gives the final implicit solution

$$y(x) = a\cosh x + b\sinh x + \int_0^x g(x - u)\sinh u\, du.$$

To compare with Example 14.10, we use $a = 1, b = -2$ and $g(x) = x$, so that

$$y(x) = \cosh x - 2\sinh x + \int_0^x (x - u)\sinh u\, du$$

Integrating by parts gives

$$\int_0^x (x - u)\sinh u\, du = [(x - u)\cosh u + \sinh u]_0^x = \sinh x - x.$$

So, finally

$$y(x) = \cosh(x) - \sinh(x) - x = e^{-x} - x,$$

in agreement with Example 14.10.

*14.3 Euler's equation

The discussion in Section 14.2 has been exclusively about linear equations with constant coefficients. However, some linear equations with variable coefficients $a_i(x)$ can be reduced to linear form with constant coefficients by a suitable transformation. Perhaps the best known of these is *Euler's equation*

$$ax^2\frac{\mathrm{d}^2y}{\mathrm{d}x^2} + bx\frac{\mathrm{d}y}{\mathrm{d}x} + cy = f(x). \tag{14.48}$$

On substituting $x = e^t$, one obtains

$$\frac{\mathrm{d}y}{\mathrm{d}x} = \frac{\mathrm{d}y}{\mathrm{d}t}\frac{\mathrm{d}t}{\mathrm{d}x} = \frac{1}{x}\frac{\mathrm{d}y}{\mathrm{d}t}$$

and

$$\frac{\mathrm{d}^2y}{\mathrm{d}x^2} = -\frac{1}{x^2}\frac{\mathrm{d}y}{\mathrm{d}t} + \frac{1}{x}\frac{1}{x}\frac{\mathrm{d}^2y}{\mathrm{d}t^2},$$

so that if $f(x)$ becomes $\tilde{f}(t)$ on changing variables, (14.48) becomes

$$a\frac{\mathrm{d}^2y}{\mathrm{d}t^2} + (b-a)\frac{\mathrm{d}y}{\mathrm{d}t} + cy = \tilde{f}(t). \tag{14.49}$$

This is a linear equation with constant coefficients and so may be solved by the methods of the last section. More generally, the nth order Euler equation

$$\sum_{i=0}^{n} a_i x^i \frac{\mathrm{d}^i y}{\mathrm{d}x^i} = f(x) \tag{14.50}$$

is reduced to a linear equation with constant coefficients by the same substitution $x = e^t$.

If $f(x) = 0$, then (14.50) can usually be solved more easily by substituting $y = x^p$, which gives

$$\left(\sum_{i=0}^{n} c_i p^i\right) x^p = 0,$$

where the constants c_i depend on the coefficients a_i. Hence if the polynomial in brackets has n distinct roots $p_1, p_2, \ldots, p_n$, the general solution to (14.50) with $f(x) = 0$ is

$$y(x) = \sum_{i=1}^{n} A_i x^{p_i},$$

where $A_1, A_2, \ldots, A_n$ are arbitrary constants. On the other hand, if the roots are not distinct, this simple method fails to give the general solution, which can still be found using the substitution $x = e^t$.

Example 14.13

In potential theory, one often meets the equation

$$\frac{d^2R}{dr^2} + \frac{2}{r}\frac{dR}{dr} - \frac{n(n+1)R}{r^2} = 0, \tag{14.51}$$

where $R(r)$ is a function of the distance from the origin r and $n \geq 0$ is a constant. Solve (14.51) subject to the boundary condition $R \to 0$ as $r \to \infty$.

Solution

On multiplying by r^2, (14.51) becomes

$$r^2\frac{d^2R}{dr^2} + 2r\frac{dR}{dr} - n(n+1)R = 0,$$

which is Euler's equation (14.49) with $a = 1, b = 2$, $c = -n(n+1)$ and $f(x) = 0$. Substituting $R = r^p$ gives

$$\left[p^2 + p - n(n+1)\right] x^p = (p-n)(p+n+1)r^p = 0$$

with two distinct solutions $p = n, p = -(n+1)$. Hence the general solution is

$$R = A_1 r^n + \frac{A_2}{r^{n+1}} = \frac{A_2}{r^{n+1}},$$

since $R \to 0$ as $r \to \infty$ implies $A_1 = 0$.

Example 14.14

Solve the equation

$$x^2\frac{d^2y}{dx^2} - x\frac{dy}{dx} + y = (\ln x)^2.$$

Solution

Substituting $x = e^t$ gives

$$\frac{d^2y}{dt^2} - 2\frac{dy}{dt} + y = t^2. \tag{1}$$

To find the complementary function, substitute $y = e^{mx}$ into the homogeneous equation

$$\frac{d^2y}{dt^2} - 2\frac{dy}{dt} + y = 0.$$

The resulting auxiliary equation is

$$m^2 - 2m + 1 = 0$$

which has a single solution $m = 1$. Hence the complementary function is [cf. (14.28b)]

$$y = (A + Bt)e^t = Ax + Bx \ln x,$$

where A and B are arbitrary constants.

Using the method of undetermined coefficients, the particular integral is of the form

$$y = b_0 + b_1 t + b_2 t^2,$$

where the constants b_0, b_1 and b_2 are to be determined. Substituting this into (1) gives

$$2b_2 - 2b_1 - 4b_2 t + b_0 + b_1 t + b_2 t^2 = t^2,$$

so that $b_0 = 6, b_1 = 4$ and $b_2 = 1$. Hence the particular integral is

$$y = t^2 + 4t + 6 = (\ln x)^2 + 4\ln x + 6$$

and the general solution is

$$y = Ax + Bx \ln x + (\ln x)^2 + 4\ln x + 6.$$

Problems 14

14.1 Find the solution of the equations

(a) $\mathrm{d}y/\mathrm{d}x = x^3 + x^2 + 2$, if $y = 2$ when $x = 1$.

(b) $\mathrm{d}y/\mathrm{d}x = y^2 - 2y + 1$, if $y = 2$ when $x = 0$.

14.2 Find the solutions of the equations

(a) $\dfrac{\mathrm{d}y}{\mathrm{d}x} = \dfrac{x + y}{x + y + 2}$, if $y = -1$, when $x = 1$,

(b) $\dfrac{\mathrm{d}y}{\mathrm{d}x} = \dfrac{x^2 e^{-y}}{x^2 + 1}$, if $y = 0$, when $x = 1$.

14.3 Show that the Fourier transform of $\exp(-\alpha x^2)$ with $\alpha > 0$ is

$$g(k) \equiv F\left[e^{-\alpha x^2}\right] = \frac{1}{2\sqrt{\pi\alpha}} e^{-k^2/4\alpha}, \tag{1}$$

by using the symmetry of $\exp(-\alpha x^2)$ to derive the equation

$$\frac{\mathrm{d}g(k)}{\mathrm{d}k} = -\frac{k}{2\alpha} g(k),$$

and then solving it subject the boundary condition

$$g(0) \equiv \frac{1}{2\pi}\int_{-\infty}^{\infty} e^{-\alpha x^2}\,\mathrm{d}x = \frac{1}{2\pi}\sqrt{\frac{\pi}{\alpha}},$$

which follows from the standard integral derived in Example 11.11. Finally, use (1) to verify (13.71).

14.4 Find the solution of the equation

$$\frac{\mathrm{d}y}{\mathrm{d}x} = \frac{y^3 - xy^2 - 7x^2y + 6x^3}{xy^2 - 2x^2y - 2x^3}.$$

14.5 Find the solution of the equation

$$\frac{\mathrm{d}y}{\mathrm{d}x} = \frac{2x + y + 1}{x - 2y + 1}.$$

14.6 Establish whether each of the following equations is exact and find the solutions of those that are:

$$(a)\ (y^3 - 3x^2) + (2xy - x^3)\frac{\mathrm{d}y}{\mathrm{d}x} = 0,$$

$$(b)\ \frac{y^2}{x} + 2y\ln x\frac{\mathrm{d}y}{\mathrm{d}x} = 0, (x > 0).$$

14.7 Find the solutions of the equations

$$\text{(a)}\ (x^2 - 2xy) - (x^2 - y + 1)\frac{\mathrm{d}y}{\mathrm{d}x} = 0,$$

$$\text{(b)}\ (8y - x^2y)\frac{\mathrm{d}y}{\mathrm{d}x} + (x - xy^2) = 0.$$

14.8 Find the solutions of the equations

$$\text{(a)}\ \frac{1}{x^3}\frac{\mathrm{d}y}{\mathrm{d}x} - \frac{2y}{x^4} = \sin x, \quad \text{(b)}\ \tan x\frac{\mathrm{d}y}{\mathrm{d}x} + y = \sin x.$$

14.9 Find the solutions of the equations

$$\text{(a)}\ 2\frac{\mathrm{d}y}{\mathrm{d}x} = y + 2\sin x, \text{ given that } y = -1 \text{ when } x = 0.$$

$$\text{(b)}\ (x+1)\frac{\mathrm{d}y}{\mathrm{d}x} = 2y + (x+1)^{5/2}\ , \text{ given that } y = 3 \text{ when } x = 0.$$

14.10 Solve the non-linear equation

$$\frac{\mathrm{d}y}{\mathrm{d}x} - y + 3x^2y^3 = 0,$$

using the substitution $z = y^{-2}$.

14.11 Find the solution of the equations

(a) $\dfrac{d^2y}{dx^2} - 2\dfrac{dy}{dx} + y = 0$, given that $y = (2,1)$ when $x = (1,0)$, respectively,

(b) $\dfrac{d^2y}{dx^2} + 6\dfrac{dy}{dx} + 13y = 0$ given that $y = 3$ and $y' = 7$ at $x = 0$.

14.12 Use the method of undetermined coefficients to find the complete solutions of the equations

(a) $\dfrac{d^2y}{dx^2} + 4\dfrac{dy}{dx} - 5y = 3e^x$, (b) $\dfrac{d^2y}{dx^2} + 2\dfrac{dy}{dx} + 5y = \sin x$.

14.13 Use the method of undetermined coefficients to find the complete solution of the equation

$$\frac{d^2y}{dx^2} - 5\frac{dy}{dx} + 6y = 2x^2e^x.$$

14.14 Find the complete solutions of the equations

(a) $\dfrac{d^2y}{dx^2} + \dfrac{dy}{dx} - 2y = e^{-2x}$, (b) $\dfrac{d^2y}{dx^2} - 3\dfrac{dy}{dx} + 2y = x^2$.

14.15 Find the general solution of the equations

(a) $\dfrac{d^2y}{dx^2} + 4\dfrac{dy}{dx} + 4y = 4e^{-2x}$, (b) $\dfrac{d^3y}{dx^3} - 3\dfrac{d^2y}{dx^2} = 36x^2$.

***14.16** Use the *D*-operator method to find the particular integrals for the equations

(a) $\dfrac{d^2y}{dx^2} - 3\dfrac{dy}{dx} + 2y = \cos 3x$, (b) $\dfrac{d^2y}{dx^2} + 3\dfrac{dy}{dx} - 2y = e^{3x}$.

***14.17** Use the *D*-operator method to find the particular integral for the equation

$$\frac{d^2y}{dx^2} + 3\frac{dy}{dx} + 2y = xe^{-2x}.$$

14.18 Find the complete solution of the equation

$$\frac{d^3y}{dx^3} + 3\frac{d^2y}{dx^2} + 3\frac{dy}{dx} + y = \sinh 2x.$$

14.19 A function ϕ satisfies the equation

$$\frac{d^2\phi}{dx^2} - C^2\phi = f(x),$$

where C is a constant and $f(x)$ is an arbitrary function such that

$$\int_{-\infty}^{\infty} |f(x)|\,dx. \qquad (1)$$

Use the relation (13.53b) to show that

$$\phi(x) = -\int_{-\infty}^{\infty} \frac{e^{ikx} h(k)}{k^2 + C^2} \, dk,$$

is a solution, where $h(k)$ is the Fourier transform of $f(x)$. What is the general solution?

***14.20** Use the Laplace transform method to find the solution of the equation

$$\frac{d^2 y(x)}{dx^2} - 3\frac{dy(x)}{dx} + 2y(x) = 3e^{2x},$$

subject to the boundary conditions $y(x) = -2$ and $y'(x) = 4$ at $x = 0$.

***14.21** A one-dimensional system undergoes forced simple harmonic motion, with an equation of motion

$$x''(t) + \omega_0^2 x(t) = A \sin(\omega t),$$

where ω_0 is the natural frequency of vibration and A is a constant. Solve for $x(t)$ with the boundary conditions $x(0) = 1$ and $x'(0) = 0$, using the Laplace transform method.

14.22 Solve Problem 14.21 using the method of undetermined coefficients.

***14.23** Find the solution of the equation

$$y''(x) - 3y'(x) + 2y(x) = h(x),$$

subject to the boundary conditions $y(0) = 1$ and $y'(0) = 0$, where $h(x)$ is an unknown function of x.

***14.24** Solve the equation

$$x^2 \frac{d^2 y}{dx^2} + 3x \frac{dy}{dx} - 3y = 4x - 3.$$

***14.25** Find the complete solution of the equation

$$x^2 \frac{d^2 y}{dx^2} + 4x \frac{dy}{dx} + 2y = \sin(3 \ln x).$$

***14.26** If $x = e^t$, show that

$$x^n \frac{d^n y}{dx^n} = \left[\frac{d}{dt} - (n-1)\right] x^{n-1} \frac{d^{n-1} y}{dx^{n-1}}.$$

Hence show that the Euler equation (14.50) is reduced to a linear equation with constant coefficients for any order n by the substitution $x = e^t$, as asserted in the text.

15

Series solutions of ordinary differential equations

In this chapter we will extend our discussion of ordinary differential equations (ODEs) to include linear second-order equations of the form

$$\frac{d^2y}{dx^2} + p(x)\frac{dy}{dx} + q(x)y = 0, \tag{15.1}$$

where the coefficients $p(x)$ and $q(x)$ are no longer restricted to constants, but may be arbitrary functions. Many ways of solving such equations apply only to a very limited range of equations, or require some prior knowledge of the solution. One such method will be mentioned at the end of Section 15.1.3. Otherwise we will confine ourselves to the most important method, which is to seek a solution in the form of a power series expansion about a particular point $x = x_0$.

This method is introduced in Section 15.1 and then, after a brief discussion of differential operators and eigenvalue equations, illustrated by applying it to two eigenvalue equations that are particularly important in physics.

15.1 Series solutions

The existence of solutions in the form of power series expansions about a particular point $x = x_0$ depends on the behaviours of $p(x)$ and $q(x)$ in the neighbourhood of x_0. Three types of behaviour need to be distinguished. If $p(x)$ and $q(x)$ are finite, single-valued and differentiable, then x_0 is called a *regular* or *ordinary point* and (15.1) is said to be *regular* at $x = x_0$. In this case, the limits of $p(x)$ and $q(x)$ as $x \to x_0$ both exist, that is, are finite. If either of these limits

Mathematics for Physicists, First Edition. B.R. Martin and G. Shaw.

Companion website: www.wiley.com/go/martin/mathsforphysicists

diverges, the point is called a *singular point* and the equation is said to be *singular* at $x = x_0$. If $x = x_0$ is a singular point, but the limits

$$\lim_{x \to x_0} (x - x_0)p(x) \quad \text{and} \quad \lim_{x \to x_0} (x - x_0)^2 q(x) \tag{15.2}$$

both exist, the ODE is said to have a *regular singularity* at $x = x_0$. If either of the limits (15.2) diverges, the equation is said to have an *essential singularity* at $x = x_0$.

The importance of this classification is that if x_0 is a regular point, $p(x)$ and $q(x)$ can be expanded as Taylor series about x_0, i.e.

$$p(x) = p_0 + p_1(x - x_0) + \cdots$$

and

$$q(x) = q_0 + q_1(x - x_0) + \cdots$$

In this case, it may be shown[1] that the general solution can be expanded in a series of the form

$$y(x) = \sum_{n=0}^{\infty} a_n (x - x_0)^n, \tag{15.3a}$$

where the radius of convergence depends on the particular equation. For regular singular points, a theorem due to Fuchs[2] shows that there exists at least one solution in the form of a *generalised power series*

$$y(x) = x^c \sum_{n=0}^{\infty} a_n (x - x_0)^n, \tag{15.3b}$$

where c is a constant to be determined. Finally, if x_0 is an essential singularity, then no infinite power series solution is possible. This does not mean that no solutions exist, just that there are no solutions of this particular form.

In what follows we will discuss the expansions about regular and regular singular points in turn, taking $x_0 = 0$ for convenience. This involves no loss of generality, because any ODE may be transformed by letting $x \to x - x_0$, so that the expansion point is $x = 0$, and vice versa.

Example 15.1

Comment on the singularity structure of the equations

$$\text{(a)}\ \frac{d^2y}{dx^2} + \frac{dy}{dx} + y = 0, \quad \text{(b)}\ x^3 \frac{d^2y}{dx^2} + x\frac{dy}{dx} + y = 0,$$

---

[1] See, for example, E. Butkov (1973), *Mathematical Physics,* Addison-Wesley, Reading, Massachusetts, pp. 146–147.

[2] E. Butkov, *loc. cit.*

(c) $\dfrac{d^2y}{dx^2} - \dfrac{2x}{1-x^2}\dfrac{dy}{dx} + \dfrac{\lambda}{1-x^2}y = 0.$

Solution

(a) Since $p(x) = q(x) = 1$, the equation is regular for all x.

(b) Written in the form (15.1), $p(x) = 1/x^2$, $q(x) = 1/x^3$ and the equation is singular at $x = 0$. Both the limit of $xp(x)$ and $x^2q(x)$ as $x \to 0$ are infinite, so $x = 0$ is an essential singularity.

(c) This is singular at $x = \pm 1$, with

$$p(x) = -\frac{2x}{1-x^2} \quad \text{and} \quad q(x) = \frac{\lambda}{1-x^2}.$$

Now,

$$\lim_{x\to\pm 1}[(1 \mp x)p(x)] = \mp 1 \quad \text{and} \quad \lim_{x\to\pm 1}[(1 \mp x)^2 q(x)] = 0,$$

so both limits exist, and thus $x = \pm 1$ are both regular singularities.

15.1.1 Series solutions about a regular point

With the regular point taken to be $x = 0$, (15.3a) becomes

$$y(x) = \sum_{n=0}^{\infty} a_n x^n, \tag{15.4a}$$

so that

$$y'(x) = \frac{dy}{dx} = \sum_{n=1}^{\infty} n a_n x^{n-1} = \sum_{n=0}^{\infty} (n+1)a_{n+1}x^n, \tag{15.4b}$$

and

$$y''(x) = \frac{d^2y}{dx^2} = \sum_{n=2}^{\infty} n(n-1)a_n x^{n-2} = \sum_{n=0}^{\infty} (n+1)(n+2)a_{n+2}x^n. \tag{15.4c}$$

If we now substitute (15.4) into the general second-order linear equation

$$y'' + p(x)y' + q(x)y = 0,$$

we can extract the coefficient of each power of x in terms of the coefficients a_n. For a general solution, the coefficient of each power of x must be equal on both sides of the equation, and so each coefficient must separately vanish. This leads to relations between the various coefficients a_n. These are called *recurrence relations* and allow values of a_n for higher values of n to be found from values of a_n for lower values of n. This is best illustrated by an example.

We will find the series solution of the equation $y'' + 7y = 0$ about the point $x = 0$. Firstly, by inspection, $x = 0$ is a regular point and we can use the results (15.4). Substituting these into $y'' + y = 0$, gives

$$\sum_{n=0}^{\infty} [(n+1)(n+2)a_{n+2} + 7a_n]x^n = 0,$$

and since each term must vanish separately, we obtain the two-term recurrence relation

$$a_{n+2} = -\frac{7a_n}{(n+1)(n+2)}, \quad n \geq 0.$$

Thus all the even coefficients may be found in terms of a_0 and all the odd coefficients may be found in terms of a_1. In the first case,

$$\begin{aligned} y_1(x) &= a_0\left[1 - \frac{7}{2!}x^2 + \frac{7^2}{4!}x^4 - \cdots\right] \\ &= a_0\sum_{n=0}^{\infty}\frac{(-1)^n 7^n}{(2n)!}x^{2n} = a_0\sum_{n=0}^{\infty}\frac{(-1)^n}{(2n)!}(\sqrt{7}x)^{2n}, \end{aligned}$$

and in the second case,

$$\begin{aligned} y_2(x) &= a_1\left[x - \frac{7}{3!}x^3 + \frac{7^2}{5!}x^5 - \cdots\right] \\ &= a_1\sum_{n=0}^{\infty}\frac{(-1)^n 7^{2n+1}}{(2n+1)!}x^{2n+1} = a_1\sum_{n=0}^{\infty}\frac{(-1)^n}{(2n+1)!}(\sqrt{7}x)^{2n+1}. \end{aligned}$$

The first result can be recognised as the series for $\cos(\sqrt{7}x)$ and the second that for $\sin(\sqrt{7}x)$, so the general solution is

$$y(x) = a_0\cos(\sqrt{7}x) + a_1\sin(\sqrt{7}x),$$

where a_0 and a_1 are arbitrary constants that would have to be fixed by suitable boundary conditions.

In this example, the solution is expressible in closed form and so may also have been found by the (easier) techniques described in Chapter 14, but it is useful to illustrate the general method. The series method is most useful for cases where no closed form for the solution exists. The following is an example of such a case.

Example 15.2

Find the general solution of the equation

$$\frac{d^2y(x)}{dx^2} + 3x\frac{dy(x)}{dx} - y(x) = 0$$

as a power series about the point $x = 0$.

Solution

The point $x = 0$ is a regular point, so we may use the expansions (15.4). Substituting them into the differential equation gives

$$\sum_{n=0}^{\infty}\Big[(n+1)(n+2)a_{n+2}x^n + 3(n+1)a_{n+1}x^{n+1} - a_n x^n\Big].$$

Then equating the coefficient of x^n to zero gives

$$a_{n+2}(n+1)(n+2) + a_n(3n-1) = 0,$$

and hence the recurrence relation is

$$a_{n+2} = -\frac{(3n-1)}{(n+1)(n+2)}a_n.$$

Thus all the even coefficients are given in terms of a_0 and all the odd coefficients are given in terms of a_1. Specifically,

$$a_2 = \frac{1}{2}a_0, \qquad a_4 = -\frac{5}{12}a_2 = -\frac{5}{24}a_0, \cdots \quad \text{and}$$

$$a_3 = -\frac{1}{3}a_1, \quad a_5 = -\frac{2}{5}a_3 = \frac{2}{15}a_1, \cdots$$

and so

$$y(x) = a_0\left(1 + \frac{1}{2}x^2 - \frac{5}{24}x^4 + \cdots\right) + a_1 x\left(1 - \frac{1}{3}x^2 + \frac{2}{15}x^4 + \cdots\right).$$

Using the d'Alembert ratio test (5.17), one sees that both the series in brackets above are convergent series in x^2 for all finite values of x.

15.1.2 Series solutions about a regular singularity: Frobenius method

If we wish to expand about a singular point x_0, we use a method due to Frobenius starting from the series (15.3b). Assuming again that $x_0 = 0$, this becomes

$$y(x) = x^c \sum_{n=0}^{\infty} a_n x^n, \tag{15.5}$$

where c is not necessarily integer and $a_0 \neq 0$. As $x = 0$ is a regular singularity, we can define new functions $s(x)$ and $t(x)$ by $s(x) \equiv xp(x)$ and $t(x) \equiv x^2 q(x)$, both of which have simple expansions

$$s(x) = \sum_{n=0}^{\infty} s_n x^n \quad \text{and} \quad t(x) = \sum_{n=0}^{\infty} t_n x^n. \tag{15.6}$$

The original ODE may now be written in terms of these new functions as

$$y'' + \frac{s(x)}{x}y' + \frac{t(x)}{x^2}y = 0, \tag{15.7}$$

where the derivatives are

$$y' = \sum_{n=0}^{\infty}(n+c)a_n x^{n+c-1}, \tag{15.8a}$$

and

$$y'' = \sum_{n=0}^{\infty}(n+c)(n+c-1)a_n x^{n+c-2}. \tag{15.8b}$$

Substituting (15.8) into (15.7), gives

$$\sum_{n=0}^{\infty}[(n+c)(n+c-1)+s(x)(n+c)+t(x)]a_n x^{n+c-2} = 0. \tag{15.9}$$

The coefficient of the lowest power of x, that is, x^{c-2} is, from (15.9) and (15.6),

$$[c(c-1)+s_0c+t_0]a_0$$

and must vanish by (15.5), so that

$$c(c-1)+cs_0+t_0 = 0. \tag{15.10}$$

This is called the *indicial equation.* It is a quadratic in c with two roots c_1 and c_2, called the *indices* of the regular singular point. Each solution, when used in (15.9), and requiring the coefficients to all vanish separately, leads to a recurrence relation between the a_n and hence to a solution of the original ODE. Again, this is best illustrated by an example.We will find the power series solution of

$$4xy'' - 3y' - y = 0,$$

about the point $x = 0$. In the standard notation used previously $s(x) \equiv xp(x) = -3/4$ and $t(x) \equiv x^2q(x) = -x/4$, and it is straightforward to show that $x = 0$ is a regular singular point. Therefore using the Frobenius series (15.5) and (15.8) in the above equation gives an analogous equation to (15.9), i.e.,

$$\sum_{n=0}^{\infty}\left[(n+c)(n+c-1) - \tfrac{3}{4}(n+c) - \tfrac{1}{4}x\right]a_n x^{n+c-2} = 0.$$

Setting the coefficient of the lowest power of x to zero, we obtain the indicial equation

$$4c(c-1) - 3c = 0,$$

with roots $c = 0, 7/4$. Demanding that the coefficients of each power x vanish separately gives the recurrence relation

$$(n+c)(n+c-1)a_n - \tfrac{3}{4}(n+c)a_n - \tfrac{1}{4}a_{n-1} = 0. \qquad (15.11)$$

Consider firstly the case, $c = 7/4$. The recurrence relation (15.11) becomes

$$a_n = \frac{a_{n-1}}{n(4n+7)}.$$

Setting $a_0 = 1$, we can calculate $a_1 = 1/11$ and from this $a_2 = 1/330$, $a_3 = 1/18810$, etc. The corresponding solution of (15.10) is

$$y_1(x) = x^{7/4}\left(1 + \tfrac{1}{11}x + \tfrac{1}{330}x^2 + \tfrac{1}{18810}x^3 + \cdots\right).$$

Similarly, for the second root $c = 0$, we find $a_1 = -1/3$, $a_2 = -1/6$, $a_3 = -1/90$, etc., and the corresponding second solution

$$y_2(x) = 1 - \tfrac{1}{3}x - \tfrac{1}{6}x^2 - \tfrac{1}{90}x^3 + \cdots.$$

The general solution is therefore

$$y(x) = c_1 y_1(x) + c_2 y_2(x),$$

where c_1 and c_2 are constants. As in Example 15.2, this solution is not in closed form, but the series again converge for all finite values of x.

In the above examples, the solutions obtained from each of the two roots of the indicial equation are linearly independent. While this is usually true, there are circumstances where it is not. An obvious example is when the two roots are equal. A second example is when the two indices differ by an integer. In this case, the recurrence relation may, or may not, lead to a second solution that is linearly independent. To illustrate this we will find a power series solution of the equation

$$x(x-1)y'' + 4xy' + 2y = 0$$

about the point $x = 0$. Using the previous notations,

$$s(x) \equiv xp(x) = \frac{4x}{x-1}, \quad t(x) \equiv x^2 q(x) = \frac{2x}{x-1},$$

and so $x = 0$ is a regular singular point. Proceeding as above, using the expansions (15.8), leads to an equation analogous to (15.9), i.e.

$$\sum_{n=0}^{\infty}\left[(n+c)(n+c-1) + \frac{4x}{x-1}(n+c) + \frac{2x}{x-1}\right]a_n x^{n+c-2} = 0.$$

If we now multiply throughout by $(x-1)$, we have

$$\sum_{n=0}^{\infty}[(n+c)(n+c-1)(x-1)+4x(n+c)+2x]a_n x^{n+c-2}=0, \tag{15.12}$$

and setting the coefficient of the lowest power of x to zero, that is, the coefficient of x^{c-2}, gives the indicial equation

$$c(c-1)=0,$$

with roots $c=0,\ 1$. It can be shown that the larger root will always give a Frobenius solution.[3] This is found by using $c=1$ in (15.12) and setting the coefficient of each power of x to zero, giving

$$-n(n+1)a_n+n(n-1)a_{n-1}+4na_{n-1}+2a_{n-1}=0,$$

and hence the recurrence relation is

$$a_n=\left(\frac{n+2}{n}\right)a_{n-1}.$$

So setting $a_0=1$, gives $a_1=3,\ a_2=6$, etc. and hence

$$y_1(x)=a_0x(1+3x+6x^2+10x^3+\cdots).$$

In the present example, the smaller root does not result in another power series solution, because repeating the procedure above for $c=0$, we find the recurrence relation

$$a_n=\left(\frac{n+1}{n-1}\right)a_{n-1},$$

and since we require $a_0\neq 0$, a_1 is infinite and the method fails.

In cases such as these, and those where the roots are equal, the Frobenius method yields a single series solution specified in terms of a single free parameter a_0. Since the general solution of a linear second order differential equation always depends on two free parameters, we need another method for finding a second independent solution. There are several ways of doing this. One is to use another result of Fuchs' theorem.[4] This states that if $y_1(x)$ is a Frobenius series, then a second solution is

$$y_2(x)=y_1(x)\ln x+z(x), \tag{15.13a}$$

[3] See, for example, P. Dennery and A. Krzywicki (1966) *Mathematics for Physicists,* Dover Publications, New York, pp. 298–301.

[4] See E. Butkov (1973), *Mathematical Physics,* Addison-Wesley, Reading, Massachusetts, p.146.

where $z(x)$ has the Frobenius form

$$z(x) = x^d \sum_{n=0}^{\infty} b_n x^n,$$

and d is the smaller of the roots of the original indicial equation. In general, the method is used by substituting $y_2(x)$ into the original differential equation and finding a solution for b_n, with $b_0 \neq 0$.

Alternatively, a more general method, which applies to any second-order linear equation where a solution $y_1(x)$ is known, is to substitute

$$y_2(x) = y_1(x)u(x) \tag{15.13b}$$

into the differential equation and solve for $u(x)$. It is illustrated in Example 15.4. Both these methods, and others, for finding a second solution are easiest to apply if the first solution is in a simple closed form.

Example 15.3

Find the series solution of the equation

$$x^2\frac{\mathrm{d}^2y}{\mathrm{d}x^2} + 2x^2\frac{\mathrm{d}y}{\mathrm{d}x} - 2y = 0$$

about the point $x = 0$.

Solution

Comparing the given equation with the general form (15.1), gives

$$p(x) = 2 \quad \text{and} \quad q(x) = -2/x^2,$$

with, in the notation of (15.6), $s(x) \equiv xp(x) = 2x \to 0$ as $x \to 0$, and $t(x) \equiv x^2q(x) = -2$ as $x \to 0$. Therefore the point $x = 0$ is a regular singularity, and we can use the expansion (15.8). Substituting into the ODE gives, from (15.9),

$$\sum_n [(n+c)(n+c-1) + 2(n+c)x - 2]a_n x^{n+c-2} = 0,$$

and equating the coefficient of the lowest power of x to zero gives the indicial equation

$$c(c-1) - 2 = 0,$$

with roots $c = -1,\ 2$. For $c = 2$, we have

$$\sum_n [a_n(n+1)(n+2)x^n + 2a_n(n+2)x^{n+1} - 2a_nx^n] = 0$$

and hence the recurrence relation is

$$a_n = -\frac{2(n+1)}{n(n+3)}a_{n-1}.$$

Thus, setting $a_0 = 1$ gives $a_1 = -1$, $a_2 = 3/5$, etc. so that

$$y_1(x) = x^2\left(1 - x + \frac{3}{5}x^2 - \cdots\right).$$

For $c = -1$, we have

$$\sum_n \left[a_n(n-1)(n-2)x^n + 2a_n(n-1)x^{n+1} - 2a_n x^n\right] = 0$$

so that the recurrence relation is

$$a_n = -\frac{2(n-2)}{n(n-3)}a_{n-1}.$$

In this case, setting $a_0 = 1$ gives $a_1 = -1$, $a_n = 0$, $n \geq 2$ etc. and

$$y_2(x) = \left(\frac{1}{x} - 1\right).$$

The general solution is therefore

$$y(x) = Ay_1(x) + By_2(x) = Ax^2\left(1 - x + \frac{3}{5}x^2 - \cdots\right) + B\left(\frac{1}{x} - 1\right),$$

where A and B are constants. In this case we have found two linearly independent solutions, even though the indices differ by an integer.

Example 15.4

Find the general solution of the equation

$$4x^2y'' + y = 0.$$

Solution

One solution, that may be found using the methods above, or by inspection, is $y_1 = \sqrt{x}$. To find a second solution, set $y_2(x) = \sqrt{x}u(x)$, with

$$y_2'' = \sqrt{x}u'' + \frac{u'}{\sqrt{x}} - \frac{u}{4x\sqrt{x}}.$$

Then, substituting into the differential equation gives

$$xu'' + u' = \frac{\mathrm{d}(xu')}{\mathrm{d}x} = 0,$$

with solution $u = A \ln x + B$, where A and B are constants. Thus the second solution is

$$y_2(x) = A\sqrt{x} \ln x + B\sqrt{x},$$

which is the general solution, with $y_1(x) = \sqrt{x}$ a special case. Note that this solution is of the form (15.13a), as required by Fuchs' theorem.

15.1.3 Polynomial solutions

Another special class of solutions using the series method is when for some value n the coefficient a_n in the recurrence relation is zero. In this case, all subsequent coefficients generated from the recurrence relation will also be zero and the infinite series actually terminates at some finite n. The solutions are then finite-order polynomials and these polynomial solutions often have a special importance in physics.As an example, consider *Hermite's equation*

$$y'' - 2xy' + \lambda y = 0, \tag{15.14}$$

where λ is a constant parameter. We can easily sec that $x = 0$ is a regular point and so an expansion about this point is

$$y = \sum_n a_n x^n.$$

Substituting into the differential equation and proceeding as in Section 15.1.2, leads to the recurrence relation

$$a_n = \frac{2(n-2) - \lambda}{n(n-1)} a_{n-2}, \quad n \geq 2.$$

Thus the even and odd coefficients are independent of each other; the even coefficients are given in terms of a_0 and the odd coefficients are given in terms of a_1. If we set $a_0 = 1$ and $a_1 = 0$, we obtain the solution

$$y_1(x) = \left[1 - \lambda\frac{x^2}{2!} - \lambda(4-\lambda)\frac{x^4}{4!} - \cdots\right], \tag{15.15a}$$

while if we set $a_0 = 0$ and $a_1 = 1$ we obtain a second solution

$$y_2(x) = \left[x + (2-\lambda)\frac{x^3}{3!} + (2-\lambda)(6-\lambda)\frac{x^5}{5!} - \cdots\right]. \tag{15.15b}$$

The general solution is then

$$y(x) = Ay_1(x) + By_2(x),$$

where A and B are arbitrary constants.

The solutions (15.15) are, in general, infinite series. To obtain a polynomial solution, we must set $\lambda = 2k$, where $k > 0$ is an integer, so that the recurrence relation gives $a_{2+k} = 0$, and one of the series (15.15a) and (15.15b) terminates. If k is even and we set $a_0 = 1$, $a_1 = 0$, the series (15.15a) terminates, giving a polynomial solution $h_k(x)$ of order k. For example,

$$h_0(x) = 1, \quad h_2(x) = 1 - 2x^2, \quad h_4(x) = 1 - 4x^2 + 4x^4/3,$$

and so on. Alternatively, If k is odd and we set $a_0 = 0$, $a_1 = 1$, the series (15.15b) terminates, again giving a polynomial solution $h_k(x)$ of order k. For example,

$$h_1(x) = x, \quad h_3(x) = x - 2x^3/3, \quad h_5(x) = x - 4x^3/3 + 4x^5/15.$$

Any polynomial of the form $H_k(x) = c_k h_k(x)$, where the c_k are constants, is called a *Hermite polynomial.* The convention for choosing the c_k is not universal, but in physics they are chosen so that the coefficient of x^k in $H_k(x)$ is 2^k, and the first six polynomials are then:

$$\begin{aligned} H_0(x) &= 1 & H_1(x) &= 2x \\ H_2(x) &= 4x^2 - 2 & H_3(x) &= 8x^3 - 12x \\ H_4(x) &= 16x^4 - 48x^2 + 12 & H_5(x) &= 32x^5 - 160x^3 + 120x \end{aligned}$$

These polynomials occur in the quantum mechanical theory of the simple harmonic oscillator.

Example 15.5

Show that the series (15.15a) and (15.15b) converge for all finite x.

Solution

The series (15.15a) is a power series of the form

$$y_1 = \sum_{m=0}^{\infty} c_m z^m,$$

where $z = x^2$ and $c_m = a_{2m}$. By (5.18) and (5.19), it converges for all $z < R$, where the radius of convergence

$$R = \lim_{m\to\infty}\left|\frac{c_m}{c_{m+1}}\right| = \lim_{m\to\infty}\left|\frac{a_{2m}}{a_{2m+2}}\right| = \lim_{m\to\infty}\left|\frac{(2m+2)(2m+1)}{(4m-\lambda)}\right| = \infty,$$

so that the series converges for $0 < z = x^2 < \infty$, and hence all finite x. A similar argument applies to (15.15b), by writing it in the form

$$y_2 = x\left(\sum_{m=0}^{\infty} c_m z^m\right),$$

where now $c_m = a_{2m+1}$.

Example 15.6

Laguerre's differential equation is

$$x\frac{\mathrm{d}^2y}{\mathrm{d}x^2} + (1-x)\frac{\mathrm{d}y}{\mathrm{d}x} + \lambda y = 0, \qquad (15.16)$$

where λ is a constant. By writing

$$y(x) = \sum_{j=0}^{\infty} a_j x^{j+c},$$

show that $c = 0$ is a solution of the indicial equation. Derive the recurrence relation for the coefficients a_j in this case, and show that when λ is a positive integer, the series terminates. Find the explicit form of the resulting solutions $y(x) = L_k(x)$ (called *Laguerre polynomials*) for fixed values of $k = 0,\ 1,\ 2,\ 3$.

Solution

From $y = \sum_j a_j x^{j+c}$, we have

$$y' = \sum_j a_j(j+c)x^{j+c-1} \quad \text{and} \quad y'' = \sum_j a_j(j+c)(j+c-1)x^{j+c-2}.$$

Substituting these into Laguerre's equation gives

$$\sum_j a_j\left[(j+c)(j+c-1)x^{j+c-1} + (j+c)x^{j+c-1} - (j+c)x^{j+c} + \lambda x^{j+c}\right].$$

The lowest power of x is x^{c-1}, and equating its coefficient to zero gives

$$a_0[c(c-1)+c] = a_0c^2 = 0,$$

that is, $c = 0$ is a solution. The general power of x^j for $c = 0$ has a coefficient

$$j(j+1)a_{j+1} + (j+1)a_{j+1} - ja_j + \lambda a_j = 0,$$

so that

$$a_{j+1} = \frac{(j-\lambda)}{(j+1)^2}a_j.$$

Thus

$$a_1 = (-\lambda)a_0, \quad a_2 = \frac{(1-\lambda)}{2^2}a_1 = \frac{(-\lambda)(1-\lambda)}{4}a_0, \cdots$$

and in general

$$a_j = a_0 \prod_{n=0}^{j-1}(n-\lambda)\left(\prod_{n=1}^{j} n^2\right)^{-1}.$$

The series for $y(x)$ with $c = 0$ is therefore

$$y(x) = a_0\left\{1 - \lambda x - [\lambda(1-\lambda)/4]x^2 + \cdots\right\}.$$

If $p = k$, where k is a positive integer, the series terminates and the resulting polynomials for $k = 0,\ 1,\ 2,\ 3$ are

$$L_0(x) = 1 \qquad L_1(x) = 1 - x$$

$$L_2(x) = 1 - 2x + \tfrac{1}{2}x^2 \qquad L_3(x) = 1 - 3x + \tfrac{3}{2}x^2 - \tfrac{1}{6}x^3$$

where, by convention, we have set $a_0 = 1$. These polynomials occur, for example, in the quantum mechanical solution of the hydrogen atom.

15.2 Eigenvalue equations

In Chapter 10, we discussed equations of the form

$$\mathbf{A}\mathbf{x} = \lambda\mathbf{x}, \tag{15.17}$$

called eigenvalue equations, where $\mathbf{A}$ was a given square matrix and $\mathbf{x}$ was a column vector to be determined; and we showed that non-trivial solutions only existed for particular values of λ. Here we shall introduce analogous eigenvalue equations for differential operators. Such equations play a central role in quantum mechanics and wave theory and, in some important cases, are solved by the methods introduced in the last section.

In Chapter 9, Section 9.3.2, we introduced the differential operator $D \equiv \mathrm{d}/\mathrm{d}x$ that transforms a function $y(x)$ into its derivative, that is, [cf. (9.46a)]

$$Df(x) \equiv \frac{\mathrm{d}y}{\mathrm{d}x} = y'(x).$$

This is a linear operator, because it satisfies the linearity condition that is, [cf. (9.46b]

$$D[ay_1(x) + by_2(x)] = aDy_1(x) + bDy_2(x), \tag{15.18}$$

where y_1, y_2 are arbitrary functions and a, b are arbitrary constants. Using this, other differential operators can be formed, for example

$$D^2y(x) \equiv D[Dy(x)] = \frac{d^2y}{dx^2},$$

or more generally,[5]

$$O = A(x)D^2 + B(x)D + C(x), \tag{15.19}$$

which transforms a function $y(x)$ to a function $z(x)$ according to

$$Oy(x) = z(x), \tag{15.20a}$$

where

$$z(x) = A(x)\frac{d^2y(x)}{dx^2} + B(x)\frac{dy(x)}{dx} + C(x)y(x). \tag{15.20b}$$

Like D, O is a linear operator,[6] i.e.

$$O[ay_1(x) + by_2(x)] = aOy_1(x) + bOy_2(x),$$

in analogy to (15.18).

In analogy to (10.1), we now define the eigenvalue equation corresponding to a given differential operator O as

$$Oy(x) = \lambda y(x), \tag{15.21a}$$

where $y(x)$ is a function subject to given boundary conditions. If O is of the form (15.19), this equation is just

$$A(x)\frac{d^2y(x)}{dx^2} + B(x)\frac{dy}{dx} + C(x)y(x) = \lambda y(x), \tag{15.21b}$$

which is a linear differential equation of the standard form (15.1) with

$$p(x) = B(x)/A(x), \quad q(x) = [C(x) - \lambda/A(x)]$$

and can be solved by series solutions about regular points or regular singularities.

Before the boundary conditions are applied, equations of the form (15.21a) and (15.21b) are linear, second-order differential equation with non-trivial solutions, that is, solutions other than $y(x) = 0$, for any value of λ. However, when boundary conditions are applied, this is not necessarily the case. The λ values for which non-trivial solutions exist are called *eigenvalues* and the corresponding solutions are

---

[5] This is not the most general form of a differential operator, but is the only form that we need consider.

[6] It is common practice to use a 'hat' over symbols to indicate that they are operators. In the case of the differential operator D this is usually omitted, so for uniformity we have also omitted it on other operators.

called *eigenfunctions.* To illustrate this, consider the simple eigenvalue equation

$$D^2y(x) = \frac{\mathrm{d}^2y}{\mathrm{d}x^2} = \lambda y(x). \tag{15.22a}$$

If $\lambda = -k^2 < 0$, where k is the wave number, this equation describes standing waves on a stretched string, provided the transverse displacement of the string is not too large. In this case, the general solution is

$$y(x) = A\cos kx + B\sin kx, \tag{15.22b}$$

where A and B are arbitrary constants. If we now impose the boundary conditions $y(0) = a$ and $y'(0) = b$, a non-trivial solution

$$y(x) = a\cos(kx) + (b/k)\sin(kx)$$

exists for any $\lambda = -k^2 < 0$. Hence, with these boundary conditions, any real $\lambda < 0$ is an eigenvalue and the set of all eigenvalues, called the *eigenvalue spectrum,* is said to be continuous. However, if the string is clamped at the points $x = 0$ and $x = L$, and is stretched between them, then the appropriate boundary conditions are

$$y(0) = y(L) = 0, \quad (L > 0).$$

which require

$$A = 0, \quad B\sin kL = 0,$$

so that non-trivial solutions only exist if $kL = \pi n$. Hence, in this case, the eigenvalues $\lambda = -k^2$ are

$$\lambda_n = -n^2\pi^2/L^2, \quad n = 1,\ 2, \ldots$$

and the eigenvalue spectrum is said to be *discrete.* The corresponding eigenfunctions are

$$y_n(x) = B\sin\left(\frac{n\pi x}{L}\right).$$

Other boundary conditions lead to other eigenvalue spectra, as illustrated in Problem 15.7 below.

Hermite's equation (15.14) and Laguerre's equation discussed in Example 15.6 are important examples of eigenvalue equations of the form (15.21b), since they play a central role in the quantum mechanical theory of the simple harmonic oscillator and the hydrogen atom, respectively. This is not the place to discuss these topics in detail, except to note that in both cases the appropriate boundary conditions as $|x| \to \infty$ are only satisfied by the polynomial solutions corresponding to eigenvalues $\lambda = 2k$ and $\lambda = k$, respectively, where k is a non-negative integer. Hence the eigenvalue spectra are discrete in both cases and it is this property that leads to quantised energy levels in these systems. Other important examples of eigenvalue equations will be discussed in the next two sections.

Example 15.7

Solve the eigenvalue equation (15.22a) with $\lambda = -k^2 < 0$ for the boundary conditions[7]

$$y(0) = y'(L) = 0, \quad (L > 0).$$

Solution

The general solution is again given by (15.22b) but now the boundary conditions require that

$$A = 0, \quad Bk\cos(kL) = 0,$$

so that non-trivial solutions only exist if $kL = (2n+1)\pi/2$. Hence the eigenvalues $\lambda = -k^2$ are

$$\lambda_n = -(2n+1)^2\pi^2/(4L^2), \quad n = 1,\ 2, \ldots$$

and the eigenfunctions are

$$y_n(x) = Bsin\left[\frac{(2n+1)\pi x}{2L}\right].$$

15.3 Legendre's equation

The *Legendre equation* is the eigenvalue equation

$$(1-x^2)\frac{\mathrm{d}^2y}{\mathrm{d}x^2} - 2x\frac{\mathrm{d}y}{\mathrm{d}x} + \lambda y = 0, \tag{15.23a}$$

where

$$\lambda = l(l+1), \tag{15.23b}$$

and l is a constant. This is an important equation for many physical systems with spherical symmetry, in which case $x = \cos\theta$, where $0 \le \theta \le \pi$ is an angular co-ordinate, and we require solutions that are finite over the range $-1 \le x \le 1$, including $x = \pm 1$.

Any solution of (15.23) is called a *Legendre function.* In the standard form, (15.23a) becomes

$$y'' + p(x)y' + q(x)y = 0, \tag{15.24}$$

where

$$p(x) = -\frac{2x}{1-x^2} \to -\infty \text{ as } x \to \pm 1,$$

[7] These are the appropriate boundary conditions for sound waves in a pipe of length L that is open at one end and closed at the other.

and

$$q(x) = \frac{\lambda}{1-x^2} \to \infty \text{ as } x \to \pm 1.$$

Hence $x = \pm 1$ are singular points of the equation. However, $x = 0$ is clearly a regular point, and so we can make a simple series expansion about $x = 0$.

$$y(x) = \sum_{n=0}^{\infty} a_n x^n.$$

Differentiating and substituting (15.4) into (15.23) gives

$$\sum_{n=0}^{\infty} \left[(n+1)(n+2)(1-x^2)a_{n+2}x^n - 2x(n+1)a_{n+1}x^n + \lambda a_n x^n\right] = 0,$$

and hence

$$(n+2)(n+1)a_{n+2} + (l^2 + l - n^2 - n)a_n = 0, \qquad (15.25)$$

where we have equated the coefficient of x^n to zero. Factorising the second term, leads to the recurrence relation

$$a_{n+2} = -\frac{(l-n)(l+n+1)}{(n+2)(n+1)}a_n. \qquad (15.26)$$

Thus, given a_0, we can find all the other even coefficients, and given a_1, we can find all the other odd coefficients. Using the ratio test, it is straightforward to show that both series converge for $|x| < 1$. The general solution is then given by the sum of the two independent linear solutions in the usual way. However, as expected, the series diverges at $x = \pm 1$, because we know these are singular points.

15.3.1 Legendre functions and Legendre polynomials

The lack of convergence at $x = \pm 1$ of the series obtained using (15.26) is an important limitation, because in many physics applications, particularly those in quantum theory, x is the cosine of an angle and l is a non-negative integer. Thus we need to find solutions that converges for *all* x, including $x = \pm 1$. This is only possible for integer values of l, as we shall show below.

The general solution of (15.23) is the sum of two series containing two constants a_0 and a_1. Using the recurrence relation (15.26) we may therefore write

$$y_l(x) = a_0\left[1 - l(l+1)\frac{x^2}{2!} + (l-2)l(l+1)(l+3)\frac{x^4}{4!} + \cdots\right]$$
$$+ a_1\left[x - (l-1)(l+2)\frac{x^3}{3!} + (l-3)(l-1)(l+2)(l+4)\frac{x^5}{5!} + \cdots\right].$$
$$(15.27a)$$

Now if, and only if, l is a non-negative integer, one of these series will terminate at $l = n$ and the other will diverge at $x = \pm 1$. This is simply seen by considering the series for $l = 0$ at $x = 1$. In this case, the even solution is simply a_0 and the odd series is

$$y_0(x = 1) = a_1\left(1 + \tfrac{1}{3} + \tfrac{1}{5} + \cdots\right)$$

which diverges. However, if $l = 1$, the odd series is just $a_1 x$, whereas the even series diverges at $x = \pm 1$.

The series that terminates defines a finite polynomial of order l, called a *Legendre polynomial* and written $P_l(x)$. The other series diverges at $x = \pm 1$ and defines a *Legendre function of the second kind*, written $Q_l(x)$. For integer l, the general solution of the Legendre equation is then

$$y_l(x) = c_1 P_l(x) + c_2 Q_l(x). \tag{15.27b}$$

The functions $Q_l(x)$ occur far less frequently in physical applications than the polynomials and we will therefore focus mainly on the latter functions. From (15.27a), if we choose the value of either a_0 or a_1 so that $y_l(1) = 1$, and hence $y_l(-1) = (-1)^l$, then the first three even-order polynomials are

$$P_0(x) = 1, \quad P_2(x) = \tfrac{1}{2}(3x^2 - 1), \quad P_4(x) = \tfrac{1}{8}(35x^4 - 30x^2 + 3), \tag{15.28a}$$

and the first three odd polynomials are

$$P_1(x) = x, \quad P_3(x) = \tfrac{1}{2}(5x^3 - 3x), \quad P_5(x) = \tfrac{1}{8}(63x^5 - 70x^3 + 15x). \tag{15.28b}$$

Choosing the constants in this way ensures that the polynomials satisfy the *normalisation condition*

$$P_0(0) = 1, \tag{15.29a}$$

while the odd and even powers in the series imply

$$P_l(-x) = (-1)^l P_l(x). \tag{15.29b}$$

The first four Legendre polynomials are plotted in Figure 15.1a. The polynomial of order l in general has l nodes, and as l increases the polynomials oscillate more and more rapidly, as illustrated in Figure 15.1b for $l = 10$.

The Legendre polynomials satisfy the *orthogonality relation*

$$\int_{-1}^{1} P_l(x) P_m(x)\, \mathrm{d}x = \frac{2}{(2l+1)}\delta_{lm}, \tag{15.30}$$

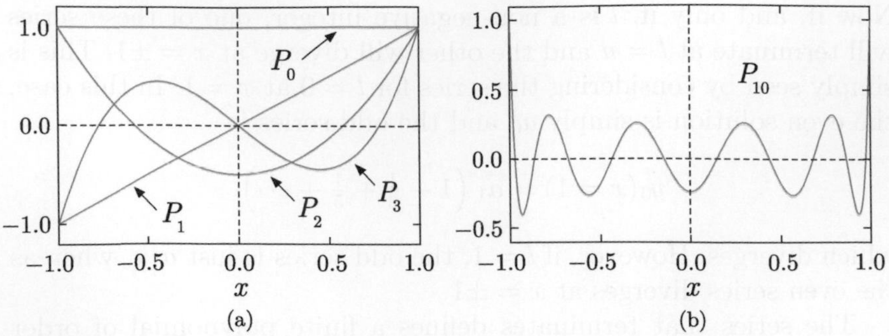

Figure 15.1 Legendre polynomials: (a) $P_l(x)$, $l = 0$, 1, 2, 3, and (b) $P_{10}(x)$.

where δ_{lm} is the Kronecker delta symbol (9.24b). For $l = m$, this reduces to

$$\int_{-1}^{1} [P_l(x)]^2 \, \mathrm{d}x = \frac{2}{(2l+1)} \tag{15.31}$$

and is a consequence of the normalisation convention (15.29a). It may be verified for individual cases using (15.28) and will be proved in general in the next section. For $l \neq m$, (15.30) may be proved by starting from the Legendre equation (15.23), which is conveniently rewritten in the form

$$\frac{\mathrm{d}}{\mathrm{d}x}[(1-x^2)y'(x)] + l(l+1)y(x) = 0.$$

Setting $y(x) = P_l(x)$, and writing this equation for two values l and m, gives

$$\frac{\mathrm{d}}{\mathrm{d}x}\left[(1-x^2)P_l'(x)\right] + l(l+1)P_l(x) = 0$$

and

$$\frac{\mathrm{d}}{\mathrm{d}x}\left[(1-x^2)P_m'(x)\right] + m(m+1)P_m(x) = 0.$$

Multiplying the first of these by $P_m(x)$ and the second by $P_l(x)$, and then subtracting one equation from the other gives

$$P_l(x)\frac{\mathrm{d}}{\mathrm{d}x}\left[(1-x^2)P_m'\right] - P_m(x)\frac{\mathrm{d}}{\mathrm{d}x}\left[(1-x^2)P_l'\right]$$
$$= [l(l+1) - m(m+1)]P_l(x)P_m(x).$$

Then integrating both sides over x from -1 to $+1$, we have

$$\int_{-1}^{1}\left\{P_l(x)\,\frac{\mathrm{d}}{\mathrm{d}x}\left[(1-x^2)P_m'\right] - P_m(x)\frac{\mathrm{d}}{\mathrm{d}x}\left[(1-x^2)P_l'\right]\right\}\mathrm{d}x$$
$$= [l(l+1) - m(m+1)]\int_{-1}^{1} P_l(x)P_m(x)\,\mathrm{d}x.$$

The left-hand side of this equation may be shown to vanish by integrating both terms by parts, and it follows that if $l \neq m$,

$$\int_{-1}^{1} P_l(x)P_m(x)\,\mathrm{d}x = 0, \quad l \neq m,$$

as required.

The orthogonality relation (15.30) is often used in conjunction with another result, which we will state without proof. This is that any function $f(x)$ that is non-singular in the range $-1 \leq x \leq 1$ can be expanded in a convergent series of the form

$$f(x) = \sum_{k=0}^{\infty} c_k P_k(x), \quad -1 \leq x \leq 1. \tag{15.32}$$

This property is called *completeness* and $P_l(x)$, $l = 0,\ 1,\ 2, \ldots$, are called a *complete set of functions*, in analogy to the definition of a complete set of basis vectors in Section 9.2.1. On multiplying (15.32) by $P_n(x)$ and integrating, one obtains

$$c_n(x) = \tfrac{1}{2}(2n+1)\int_{-1}^{1} f(x)P_n(x)\,\mathrm{d}x, \quad n = 0,\ 1, \ldots \tag{15.33}$$

for the coefficients in (15.32). This expansion is called a *Legendre series* and is closely analogous to a Fourier expansion, as can be seen by comparing (15.30), (15.32) and (15.33) with (13.38), (13.39a) and (13.39b) respectively.[8] The expansion (5.32) is often used in numerical work where one has a large number of measurements of a quantity f as a function of angle, that is, an angular distribution $f(\cos\theta)$, and requires a convenient approximate representation of them.

We conclude this section with a brief account of the Legendre functions of the second kind. As discussed earlier, these are defined by the first series in (15.27a) for odd l, where by convention we take $a_0 = 1$; and by the second series in (15.27a) for even l, where we take $a_1 = 1$. For integer l, the resulting series can be conveniently summarised by introducing the *double factorial*

$$\begin{aligned} k!! &\equiv k(k-2)\ldots 3.1 \quad (k \text{ odd}), \\ k!! &\equiv k(k-2)\ldots 4.2 \quad (k \text{ even}), \end{aligned} \tag{15.34a}$$

which satisfy the identities

$$(2n)!! = 2^n n!, \quad (2n-1)!! = \frac{(2n)!}{(2n)!!}, \tag{15.34b}$$

[8] In addition, both are analogous to the expansion of a vector in terms of a set of basis vectors, as discussed in Section *13.2.1.

where $0!! = 1$, by definition. Using relations like

$$(l+2)(l+4)\ldots(l+2n) = (l+2n)!!/l!!,$$

one finds from (15.27a) and (15.27b) that

$$\begin{aligned} Q_l(x) &= \sum_{n=0}^{\infty}\left[(-1)^n\frac{(l+2n)!!}{l!!}\frac{(l-1)!!}{(l-2n-1)!!}\right]\frac{x^{2n+1}}{(2n+1)!}, \quad \text{(even } l\text{)} \\ &= \sum_{n=0}^{\infty}\left[(-1)^n\frac{(l+2n-1)!!}{(l-1)!!}\frac{l!!}{(l-2n)!!}\right]\frac{x^{2n}}{(2n)!}. \quad \text{(odd } l\text{)} \end{aligned} \tag{15.35}$$

These series diverge at $x = \pm 1$, as shown by expressing them in closed form for $l = 0, 1$ in Example 15.8, and generalising the result to all integer l in Problem 15.14.

Example 15.8

Verify that the lowest-order Legendre functions of the second kind are

$$Q_0(x) = \frac{1}{2}\ln\left(\frac{1+x}{1-x}\right) \quad \text{and} \quad Q_1(x) = \frac{x}{2}\ln\left(\frac{1+x}{1-x}\right) - 1.$$

Solution

For $l = 0$, the first series in (15.27) is $P_0(x) = 1$, while the second gives

$$Q_0(x) = x + x^3/3 + \cdots + a_n x^n + \cdots,$$

where $a_n = 0$ (n even) and $a_{n+2} = na_n/(n+2)$ (n odd), using (15.26). On the other hand, using the Maclaurin series of Table 5.1 for $\ln(1+x)$, one obtains

$$\begin{aligned} \frac{1}{2}\ln\left(\frac{1+x}{1-x}\right) &= \frac{1}{2}\left[\ln(1+x) - \ln(1-x)\right] \\ &= x + x^3/3 + \cdots + c_n x^n + \cdots, \end{aligned}$$

where $c_n = 0$ (n even) and $c_n = n^{-1}$(n odd). Since $c_0 = a_0 = 1$, and c_n satisfy the same recurrence relation as the a_n, the series are identical, so that

$$Q_0(x) = \frac{1}{2}\ln\left(\frac{1+x}{1-x}\right),$$

as required. A similar argument leads to the result for $Q_1(x)$. We see that both are singular at $x = \pm 1$.

*15.3.2 The generating function

A useful technique for deriving properties of Legendre polynomials is to use the *generating function*

$$G(x,h) = (1 - 2xh + h^2)^{-1/2}, \tag{15.36}$$

where h is a dummy variable and

$$G(x,h) = \sum_{l=0}^{\infty} P_l(x)h^l. \tag{15.37}$$

To prove (15.37), we have to show that the functions $P_l(x)$ on the right-hand side really do satisfy the Legendre equation and that they have the property $P_l(1) = 1$. The latter follows simply by putting $x = 1$ in (15.36) so that

$$G(1,h) = (1 - 2h + h^2)^{-1/2} = \frac{1}{1-h} = 1 + h + h^2 + \cdots,$$

and then equating this to the right-hand side of (15.37) to give

$$1 + h + h^2 + \cdots \equiv P_0(1) + P_1(1)h + P_2(1)h^2 + \cdots,$$

Since this relation is an identity in h, the coefficients of h^n on both sides must be equal and so $P_l(1) = 1$. To show that the $P_l(x)$ in (15.37) satisfy the Legendre equation, we use the identity

$$(1 - x^2)\frac{\partial^2 G}{\partial x^2} - 2x\frac{\partial G}{\mathrm{d}x} + h\frac{\partial^2}{\partial h^2}(hG) = 0, \tag{15.38}$$

that may be verified from the definition (15.36). Substituting (15.37) into (15.38) gives

$$\sum_{l=0}^{\infty} \left[(1 - x^2)P_l''(x) - 2xP_l'(x) + l(l+1)P_l(x)\right] h^l = 0.$$

Since this is an identity in h, the coefficient of each power of h must vanish, and hence

$$(1 - x^2)P_l''(x) - 2xP_l'(x) + l(l+1)P_l(x) = 0.$$

But this is the Legendre equation, and so the P_l of (15.37) are indeed Legendre functions.

The generating function is useful in deriving recurrence relations for Legendre polynomials. These are relations that relate two or more polynomials of different orders, that is, with different values of l, and by analogy to the recurrence relations discussed earlier, they provide a simple way of evaluating higher-order polynomials from polynomials of lower order.[9] Some examples of recurrence relations are:

$$lP_l(x) = (2l-1)xP_{l-1}(x) - (l-1)P_{l-2}(x), \tag{15.39a}$$

$$xP_l'(x) - P_{l-1}'(x) = lP_l(x), \tag{15.39b}$$

$$P_l'(x) - xP_{l-1}'(x) = lP_{l-1}(x), \tag{15.39c}$$

$$(1-x^2)P_l'(x) = lP_{l-1}(x) - lxP_l(x), \tag{15.39d}$$

$$(2l+1)P_l(x) = P_{l+1}'(x) - P_{l-1}'(x). \tag{15.39e}$$

As an example of how these are derived using the generating function, we will prove (15.39b). Differentiating (15.36) and (15.37) partially with respect to x, keeping h constant, gives

$$h(1-2xh+h^2)^{-3/2} = \sum_{l=0}^{\infty} P_l'(x)h^l,$$

while differentiating with respect to h, keeping x constant, gives

$$(x-h)(1-2xh+h^2)^{-3/2} = \sum_{l=0}^{\infty} lP_l(x)h^{l-1}.$$

Comparing these two equations gives

$$(x-h)\sum_{l=0}^{\infty} P_l'(x)h^l = h\sum_{l=1}^{\infty} lP_l(x)h^{l-1},$$

and equating the coefficients of h^l gives

$$lP_l(x) = xP_l'(x) - P_{l-1}'(x),$$

which is (15.39b). Proofs of some of the other relations are left to the Examples and Problems. One can show that exactly the same recurrence relations apply to the Legendre functions of the second kind (see Problem 15.12).

The generating function also yields an elegant derivation of the normalisation formula (15.31). To do this, we evaluate

$$f(h) = \int_{-1}^{1} G^2(x,h)\,\mathrm{d}x$$

[9] For large values of l, direct evaluation of the polynomials must be done with care, because cancellations between different terms can lead to rounding errors. The latter are greatly reduced by using the recurrence relations.

using (15.36) and (15.37). From (15.36) we obtain

$$f(h) = \int_{-1}^{1} \frac{1}{1 - 2xh + h^2} \mathrm{d}x = \frac{1}{h} \ln\left(\frac{1+h}{1-h}\right) = \sum_{l=0}^{\infty} \frac{2}{(2l+1)} h^{2l}, \tag{15.40a}$$

where we have used the Maclaurin expansion of Table 5.1 to expand the logarithms. On the other hand, using (15.37) gives

$$f(h) = \sum_{l,m=0}^{\infty} h^{l+m} \int_{-1}^{1} P_l(x) P_m(x)\, \mathrm{d}x = \sum_{l=0}^{\infty} h^{2l} \int_{-1}^{1} [P_l(x)]^2\, \mathrm{d}x, \tag{15.40b}$$

where we have used the orthogonality relation (15.33). Equating powers of $2l$ in (15.40a) and (15.40b) yields (15.31) as required.

Finally, a well-known physical application of (15.36) and (15.37) is the expansion of a potential $V(\mathbf{r})$ due to a point charge, or mass, at $\mathbf{r} = \mathbf{a}$, in powers of $1/r$, where $r = |\mathbf{r}|$. From Figure 15.2, we have in the electrostatic case

$$V(\mathbf{r}) = \frac{q}{4\pi\varepsilon_0} \frac{1}{|\mathbf{r} - \mathbf{a}|},$$

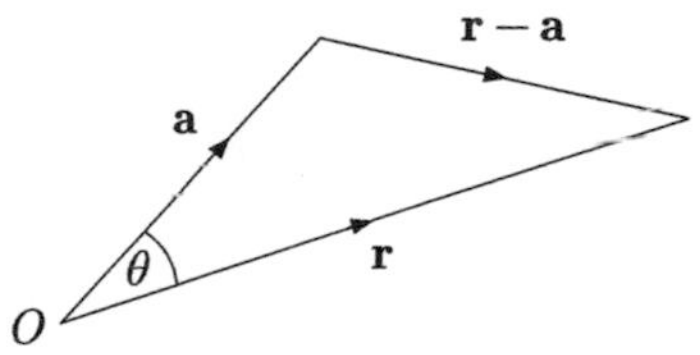

Figure 15.2 Construction for the multipole expansion.

where q is the charge and ε_0 is the permittivity of the vacuum. Writing, $a = |\mathbf{a}|$, and $h = a/r$, for $r > a$ we have

$$V(\mathbf{r}) = \frac{q}{4\pi\varepsilon_0 r} \frac{1}{(1 - 2h\cos\theta + h^2)^{1/2}},$$

and expanding this using (15.36) and (15.37) gives

$$V(\mathbf{r}) = \frac{q}{4\pi\varepsilon_0 r} \sum_{l=0}^{\infty} \left(\frac{a}{r}\right)^l P_l(\cos\theta). \quad r > a \tag{15.41a}$$

For $r < a$, the corresponding result is

$$V(\mathbf{r}) = \frac{q}{4\pi\varepsilon_0 a} \sum_{l=0}^{\infty} \left(\frac{r}{a}\right)^l P_l(\cos\theta). \quad r < a \tag{15.41b}$$

Equations (15.41) are called the axial multipole expansions. Using them, the potential due to any linear distribution of point charges can then be obtained by adding the contributions of each point charge using (15.41). For example, for a dipole with $-e$ at $r = 0$ and $+e$ at $r = a$, one obtains the simple result

$$V(\mathbf{r}) = \frac{\mu\cos\theta}{4\pi\varepsilon_0 r^2}$$

in the limit $r \gg a$, where $\mu = ea$ is the dipole moment.

Example 15.9

Evaluate $P_n(0.5)$ to 4 significant figures for $0 \le n \le 10$.

Solution

Using (15.28a) and (15.28b) we can evaluate the first five polynomials to be:

$$P_0(0.5) = 1 \qquad P_3(0.5) = -0.4375$$

$$P_1(0.5) = 0.5 \qquad P_4(0.5) = -0.28906 = -0.2891 \text{ (4 sf)}$$

$$P_2(0.5) = -0.125 \quad P_5(0.5) = -0.089844 = -0.08984 \text{ (4 sf)}.$$

Then, using (15.39a),

$$P_l(0.5) = \frac{(2l-1)}{2l} P_{l-1}(0.5) - \frac{(l-1)}{l} P_{l-2}(0.5),$$

we can generate the values of the other polynomials. They are:

$$P_6(0.5) = \tfrac{11}{12} P_5(0.5) - \tfrac{5}{6} P_4(0.5) = 0.32324 = 0.3232 \text{ (4 sf)},$$
$$P_7(0.5) = \tfrac{13}{14} P_6(0.5) - \tfrac{6}{7} P_5(0.5) = 0.22314 = 0.2231 \text{ (4 sf)},$$
$$P_8(0.5) = \tfrac{15}{16} P_7(0.5) - \tfrac{7}{8} P_6(0.5) = -0.073639 = -0.07364 \text{ (4 sf)},$$
$$P_9(0.5) = \tfrac{17}{18} P_8(0.5) - \tfrac{8}{9} P_6(0.5) = -0.26789 = -0.2679 \text{ (4 sf)},$$
$$P_{10}(0.5) = \tfrac{19}{20} P_9(0.5) - \tfrac{9}{10} P_7(0.5) = -0.18822 = -0.1882 \text{ (4 sf)}.$$

*15.3.3 Associated Legendre equation

Another equation that is closely associated with the Legendre equation is the *associated Legendre equation*

$$(1-x^2)y'' - 2xy' + \left[l(l+1) - \frac{m^2}{1-x^2} \right] y = 0, \qquad (15.42)$$

where m and l are integers and in physical situations $-l \le m \le l$. This equation reduces to the Legendre equation if $m = 0$, but in physical applications it is often the family of equations (15.42) that occurs, rather than just the Legendre equation itself. However, the solutions of (15.42), called the *associated Legendre functions,* are easily obtained from the Legendre functions already derived, as we now show.

To do this, we substitute

$$y(x) = (-1)^m (1-x^2)^{m/2} u(x), \qquad m \ge 0 \qquad (15.43)$$

into (15.42) to obtain, after some simplification,

$$(1-x^2)\frac{d^2u}{dx^2} - 2(m+1)x\frac{du}{dx} + (l-m)(l+m+1)u = 0. \tag{15.44}$$

On the other hand, on differentiating Legendre's equation (15.23) m times, we obtain

$$(1-x^2)\frac{d^{m+2}y}{dx^{m+2}} - 2(m+1)x\frac{d^{m+1}y}{dx^{m+1}} + (l-m)(l+m+1)\frac{d^m y}{dx^m} = 0. \tag{15.45}$$

Comparing (15.44) and (15.45), we see that $u = d^m y/dx^m$, where y is a solution of Legendre's equation. Hence from (15.42) and (15.27b) the general equation for $m \geq 0$ is

$$y_{lm}(x) = (-1)^m (1-x^2)^{m/2} \frac{d^m}{dx^m}[c_1 P_l(x) + c_2 Q_l(x)], \quad m \geq 0. \tag{15.46}$$

In applications, we are mostly interested in the *associated Legendre polynomials*

$$P_l^m(x) = (-1)^m (1-x^2)^{m/2} \frac{d^m P_l(x)}{dx^m}, \quad m \geq 0 \tag{15.47a}$$

and since the associated Legendre equation depends only on m^2, we can define

$$P_l^{-m}(x) \equiv c_{lm} P_l^m(x), \quad m < 0$$

where c_{lm} is a constant. The usual convention is to define (cf. Section 15.3.4 below)

$$P_l^{-m}(x) \equiv (-1)^m \frac{(l-m)!}{(l+m)!} P_l^m(x), \tag{15.47b}$$

when the orthogonality relation analogous to (15.30) for given m is[10]

$$\int_{-1}^{1} P_l^m(x) P_{l'}^m(x)\, dx = \frac{2}{(2l+1)} \frac{(l+m)!}{(l-m)!} \delta_{ll'}. \tag{15.48}$$

[10]For a derivation of this result, see G.B. Arfken and H.J. Weber (2005) *Mathematical Methods for Physicists,* 6th edn., Academic Press, San Diego, California, Section 12.5.

Example 15.10

In most applications $x = \cos\theta$, where θ is an angular co-ordinate. Evaluate $P_l^m(\cos\theta)$ for $l \leq 2$ and $m \geq 0$.

Solution

Writing $x = \cos\theta$, (15.47a) becomes

$$P_l^m(\cos\theta) = (-1)^m(\sin\theta)^m \frac{\mathrm{d}^m P_l(\cos\theta)}{\mathrm{d}\cos^m\theta}.$$

Using the explicit forms for $P_l(x)$ given in (15.28), one easily obtains

$$P_0^0(\cos\theta) = 1 \qquad P_2^2(\cos\theta) = 3\sin^2\theta$$

$$P_1^1(\cos\theta) = -\sin\theta \qquad P_2^1(\cos\theta) = -3\sin\theta\cos\theta$$

$$P_1^0(\cos\theta) = \cos\theta \qquad P_2^0(\cos\theta) = \tfrac{1}{2}(3\cos^2\theta - 1)$$

*15.3.4 Rodrigues' formula

In the previous sections we have derived the properties of Legendre polynomials from the properties of Legendre's equation, or by using the generating function (15.36) and (15.37). An alternative approach is to exploit, or even define, the polynomials by using *Rodrigues' formula*,

$$P_l(x) = \frac{1}{2^l l!}\frac{\mathrm{d}^l}{\mathrm{d}x^l}(x^2-1)^l. \qquad (15.49)$$

To derive this result, we note that for even l, the Legendre polynomials can be written in the compact form [cf. Problems (15.12) and (15.13)]

$$P_l(x) = \sum_{k=0}^{l/2} \frac{(-1)^k(2l-2k)!}{2^l k!(l-2k)!(l-k)!}x^{l-2k}, \quad (l \text{ even}). \quad (15.50a)$$

Since

$$\left(\frac{\mathrm{d}}{\mathrm{d}x}\right)^l x^{2l-2k} = \frac{(2l-2k)!x^{l-2k}}{(l-k)!}, \quad 2k \leq l$$

$$= 0, \quad 2k > l,$$

this becomes

$$P_l(x) = \frac{1}{2^l l!}\left(\frac{\mathrm{d}}{\mathrm{d}x}\right)^l \sum_{k=0}^{l} \frac{(-1)^k l!}{k!(l-k)!}x^{2l-2k}, \qquad (15.51)$$

where the sum has been extended to all $k \leq l$. The reason for this is that we can now use the binomial theorem (1.23) and (1.24), to write

$$(x^2-1)^l = \sum_{k=0}^{l} \frac{l!}{k!(l-k)!}x^{2l-2k}(-1)^k,$$

and Rodrigues' formula (15.49) follows. A similar argument, starting from the expansion

$$P_l(x) = \sum_{k=0}^{(l-1)/2} \frac{(-1)^k (2l-2k)!}{2^l k!(l-2k)!(l-k)!} x^{l-2k}, \quad (l \text{ odd}) \qquad (15.50\text{b})$$

establishes Rodrigues' formula for odd l also.

Rodrigues' formula can be used to derive many useful results on Legendre polynomials as illustrated in Example 15.10. It is also easily extended to associated Legendre polynomials by substituting (15.49) into (15.47a) to give

$$P_l^m(x) = \frac{(-1)^m}{2^l l!}(1-x^2)^{m/2} \frac{\mathrm{d}^{l+m}}{\mathrm{d}x^{l+m}}(x^2-1)^l \qquad (15.52)$$

for $m \geq 0$. However, although this formula is derived from (15.47a) for $m \geq 0$, the right-hand side is defined for negative $m \geq -l$; and if it is used to define $P_l^{-m}(x)$, it can be shown to automatically lead to the normalisation (15.47b) for $P_l^{-m}(x)$ adopted in the previous section.

Example 15.11

Use Rodrigues' formula to prove that

$$I = \int_{-1}^{1} x^n P_l(x)\mathrm{d}x = 0, \quad l > n.$$

Solution

From Rodriques' formula we have

$$2^l l! I = \int_{-1}^{1} x^n \frac{\mathrm{d}^l}{\mathrm{d}x^l}(x^2-1)^l.$$

Now

$$\frac{\mathrm{d}^k}{\mathrm{d}x^k}(x^2-1)^l$$

always contains a factor (x^2-1) if $k \leq l$, so that

$$\left[\frac{\mathrm{d}^k}{\mathrm{d}x^k}(x^2-1)^l\right]_{-1}^{1} = 0 \quad \text{for} \quad k \leq l.$$

Hence, by repeated partial integration, we find

$$2^l l! I = -n \int_{-1}^{1} x^{n-1} \frac{\mathrm{d}^{l-1}}{\mathrm{d}x^{l-1}} (x^2-1)^l \mathrm{d}x$$
$$= (-1)^2 n(n-1) \int_{-1}^{1} x^{n-2} \frac{\mathrm{d}^{l-2}}{\mathrm{d}x^{l-2}} (x^2-1)^l \mathrm{d}x$$
$$= (-1)^n n! \int_{-1}^{1} \frac{\mathrm{d}^{l-n}}{\mathrm{d}x^{l-n}} (x^2-1)^l \mathrm{d}x$$
$$= (-1)^n n! \left[\frac{\mathrm{d}^{l-n+1}}{\mathrm{d}x^{l-n+1}} (x^2-1)^l \right]_{-1}^{1} = 0,$$

since $k = l - n + 1 \le l$ if $l > n$. Hence $I = 0$ as required.

15.4 Bessel's equation

Bessel's equation is

$$x^2 y'' + xy' + (x^2 - \nu^2)y = 0, \tag{15.53a}$$

where ν is a number and we can take $\nu \ge 0$ with no loss of generality. It is an eigenvalue equation of the form (15.21), with eigenvalues ν^2. Bessel's equation frequently occurs in studying systems with cylindrical symmetry, when $x = \rho$, the shortest distance from a point to the axis of symmetry. Such applications are extremely varied, encompassing for example, heat flow and diffusion problems, cylindrical waveguides (e.g. propagation of signals in optical fibres) and vibrating drums. In such examples, we are usually interested in solutions that are finite and well-defined for $0 \le x < \infty$, including at the end point $x = 0$.

In the standard form (15.1), Bessel's equation becomes

$$y'' + \frac{1}{x} y' + \left(1 - \frac{\nu^2}{x^2}\right) y = 0. \tag{15.53b}$$

One easily shows that $x = 0$ is a regular singular point and so we can use the Frobenius method of Section 15.1.3 to find a solution of the form

$$y = x^c \sum_{n=0}^{\infty} a_n x^n. \tag{15.54}$$

Substituting (15.54) into (15.53b) and using (15.8), gives after some simplification,

$$\sum_{n=0}^{\infty} a_n x^{n+c-2}\left\{\left[(c+n)^2-\nu^2\right]+x^2\right\}=0. \qquad (15.55)$$

Setting $n = 0$ and demanding that the coefficient of x^{c-2} vanishes yields,

$$c = \pm\nu \qquad (15.56)$$

and by considering the coefficients of higher powers of x,

$$\left[(c+1)^2-\nu^2\right] a_1 = 0,$$

and

$$\left[(c+n)^2-\nu^2\right] a_n + a_{n-2} = 0, \qquad n \geq 2$$

which, using (15.56), become

$$(1 \pm 2\nu)\, a_1 = 0, \qquad (15.57)$$

and

$$a_n = -\frac{a_{n-2}}{(n+c)^2-\nu^2}. \qquad (15.58)$$

Hence all the odd coefficients vanish and the even coefficients can be obtained in terms of a_0.

15.4.1 Bessel functions

We start by considering the case where $c = \nu$. From (15.58),

$$a_n = -\frac{a_{n-2}}{(n+c)^2-\nu^2} = -\frac{a_{n-2}}{n(n+2\nu)},$$

or, equivalently,

$$a_{2n} = -\frac{a_{2n-2}}{4n(n+\nu)}. \qquad (15.59)$$

Using this recurrence relation we find

$$\begin{aligned} a_2 &= -\frac{a_0}{2^2(1+\nu)}, \\ a_4 &= -\frac{a_2}{2^3(2+\nu)} = \frac{a_0}{2!2^4(1+\nu)(2+\nu)}, \qquad (15.60) \\ a_6 &= -\frac{a_4}{(3\times 2^2)(3+\nu)} = -\frac{a_0}{3!2^6(1+\nu)(2+\nu)(3+\nu)}, \end{aligned}$$

and so on. If ν is a positive integer, it can be seen that the denominator can be written compactly in terms of factorials, but in the general case where ν is not an integer, we need to use a notation that reduces to factorials for integral ν. The required function is called a *gamma function* $\Gamma(\nu)$ and is defined for positive ν by

$$\Gamma(\nu) \equiv \int_0^\infty x^{\nu-1} e^{-x} \mathrm{d}x, \quad \nu > 0. \tag{15.61}$$

It can be shown from this definition, by integrating by parts (see Problem 4.12), that

$$\Gamma(\nu+1) = \nu\Gamma(\nu). \tag{15.62a}$$

This is a recurrence relation for the gamma function and can be used, together with (15.61), to extend the definition from $\nu > 0$ to all ν, including $\nu \le 0$. For integer $n \ge 0$, together with $\Gamma(1) = 1$ obtained directly from (15.61), it leads to

$$\Gamma(n) = (n-1)!, \quad n \ge 1 \tag{15.62b}$$

with $0! \equiv \Gamma(1) = 1$, while for integers $n \le 0$ one has

$$\lim_{\nu \to n} \Gamma(\nu) = \pm\infty, \quad n \le 0, \tag{15.62c}$$

where the sign depends on the direction of approach to the limit. The resulting behaviour of the gamma function for $-5 \le \nu \le 4$ is shown in Figure 15.3.

Returning to the series defined by (15.59), we see that the relations (15.60), written in terms of gamma functions, are

$$a_2 = -\frac{a_0\Gamma(1+\nu)}{2^2\Gamma(2+\nu)}, \quad a_4 = \frac{a_0\Gamma(1+\nu)}{2!2^4\Gamma(3+\nu)}, \quad a_6 = -\frac{a_0\Gamma(1+\nu)}{3!2^6\Gamma(4+\nu)}$$

and in general,

$$a_{2n} = (-1)^n a_0 \frac{\Gamma(1+\nu)}{n!2^{2n}\Gamma(n+\nu+1)}. \tag{15.63}$$

It is usual to set

$$a_0 = \frac{1}{2^\nu\Gamma(1+\nu)} \tag{15.64}$$

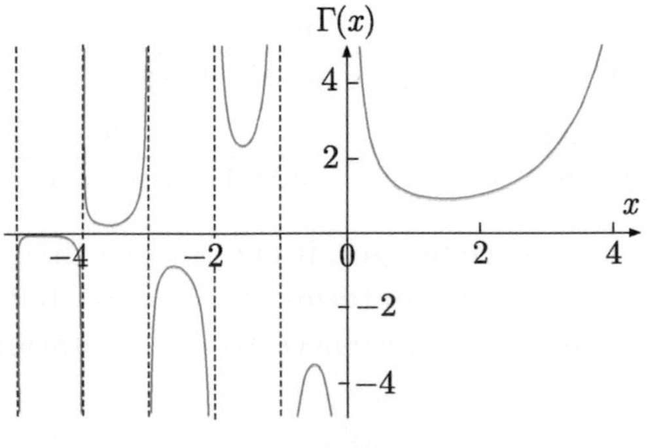

Figure 15.3 The gamma function $\Gamma(x)$.

and the function $y(x)$ is then called the *Bessel function of the first kind of order* ν, written $J_\nu(x)$. Using (15.54), (15.62), (15.63) and (15.64), we find

$$J_\nu(x) = \sum_{n=0}^{\infty} \frac{(-1)^n}{\Gamma(n+1)\Gamma(n+\nu+1)} \left(\frac{x}{2}\right)^{2n+\nu}. \tag{15.65}$$

We next consider the case where $c = -\nu$. It is not necessary to repeat all the steps that led to the derivation of (15.65). All we have to do is replace ν by $-\nu$ in that equation. This gives

$$J_{-\nu}(x) = \sum_{n=0}^{\infty} \frac{(-1)^n}{\Gamma(n+1)\Gamma(n-\nu+1)} \left(\frac{x}{2}\right)^{2n-\nu} \tag{15.66}$$

The series (15.65) and (15.66) are easily shown to converge for $0 < x < \infty$ using the ratio test and $J_{-\nu}(x)$, like $J_\nu(x)$, is also called a Bessel function of the first kind.

At this point, we distinguish between integer and non-integer ν. For *non-integer* ν, $J_\nu(x)$ and $J_{-\nu}(x)$ are independent solutions, as is the linear combination

$$y(x) = c_1 J_\nu(x) + c_2 J_{-\nu}(x), \tag{15.67}$$

where c_1 and c_2 are arbitrary constants. However, as can be seen from the first terms in (15.65) and (15.66), only $J_\nu(x)$ with $\nu \geq 0$ is non-singular as $x \to 0$.

For integer $\nu = m > 0$, the situation is somewhat different. This is because the first terms in (15.66) vanish by (15.62c), so that

$$\begin{aligned} J_{-m}(x) &= \sum_{n=m}^{\infty} \frac{(-1)^n}{\Gamma(n+1)\Gamma(n-m+1)} \left(\frac{x}{2}\right)^{2n-m} \\ &= \sum_{k=0}^{\infty} \frac{(-1)^{k+m}}{\Gamma(k+m+1)\Gamma(k+1)} \left(\frac{x}{2}\right)^{2k+m} = (-1)^m J_m(x), \end{aligned}$$

where we have defined $k = n - m$. Hence for integer m, $J_m(x)$ and $J_{-m}(x)$ are not independent solutions and another solution must be found. For this reason, it is conventional to replace $J_{-\nu}(x)$ by the function

$$N_\nu(x) \equiv \frac{J_\nu(x)\cos(\nu\pi) - J_{-v}(x)}{\sin(\nu\pi)}. \tag{15.68}$$

These functions are called *Bessel functions of the second kind.*[11] For non-integer ν, they are obviously solutions of Bessel's equation, since

[11] The use of the notation N for the function is because it is also known as a *Neumann function*; some authors refer to it as a Weber function and use the letter Y.

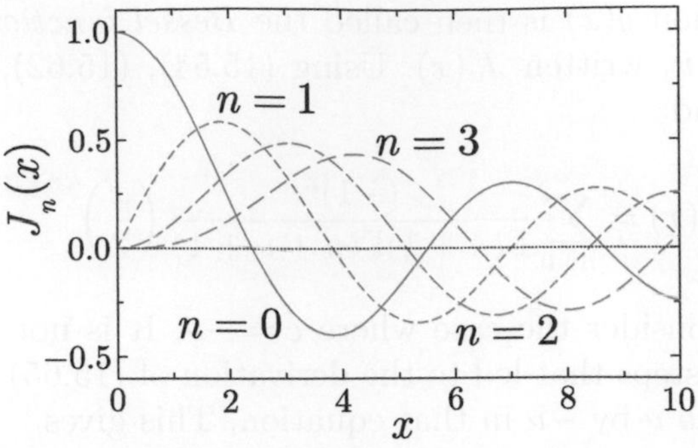

Figure 15.4 Bessel functions $J_n(x)$: $n = 0$ (solid), $n = 1$ (short dash), $n = 2$ (long dash-short dash), $n = 3$ (long dash).

they are just well-defined linear combinations of $J_\nu(x)$ and $J_{-\nu}(x)$. However, it can be shown that they are also solutions for integer m, provided we interpret (15.68) as

$$N_m(x) = \lim_{\nu \to m} \left[\frac{J_\nu(x)\cos(\nu\pi) - J_{-\nu}(x)}{\sin(\nu\pi)} \right]. \tag{15.69}$$

The general solution of Bessel's equation is then written

$$y(x) = AJ_\nu(x) + BN_\nu(x) \tag{15.70}$$

for both integer and non-integer ν, where A and B are arbitrary constants.

We will not discuss the functions $N_\nu(x)$ further, because only Bessel functions of the first kind $J_\nu(x)$ with $\nu \geq 0$ are non-singular as $x \to 0$. These, and especially those with integer ν, are the most important in applications, and the behaviour of $J_n(x)$ are shown in Figure 15.4 for $n = 0,\ 1,\ 2$ and 3, and $0 \leq x \leq 10$. As seen from (15.65), $J_n(0) = 0$ for $n > 0$. The positions of the zeros for $x > 0$ are also important in applications. The values of the first five zeros of the Bessel functions $J_n(x)$, $n = 1,\ 2, \ldots, 5$ are given in Table 15.1.

Further properties of Bessel functions that are useful in applications are discussed in the next subsection. However, before doing so, we warn the reader that $J_\nu(x)$ and $N_\nu(x)$ are not the only forms referred to as Bessel functions. There are others, such as *spherical*

Table 15.1 Values of the first five zeros of the Bessel functions $J_n(x)$, for $n = 0, \ldots, 5$

	$J_0(x)$	$J_1(x)$	$J_2(x)$	$J_3(x)$	$J_4(x)$	$J_5(x)$
1	2.4048	3.8317	5.1356	6.3802	7.5883	8.7715
2	5.5201	7.0156	8.4172	9.7610	11.0647	12.3386
3	8.6537	10.1735	11.6198	13.0152	14.3725	15.7002
4	11.7915	13.3237	14.7960	16.2235	17.6160	18.9801
5	14.9309	16.4706	17.9598	19.4094	20.8269	22.2178

Bessel functions (that arise in scattering problems) and *Hankel functions.* We will not discuss these other forms here.

Example 15.12

Show that when $x^2 \gg \max\{1, \nu^2\}$, the general solution to Bessel's equation (15.53) reduces to

$$y = \frac{1}{\sqrt{x}}(A \sin x + B \cos x).$$

Solution

Neglecting the term in ν^2/x^2, Bessel's equation reduces to

$$y'' + \frac{1}{x}y' + y = 0.$$

Then on substituting $y = u(x)/\sqrt{x}$, we obtain, after a little algebra,

$$\frac{\mathrm{d}^2u}{\mathrm{d}x^2} + \left(1 + \frac{1}{4x^2}\right) u = 0,$$

which reduces to

$$\frac{\mathrm{d}^2u}{\mathrm{d}x^2} + u = 0$$

when $x^2 \gg 1$. The solution of this equation is

$$u(x) = A \sin x + B \cos x,$$

so that finally

$$y(x) = \frac{u(x)}{\sqrt{x}} = \frac{1}{\sqrt{x}}(A \sin x + B \cos x),$$

as required.

*15.4.2 Properties of non-singular Bessel functions $J_\nu(x)$

The values and properties of the various types of Bessel functions are extensively listed in reference books and on the web.[12] Here we restrict ourselves to just some of the properties of Bessel functions that are non-singular at $x = 0$, that is, Bessel functions of the first kind $J_\nu(x)$ $(\nu > 0)$.

Bessel functions obey recurrence relations that are somewhat similar to those obtained in Section 15.3.2 for Legendre polynomials.

[12]See, for example, G.B. Arfken and H.J. Weber (2005) *Mathematical Methods for Physicists,* 6th edn., Academic Press, San Diego, California, Chapter 11 and references therein.

Some of these recurrence relations, which hold for positive and negative ν, are

$$\frac{\mathrm{d}}{\mathrm{d}x}[x^{\nu}J_{\nu}(x)] = x^{\nu}J_{\nu-1}(x), \tag{15.71a}$$

$$\frac{\mathrm{d}}{\mathrm{d}x}[x^{-\nu}J_{\nu}(x)] = -x^{-\nu}J_{\nu+1}(x), \tag{15.71b}$$

$$J_{\nu-1}(x) + J_{\nu+1}(x) = \frac{2\nu}{x}J_{\nu}(x), \tag{15.71c}$$

$$J_{\nu-1}(x) - J_{\nu+1}(x) = 2J'_{\nu}(x), \tag{15.71d}$$

$$xJ'_{\nu}(x) = -\nu J_{\nu}(x) + xJ_{\nu-1}(x) = \nu J_{\nu}(x) - xJ_{\nu+1}(x). \tag{15.71e}$$

Such relations are easily confirmed using the series representation (15.65). For example, if we differentiate the product $x^{\nu}J_{\nu}(x)$ using (15.65), we obtain

$$\frac{\mathrm{d}}{\mathrm{d}x}[x^{\nu}J_{\nu}(x)] = x^{\nu}\sum_{n=0}^{\infty}\frac{(-1)^n x^{(\nu-1)+2n}}{2^{(\nu-1)+2n}n!\Gamma[(\nu-1)+n+1]} = x^{\nu}J_{\nu-1}(x),$$

which is (15.71a). Expanding the left-hand side of this expression and dividing by $x^{\nu-1}$, gives

$$xJ'_{\nu}(x) + \nu J_{\nu}(x) = xJ_{\nu-1}. \tag{15.72a}$$

In a similar way we may show that

$$xJ'_{\nu}(x) - \nu J_{\nu}(x) = -xJ_{\nu+1}. \tag{15.72b}$$

These relations are equivalent to (15.71e), and adding them and dividing by x gives (15.71d).

In Section 15.3.1, we saw an arbitrary function that is non-singular in the range $-1 < x < 1$ could be expanded in terms of Legendre polynomials [cf. (15.32)]. If $a_{\nu n} > 0$ are the zeroes of the Bessel function $J_{\nu}(x)$, i.e.

$$J_{\nu}(a_{\nu n}) = 0, \quad n = 0,\ 1,\ 2, \ldots, \tag{15.73}$$

a similar expansion in terms of Bessel functions in the range $0 < x < 1$ can be obtained by considering the functions $J_{\nu}(a_{\nu n}x)$, which satisfy the relations[13]

$$\int_0^1 x\ J_{\nu}(a_{\nu m}x)J_{\nu}(a_{\nu n}x)\mathrm{d}x = \tfrac{1}{2}[J'_{\nu}(a_{\nu n})]^2\delta_{nm}. \tag{15.74}$$

[13] The proof of this relation is long and will not be reproduced here. It may be found in, for example, G.B. Arfken and H.J. Weber (2005) *Mathematical Methods for Physicists,* 6th edn., Academic Press, San Diego, California, Section 11.2.

For $m \neq n$, this differs from the orthogonality relations (15.33) obtained for Legendre polynomials by the presence of the factor x and in the range of integration. For this reason, the functions $J_\nu(a_{\nu n}x)$ are said to be orthogonal with weight function x in the domain $0 \leq x \leq 1$. In analogy to (15.32), it can be shown that an arbitrary function $f(x)$ that is non-singular in the domain $0 \leq x \leq 1$ can be expanded in the form

$$f(x) = \sum_{n=1}^{\infty} c_{\nu n} J_\nu(a_{\nu n}x), \quad 0 \leq x \leq 1, \tag{15.75}$$

i.e.$J_\nu(a_{\nu n}x)$, $n = 1, 2, 3, \ldots$, form a complete set of functions in this range. The coefficients $c_{\nu k}$ are then obtained by multiplying (15.75) by $xJ_\nu(a_{\nu k}x)$ and integrating using (15.74) to give

$$c_{\nu k} = \frac{2}{[J'_\nu(a_{\nu k})]^2} \int_0^1 x\ f(x) J_\nu(a_{\nu k}x) \mathrm{d}x. \tag{15.76}$$

The expansion (15.75) is called a *Fourier-Bessel series* and is often used in the solution of partial differential equations in cylindrical polar co-ordinates.

Example 15.13

By analogy with the generating function for Legendre polynomials, the generating function for Bessel functions of integer order n can be shown to be

$$G(x, h) = \exp\left[\frac{x}{2}\left(h - \frac{1}{h}\right)\right] = \sum_{n=-\infty}^{\infty} J_n(x)h^n.$$

Use this relation to prove the recurrence relation (15.72c), for integer n.

Solution

Differentiating $G(x, h)$ partially with respect to h we obtain

$$\frac{\partial G(x, h)}{\partial h} = \frac{x}{2}\left(1 + \frac{1}{h^2}\right) G(x, h) = \sum_{n=-\infty}^{\infty} nJ_n(x)h^{n-1}.$$

Substituting for $G(x, h)$ using the definition above again gives

$$\frac{x}{2}\left(1 + \frac{1}{h^2}\right) \sum_{n=-\infty}^{\infty} J_n(x)h^n = \sum_{n=-\infty}^{\infty} nJ_n(x)h^{n-1}.$$

Then equating coefficients of h^n, we have

$$\frac{x}{2}[J_n(x) + J_{n+2}(x)] = (n + 1)J_{n+1}(x),$$

and finally replacing n by $n - 1$ yields (15.72c).

Problems 15

15.1 Discuss the feasibility of finding power series solutions of the equations

$$\text{(a) } (1-x^2)y'' - 3x^2y' + (1-x)y = 0, \quad \text{(b) } x^3y'' + y' + x^2y = 0.$$

15.2 Find the complete series solution of the equation

$$(1+x^2)y''(x) + xy'(x) + y(x) = 0,$$

about the point $x = 0$.

15.3 Find the general solution of the equation

$$x(2-x)\frac{\mathrm{d}^2y}{\mathrm{d}x^2} + 3(1-x)\frac{\mathrm{d}y}{\mathrm{d}x} - y = 0$$

as a power series about $x = 1$.

15.4 Confirm that $x = 0$ is a singular point of the equation

$$x^2\frac{\mathrm{d}^2y}{\mathrm{d}x^2} + x\frac{\mathrm{d}y}{\mathrm{d}x} - 9y = 0$$

and deduce the nature of the singularity. Hence solve for $y(x)$ as a power series.

15.5 Show that the solutions of the indicial equation for the ODE

$$2x\frac{\mathrm{d}^2y}{\mathrm{d}x^2} - \frac{\mathrm{d}y}{\mathrm{d}x} + 2y = 0$$

are 0 and 3/2, and for the latter case find the power series solution of the equation.

15.6 One solution of the equation

$$x^2y''(x) + 3xy'(x) + y(x) = 0,$$

is $y_1(x) = 1/x$. Find a second independent solution $y_2(x)$ by writing $y_2(x) = y_1(x)u(x)$ and solving for $u(x)$. Hence find the general solution.

15.7 Show that the indicial equation for the ODE

$$x(x-1)^2y''(x) - 2y(x) = 0$$

has solutions $c = 0$ and 1, and find the explicit form of the solution for the larger of the two values. Assuming the smaller value does not lead to an independent solution of the ODE, use Fuchs' theorem to find a second independent solution, and hence the complete solution of the equation.

15.8 Show that $y_n(x) = e^{-x^2/2}H_n(x)$, $n = 0, 1, 2, \ldots$, where $H_n(x)$ is a Hermite polynomial, is a solution of the equation

$$\frac{\mathrm{d}^2y}{\mathrm{d}x^2} - x^2y + (2n+1)y = 0. \tag{1}$$

Hence show that these functions satisfy the orthogonality relation

$$\int_{-\infty}^{\infty} e^{-x^2} H_n(x) H_m(x)\, dx = 0, \quad n \neq m.$$

15.9 If the series solution of the equation

$$2x\frac{d^2y}{dx^2} + (3-x)\frac{dy}{dx} + \alpha y = 0$$

is

$$y(x) = \sum_{n=0}^{\infty} a_n x^{n+c}, \quad (a_0 \neq 0),$$

where α is a constant, show that $c = 0$ or $-1/2$, and in the former case deduce the recurrence relation

$$(2n+3)(n+1)a_{n+1} = (n-\alpha)a_n.$$

Show also that if $\alpha = m$, where m is a positive integer, a polynomial solution results and deduce its form.

15.10 A real function $y(x)$ satisfies the equation

$$y'' + 2y' + \lambda y = 0,$$

and is subject to the boundary conditions $y = 0$ at $x = 0$ and $x = 1$. Find the eigenvalues $\lambda = \lambda_n$ and the corresponding eigenfunctions.

15.11 Show that the substitution $x = e^t$ reduces the equation

$$x^2\frac{d^2y(x)}{dx^2} + 2x\frac{dy(x)}{dx} + \frac{1}{4}y(x) + \lambda_n y(x) = 0. \quad (1)$$

to a linear second-order equation in t with constant coefficients. Hence find the eigenvalues $\lambda_n > 0$ and the corresponding normalised eigenfunctions $y_n(x)$ subject to the boundary conditions $y(x=1) = y(x=e) = 0$.

***15.12** Use (15.35) to show that Legendre functions of the second kind for integer l satisfy the recurrence relation

$$lQ_l(x) = (2l-1)xQ_{l-1}(x) - (l-1)Q_{l-2}(x). \quad (1)$$

Use this result, together with the expressions for $Q_0(x)$ and $Q_1(x)$ given in Example (15.8), to prove that

$$Q_l(x) = \frac{P_l(x)}{2}\ln\left(\frac{1+x}{1-x}\right) - q_l(x)$$

for all integer l, where $P_l(x)$ is the corresponding Legendre polynomial and $q_l(x) = 0$ $(l = 0)$, or is a polynomial of order $l-1$ $(l \geq 1)$. Find the form of $q_l(x)$ for $l = 2,\ 3,\ 4$.

15.13 A function $f(x)$ that is non-singular in the range $-1 \le x \le 1$ is expanded in a Legendre series

$$f(x) = \sum_{k=0}^{\infty} c_k P_k(x), \quad -1 \le x \le 1. \tag{1}$$

Show that the coefficients are unique and evaluate the integral

$$\int_{-1}^{1} [f(x)]^2 \, \mathrm{d}x \tag{2}$$

in terms of the coefficients c_k.

***15.14** Use the generating function

$$G(x, h) = (1 - 2xh + h^2)^{-1/2}$$

for the Legendre polynomials to derive the recurrence relation

$$(2l + 1)xP_l(x) = (l + 1)P_{l+1}(x) + lP_{l-1}(x).$$

***15.15** A linear electric quadrupole is composed of a charge e at $\mathbf{r} = \mathbf{p}$, a charge e at $\mathbf{r} = -\mathbf{p}$, and a charge $-2e$ at the origin. Expand the resulting electrostatic field $V(\mathbf{r})$ in powers of p/r, where $p = |\mathbf{p}|$ and $r = |\mathbf{r}|$, and hence obtain its form for $r \gg p$.

***15.16** Use suitable recurrence relations, or otherwise, to show that

$$P_{2n}(0) = \frac{(-1)^n (2n-1)!!}{(2n)!!} \quad \text{and} \quad P_{2n+1}{}'(0) = \frac{(-1)^n (2n+1)!!}{(2n)!!}, \tag{1}$$

where $P_k(x)$ is a Legendre polynomial, $k = 0, 1, 2, \ldots$.

***15.17** Verify (15.50a) for Legendre polyomials $P_l(x)$ of even order l by showing that it leads to the correct recurrence relation (15.26) and the normalisation conditions (1) of Question 15.12.

***15.18** Use Rodrigues' formula to deduce the coefficient c_n in the expansion

$$x^n = \sum_{l=0}^{n} c_l P_l(x).$$

The standard integral

$$B(n+1, \; n+1) \equiv \int_0^1 t^n (1-t)^n \mathrm{d}t = \frac{n!n!}{(2n+1)!}$$

may be useful.

15.19 A function $y(x)$ satisfies the equation

$$\frac{\mathrm{d}^2 y(x)}{\mathrm{d}x^2} + \frac{1}{x}\frac{\mathrm{d}y(x)}{\mathrm{d}x} - \frac{1}{x^2}y(x) + k^2 y(x) = 0,$$

together with the boundary conditions $y = 0$ at $x = 0$ and $x = 1$. By using the variable $z = kx$, find the four lowest allowed values of k.

***15.20** Figure 15.4 suggests that, if $J_n(x)$ $(n = 0,\ 1,\ 2,\ 3)$ are Bessel functions of the first kind, then for $n \geq 1$, $J_{n+1}(x) \approx J_{n-1}(x)$ at the maximum of $J_n(x)$; and $J_{n+1}(x) \approx -J_{n-1}(x)$ at a zero of $J_n(x)$. Show that these relations are both exact for all $n \geq 1$.

15.21 Consider the use of the expansion (15.65) to evaluate the Bessel function $J_2(x)$ at $x = 2$. Use (5.62) to determine how many terms must be retained to ensure that the error in truncating the series is less than 10^{-5}, and hence evaluate $J_2(2)$ to 5 decimal places. How many extra terms would be required to evaluate it to 7 decimal places?

15.22 Bessel functions of the second kind $N_\nu(x)$ are singular at $x = 0$.
(a) By using (15.68) and (15.69), show that for small x,

$$N_0(x) = \frac{2}{\pi}[\ln x + \gamma - \ln 2] + O(x).$$

(b) By letting $\nu = 1 + \varepsilon$ and considering the behaviour of the Bessel functions $J_{\pm\nu}(x)$ as $\varepsilon \to 1$, derive the relation

$$N_1(x) = \frac{1}{\pi}\left[-\left(\frac{2}{x}\right) + (2\gamma - 1)\left(\frac{x}{2}\right) + 2\left(\frac{x}{2}\right)\ln\left(\frac{x}{2}\right)\right] + O(x^2),$$

where $\gamma = 0.57721\ldots$ is the *Euler-Mascheroni constant.* You may assume, that for small ν,

$$\Gamma(\nu) = \frac{1}{\nu} - \gamma + \frac{1}{6}\left(3\gamma^2 + \frac{\pi^2}{2}\right)\nu + O(\nu^2). \qquad (1)$$

***15.23** Use the series (15.65) to show that the Bessel function

$$J_{1/2}(x) = \left(\frac{2}{\pi x}\right)^{1/2} \sin x,$$

given that $\Gamma(1/2) = \sqrt{\pi}$. Use this result to express $J_{-1/2}(x)$ and $J_{3/2}(x)$ in terms of trigonometric functions.

16

Partial differential equations

In Chapters 14 and 15 we discussed ordinary differential equations and their solutions. These are equations that contain a dependent variable y, which is a function of a single variable x, and derivatives of y with respect to x. In this chapter, we extend the discussion to similar equations that involve functions of two or more variables $x_1, x_2, \ldots, x_n$. These are called *partial differential equations (PDEs)* because the functional form analogous to (14.2) in general contains partial differentials with respect to several variables, including mixed derivatives. If we consider a function u of just two variables $x_1 = x$ and $x_2 = y$, then examples of partial differential equations are

$$\frac{\partial u}{\partial x} + y^2 \frac{\partial u}{\partial y} = 3u, \tag{16.1a}$$

$$\frac{\partial^2 u}{\partial x^2} + f(x, y) \frac{\partial^2 u}{\partial y^2} = u^2, \tag{16.1b}$$

and

$$\tan y \frac{\partial^2 u}{\partial x^2} + \cos x \frac{\partial u}{\partial y} = 0, \tag{16.1c}$$

where $f(x, y)$ is an arbitrary function of x and y.

By analogy with the definitions in Chapter 14, the *degree* of the equation is defined as the power to which the highest order derivative is raised after the equation is rationalised, if necessary; and the *order* of a partial differential equation is the order of the highest derivative in the equation. Thus (16.1a) is first-order and (16.1b) and (16.1c) are second-order equations. In addition, all three equations are *linear*, in the sense that they contain only u and its derivatives to first degree; products between them are absent.

Mathematics for Physicists, First Edition. B.R. Martin and G. Shaw.

Companion website: www.wiley.com/go/martin/mathsforphysicists

Non-linear equations, such as

$$u^2\frac{\partial^2 u}{\partial y^2} + \left(\frac{\partial u}{\partial x}\right)^3 = u,$$

will not be discussed.

A linear equation is said to be *homogeneous* if each term contains either the dependent variable or one of its derivatives, so that if u is a solution, so is λu, where λ is a constant. Thus (16.1a) and (16.1c) are homogeneous, while (16.1b) is inhomogeneous. An important property of linear homogeneous equations is that, if u_1 and u_2 are solutions, then so is any arbitrary linear combination of them $(\alpha u_1 + \beta u_2)$, where α and β are arbitrary constants. This is called the *superposition principle* and is widely used in finding the general solution of some PDEs using a method known as 'separation of variables' that we will discuss in Sections 16.2 and 16.3.

One important difference between PDEs and ODEs needs to be mentioned. As discussed in Chapter 14, a linear combination of two or more solutions for an nth order linear homogeneous ODE will in general have n arbitrary constants, which can only be determined by suitable boundary conditions, for example, the value of the function $y(x)$ can be specified for n values of x to determine the constants. The solution of a PDE, however, usually contains a number of arbitrary *functions*. For example, it is easily verified by direct substitution that the simple first-order PDE

$$y\frac{\partial u(x,y)}{\partial y} = u(x,y),$$

has solutions of the form $u(x,y) = y\,f(x)$, where $f(x)$ is an arbitrary function of x. Similarly, the simple second-order equation

$$\frac{\partial^2 u}{\partial x \partial y} = 0 \tag{16.2a}$$

has a general solution of the form

$$u(x,y) = f(x) + g(y), \tag{16.2b}$$

where $f(x)$ and $g(y)$ are arbitrary differentiable functions, as may be found by successive integration with respect to x (at fixed y) and y (at fixed x). However, unlike the analogous situation for ODEs, it does not in general follow that an nth order PDE always contains n arbitrary functions, but it is true for linear equations with constant coefficients, which is the class of equations discussed in the rest of this chapter. As for ODEs, these arbitrary functions must be determined by imposing boundary conditions. In this case, these will have to take the form of specifying u along a continuum of points, for example along a line in the (x, y) or (x, t) plane, but the appropriate form of the boundary conditions depends on the type of PDE. This will be discussed in more detail in Section 16.6.

Example 16.1

(a) Show that $z(x, y) = f(x + 3y) + g(x - 3y)$, where f and g are arbitrary differentiable functions, is a general solution of the partial differential equation

$$9\frac{\partial^2 z}{\partial x^2} = \frac{\partial^2 z}{\partial y^2}.$$

(b) Find the PDE that has a general solution $z(x, y) = e^x f(3y + 2x)$, where f is an arbitrary differentiable function.

Solution

(a) Let $u = x + 3y$ and $w = x - 3y$, so that $z = f(u) + g(w)$. Then

$$\frac{\partial z}{\partial x} = \frac{\partial f}{\partial u}\frac{\partial u}{\partial x} + \frac{\partial g}{\partial w}\frac{\partial w}{\partial x} = f'(u) + g'(w),$$

and

$$\frac{\partial^2 z}{\partial x^2} = \frac{\partial f'(u)}{\partial u}\frac{\partial u}{\partial x} + \frac{\partial g'(w)}{\partial w}\frac{\partial w}{\partial x} = f''(u) + g''(w).$$

Also,

$$\frac{\partial^2 z}{\partial y^2} = 9f''(u) + 9g''(w),$$

and so z is a solution of the PDE

$$9\frac{\partial^2 z}{\partial x^2} = \frac{\partial^2 z}{\partial y^2}.$$

Moreover, since this equation is linear with constant coefficients and of order two, the solution contains two arbitrary functions and is a general solution.

(b) Set $u = 3y + 2x$, so that $z = e^x f(u)$. Then

$$\frac{\partial z}{\partial x} = e^x\frac{\partial f}{\partial x} + e^x f = 2e^x f' + e^x f,$$

and

$$\frac{\partial z}{\partial y} = 3e^x f'.$$

So, eliminating f' gives the PDE

$$3\frac{\partial z}{\partial x} - 2\frac{\partial z}{\partial y} = 3z.$$

16.1 Some important PDEs in physics

Partial differential equations are in general far more difficult to solve than ordinary differential equations, and except for certain special types of equation, no general method of solution exists. In this chapter we will concentrate on discussing some specific second-order equations, since these include many of the most important equations in physics. These include the following, where u is a finite physical quantity that can depend on three space co-ordinates x, y, z and the time t.

(i) *The wave equation*

$$\boldsymbol{\nabla}^2 u(\mathbf{r},t) = \frac{1}{v^2}\frac{\partial^2 u(\mathbf{r},t)}{\partial t^2}. \tag{16.3a}$$

In mechanics, u could be a displacement of a vibrating medium, or in electromagnetism, the component of an electromagnetic wave, etc., and v is the speed of propagation of the associated wave.

(ii) *The diffusion equation*

$$\boldsymbol{\nabla}^2 u(\mathbf{r},t) = \frac{1}{\kappa}\frac{\partial u(\mathbf{r},t)}{\partial t}. \tag{16.4}$$

This describes the diffusion of material particles, where ρ is the density of diffusing particles and the constant κ is called the diffusivity. It also describes heat conduction in a region that contains no heat sources or sinks, where it is referred to as the *heat conduction equation.* In this case, u is the temperature and κ is called the *thermal diffusivity.* It is given by $\kappa = k/(s\rho)$, where k is the thermal conductivity, s is the specific heat capacity, and ρ is the density of the material.

(iii) *Laplace's equation*

$$\boldsymbol{\nabla}^2 u(\mathbf{r}) = 0. \tag{16.5}$$

Here u could be, for example, a steady-state temperature distribution, or the electrostatic or gravitational potential in free space.

(iv) *Poisson's equation*

$$\boldsymbol{\nabla}^2 u(\mathbf{r}) = \rho(\mathbf{r}). \tag{16.6}$$

The quantity u describes the same quantities as in Laplace's equation, but now in a region containing an appropriate source $\rho(\mathbf{r})$. For the electrostatic potential $u = \phi$, Poisson's equation takes the form (12.60), so that the force is proportional to the electric charge density; for the gravitational potential ψ, (12.58) and (12.59) give

$$\boldsymbol{\nabla}^2 \psi = 4\pi G \rho_m(\mathbf{r}),$$

where $\rho_m(\mathbf{r})$ is the mass density and G is Newton's gravitational constant.

(v) *Schrödinger's equation*

$$-\frac{\hbar^2}{2m}\nabla^2 u(\mathbf{r},t) + V(\mathbf{r})u(\mathbf{r},t) = i\hbar\frac{\partial u(\mathbf{r},t)}{\partial t}. \tag{16.7}$$

This is the equation of motion for a particle of mass m in a potential $V(\mathbf{r})$ in non-relativistic quantum mechanics, where $\hbar$ is Planck's constant h divided by 2π. The quantity u is the *Schrödinger wave function* and is usually complex.

An important feature of all these equations is that the spatial derivatives enter via the scalar differential operator

$$\nabla^2 = \frac{\partial^2}{\partial x^2} + \frac{\partial^2}{\partial y^2} + \frac{\partial^2}{\partial z^2},$$

which reflects the rotational invariance of the laws of physics. In particular, they do not include mixed partial derivatives of the forms $\partial^2 u/\partial x\partial y, \partial^2 u/\partial x\partial t$, etc. Because of this, and subject to the forms of $\rho(\mathbf{r})$ and $V(\mathbf{r})$, these equations can be solved by the method of *separation of variables,* in which the problem is converted into one of solving a set of ordinary differential equations, each in a single variable. This method is discussed in the next two sections; other methods of solution will be treated in later sections.

16.2 Separation of variables: Cartesian co-ordinates

In this method, given a PDE for a function $u(x, y, z, t)$, we seek solutions of the form

$$u(x, y, z, t) = X(x)Y(y)Z(z)T(t), \tag{16.8}$$

in which u is given as the product of functions of single variables. Here we have used the convention of denoting the single-variable functions by an upper-case letter and its argument by the corresponding lower-case letter. There could also be common parameters that occur in each function, but each depends on only one independent variable. We then try to rewrite the original PDE in the form of four separate ODEs for each of the functions X, Y, Z, T using a procedure to be explained below. If this is possible, then the original PDE is said to be *separable,* and one can seek solutions to the individual ODEs using the methods discussed in Chapters 14 and 15. If it is not possible, then the equation is not separable, and other methods must be used.

If a function can be written in the form (16.8), it is also said to be *separable*. Thus $xy^2z^3\tan(at)$ is separable in all four variables, and is said to be *completely separable*; whereas $(x^2+y^{-2})z^3\sin(\theta t)$ is only separable in z and t and is therefore *partially separable*; but $xz+ayt$ cannot be separated in any variable and so is *inseparable*. Obviously the individual solutions initially found by separation of variables are, by construction, completely separable functions. However, this is not as restrictive as it might seem, since, as we shall see, the method leads to many such solutions, usually an infinite number, and the general solution, which is rarely separable, can usually be constructed from them. This is particularly easy in the important case of linear homogeneous equations, when any linear combination of a given set of solutions is itself a solution, and for such equations it is also known as the *Fourier method* of solution.

In this section, we shall introduce the method using Cartesian co-ordinates, leaving the important but more complicated case of polar co-ordinates until Section 16.3. In both cases, we will use the physical equations introduced in the previous section to do this, applying boundary conditions and constraints appropriate to the function u being a physical quantity.

16.2.1 The wave equation in one spatial dimension

To illustrate the general method, we will solve the wave equation (16.3a) for a single spatial variable x, when it reduces to

$$\frac{\partial^2 u}{\partial x^2}=\frac{1}{v^2}\frac{\partial^2 u}{\partial t^2}. \tag{16.3b}$$

This equation could describe, for example, a vibrating string undergoing small transverse displacements $u(x,t)$, where the wave velocity is v. For definiteness, we will assume the string is clamped at $x=0$ and $x=L$, so that the boundary conditions include the constraints

$$u(0,t)=0, \quad u(L,t)=0 \tag{16.9}$$

at all times t.

To solve (16.3b), we assume a separable form

$$u(x,t)=X(x)T(t), \tag{16.10}$$

which when substituted into (16.3b) gives,

$$X''T=v^{-2}XT''. \tag{16.11}$$

Here the primes indicate differentiation with respect to the single variable appropriate to the symbol for the function, that is,

$X'' = \mathrm{d}^2X/\mathrm{d}x^2$ and $T'' = \mathrm{d}^2T/\mathrm{d}t^2$. If we now divide (16.11) by $u = XT$, we have

$$\frac{X''}{X} = \frac{1}{v^2}\frac{T''}{T}. \tag{16.12}$$

This equation is of a very special form, where each term that appears is a function of only one variable; X''/X is a function of x only and $T''/(v^2T)$ is a function of t only. It can therefore only be satisfied if each term is equal to the same constant, called the *separation constant*, which will be denoted by $-k^2$ for later convenience. Thus we can write

$$X'' = -k^2X, \tag{16.13a}$$

and

$$T'' = -\omega^2T \tag{16.13b}$$

where $\omega = kv$.

The first of these equations, together with the boundary conditions $X(0) = X(L) = 0$, which follow from (16.9), is just the eigenvalue problem that was solved earlier in Section 15.2. There we found that solutions only exist for

$$k = k_n \equiv n\pi/L, \quad n = 1, 2, \ldots \tag{16.14}$$

and are

$$X(x) = \sin(n\pi x/L), \quad 0 < x < L. \tag{16.15}$$

The corresponding values of ω are then $\omega_n = vk_n = n\pi vx/L$, and the general solution to (16.13b) is therefore

$$T(t) = A_n\cos(n\pi vt/L) + B_n\sin(n\pi vt/L), \tag{16.16}$$

where A_n and B_n are arbitrary constants, so that we arrive at a set of solutions

$$u_n(x,t) = \sin(n\pi x/L)\left[A_n\cos(n\pi vt/L) + B_n\sin(n\pi vt/L)\right], \tag{16.17}$$

where $n = 1,\ 2, \ldots$ is any positive integer. Each of these solutions is a 'normal mode' in which $u(x,t)$ oscillates with a single angular frequency $\omega_n = n\pi v/L$ for all x-values[1]. At this point, we recall that the wave equation is linear and homogeneous, so that any linear combinations of solutions of the form (16.17) is also a solution of the original equation. The general solution is a linear combination of the normal modes, that is,

$$u(x,t) = \sum_{n=1}^{\infty} u_n(x,t) = \sum_{n=1}^{\infty} \sin(n\pi x/L)\,[B_n\cos(n\pi vt/L) + A_n\sin(n\pi vt/L).] \tag{16.18}$$

[1] Compare the discussion of normal modes for mechanical systems in Section 10.2.1.

Finally, to determine the arbitrary constants A_n and B_n and obtain a unique solution requires imposing further boundary conditions that have to be specified according to the problem. For the case of a vibrating string, these might be that the string is released from rest at $t = 0$ from its initial state $u(x, 0)$. The first of these conditions, that is,

$$\dot{u}(x, 0) \equiv \left.\frac{\partial u(x,t)}{\partial t}\right|_{t=0} = 0,$$

leads to the result $A_n = 0$ (all n) on substituting into (16.18), so that the solution becomes

$$u(x,t) = \sum_{n=1}^{\infty} u_n(x,t) = \sum_{n=1}^{\infty} B_n \sin(n\pi x/L)\cos(n\pi vt/L)\,. \quad (16.19)$$

In addition, the coefficients B_n can be determined, given the initial displacement $u(x, 0)$, since from (16.19)

$$u(x,0) = \sum_{n=1}^{\infty} B_n \sin(n\pi x/L)\,,$$

which is just the Fourier expansion of the initial configuration of $u(x, 0)$ extended, as an odd function, to the range $-L < x < L$ (cf. Section 13.1.4). To invert it, we multiply both sides by $\sin(m\pi z/L)$ and integrate from 0 to L using the orthonormality relation

$$\int_0^L \sin\left(\frac{n\pi x}{L}\right)\sin\left(\frac{m\pi x}{L}\right)\mathrm{d}x = \frac{L}{2}\delta_{nm}. \quad (16.20)$$

This gives

$$B_n = \frac{2}{L}\int_0^L u(x,0)\sin(n\pi x/L)\,\mathrm{d}x \quad (16.21)$$

and the solution (16.19) is completely determined.

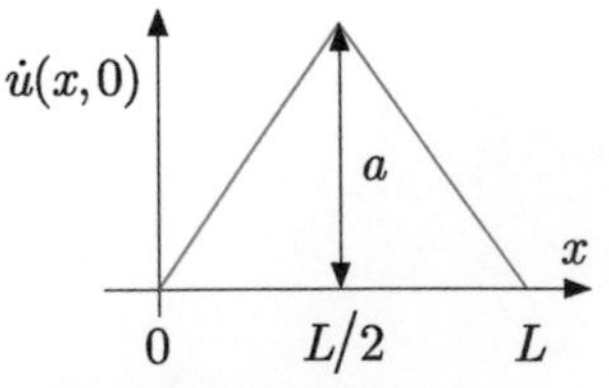

Figure 16.1

Example 16.2

A stretched string of length L lies in the x-direction and is clamped at both ends. At time $t = 0$, the transverse displacement $u(x, 0) = 0$, but the transverse velocity $\dot{u}(x, 0)$ is as shown in Figure 16.1. If the velocity of waves along the string is v, what is the function $u(x, t)$ that describes the shape of the string at subsequent times?

Solution

We again solve the wave equation (16.3b) subject to the boundary conditions (16.9) corresponding to both ends of the string being clamped, so the general solution is given by (16.18). However, the condition $u(x,0) = 0$ implies $B_n = 0$, so that the general solution becomes

$$u(x,t) = \sum_{n=1}^{\infty} A_n \sin(n\pi x/L)\sin(n\pi vt/L) \tag{1}$$

and thus

$$\dot{u}(x,0) = \frac{\pi v}{L}\sum_{n=1}^{\infty} nA_n \sin\left(\frac{n\pi x}{L}\right), \tag{2}$$

where from Figure 16.1 we have

$$\dot{u}(x,0) = \begin{cases} 2ax/L & 0 < x < L/2, \\ 2a - 2ax/L & L/2 < x < L. \end{cases}$$

Equation (2) is just the Fourier expansion of $\dot{u}(x,0)$ extended as an odd function, so that in analogy to (16.21), we now have

$$\frac{n\pi vA_n}{L} = \frac{2}{L}\int_0^L \dot{u}(x,0)\sin(n\pi x/L)\,dx.$$

On substituting for $\dot{u}(x,0)$ and integrating by parts, one obtains

$$A_n = \begin{cases} \dfrac{8aL}{n^3\pi^3 v}(-1)^{(n-1)/2} & n \text{ odd} \\ 0 & n \text{ even} \end{cases}$$

so that, finally,

$$u(x,t) = \frac{8aL}{\pi^3 v}\sum_{n\text{ odd}} \frac{(-1)^{(n-1)/2}}{n^3}\sin\left(\frac{n\pi x}{L}\right)\sin\left(\frac{n\pi vt}{L}\right).$$

16.2.2 The wave equation in three spatial dimensions

The method of Section 16.2.1 is easily extended to more than one spatial dimension. To illustrate this, we shall consider the wave equation in three dimensions with $0 < x,\ y,\ z < L$, that is, for waves confined to a cubic box. We then find the form of the waves in this box by solving the equation

$$\frac{\partial^2 u}{\partial x^2} + \frac{\partial^2 u}{\partial y^2} + \frac{\partial^2 u}{\partial z^2} = \frac{1}{v^2}\frac{\partial^2 u}{\partial t^2}, \tag{16.22}$$

assuming that u vanishes at the walls of the box, that is,

$$\left.\begin{aligned} u(0,\ y,\ z,\ t) = u(L,\ y,\ z,\ t) = 0 \\ u(x,\ 0,\ z,\ t) = u(x,\ L,\ z,\ t) = 0 \\ u(x,\ y,\ 0,\ t) = u(x,\ y,\ L,\ t) = 0 \end{aligned}\right\}. \tag{16.23}$$

In this case, substituting (16.8) into (16.22) and dividing by $a = XYZT$, gives

$$\frac{1}{X}\frac{\partial^2 X}{\partial x^2} + \frac{1}{Y}\frac{\partial^2 Y}{\partial y^2} + \frac{1}{Z}\frac{\partial^2 Z}{\partial z^2} = \frac{1}{v^2 T}\frac{\partial^2 T}{\partial t^2}.$$

As before, since the left-hand side is independent of t and the right-hand side depends solely on t, both sides must equal a constant, which we will denote by $-k^2$. We thus obtain

$$T'' + \omega^2 T = 0, \quad \omega = kv \tag{16.24a}$$

and

$$\frac{X''}{X} + \frac{Y''}{Y} = -k^2 - \frac{Z''}{Z},$$

where we have transferred the term in Z to the right-hand side. Since the left-hand side of this equation is independent of Z and the right-hand side depends solely on Z, both sides again must be a constant, which we write as $-k^2 + k_z^2$. We then obtain

$$Z'' + k_z^2 Z = 0 \tag{16.24b}$$

and

$$\frac{X''}{X} + \frac{Y''}{Y} = -k^2 + k_z^2.$$

Taking Y''/Y to the right-hand side and repeating the argument then gives

$$Y'' + k_y^2 Y = 0, \tag{16.24c}$$

and

$$X'' + k_x^2 X = 0, \tag{16.24d}$$

where

$$k^2 = k_x^2 + k_y^2 + k_z^2. \tag{16.25}$$

We have now converted the wave equation (16.22) into four ODEs (16.24a)–(16.24d). Furthermore, (16.24b)–(16.24d) have the same form as (16.13a) with the same boundary conditions, and (16.24a) is identical to (16.13a). Solving in the same way, one finds that solutions satisfying the boundary conditions (16.23) only exist for

$$k_x = n_x\pi/L, \quad k_y = n_y\pi/L, \quad k_z = n_z\pi/L, \tag{16.26}$$

in analogy to (16.14), where $n_x,\ n_y,\ n_z$ are positive integers; and the form of the solution is

$$u_{n_x n_y n_z} = \sin(n_x \pi x/L)\sin(n_y \pi y/L)\sin(n_z \pi z/L) \times [A\cos(\omega t) + B\sin(\omega t)], \tag{16.27}$$

in analogy to (16.17), where

$$\omega = kv = \left(n_x^2 + n_y^2 + n_z^2\right)^{1/2} \pi v/L \tag{16.28}$$

and the arbitrary constants A and B can also depend on $n_x,\ n_y,\ n_z$.

The solutions (16.27) are the normal modes for waves confined to a cubic box and the general solution is a linear combination

$$u = \sum_{n_x, n_y, n_z} u_{n_x n_y n_z}. \tag{16.29}$$

These results apply to any type of wave confined to a cubic box, for example, sound waves or electromagnetic waves, and are easily generalised to any rectangular box with sides $L_x,\ L_y,\ L_z$.

Example 16.3

For a cube of sides L, with $V = L^3$, on average, how many normal modes (16.27) characterised by the positive integers $n_x,\ n_y,\ n_z$ are such that $(k_x^2 + k_y^2 + k_z^2)^{1/2}$ lies in the range k to $k + \mathrm{d}k$?

Solution

This can be calculated by representing the solutions as dots in 'k-space', with axes $k_x,\ k_y,\ k_z$, as illustrated in Figure 16.2 for the simpler case of two dimensions. The dots correspond to the allowed values (16.26) and are separated by π/L, so that the average density of normal modes in k-space is V/π^3; and the volume of k-space from k to $k + \mathrm{d}k$ for small $\mathrm{d}k$ is

$$\tfrac{1}{8} 4\pi k^2 \mathrm{d}k,$$

where the factor 1/8 occurs because $k_x,\ k_y,\ k_z > 0$. Hence the average number of normal modes between k and $k + \mathrm{d}k$ is

$$n(k)\mathrm{d}k \equiv \frac{V}{\pi^3}\frac{1}{8}4\pi k^2 \mathrm{d}k = \frac{Vk^2 \mathrm{d}k}{2\pi^2}. \tag{16.30}$$

This formula, which also holds for rectangular boxes of unequal sides, is called the 'density of states' formula and occurs in many contexts in statistical physics.

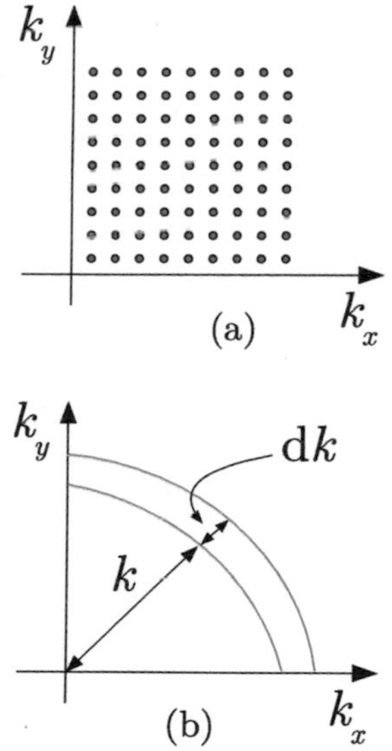

Figure 16.2 Diagrams used in the derivation of the density-of-states formula (16.30), where only the (k_x, k_y) plane is shown. The diagrams are not to the same scale, and in practical application the spacing between the points in the left-hand diagram is normally extremely small compared to typical k-values.

16.2.3 The diffusion equation in one spatial dimension

We next consider the diffusion equation (16.4), which is again second order in spatial derivatives but, in contrast to the wave equation, only first order in the time derivative. To focus on this difference, we shall just consider it in one spatial dimension. The generalisation to three spatial dimensions is very similar to that discussed in detail for the wave equation and is left as an exercise for the reader.

In one spatial dimension, the diffusion equation becomes

$$\frac{\partial^2 u}{\partial x^2} = \frac{1}{\kappa}\frac{\partial u}{\partial t}, \tag{16.31}$$

and, as an example, we will find solutions subject to the boundary conditions

$$u(0,t) = u(L,t) = 0 \text{ for } t \geq 0 \quad \text{and} \quad u(x,0) = f(x) \text{ for } 0 \leq x \leq L, \tag{16.32}$$

where L is a constant and $f(x)$ is a given function that must vanish at the end points 0 and L for the boundary conditions to be consistent. We start by again assuming a solution of the separable form (16.10), and after substituting into (16.31) and dividing by u we have

$$\frac{1}{X}\frac{\mathrm{d}^2 X}{\mathrm{d}x^2} = \frac{1}{\kappa T}\frac{\mathrm{d}T}{\mathrm{d}t}.$$

Each side now has to be a constant, which we will denote by α. For $\alpha \geq 0$, it may easily be shown that the only solution consistent with the first boundary condition is $u(x,t) = 0$. However, for $\alpha \equiv -\lambda^2 < 0$, the solutions for $X(x)$ and $T(t)$ are

$$X(x) = A\cos\lambda x + B\sin\lambda x, \quad T(t) = Ce^{-\lambda^2\kappa t},$$

where A, B and C are constants. The solution is therefore

$$u(x,t) = [A\cos\lambda x + B\sin\lambda x]\,e^{-\lambda^2\kappa t}, \tag{16.33}$$

where the constant C has been absorbed into A and B.

Using the first boundary condition with $x = 0$ gives

$$0 = Ae^{-\lambda^2\kappa t}$$

for all t, and hence $A = 0$. Similarly putting $x = L$ gives

$$0 = (B\sin\lambda L)e^{-\lambda^2\kappa t}$$

and hence only gives non-trivial solutions provided

$$\sin\lambda L = 0, \text{ i.e. } \lambda = n\pi/L,\ n = 1,\ 2, \ldots.$$

Thus the general solution is the superposition

$$u(x,t) = \sum_{n=1}^{\infty} B_n \exp\left(-n^2\pi^2\kappa t\big/L^2\right) \sin\left(n\pi x/L\right). \qquad (16.34)$$

As in the corresponding result (16.19) for the wave equation, at any fixed time t the solution takes the form of a Fourier sine series, but now each term is associated with an exponential rather than oscillating time dependence.

Finally, imposing the boundary condition at $t = 0$ gives

$$f(x) = \sum_{n=1}^{\infty} B_n \sin\left(\frac{n\pi x}{L}\right).$$

Multiplying by $\sin(m\pi x/L)$ and integrating from 0 to L using the orthonormality relation (16.20) then gives

$$B_m = \frac{2}{L}\int_0^L f(x)\sin\left(m\pi x/L\right)\,\mathrm{d}x, \quad m = 1,\ 2, \ldots \qquad (16.35)$$

which can be evaluated for any initial distribution $u(x,0) = f(x)$ to give the coefficients in (16.34).

Example 16.4

A thin metal bar has its ends at $x = 0$ and $x = L$. At $x = 0$ its temperature u is kept at a constant value u_0 at all times t, and at $x = L$ it is insulated so that $\partial u/\partial t$ is always zero. Initially, the temperature distribution in the bar is given by

$$u(x,0) = u_0 + \sin\left(\frac{3\pi x}{2L}\right), \quad 0 < x < L.$$

Assuming the heat flow is determined by the one-dimensional diffusion equation, find the general expression for $u(x,t)$ valid for $0 \le x \le L$ and all t.

Solution

Define $T \equiv u - u_0$, so that the boundary conditions become

(a) $T = 0$ at $x = 0$, (b) $\partial T/\partial t = 0$ at $x = L$, and
(c) $T = \sin\left(3\pi x/2L\right)$ at $t = 0$.

The general solution to the diffusion equation in one dimension was found in (16.33) and is

$$T(x,t) = [A\cos(\lambda x) + B\sin(\lambda x)]e^{-\lambda^2\kappa t},$$

where A, B and λ are constants. Using (a) gives

$$0 = Ae^{-\lambda^2 \kappa t} \text{ which implies } A = 0.$$

Then using (b),

$$\left.\frac{\partial T}{\partial t}\right|_{x=L} = \lambda B \cos(\lambda L) e^{-\lambda^2 \kappa t},$$

which is only possible if $\lambda = (n+1)\pi/(2L)$. So the solution is

$$T(x,t) = \sum_{n=0}^{\infty} B_n \sin\left[\frac{(2n+1)\pi x}{2L}\right] e^{-\lambda^2 \kappa t}.$$

The coefficients B_n are found by using (c). Thus

$$T(x,0) = \sum_{n=0}^{\infty} B_n \sin\left[\frac{(2n+1)\pi x}{2L}\right] = \sin\left(\frac{3\pi x}{2L}\right),$$

from which we deduce that $B_1 = 1$ and $B_{n\neq 1} = 0$. So, finally,

$$u = T + u_0 = u_0 + \sin\left(\frac{3\pi x}{2L}\right) e^{-\lambda^2 \kappa t}.$$

16.3 Separation of variables: polar co-ordinates

In this section, we shall explore separation of variables using polar co-ordinates, rather than Cartesian co-ordinates. This is particularly useful for systems with circular, cylindrical or spherical symmetry. Throughout the discussion, which will again proceed via examples, we shall have in mind that the dependent variable is a physical quantity, and therefore will emphasise solutions that are finite and single-valued.

16.3.1 Plane-polar co-ordinates

Plane-polar co-ordinates were introduced in Section 2.2.1, where from Figure 2.4 we see that the co-ordinates (r,θ) and $(r,\theta+2\pi)$ represent the same point in the plane. Hence if $u(r,\theta)$ represents a measurable quantity, it must not only be finite and continuous, but must also satisfy the boundary condition

$$u(r,\theta) = u(r,\theta+2\pi), \tag{16.36a}$$

which for separable solutions

$$u(r,\theta) = R(r)\Theta(\theta) \tag{16.37}$$

implies

$$\Theta(\theta) = \Theta(\theta+2\pi). \tag{16.36b}$$

To illustrate the consequences of this, we will consider solutions of the Laplace and wave equations in two spatial dimensions.

In plane–polar co-ordinates, Laplace's equation is given by [cf. (7.29)]

$$\frac{\partial^2 u}{\partial r^2} + \frac{1}{r}\frac{\partial u}{\partial r} + \frac{1}{r^2}\frac{\partial^2 u}{\partial \theta^2} = 0, \tag{16.38a}$$

or equivalently,

$$\frac{1}{r}\frac{\partial}{\partial r}\left(r\frac{\partial u}{\partial r}\right) + \frac{1}{r^2}\frac{\partial^2 u}{\partial \theta^2} = 0. \tag{16.38b}$$

Substituting (16.37) into (16.38b) and multiplying by r^2 gives

$$\frac{r}{R}\frac{\mathrm{d}}{\mathrm{d}r}\left(r\frac{\mathrm{d}R}{\mathrm{d}r}\right) = -\frac{1}{\Theta}\frac{\mathrm{d}^2\Theta}{\mathrm{d}\theta^2}, \tag{16.39}$$

where the left-hand side is independent of θ and the right-hand side is independent of r, so that both must be constant. Denoting the constant by m^2, we then have

$$\Theta'' + m^2\Theta = 0 \tag{16.40a}$$

and

$$r\frac{\mathrm{d}}{\mathrm{d}r}\left(r\frac{\mathrm{d}R}{\mathrm{d}r}\right) - m^2 R = 0. \tag{16.40b}$$

In solving these equations, it will be convenient to treat the cases $m = 0$ and $m \neq 0$ separately.

(i) $m = 0$ In this case (16.40a) has the trivial solution $\Theta = A + B\theta$, where A, B are arbitrary constants. Similarly, (16.40b) reduces to

$$\frac{\mathrm{d}}{\mathrm{d}r}\left(r\frac{\mathrm{d}R}{\mathrm{d}r}\right) = 0,$$

with the general solution

$$R = C\ln r + D.$$

Hence the general solution is

$$u = (A + B\theta)(C\ln r + D),$$

which reduces to

$$u_0 = C\ln r + D$$

on imposing the boundary condition (16.36a) and absorbing the constant A into the arbitrary constants C and D.

(ii) $m \neq 0$ In this case (16.40a) has the general solution

$$\Theta = A_m \cos m\theta + B_m \sin m\theta,$$

while the corresponding general solution of (16.40b) is

$$R = C_m r^m + D_m r^{-m},$$

as is easily verified by substitution. In addition, we note that m must be an integer if the boundary condition (16.36b) is to be satisfied, so that for $m \neq 0$, the separable solutions of Laplace's equation are:

$$u_m = (C_m r^m + D_m r^{-m})(A_m \cos m\theta + B_m \sin m\theta), \ m = 1, 2, \ldots \tag{16.41b}$$

where A_m, B_m, C_m, D_m are arbitrary constants. Finally, the general solution is given by

$$u(r, \theta) = \sum_{m=0}^{\infty} u_m(r, \theta), \tag{16.42}$$

where the various constants must be determined from boundary conditions.

The angular dependence of the separable solutions (16.41) is characteristic of the separable solutions of many other equations with circular symmetry, including the wave equation (16.3); the diffusion equation (16.4a); and the Schrödinger equation (16.7), provided the potential $V(\mathbf{r}) = V(r)$ is independent of angle. To illustrate this, we consider the wave equation in two spatial dimensions, which is given by

$$\frac{1}{r}\frac{\partial}{\partial r}\left(r\frac{\partial u}{\partial r}\right) + \frac{1}{r^2}\frac{\partial^2 u}{\partial \Theta^2} = \frac{1}{v^2}\frac{\partial^2 u}{\partial t^2} \tag{16.43}$$

in planar co-ordinates. Assuming a separable solution

$$u(r, \theta, t) = R(r)\Theta(\theta)T(t) \tag{16.44}$$

gives

$$\frac{1}{Rr}\frac{\mathrm{d}}{\mathrm{d}r}\left(r\frac{\mathrm{d}R}{\mathrm{d}r}\right) + \frac{1}{r^2\theta}\frac{\mathrm{d}^2\Theta}{\mathrm{d}\theta^2} = \frac{1}{v^2 T}\frac{\mathrm{d}^2 T}{\mathrm{d}t^2}$$

after dividing by u, and hence

$$T'' + k^2 v^2 T = 0 \tag{16.45a}$$

and

$$\frac{1}{Rr}\frac{\mathrm{d}}{\mathrm{d}r}\left(r\frac{\mathrm{d}R}{\mathrm{d}r}\right) + \frac{1}{r^2\theta}\frac{\mathrm{d}^2\Theta}{\mathrm{d}\theta^2} = -k^2 \tag{16.46}$$

on separating off the right-hand side and denoting the separation constant by $-k^2$. After multiplying by r^2, (16.46) can also be separated in a similar way to Laplace's equation to give

$$\Theta'' + m^2\Theta = 0 \tag{16.45b}$$

and

$$r\frac{\mathrm{d}}{\mathrm{d}r}\left(r\frac{\mathrm{d}R}{\mathrm{d}r}\right) + (k^2 r^2 - m^2)R = 0. \tag{16.45c}$$

Here, equation (16.45b) is identical to (16.40a) and has the same solution, while on expanding the differential in (16.45c) and substituting $z = kr$, we obtain

$$z^2\frac{\mathrm{d}^2R}{\mathrm{d}z^2} + z\frac{\mathrm{d}R}{\mathrm{d}z} + (z^2 - m^2)R = 0,$$

which is Bessel's equation (15.53a) and has the general solution

$$R(r) = C_m J_m(kr) + D_m N_m(kr), \qquad (16.47a)$$

where C_m and D_m are arbitrary constants and J_m and N_m are the Bessel functions discussed in Section 15.4.1. In particular, $N_m(kr)$ is singular at $r = 0$, so if we require u to be finite at $r = 0$, (16.47a) reduces to

$$R(r) = C_m J_m(kr). \qquad (16.47b)$$

Finally, the general solution of (16.45) is

$$T(t) = E\cos\omega t + F\sin\omega t, \qquad (16.48)$$

where the angular frequency $\omega = kv$. Hence the separable solutions that are finite at $r = 0$ are

$$u_0 = J_0(kr)[E_0\cos\omega t + F_0\sin\omega t], \quad m = 0, \qquad (16.49a)$$

and

$$u_m = J_m(kr)[E_m\cos\omega t + F_m\sin\omega t][A_m\cos m\theta + B_m\sin m\theta], \quad m \neq 0, \qquad (16.49b)$$

where $m = 1,\ 2, \ldots$ must be an integer if the boundary condition (16.36b) is to be satisfied, and where C_m in (16.47b) has been absorbed into the other constants. In particular, the angular dependence of the separable solutions is the same as for Laplace's equation and is characteristic of the separable solutions of many equations with circular symmetry.

Example 16.5

A circular drum has radius a and is fixed at the circumference so that the transverse displacement $u(r = a,\ \theta, t) = 0$. For $r < a$, the displacement obeys the wave equation (16.43), where the velocity v of the waves is determined by the tension and mass per unit area of the skin. Derive expressions for the frequencies of the four normal modes with the lowest frequencies.

Solution

The displacement u must be single-valued and be finite at $r = 0$. Hence the separable solutions are given by (16.49), and each corresponds to a normal mode since u oscillates with a fixed angular frequency $\omega = kv$ independent of r and θ. The possible values of k and hence ω, are determined by the boundary condition $u = 0$ at $r = a$, which requires $J_m(ka) = 0$. From Table 15.1 we see that for any m the three lowest occur at

$$\alpha \equiv ka = 2.4048,\ 3.8317,\ 5.1356,\ 5.5201,$$

corresponding to $m = 0,\ 1,\ 2,\ 0$, respectively. Hence the four lowest values for the frequency $\nu = \omega/2\pi = kv/2\pi$ are

$$v = \alpha v/2\pi a,$$

where α has the values given above.

16.3.2 Spherical polar co-ordinates

The above discussion of plane polar co-ordinates is easily extended to the spherical polar co-ordinates (r, θ, ϕ) discussed in Section 11.3.1. However in this case, (r, θ, ϕ) and $(r, \theta, \phi + 2\pi)$ correspond to the same point in space, so that if u is a physical quantity, we must impose the boundary condition

$$u(r, \theta, \phi + 2\pi) = u(r, \theta, \phi) \tag{16.50}$$

in addition to requiring that u is finite and continuous.

Again, we shall illustrate this by considering Laplace's equation (16.2). We start by using the result given in Table 12.1 for $\boldsymbol{\nabla}^2$ in spherical polar co-ordinates to express (16.2) as

$$\boldsymbol{\nabla}^2 u(r, \theta, \phi) = \frac{1}{r^2}\frac{\partial}{\partial r}\left(r^2\frac{\partial u}{\partial r}\right) + \frac{1}{r^2 \sin\theta}\frac{\partial}{\partial \theta}\left(\sin\theta\frac{\partial u}{\partial \theta}\right) + \frac{1}{r^2\sin^2\theta}\frac{\partial^2 u}{\partial \phi^2} = 0, \tag{16.51}$$

and then substitute the decomposition

$$u(r,\ \theta,\ \phi) = R(r)\Theta(\theta)\Phi(\phi) \tag{16.52}$$

into (16.51). After dividing through by u and multiplying by r^2, we obtain the equation

$$\frac{\sin^2\theta}{R}\frac{\mathrm{d}}{\mathrm{d}r}\left(r^2\frac{\mathrm{d}R}{\mathrm{d}r}\right) + \frac{1}{\Theta}\sin\theta\frac{\mathrm{d}}{\mathrm{d}\theta}\left(\sin\theta\frac{\mathrm{d}\Theta}{\mathrm{d}\theta}\right) + \frac{1}{\Phi}\left(\frac{\mathrm{d}^2\Phi}{\mathrm{d}\phi^2}\right) = 0. \tag{16.53}$$

The first two terms are functions of both r and θ, whereas the third term is a function of ϕ alone. Since the three variables are independent, this means that the third term must be a constant, and so we can set

$$\frac{\mathrm{d}^2\Phi}{\mathrm{d}\phi^2} = -m^2\Phi. \tag{16.54}$$

The rest of the equation can then be manipulated into the form

$$\frac{1}{R}\frac{\mathrm{d}}{\mathrm{d}r}\left(r^2\frac{\mathrm{d}R}{\mathrm{d}r}\right) = \frac{m^2}{\sin^2\theta} - \frac{1}{\Theta}\frac{1}{\sin\theta}\frac{\mathrm{d}}{\mathrm{d}\theta}\left(\sin\theta\frac{\mathrm{d}\Theta}{\mathrm{d}\theta}\right).$$

Again we use the fact that each side is a function of a single independent variable and hence must be a constant, which we will denote by $l(l+1)$ for later convenience. We thus have the two equations

$$\frac{\mathrm{d}}{\mathrm{d}r}\left(r^2\frac{\mathrm{d}R}{dr}\right) = l(l+1)R, \tag{16.55}$$

and

$$\frac{\mathrm{d}}{\mathrm{d}\theta}\left(\sin\theta\frac{\mathrm{d}\Theta}{\mathrm{d}\theta}\right) + \left(l(l+1)\sin\theta - \frac{m^2}{\sin\theta}\right)\Theta = 0. \tag{16.56}$$

The radial equation (16.55) may be written as

$$r^2\frac{\mathrm{d}^2R}{\mathrm{d}r^2} + 2r\frac{\mathrm{d}R}{dr} - l(l+1)R = 0,$$

and has the general solution

$$R_l(r) = A_l r^l + B_l\big/r^{l+1}, \tag{16.57}$$

where A_l and B_l are arbitrary constants, and the possible values of l are restricted by consideration of the angular dependence, as we shall soon see.

We start with the ϕ-dependence. The only solutions of (16.54) compatible with the boundary conditions (16.50) are

$$\Phi = \begin{cases} \bar{A}_0 & m = 0 \\ \bar{A}_m\cos m\phi + \bar{B}_m\sin m\phi & m = 1,\ 2,\ldots \end{cases} \tag{16.58}$$

Alternatively, and equivalently, we can choose a set of solutions

$$\Phi = A_m e^{imx},\ m = 0, \pm 1, \ldots \tag{16.59}$$

where the coefficients A_m are arbitrary constants.

We next turn to the θ dependence. On setting $\mu = \cos\theta$ and expanding the first term in (16.56), it becomes

$$\frac{\mathrm{d}}{\mathrm{d}\mu}\left[(1-\mu^2)\frac{\mathrm{d}\Theta}{\mathrm{d}\theta}\right] + \left[l(l+1) - \frac{m^2}{(1-\mu^2)}\right]\Theta = 0. \tag{16.60}$$

This is the *associated Legendre equation*, which has solutions[2] that are finite for $\mu = \cos\theta = -1$ (i.e. on the negative z-axis) only if $l = 0, 1, 2, \ldots$ and the m values are restricted to $-l \le m \le l$. They are called *associated Legendre polynomials* and denoted $P_l^m(\cos\theta)$, so that we write

$$\Theta = C_{lm} P_l^m(\cos\theta) \tag{16.61}$$

and from (16.57), (16.59) and (16.61), we find that the finite separable solutions which satisfy the boundary conditions (16.50) are

$$u_{lm} = \left[A_{lm} r^l + B_{lm} r^{-(l+1)}\right] P_l^m(\cos\theta) e^{im\phi}, \tag{16.62}$$

where $l = 0, 1, 2, \ldots$, $-l \le m \le l$ and A_{lm}, B_{lm} are arbitrary constants. The explicit forms of the first few polynomials are given in Example 15.10. For $m = 0$, when (16.60) reduces to Legendre's equation (15.23), they reduce to the Legendre polynomials $P_l(\cos\theta)$ of order l discussed in detail in Section 15.3.1. Hence for $m = 0$ the solutions (16.62) reduce to

$$u_{l0} = \left[A_{l0} r^l + B_{l0} r^{-(l+1)}\right] P_l(\cos\theta), \tag{16.63}$$

which are independent of the azimuthal angle ϕ and unchanged by rotations about the z-axis. For $l = m = 0$, since $P_0(\cos\theta) = 1$, (16.63) reduces to

$$u_{00} = \left[A_{00}^l + B_{00} r^{-1}\right], \tag{16.64}$$

which is the most general spherically symmetric solution to Laplace's equation. The most general solution, without symmetry constraints, is obtained by taking linear combinations of the separable solutions (16.63) and is

$$u(r,\ \theta, \phi) = \sum_{l=0}^{\infty} \sum_{m=-l}^{l} \left[A_{lm} r^l + B_{lm} r^{-(l+1)}\right] P_l^m(\cos\theta) e^{im\phi}. \tag{16.65}$$

Finally, we note that the angular dependence of the separable solutions (16.62) is not specific to Laplace's equation, but is shared by the separable solutions of other important equations that have spherical symmetry. For example, the separable solutions of the wave equation and the diffusion equation also take the form

$$R(r) T(t) P_l^m(\cos\theta) e^{im\phi} \tag{16.66}$$

for appropriate choices of $R(r)$ and $T(t)$, as the reader may verify.

[2] The associated Legendre equation is solved in the starred Sections 15.3.3. Here the results are stated without proof for the benefit of readers who have not studied that section.

In practice, it is common to rewrite equations like (16.62) and (16.65) in terms of the so-called *spherical harmonics* defined by

$$Y_l^m(\theta,\phi) \equiv \sqrt{\frac{(2l+1)(l-m)!}{4\pi(l+m)!}}\,P_l^m(\cos\theta)e^{im\phi}, \qquad (16.67a)$$

where the constant is chosen so that the normalisation condition

$$\int_0^{2\pi} d\phi \int_{-1}^{1} Y_l^m(\theta,\phi)\left[Y_{l'}^{m'}(\theta,\phi)\right]^* d(\cos\theta) = \delta_{ll'}\delta_{mm'} \qquad (16.67b)$$

is satisfied. With this convention, the first few spherical harmonics are given by

$$Y_0^0 = \sqrt{\tfrac{1}{4\pi}}$$
$$Y_1^0 = \sqrt{\tfrac{3}{4\pi}}\cos\theta \qquad Y_1^{\pm1} = \mp\sqrt{\tfrac{3}{8\pi}}\sin\theta e^{\pm i\phi}$$
$$Y_2^0 = \sqrt{\tfrac{5}{16\pi}}(3\cos^2\theta - 1) \quad Y_2^{\pm1} = \mp\sqrt{\tfrac{15}{8\pi}}\sin\theta\cos\theta e^{\pm i\phi}$$
$$Y_2^{\pm2} = \sqrt{\tfrac{15}{32\pi}}\sin^2\theta e^{\pm 2i\phi}.$$

Example 16.6

Show that the only solution of Laplace's equation that is finite at all points in space, single-valued and satisfies $u \to 0$ as $r \to \infty$ is $u = 0$. Hence show that for a given $\rho(\mathbf{r})$, the solution of Poisson's equation $\boldsymbol{\nabla}^2 u = 4\pi\rho(\mathbf{r})$ satisfying the same condition $u \to 0$ as $r \to \infty$ is unique.

Solution

If $u(r,\theta,\phi)$ is finite at $r = 0$, the coefficients B_l must vanish in the general solution (16.65) of Laplace's equation; and if $u \to 0$ as $r \to \infty$, the coefficients A_l also vanish. Thus the general solution reduces to the trivial solution $u = 0$, as required.

Suppose we have two distinct solutions u_1 and u_2 of Poisson's equation, both satisfying the boundary conditions. Then from Poisson's equation we have

$$\boldsymbol{\nabla}^2(u_1 - u_2) = 0.$$

But we have just shown that the only solution to this equation is $u_1 - u_2 = 0$, so $u_1 = u_2$, contradicting the assumption $u_1 \neq u_2$. Hence there is only one unique solution.

Example 16.7

A sphere of radius r_0 centred on the origin has a surface temperature

$$u(r_0, \theta, \phi) = u_0(\cos\theta - \cos^3\theta)$$

where u_0 is a constant and (θ, ϕ) are polar angles. Find the temperature at a point within the sphere, assuming it is at thermal equilibrium.

Solution

At thermal equilibrium, the temperature is independent of time, so that the heat conduction equation (16.4) reduces to Laplace's equation; and since we are dealing with a sphere, we use spherical polar co-ordinates. Laplace's equation is then given by (16.51) and the general single-valued solution is given by (16.65), that is,

$$u(r,\ \theta, \phi) = \sum_{l=0}^{\infty} \sum_{m=-l}^{l} \left[A_{lm} r^l + B_{lm} r^{-(l+1)}\right] P_l^m(\cos\theta) e^{im\phi}.$$

The requirement that the temperature at $r = 0$ is finite implies $B_{lm} = 0$, and the requirement that the temperature at r_0 is independent of ϕ implies that only terms with $m = 0$ contribute, so the solution reduces to

$$u(r, \theta, \phi) = \sum_{l=0}^{\infty} c_l r^l P_l(\cos\theta),$$

where the constant coefficients $c_l \equiv A_{l0}$. To find these we use the boundary condition

$$T(r_0, \theta, \phi) = \sum_{l=0}^{\infty} c_l r_0^l P_l(\cos\theta) = T_0(\cos\theta - \cos^3\theta).$$

But,

$$P_1(\cos\theta) = \cos\theta, \quad P_3(\cos\theta) = \tfrac{1}{2}(5\cos^3\theta - 3\cos\theta),$$

so that

$$\cos\theta - \cos^3\theta = \tfrac{2}{5}P_1(\cos\theta) - \tfrac{2}{5}P_3(\cos\theta),$$

and equating coefficients of $P_l(\cos\theta)$ gives

$$c_1 = \frac{2u_0}{5r_0}; \quad c_3 = -\frac{2u_0}{5r_0^3}; \quad c_l = 0,\ l \neq 1,\ 3.$$

Finally, substituting into the formula for the general solution gives

$$u(r, \theta, \phi) = \frac{2}{5}u_0 \left[\left(\frac{r}{r_0}\right) P_1(\cos\theta) - \left(\frac{r}{r_0}\right)^3 P_3(\cos\theta)\right].$$

16.3.3 Cylindrical polar co-ordinates

We conclude the discussion of polar co-ordinates by considering cylindrical polar co-ordinates defined in Section 11.3.1. In this case, the co-ordinates (ρ, ϕ, z) and $(\rho, \phi + 2\pi, z)$ represent the same point, so the boundary condition corresponding to (16.50) is

$$u(\rho, \phi, z) = u(\rho, \phi + 2\pi, z). \tag{16.69}$$

We shall again take as our example the Laplace equation, which in cylindrical polar co-ordinates takes the form

$$\frac{1}{\rho}\frac{\partial}{\partial r}\left[\rho\left(\frac{\partial u}{\partial \rho}\right)\right] + \frac{1}{\rho^2}\frac{\partial^2 u}{\partial \phi^2} + \frac{\partial^2 u}{\partial z^2} = 0, \tag{16.70}$$

where we have used the expression for $\boldsymbol{\nabla}^2$ given in Table 12.1. Assuming a separable solution

$$u(\rho, \phi, z) = \mathrm{P}(\rho)\Phi(\phi)Z(z) \tag{16.71}$$

then gives

$$\frac{1}{\mathrm{P}\rho}\frac{\mathrm{d}}{\mathrm{d}\rho}\left(\rho\frac{\mathrm{dP}}{\mathrm{d}\rho}\right) + \frac{1}{\rho^2\Phi}\frac{\mathrm{d}^2\Phi}{\mathrm{d}\phi^2} + \frac{1}{Z}\frac{\mathrm{d}^2 Z}{\mathrm{d}z^2} = 0 \tag{16.72}$$

on substituting into (16.70) and dividing by u. In this equation, the first two terms are independent of z, while the third term depends only on z, and so must be a constant. Hence, separating the third term, we obtain

$$\frac{\mathrm{d}^2 Z}{\mathrm{d}z^2} = k^2 Z \tag{16.73a}$$

and

$$\frac{1}{\mathrm{P}\rho}\frac{\mathrm{d}}{\mathrm{d}\rho}\left(\rho\frac{\mathrm{dP}}{\mathrm{d}\rho}\right) + \frac{1}{\rho^2\Phi}\frac{\mathrm{d}^2\Phi}{\mathrm{d}\phi^2} + k^2 = 0,$$

where we have taken the separation constant to be k^2. The second of these equations is not separable as it stands, but multiplying through by ρ^2 gives

$$\frac{\rho}{R}\frac{\mathrm{d}}{\mathrm{d}\rho}\left(\rho\frac{\mathrm{d}R}{\mathrm{d}\rho}\right) + k^2\rho^2 + \frac{1}{\Theta}\frac{\mathrm{d}^2\Theta}{\mathrm{d}\theta^2} = 0,$$

which is separable. Separating the third term and denoting the separation constant by m^2 then gives

$$\frac{\mathrm{d}^2\Phi}{\mathrm{d}\phi^2} + m^2\Phi = 0 \tag{16.73b}$$

and

$$\rho^2\frac{\mathrm{d}^2\mathrm{P}}{\mathrm{d}\rho^2} + \rho\frac{\mathrm{dP}}{\mathrm{d}\rho} + (k^2\rho^2 - m^2)\mathrm{P} = 0, \tag{16.73c}$$

where we have expanded the ρ-derivative.

It remains to solve the three ODEs (16.73a)–(16.73c). For any k, the general solution of the first of these is

$$Z(z) = A_k e^{kz} + B_k e^{-kz} \tag{16.74}$$

where A_k and B_k are arbitrary constants. If we impose the boundary conditions (16.69), the general solutions of (16.73b) are

$$\Phi(\phi) = \begin{cases} C_0 & m = 0 \\ C_m \cos m\phi + D_m \sin m\phi & m \neq 0 \end{cases} \tag{16.75}$$

where $m \geq 0$ is an integer, and C_m, D_m are arbitrary constants. Finally, on setting $\eta = k\rho$ in (16.73c) we obtain

$$\eta^2 \frac{\mathrm{d}^2 \mathrm{P}}{\mathrm{d}\eta^2} + \eta \frac{\mathrm{d}\mathrm{P}}{\mathrm{d}\eta} + (\eta^2 - m^2)\mathrm{P} = 0.$$

This is Bessel's equation (15.53), with general solutions of the form (15.70), i.e.

$$\mathrm{P}(\rho) = E_m J_m(k\rho) + F_m N_m(k\rho), \tag{16.76}$$

where E_m and F_m are arbitrary constants and $J_m(k\rho)$ and $N_m(k\rho)$ are Bessel functions of the first and second kind respectively. These Bessel functions were discussed in Section 15.4.1, where we saw that $N_m(k\rho)$ was singular at $k\rho = 0$. If we require solutions that are finite at $\rho = 0$, we must therefore set $F_m = 0$ in (16.71). Hence those separable solutions (16.66) that are both finite and single-valued are

$$u(\rho, \phi, z) = J_0(k\rho)\left[A_k e^{kz} + B_k e^{-kz}\right], \quad m = 0 \tag{16.77a}$$

and

$$u(\rho, \phi, z) = J_m(\rho\rho)\left[A_k e^{kz} + B_k e^{-kz}\right]\left[C_m \cos m\phi + D_m \sin m\phi\right], \quad m = 1, 2, \ldots \tag{16.77b}$$

where we have absorbed C_0 and E_m into the other constants. Since Laplace's equation is homogeneous, more general solutions may then be formed by linear superposition of the solutions (16.77a) and (16.77b), where in applications the possible values of k and the various constants must be determined by boundary conditions, as we shall illustrate by an example.

Example 16.8

A solid cylinder of radius R and length L has its top and bottom faces maintained at zero temperature. If the curved surfaces have a temperature $u = u_0 z^2(L - z)$, where u_0 is a constant, find the equilibrium temperature at an arbitrary point within its volume. Note the integral (which may be done by parts)

$$\int_0^L z^2(L-z)\sin\left(\frac{n\pi z}{L}\right)\mathrm{d}z = -\frac{2L^4}{n^3\pi^3}[1+2(-1)^n].$$

Solution

We need to solve Laplace's equation, and because the temperature does not depend on θ, we need only the general solution (16.77a) for $m = 0$. Setting $k = in\pi/L$ so that the temperature at $z = 0$ and $z = L$ are equal, we may write this as

$$u(\rho, z) = \sum_{n=1}^{\infty} I_0(n\pi\rho/L)[C_n\cos(n\pi z/L) + D_n\sin(n\pi z/L),$$

where $I_0(x) \equiv J_0(ix)$ is called the *modified Bessel function* of zero order.[3] Imposing the boundary conditions at the top and bottom faces implies $C_n = 0$ for all n and so

$$u(\rho, z) = \sum_{n=1}^{\infty} D_n I_0(n\pi\rho/L)\sin(n\pi z/L). \qquad (1)$$

Next we impose the boundary condition on the curved surface, so that

$$u_0 z^2(L-z) = \sum_{n=1}^{\infty} D_n I_0(n\pi R/L)\sin(n\pi z/L).$$

To find the coefficients, we multiply both sides by $\sin(m\pi z/L)$ and integrate from 0 to L, using the orthogonality properties (16.20). This gives

$$D_n\frac{L}{2}I_0\left(\frac{n\pi R}{L}\right) = u_0\int_0^1 z^2(L-z)\sin\left(\frac{n\pi z}{L}\right)\mathrm{d}z,$$

and using the given integral, we obtain

$$D_n = -\frac{4u_0L^3}{n^3\pi^3}\frac{[1+2(-1)^n]}{I_0(n\pi R/L)}.$$

Then substituting into (1) gives the solution

$$u(\rho, z) = -\frac{4L^3u_0}{\pi^3}\sum_{n=1}^{\infty}\frac{[1+2(-1)^n]}{n^3}\frac{I_0(n\pi\rho/L)}{I_0(n\pi R/L)}\sin(n\pi z/L).$$

---

[3]For the properties of these functions, including a plot of I_0, see G.B. Arfken and H.J. Weber (2005) *Mathematical Methods for Physicists,* 6th edn., Academic Press, San Diego, California, Section 11.5.

*16.4 The wave equation: d'Alembert's solution

In the next three sections, we shall consider other methods of solution of PDEs, mainly applied to functions of two variables. We start with the wave equation (16.3b), which we will solve by introducing the new variables

$$\xi = x - vt \quad \text{and} \quad \eta = x + vt. \tag{16.78}$$

On changing variables using (7.24), we obtain

$$\frac{\partial^2 u}{\partial x^2} = \frac{\partial^2 u}{\partial \xi^2} + 2\frac{\partial^2 u}{\partial \xi \partial \eta} + \frac{\partial^2 u}{\partial \eta^2},$$

and

$$\frac{\partial^2 u}{\partial t^2} = v^2\left(\frac{\partial^2 u}{\partial \xi^2} - 2\frac{\partial^2 u}{\partial \xi \partial \eta} + \frac{\partial^2 u}{\partial \eta^2}\right),$$

so that the wave equation becomes

$$\frac{\partial^2 u}{\partial \xi \partial \eta} = 0, \tag{16.79}$$

with the general solution [cf. (16.2a) and (16.2b)]

$$u = f(\xi) + g(\eta) = f(x - vt) + g(x + vt), \tag{16.80}$$

where f and g are arbitrary differentiable functions.

Equation (16.80) is the general solution of the wave equation and each of the two terms has a simple interpretation. Let us suppose

$$u(x,t) = f(x - vt). \tag{16.81}$$

Then we have

$$u(x + vt, t) = u(x, 0)$$

for all t, so that the solution moves as illustrated in Figure 16.3 for a simple choice of $u(x,0)$. It represents a travelling wave moving in the positive x-direction with speed v. Thus, for example, a simple harmonic wave with wavelength λ, wave number $k = 2\pi/\lambda$ and angular frequency $\omega = kv$, can be written in the form

$$f = A\sin[kx - \omega t + \alpha] = A\sin[k(x - vt) + \alpha]$$

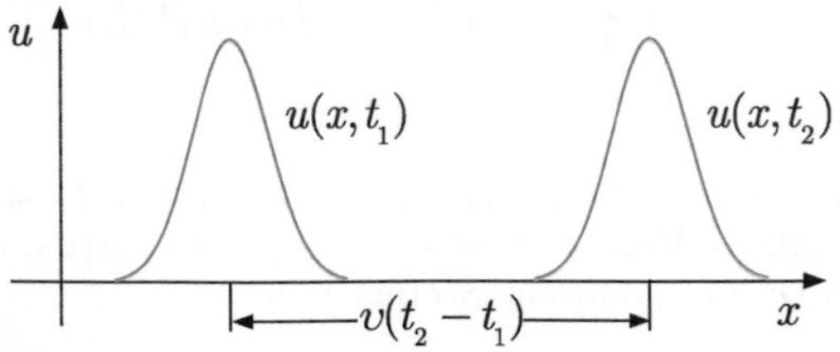

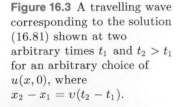

Figure 16.3 A travelling wave corresponding to the solution (16.81) shown at two arbitrary times t_1 and $t_2 > t_1$ for an arbitrary choice of $u(x,0)$, where $x_2 - x_1 = v(t_2 - t_1)$.

where A and α are arbitrary constants. Similarly, $g(x + vt)$ is a wave travelling in the minus x-direction and (16.80) shows that any other non-trivial solution of the wave equation may be written as a sum of a wave travelling to the right and one travelling to the left.

At this point, we digress briefly to indicate how this description can be extended to three dimensions. Denoting a point in space by its position vector $\mathbf{r}$, one easily shows that the functions

$$f(\hat{\mathbf{k}} \cdot \mathbf{r} - vt) \quad \text{and} \quad g(\hat{\mathbf{k}} \cdot \mathbf{r} + vt) \tag{16.82}$$

are solutions of the three-dimensional wave equation, where the unit vector $\hat{\mathbf{k}}$ indicates a chosen direction in space. Since the equation of a plane perpendicular to the unit vector $\hat{\mathbf{k}}$ is $\hat{\mathbf{k}} \cdot \mathbf{r} = c$, where c is a constant, then at any fixed time t, the functions f and g are constant over the whole plane. They are therefore *plane waves* travelling in the positive and negative $\hat{\mathbf{k}}$ directions, respectively. This may be more familiar if we note that a simple harmonic wave in the $\hat{\mathbf{k}}$ direction analogous to (16.80) is given by

$$A \sin(\mathbf{k} \cdot \mathbf{r} - \omega t + \alpha) = A \sin[k(\hat{\mathbf{k}} \cdot \mathbf{r} - vt) + \alpha],$$

where A and α are again constants and the wave vector $\mathbf{k} = k\,\hat{\mathbf{k}}$.

We now return to the one-dimensional case and consider its solution subject to the initial conditions

$$u(x, 0) = \alpha(x) \tag{16.83a}$$

and

$$\dot{u}(x, 0) \equiv \left.\frac{\partial}{\partial t} u(x, t)\right|_{t=0} = \beta(x), \tag{16.83b}$$

where $\alpha(x)$ and $\beta(x)$ are given functions. Substituting the general solution (16.80) into these boundary conditions then gives

$$\alpha(x) = f(x) + g(x) \tag{16.84a}$$

and

$$\beta(x) = -vf'(x) + vg'(x), \tag{16.84b}$$

where

$$f'(x) \equiv \left.\frac{\mathrm{d}f(q)}{\mathrm{d}q}\right|_{q=x}$$

and similarly for $g'(x)$. We now integrate (16.84b) to get

$$f(x) - g(x) + c = -\frac{1}{v} \int_a^x \beta(q) \mathrm{d}q,$$

where the integration constant c will depend on the arbitrarily chosen lower limit a of the integration. From this equation, together with (16.84a), we obtain

$$f(x) = \frac{1}{2}\alpha(x) - \frac{1}{2v}\int_a^x \beta(q)\mathrm{d}q - \frac{c}{2}$$

and

$$g(x) = \frac{1}{2}\alpha(x) + \frac{1}{2v}\int_a^x \beta(q)\mathrm{d}q + \frac{c}{2},$$

and hence

$$f(x-vt) = \frac{1}{2}\alpha(x-vt) - \frac{1}{2v}\int_a^{x-vt} \beta(q)\mathrm{d}q - \frac{c}{2}$$

and

$$g(x+vt) = \frac{1}{2}\alpha(x+vt) + \frac{1}{2v}\int_a^{x+vt} \beta(q)\mathrm{d}q + \frac{c}{2}.$$

Finally, adding we find

$$u(x,t) = \frac{1}{2}\alpha(x-vt) + \frac{1}{2}\alpha(x+vt) + \frac{1}{2v}\int_{x-vt}^{x+vt} \beta(q)\mathrm{d}q. \qquad (16.85)$$

This is *d'Alembert's solution* to the wave equation (16.3b) subject to the boundary conditions (16.83). It is unique and independent of the intermediate constants c and a introduced in its derivation. Furthermore, at any point $x = x_0$, $u(x_0, t)$ is dependent only on the initial values $u(x, 0)$ and $\dot{u}(x, 0)$ in the range $x_0 - vt < x < x_0 + vt$, and is independent of u and $\dot{u}$ outside this range. This embodies the idea of 'causality' for the wave equation, since it is just the range from within which a signal emitted at $t = 0$ and travelling with speed v can reach x_0 in a time less than or equal to t.

Example 16.9

In Section 16.2.1 we derived the separable solutions (16.17) for the wave equation. Write (16.17) for the case $B_n = 0$ as a special case of the general solution (16.80), identifying the forms $f(x - vt)$ and $g(x + vt)$.

Solution

For $B_n = 0$ we have

$$u_n(x) = A_n \sin(n\pi x/L)\cos(n\pi vt/L) \tag{1}$$

$$= \frac{A_n}{4i}\left[e^{in\pi x/L} - e^{-in\pi x/L}\right]\left[e^{in\pi vt/L} + e^{-in\pi vt/L}\right]$$

$$= f(x - vt) + g(x + vt),$$

where

$$f(x - vt) = \frac{A_n}{2}\sin[n\pi(x - vt)/L] \tag{2a}$$

and

$$g(x + vt) = \frac{A_n}{2}\sin[n\pi(x + vt)/L]\,. \tag{2b}$$

This illustrates the fact that a 'standing wave' like (1) in which the nodes in x remain stationary, can be written as the sum of two 'equal but opposite' waves (2a, 2b) travelling in the positive and negative x directions, respectively.

*16.5 Euler Equations

The method used in the previous section to obtain (16.80) as the general solution of the wave equation can be extended to solve any equation of the form

$$A\frac{\partial^2 u}{\partial x^2} + 2B\frac{\partial^2 u}{\partial x \partial y} + C\frac{\partial^2 u}{\partial y^2} = 0, \tag{16.86}$$

where A, B, C, are given constants and x, y are any variables, not necessarily Cartesian co-ordinates. Such equations are often called Euler's equations.[4] To solve them we introduce the new variables

$$\xi \equiv x + \lambda_1 y, \quad \eta \equiv x + \lambda_2 y, \tag{16.87}$$

where λ_1 and λ_2 are constants. We then try to find values of λ_1 and λ_2 such that (16.86) reduces to an equation of the form (16.79), with a general solution

$$u = f(\xi) + g(\eta) = f(x + \lambda_1 y) + g(x + \lambda_2 y) \tag{16.88}$$

analogous to (16.80) for the wave equation.

[4] They are, however, not the same as the Euler equations discussed in Section 14.3.

To see whether this is possible, we change variables using (7.24) to obtain

$$\frac{\partial u}{\partial x} = \frac{\partial u}{\partial \xi}\frac{\partial \xi}{\partial x} + \frac{\partial u}{\partial \eta}\frac{\partial \eta}{\partial x} = \frac{\partial u}{\partial \xi} + \frac{\partial u}{\partial \eta},$$

$$\frac{\partial u}{\partial y} = \frac{\partial u}{\partial \xi}\frac{\partial \xi}{\partial y} + \frac{\partial u}{\partial \eta}\frac{\partial \eta}{\partial y} = \lambda_1\frac{\partial u}{\partial \xi} + \lambda_2\frac{\partial u}{\partial \eta},$$

$$\frac{\partial^2 u}{\partial x^2} = \frac{\partial}{\partial x}\left(\frac{\partial u}{\partial x}\right) = \left(\frac{\partial}{\partial \xi} + \frac{\partial}{\partial \eta}\right)\left(\frac{\partial u}{\partial \xi} + \frac{\partial u}{\partial \eta}\right)$$
$$= \frac{\partial^2 u}{\partial \xi^2} + 2\frac{\partial^2 u}{\partial \xi \partial \eta} + \frac{\partial^2 u}{\partial \eta^2},$$

$$\frac{\partial^2 u}{\partial y^2} = \frac{\partial}{\partial y}\left(\frac{\partial u}{\partial y}\right) = \left(\lambda_1\frac{\partial}{\partial \xi} + \lambda_2\frac{\partial}{\partial \eta}\right)\left(\lambda_1\frac{\partial u}{\partial \xi} + \lambda_2\frac{\partial u}{\partial \eta}\right)$$
$$= \lambda_1^2\frac{\partial^2 u}{\partial \xi^2} + 2\lambda_1\lambda_2\frac{\partial^2 u}{\partial \xi \partial \eta} + \lambda_2^2\frac{\partial^2 u}{\partial \eta^2},$$

$$\frac{\partial^2 u}{\partial x \partial y} = \frac{\partial}{\partial x}\left(\frac{\partial u}{\partial y}\right) = \left(\frac{\partial}{\partial \xi} + \frac{\partial}{\partial \eta}\right)\left(\lambda_1\frac{\partial u}{\partial \xi} + \lambda_2\frac{\partial u}{\partial \eta}\right)$$
$$= \lambda_1\frac{\partial^2 u}{\partial \xi^2} + (\lambda_1 + \lambda_2)\frac{\partial^2 u}{\partial \xi \partial \eta} + \lambda_2\frac{\partial^2 u}{\partial \eta^2}.$$

Substituting these expressions into (16.86) gives

$$\left(A + 2B\lambda_1 + \lambda_1^2 C\right)\frac{\partial^2 u}{\partial \xi^2} + \left(A + 2B\lambda_2 + \lambda_2^2 C\right)\frac{\partial^2 u}{\partial \eta^2}$$
$$+2\left[A + C\lambda_1\lambda_2 + B(\lambda_1 + \lambda_2)\right]\frac{\partial^2 u}{\partial \xi \partial \eta} = 0. \tag{16.89}$$

If we now choose λ_i $(i = 1, 2)$ to be the roots of

$$(A + 2B\lambda_i + \lambda_i^2 C) = 0, \tag{16.90}$$

then (16.89) reduces to

$$[A + C\lambda_1\lambda_2 + B(\lambda_1 + \lambda_2)]\frac{\partial^2 u}{\partial \xi \partial \eta} = 0. \tag{16.91}$$

So, provided the term in square brackets does not vanish,

$$\frac{\partial^2 u}{\partial \xi \partial \eta} = 0,$$

and the solution (16.88) follows by successive integrations.

The condition that the square bracket in (16.91) does not vanish is easily found by noting from Equation (2.7) that

$$\lambda_1 + \lambda_2 = -2B/C \quad \text{and} \quad \lambda_1\lambda_2 = A/C,$$

so that

$$[A + C\lambda_1\lambda_2 + B(\lambda_1 + \lambda_2)] = \frac{2}{C}(AC - B^2),$$

that is, the equation must be such that $AC \neq B^2$.

If $AC = B^2$, the square bracket in (16.91) does vanish and (16.90) has only one solution, which is a repeated root given by $\lambda = -B/C$. In this case, we choose, $\lambda_1 = -B/C$, $\lambda_2 = 0$, and substituting these in (16.87) we find that the first and third terms vanish and the equation reduces to

$$\frac{\partial^2 u}{\partial \eta^2} = 0.$$

Direct integration then gives

$$u(\xi,\ \eta) = f(\xi) + \eta g(\xi), \tag{16.92}$$

where f and g are again arbitrary functions of ξ. The solution of the PDE when $AC = B^2$ is therefore

$$u(x,\ y) = f(x - By/C) + x\, g(x - By/C), \tag{16.93}$$

where f and g are arbitrary functions of ξ.

To summarise, we have to distinguish between the cases $AC = B^2$, when the general solution is given by (16.93); and $AC \neq B^2$, when the general solution is given by (16.88), where λ_1, λ_2 are the roots of (16.90).

Example 16.10

Find the general form of the solutions to the following equations:

$$\text{(a)}\ \frac{\partial^2 u}{\partial x^2} + \frac{\partial^2 u}{\partial y^2} = 0, \ \text{(b)}\ \frac{\partial^2 u}{\partial x^2} + 4\frac{\partial^2 u}{\partial x \partial y} + 4\frac{\partial^2 u}{\partial y^2} = 0$$

Solution

(a) This is just Laplace's equation in two dimensions. Since $A = C = 1$, $B = 0$, the condition $AC \neq B^2$ is satisfied and (16.90) reduces to $1 + \lambda_1^2 = 0$, with roots $\lambda_1 = -\lambda_2 = i$. Hence the general solution (16.88) is

$$u(x, y) = f(x + iy) + g(x - iy),$$

where f and g are arbitrary functions.

(b) In this, $A = 1,\ B = 2,\ C = 4$, so that $AC = B^2$, and the general solution is given by (16.93), that is,

$$u(x,y) = f(x - \tfrac{1}{2}y) + x\,g(x - \tfrac{1}{2}y) = \tilde{f}(2x - y) + x\,\tilde{g}(2x - y),$$

where again $\tilde{f}$ and $\tilde{g}$ are arbitrary functions.

*16.6 Boundary conditions and uniqueness

So far, we have focussed on problems that can be solved exactly. However, it is often not possible to do this in practice, and then one must resort to numerical methods to find approximate solutions of a given PDE that satisfy specific boundary conditions. In these cases, especially, it is very useful to know in advance what boundary conditions result in a unique, stable solution of the PDE, where by stable we mean that very small changes in the boundary conditions do not lead to very large changes in the solution. Here we will simply state the main results without proof, since the derivations are often difficult.[5]

The boundary conditions take the form of information about the dependent variable u specified on a continuous boundary, which may be open, like a plane in three-dimensional space, or closed, like the surface of a sphere. The main types of boundary conditions are classified as follows:

Dirichlet.
The value of u is specified at each point of the boundary.

Neumann
The value of the normal derivative $\partial u/\partial n = \nabla u \cdot \hat{\mathbf{n}}$, where $\hat{\mathbf{n}}$ is the unit normal to the boundary, is specified at each point of the boundary.

Cauchy
The values of both u and $\partial u/\partial n$ are specified at each point of the boundary.

The next step is to classify PDEs into three types, called elliptic, hyperbolic and parabolic. In doing so, we shall focus on second-order linear equations that contain only the derivatives

$$\partial^2 u/\partial t^2,\ \partial u/\partial t \quad \text{and/or} \quad \boldsymbol{\nabla}^2 u \qquad (16.94a)$$

[5] For a discussion of these questions see, for example, P. M. Morse and H. Feshbach (1953) *Methods of Theoretical Physics, Volume 1,* McGraw-Hill Book Company, New York, Chapter 6.

with constant coefficients, where $\nabla^2 u$ is replaced by $\partial^2 u/\partial x^2$ if there is only a single spatial variable. Thus, we consider PDEs of the form

$$A\nabla^2 u + B\frac{\partial^2 u}{\partial t^2} + C\frac{\partial u}{\partial t} + Du = \rho, \tag{16.94b}$$

where A, B, C and D are constants and ρ is a given function. This form includes many equations of physical interest, including all those listed in Section 6.1. They are then classified according to which of the partial derivatives (16.94a) occur, and by the relative sign of their coefficients. We consider each in turn.

Elliptic equations
These are defined as those containing $\nabla^2 u$ and $\partial^2 u/\partial t^2$ with coefficients of the same sign, that is, $AB > 0$; or just $\nabla^2 u$ with no time derivatives. The latter are of most interest and include Laplace's equation, Poisson's equation and the Helmholtz equation

$$(\nabla^2 + k^2)u(\mathbf{r}) = \rho(\mathbf{r}) \tag{16.95}$$

both with a source ($\rho \neq 0$) and without ($\rho = 0$), where k^2 is a positive real constant.

Elliptic equations have the property that if either Dirichlet or Neumann boundary conditions are applied on a closed boundary, then the equation has a unique and stable solution within the boundary. Consequently, if Cauchy boundary conditions are applied, the equation in general has no solutions, and the equation is said to be over-constrained. The closed boundary may be finite, as illustrated for the Laplace equation in Examples 16.7 and 16.8, which used Dirichlet conditions on a finite closed surface; or the surface may be at infinity, as illustrated for Poisson's equation in Example 16.6. Alternatively, if one wishes to determine the function outside a finite closed boundary, then either Dirichlet or Neumann conditions must be applied both on the finite boundary and at infinity.

Hyperbolic equations
These are defined as those containing $\nabla^2 u$ and $\partial^2 u/\partial t^2$ with coefficients of the opposite sign, that is, $AB < 0$. They may in principle also contain terms in $\partial u/\partial t$, but in practice such terms are usually absent in physical applications. Examples of hyperbolic equations are the wave equations (16.3a) and (16.3b) and the Klein-Gordon equation.

$$\frac{1}{c^2}\frac{\partial^2 u}{\partial t^2} - \nabla^2 u + m^2 u = 0,$$

which plays an important role in relativistic quantum mechanics, where c is the speed of light and m is a particle mass.

Hyperbolic equations have unique and stable solutions if Cauchy boundary conditions are applied on an open boundary. In physical applications, this is usually taken to correspond to a constant time, which can always be chosen to be $t = 0$. One thus has to specify $u(r,\ t = 0)$ and the time derivative of $u(r, t)$ at $t = 0$ to obtain a unique and stable solution, as illustrated in the wave equation in Section 16.2.1 [cf. (16.19)] and Example 16.2 for standing waves constrained to vanish at $x = 0,\ L$; and for travelling waves in one dimension in Section 16.4 [cf. (16.83, 85)].

Parabolic equations
These are defined as those containing terms in $\nabla^2 u$ and $\partial u/\partial t$, but not terms in $\partial^2 u/\partial t^2$, that is, $AB = 0$. Examples are the diffusion equation (16.4) and the Schrödinger equation (16.7). In this case, unique and stable solutions are obtained if Dirichlet or Neumann conditions are imposed on an open boundary. This is almost always chosen to be constant time $t = t_0$, and unique and stable solutions for $t > t_0$ are obtained given either $u(r,\ t_0)$ or $\dot{u}(r,\ t_0)$. This is illustrated for the case of Dirichlet boundary conditions for the examples given in Section 16.2.3 and Section 16.6.1 below.

Finally, we stress again that we have only considered equations containing the partial derivatives (16.89), since this covers many of the most important PDEs in physical applications. The discussion can, however, be extended to all linear second-order PDEs with constant coefficients.[6]

*16.6.1 Laplace transforms

In Section 14.2.4, we introduced Laplace transforms (14.44) and showed how they could be used to obtain solutions of ODEs that automatically incorporated given boundary conditions. This method can be extended in principle to PDEs in which the boundary conditions are given at an initial time $t = 0$, although, as in ODEs, it may be difficult to perform the inverse Laplace transform required to obtain the final solution.

[6] The case of two independent variables, which gives rise to the nomenclature elliptic, hyperbolic and parabolic, is discussed in, for example, Chapter 2 of G. Stephenson (1985) *Partial Differential Equations for Scientists and Engineers,* 3rd edn, Longman, London, while the more complicated case of any number of independent variables is summarised, for example, in Section 23, Chapter 4, of P. Dennery and A. Krzywicki (1966) *Mathematics for Physicists,* Dover Publications, New York.

To illustrate this, we shall consider the bounded solution of the diffusion equation in one spatial dimension (16.31) in the range $0 < x < \infty$ for times $t > 0$, with boundary conditions

$$u(x,t) = 0, \quad t < 0; \quad u(0,t) = u_0, \quad t \geq 0. \tag{16.96}$$

This could, for example, describe the temperature distribution due to heat flow along a long rod with one end at $x = 0$, if the rod is initially at zero temperature $u = 0$, but is in contact at $t > 0$ with a heat bath of constant temperature u_0 at the end $x = 0$, and the sides of the bar are perfectly lagged so that heat flow from the sides can be neglected.

To solve this problem, we take the Laplace transform of both sides of (16.31) with respect to time t from (14.44) and (14.45a). We then have, with an appropriate change of notation:

$$L[u(x,t)] \equiv \int_0^\infty u(x,t)e^{-pt}\mathrm{d}t = F(x,p) \tag{16.97a}$$

and

$$L\left[\frac{\partial u(x,t)}{\partial t}\right] = -u(x,0) + pF(x,p) = pF(x,p), \tag{16.97b}$$

where we have used (16.96) to set $u(x,0) = 0$ in (16.97b). Hence (16.31) becomes

$$\frac{\partial^2 F(x,p)}{\partial x^2} = \frac{p}{\kappa}F(x,p) \tag{16.98}$$

with the general solution

$$F(x,p) = A\exp(\alpha x) + B\exp(-\alpha x),$$

where $\alpha = (p/\kappa)^{1/2} > 0$ and A and B are arbitrary constants. If $u(x,t)$ is bounded as $x \to \infty$, which is an obvious requirement if it represents a temperature distribution, then it follows that $F(x,p)$ must also be bounded as $x \to \infty$, so that $A = 0$. The value of B is then found by imposing the boundary condition $u(0,t) = u_0$, which from (16.97a) gives

$$A = F(0,p) = \int_0^\infty u_0 e^{-pt}\mathrm{d}t = u_0\, p^{-1}.$$

Hence,

$$F(x,p) = \frac{u_0}{p}\exp\left[-\left(\frac{p}{\kappa}\right)^{1/2} x\right] \tag{16.99a}$$

and the final solution is given by the inverse transform

$$u(x,p) = L^{-1}[F(x,p)]. \tag{16.99b}$$

At this point, we remind the reader that, as discussed in Section (14.2.4), finding inverse Laplace transforms is difficult and often impossible to do in closed form. One frequently has to resort to tables of such transforms like that of Table 14.1, or the more extensive tables available in the literature.[7] In the case above, the required inverse transform can be expressed in terms of the *error function*

$$\mathrm{erf}(t) \equiv \frac{2}{\sqrt{\pi}} \int_0^t \exp(-u^2)\mathrm{d}u \qquad (16.100\mathrm{a})$$

and the associated *complementary error function*

$$\mathrm{erfc}(t) \equiv 1 - \mathrm{erf}(t) = \frac{2}{\sqrt{\pi}} \int_t^\infty \exp(-u^2)\mathrm{d}u. \qquad (16.100\mathrm{b})$$

The error function is normalised so that it tends to unity as $t \to \infty$, since (cf. Example 11.11)

$$\int_0^\infty \exp(-u^2)\mathrm{d}u = \frac{1}{2}\int_{-\infty}^\infty \exp(-u^2)\mathrm{d}u = \frac{\sqrt{\pi}}{2}.$$

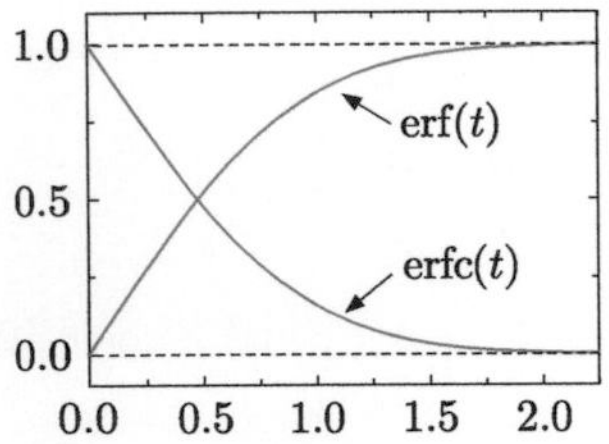

Figure 16.4 The error function (16.100a) and the complementary error function (16.100b).

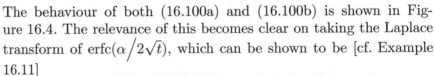

The behaviour of both (16.100a) and (16.100b) is shown in Figure 16.4. The relevance of this becomes clear on taking the Laplace transform of $\mathrm{erfc}(\alpha/2\sqrt{t})$, which can be shown to be [cf. Example 16.11]

$$L\left[\mathrm{erfc}\left(\frac{\alpha}{2\sqrt{t}}\right)\right] = \frac{1}{p}\exp\left(-\alpha\sqrt{p}\right). \qquad (16.101\mathrm{a})$$

This obviously implies

$$L^{-1}\left[\frac{1}{p}\exp\left(-\alpha\sqrt{p}\right)\right] = \mathrm{erfc}\left(\frac{\alpha}{2\sqrt{t}}\right), \qquad (16.101\mathrm{b})$$

which, together with (16.99a) and (16.99b), gives

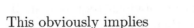

$$u(x,t) = u_0 \mathrm{erfc}\left(\frac{x}{2\sqrt{\kappa t}}\right) \qquad (16.102)$$

as the final solution. Hence $u(x,t) \to u_0$ as $t \to \infty$, but more slowly as x increases, which is what one intuitively expects if u represents the temperature of a long bar, as in this example.

---

[7] For example: Alan Jeffreys and Hui-Hui Dai (2008) *Handbook of Mathematical Formulas and Integrals,* 4th edn., Academic Press., New York, pp. 342–352.

Example 16.11

Derive the relation (16.101a).

Solution

From the definition of the complementary error function (16.100b) and the Laplace transform (14.44), we have

$$L\left[\operatorname{erfc}\left(\frac{\alpha}{2\sqrt{t}}\right)\right] = \frac{2}{\sqrt{\pi}}\int_0^\infty e^{-pt}\left[\int_{\alpha/2\sqrt{t}}^\infty \exp(-u^2)\mathrm{d}u\right]\mathrm{d}t.$$

We then change the orders of integration in the double integral using the method of Section 11.2.2. This gives

$$L\left[\operatorname{erfc}\left(\frac{\alpha}{2\sqrt{t}}\right)\right] = \frac{2}{\sqrt{\pi}}\int_0^\infty \exp(-u^2)\left[\int_{\alpha^2/4u^2}^\infty e^{-pt}\mathrm{d}t\right]\mathrm{d}u,$$

which after the evaluation of the integral over t is

$$L\left[\operatorname{erfc}\left(\frac{\alpha}{2\sqrt{t}}\right)\right] = \frac{2}{p\sqrt{\pi}}\int_0^\infty \exp\left[-\left(u^2+\frac{\alpha^2 p}{4u^2}\right)\right]\mathrm{d}u.$$

It remains to evaluate the integral over u. This may be done by defining a new variable

$$w \equiv u - \frac{\alpha\sqrt{p}}{2u}$$

where $0 < u < \infty$ implies $-\infty < w < \infty$. Solving for u gives

$$u = \frac{1}{2}\left[w + \left(w^2 + 2\alpha\sqrt{p}\right)^{1/2}\right],$$

and hence

$$\frac{\mathrm{d}u}{\mathrm{d}w} = \frac{1}{2} + \frac{w}{(w^2+2\alpha\sqrt{p})^{1/2}}.$$

We can now change the variable of integration to w to give

$$L\left[\operatorname{erfc}\left(\frac{\alpha}{2\sqrt{t}}\right)\right] = \frac{e^{-\alpha\sqrt{p}}}{p\sqrt{\pi}}\left[\int_{-\infty}^\infty \exp(-w^2)\mathrm{d}w + 2\int_{-\infty}^\infty \frac{w\exp(-w^2)}{(w^2+2\alpha\sqrt{p})^{1/2}}\mathrm{d}w\right].$$

The first of the integrals in brackets is $\sqrt{\pi}$ (cf. Example 11.11) and the second is zero because the integrand is an odd function of w. So finally,

$$L\left[\operatorname{erfc}\left(\frac{\alpha}{2\sqrt{t}}\right)\right] = \frac{e^{-\alpha\sqrt{p}}}{p},$$

as required.

Problems 16

16.1 A function $u(x, y, z)$ satisfies the Helmholtz equation

$$(\nabla^2 + k^2)\, u = 0$$

in the range $0 \le x \le L$, where k is a constant. Find the values of k^2 such that u satisfies the boundary conditions

$$u(0, y, z) = u(L, y, z) = 0; \quad u(x, 0, z) = u(x, L, z) = 0;$$
$$u(x, y, 0) = u(x, y, L) = 0,$$

and give the corresponding solutions.

16.2 In quantum mechanics, the wave function $u(x, t)$ of a particle of mass m moving freely in one dimension is described by the Schrödinger equation

$$-\frac{\hbar^2}{2m}\frac{\partial^2 u}{\partial x^2} = i\hbar\frac{\partial u}{\partial t},$$

where $\hbar \equiv h/2\pi$ and h is Planck's constant. Show that separable solutions of the form

$$u(x, t) = X(x)\exp\left(-iEt/\hbar\right)$$

exist, where E is an arbitrary real constant. What are the possible values of E if u satisfies the periodic boundary conditions

$$u(x + L, t) = u(x, t)$$

for any x?

16.3 A rectangular plate with sides of length a and b is oriented so that $0 \le x \le a$, $0 \le y \le b$. The edges corresponding to $x = 0$, $x = a$, $y = 0$ are each kept at temperature zero, and the other edge has a temperature distribution along its length given by $u(x, b) = u_0 x^2 (a - x)/a$, where u_0 is a constant. Find an expression for the temperature at an arbitrary point on the plate. Note the integral

$$\int_0^a x^2 (a - x) \sin\left(\frac{n\pi x}{a}\right) \mathrm{d}x = \frac{-2a^4}{n^3\pi^3}\left[1 + 2(-1)^n\right].$$

16.4 A thin rectangular plate, defined by $0 \le x \le a$, $0 \le y \le b$, is clamped along its perimeter. By solving the two-dimensional wave equation with velocity v, show that its vibrational modes are given by

$$u(x, y, t) = \sum_{n=1}^{\infty}\sum_{m=1}^{\infty} u_{nm}(x, y, t),$$

where

$$u_{nm}(x, y, t) = [a_{nm}\cos(\omega_{nm} t) + b_{nm}\sin(\omega_{nm} t)]\sin\left(\pi n x/a\right)\sin(\pi m x/b),$$

with

$$\omega_{nm} = v\pi \left(\frac{n^2}{a^2} + \frac{m^2}{b^2} \right)^{1/2},$$

and a_{nm} and b_{nm} are arbitrary constants. Hence show that if the plate is released from rest with an initial profile

$$u(x, y, 0) = \sin(x\pi/a)\sin(y\pi/b),$$

its subsequent motion is described by

$$u(x, y, t) = \sin\left(\frac{\pi x}{a}\right)\sin\left(\frac{\pi y}{b}\right)\cos\left[\left(\frac{a^2+b^2}{a^2b^2}\right)^{1/2} v\pi t\right].$$

16.5 A thin insulated rod of length L, with ends at $x = 0$ and $x = L$, has an initial temperature along its length given by

$$u(x, 0) = u_0\, x(L - x),$$

where u_0 is a constant. If the ends of the rod are kept at temperature zero, find an expression for $u(x, t)$ for $t > 0$.

16.6 Find the function $u(x, y)$ that describes the steady-state distribution of temperature through a two-dimensional semi-infinite slab, where $0 \le x \le d$ and $0 \le y < \infty$, if the long edges of the slab are kept at zero temperature. Assume that for $0 < x < d$ and $y = 0$, $u(x, y) = f(x)$, where $f(x)$ is a given function (the form of which would have to be such as to satisfy the boundary condition $u(0, y) = u(d, y) = 0$) and $u(x, y) \to 0$ as $y \to \infty$ for $0 \le x \le d$.

16.7 Find the single-valued solution $u(r, \theta)$ of the two-dimensional Laplace equation within a circle of radius R, subject to the boundary condition $u(R, \theta) = f(\theta)$, where $f(\theta)$ is an arbitrary positive function of θ.

16.8 Show that a spherically symmetric potential u that obeys the Laplace equation and vanishes at infinity may be written $u(r) = a/r$, where a is a constant.

16.9 A neutral conducting sphere of radius a is centred at the origin and is exposed to a uniform electric field $\mathbf{E}$ in the z-direction. Find the electrostatic potential u satisfying Laplace's equation outside the sphere if the potential on the sphere is set, by convention, to zero.

16.10 Show that the differential equation

$$\left[\boldsymbol{\nabla}^2 + f(r) + \frac{g(\theta)}{r^2} + \frac{h(\phi)}{r^2\sin^2\theta}\right] u(r, \theta, \phi) = 0$$

has separable solutions of the form $u(r, \theta, \phi) = R(r)\Theta(\theta)\Phi(\phi)$, where r, θ, ϕ are spherical polar co-ordinates, and f, g and h are arbitrary functions.

16.11 On substituting $u = \psi(r, \theta, \phi)\exp(-iEt/\hbar)$ into the Schrödinger equation (16.7), one obtains the so-called *time independent Schrödinger equation*

$$-\frac{\hbar^2}{2m}\nabla^2\psi + V(r)\psi = E\psi. \quad (1)$$

Show that for spherical potentials $V(\mathbf{r}) = V(r)$, this equation has separable solutions of the form

$$\psi(r, \theta, \phi) = r^{-1}R(r)Y_{lm}(\theta, \phi), \quad (2)$$

where Y_{lm} are spherical harmonics; and find the ODE satisfied by the radial function $R(r)$.

16.12 Verify that

$$u(r, t) = \frac{1}{t^{3/2}}\exp\left(-\frac{r^2}{4\kappa t}\right),$$

where κ is the diffusivity, is a solution of the diffusion equation in spherical co-ordinates.

16.13 A sphere of radius R has the surface of its upper $(0 \le \theta < \pi/2)$ hemisphere held at a constant temperature T_U, and the surface of its lower $(\pi/2) \le \theta < \pi)$ hemisphere held at a constant temperature T_L. Assuming it is in thermal equilibrium, find an expansion for the temperature $u(r, \theta)$ within the sphere, accurate to terms of order $(r/R)^3$.

16.14 Find separable solutions of the Helmholtz equation

$$(\nabla^2 + k^2)u(r, \theta, z) = 0,$$

in cylindrical polar co-ordinates, when $k^2 > 0$, if u is single-valued, finite and tends exponentially to zero as $z \to \infty$.

16.15 A solid semi-infinite cylinder of unit radius is in thermal equilibrium. Show that the temperature distribution $u(\rho, \phi, z)$ in the cylinder, subject to the boundary conditions (1) $u = \rho\sin\phi$ on the base $z = 0$, and (2) $u = 0$ on the curved surface, is

$$u(\rho, \phi, z) = \sum_{n=1}^{\infty}\frac{2}{k_n J_2(k_n)}J_1(k_n\rho)e^{-k_n z}\sin\phi,$$

where J_ν is a Bessel function of the first kind of order ν, and k_n are the zeros of $J_1(k)$.

16.16 If in question 16.15 the base is kept at a constant temperature u_0, then the resulting temperature distribution is

$$u(\rho, \phi, z) = \sum_{n=1}^{\infty}\frac{2u_0}{k_n J_1(k_n)}J_0(k_n\rho)e^{-k_n z}.$$

Use this expansion to calculate the value of the temperature at $\rho = 1/2$ and $z = 1$ to three decimal places if $u_0 = 50$. [Note: the positions of the zeros of the Bessel functions are given in Table 15.1 and values of the Bessel function $J_n(x)$ may be found from a number of widely available sources, for example, the function BESSEL(x, n) in a Microsoft Excel spreadsheet, or the website www.wolframalpha.com.]

***16.17** Find the solution to the equation

$$\frac{\partial^2 u}{\partial x^2} = \frac{1}{v^2}\frac{\partial^2 u}{\partial t^2}$$

subject to the initial conditions

(a) $u(x,0) = \exp(-x^2); \quad \dot{u}(x,0) = 0;$
(b) $u(x,0) = 0; \quad \dot{u}(x,0) = x\exp(-x^2).$

***16.18** Find the general form of the solution to the following equations:

$$\text{(a) } 2\frac{\partial^2 u}{\partial x^2} + 5\frac{\partial^2 u}{\partial x \partial y} + 2\frac{\partial^2 u}{\partial y^2} = 0, \quad \text{(b) } 9\frac{\partial^2 u}{\partial x^2} - 6\frac{\partial^2 u}{\partial x \partial y} + \frac{\partial^2 u}{\partial y^2} = 0,$$

$$\text{(c) } \frac{\partial^2 u}{\partial x^2} - 4\frac{\partial^2 u}{\partial x \partial y} + 5\frac{\partial^2 u}{\partial y^2} = 0, \quad \text{(d) } \frac{\partial^2 u}{\partial x^2} + 2\frac{\partial^2 u}{\partial x \partial y} = 0.$$

***16.19** Solve the equation

$$\frac{\partial^2 u}{\partial x^2} + 2\frac{\partial^2 u}{\partial x \partial y} + \frac{\partial^2 u}{\partial y^2} = 0,$$

subject to the boundary conditions $u(0,y) = y^2, u(x,0) = \sin x$.

***16.20** A thin insulated metal rod lies horizontally in the semi-infinite region $x \geq 0$ and is initially at zero temperature. At time $t > 0$, the end at $x = 0$ is placed in contact with a heat bath with fixed temperature u_0. If $F(x,p) \equiv L[u(x,t)]$ is the Laplace transform of $u(x,t)$ show that the distribution of temperature along the rod at time t may be written as

$$u(x,t) = u_0 L^{-1}\left[\frac{\exp(-x\sqrt{p/\kappa})}{p}\right],$$

where L^{-1} denotes an inverse Laplace transform and κ is the thermal diffusivity.

***16.21** Show that the solution of the equation

$$\frac{\partial^2 u}{\partial x^2} = \frac{\partial u}{\partial t},$$

for $t > 0$ and $0 < x < a$, subject to the boundary conditions

$$u(x,0) = 0, \quad 0 < x < a \tag{1}$$

$$(\partial u/\partial x)|_{x=a} = 0, \quad t > 0 \tag{2}$$

and

$$u(0,t) = u_0, \quad t > 0 \tag{3}$$

may be written

$$u(x,t) = u_0 L^{-1}\left[\frac{\cosh\{(x-a)p^{1/2}\}}{p\cosh(ap^{1/2})}\right],$$

where L^{-1} is the inverse Laplace transform.

***16.22** Show that the solution of the wave equation with unit velocity for $t > 0$ and $0 < x < a$, where a is a constant, subject to the boundary conditions

$$u(x,0) = 0, \quad (\partial u/\partial t)|_{t=0} = 0; \quad 0 < x < a \tag{1}$$

and

$$u(0,t) = 0, \quad (\partial u/\partial x)|_{x=a} = f; \quad t > 0 \tag{2}$$

where f is a constant, may be written

$$u(x,t) = f\, L^{-1}\left[\frac{\sinh(px)}{p^2\cosh(pa)}\right],$$

where L^{-1} is the inverse Laplace transform.

Answers to selected problems

Problems 1

1.1 (a) $p = 124.68$; (b) $q = 400$; (c) $r = 0.04$.
1.2 (a) $756 = 2^2 3^3 7$; (b) $4/7 = 0.571$.
1.3 (a) $p = 1.246 \times 10^3$; (b) -1.57×10^3; (c) 4.5×10^{-3}.
1.5 (a) $2^{5/2} 3^{11/12}$; (b) $3^{-1/6} 5^{12/5}$; (c) $7^{7/3} 2^{-1/2}$.
1.6 (a) $(8\sqrt{5} - 14)/31$; (b) $(4 - 3\sqrt{3} + \sqrt{5} + 2\sqrt{15})/22$.
***1.7** (a) $10.31 = 1010.0101_2 \ldots$; (b) $1101.01_2 = 13.25$.
***1.8** (a) $p \times q = 32130_4$; (b) $(p - q) = 11_4$.
1.9 (a) $1/x^3$; (b) $(3 + 8\sqrt{x} - 3x)/(9 - x)$; (c) $3^{5/6} a^{-5/3}$.
1.10 (a) identity; (b) equation.
1.11 (a) $-1 < x < 4$; (b) $-4 < x < -2$ or $2 < x < 4$; (c) $x > -2$ or $x < -4$.
1.12 31.681.
1.13 $-9828a^3b^{11}$.
1.17 (a) $x = \left[2(y-3)/y^2 - 1\right]^{1/2}$; (b) $x = (2y+1)/(1-y)$; (c) $x = y^6 + 2$.
1.18 $f_S(x) = -(x^2 - 12)/(x^2 - 9)$; $f_A(x) = x/(x^2 - 9)$.
1.19 (a) $f^{-1}(x) = 2(x+1)/(1-x)$, $(x \neq 1)$; (b) $f^{-1}(x) = 27x^3 - 4$.
1.20 (a) $y = -x + 3$; (b) $3y = 2x - 7$; (c) $2y + 5x = 13$; (d) $5y + 2x = 22$.
1.21 $\sqrt{34}$.
1.22 Area = 12.
1.23 (a) $(x-1)^2 + (y-3)^2 = 4$; (b) $(x-1)^2 + (y-3)^2 = 13$.
1.24 Centre (1, 3/2), radius $r = 3/2$.

Problems 2

2.1 $3x^2 - 2x + 1 = 0$.
2.3 Gradients $m = \pm\left[(c^2 - r^2)/r^2\right]^{1/2}$.
2.4 Points of intersection: $(x, y) = \left(\frac{1}{2} + \frac{\sqrt{7}}{2},\ -\frac{1}{2} + \frac{\sqrt{7}}{2}\right)$ and $\left(\frac{1}{2} - \frac{\sqrt{7}}{2},\ -\frac{1}{2} - \frac{\sqrt{7}}{2}\right)$ Angle at centre 2.42 rad = 139°.
2.5 (a) $(x-1)(x^2 + 2x + 1) - 3$.
2.6 $x = 1,\ 2,\ (-1 + \sqrt{5})/2, (-1 - \sqrt{5})/2$.
2.7 1.526.
2.8 (a) $\dfrac{1}{(x-2)} - \dfrac{2}{(x-3)} + \dfrac{3}{(x+4)}$; (b) $\dfrac{2x-1}{x^2+2x-4} + \dfrac{3}{2x+1}$; (c) $\dfrac{2}{(x-1)} - \dfrac{2}{(2x+1)} + \dfrac{3}{(2x+1)^2}$.
2.9 (a) $(x-3) + \dfrac{3}{(x-1)} + \dfrac{2}{(x+2)}$; (b) $\dfrac{2}{(x+2)} - \dfrac{3x+1}{(3x^2+x-1)}$;
(c) $-\dfrac{1}{2(x-1)} + \dfrac{1}{2(x-3)} + \dfrac{2}{(x-3)^2} + \dfrac{13}{(x-3)^3}$.

Mathematics for Physicists, First Edition. B.R. Martin and G. Shaw.

Companion website: www.wiley.com/go/martin/mathsforphysicists

2.11 (a) $\theta = 0.666$ and 2.475 radians; $\theta = \pi/2, 3\pi/2$ radians. (b) $\theta = \pi/2, 3\pi/2$; $\theta = \pi/3,\ 5\pi/3,\ \pi$.
2.12 $\theta = \pm 2n\pi/(k-1)$ for all integer n and $k \neq 1$, and $\theta = \pm(2n+1)\pi/(k+1)$ for all integer n and $k \neq -1$.
2.13 $x = a^2 \sin\theta/p$, $y = -b^2\cos\theta/p$.
2.15 $A = 0.643$ radians, $C = 1.999$ radians, $c = 7.59$ cm.
2.16 $A = 0.809$ radians, $B = 1.391$ radians, $C = 0.942$ radians.
2.18 (a) $\log(x^2)$; (b) 0.
2.19 (a) $x = 31.765$; (b) $x = 1.122$.
2.20 (a) $x = 32.66$; (b) $x = 3.401$.
2.22 (a) $x = -\frac{1}{2}\ln 5$ or $x = 0$; (b) $x = 12/13$; (c) $c \geq \sqrt{3/2}$ or $c \leq -\sqrt{3/2}$.
2.23 $x = \pm 1.317$.
2.24 $t_1 t_2 = -1$.
2.25 tangent: $y = x + 1$; normal: $y = -x + 3$.

Problems 3

3.1 (a) 1; (b) 0; (c) 4.
3.2 (a) 10; (b) 1/2; (c) 1.
3.5 (a) Removable discontinuity at $x = 0$; (b) Non-removable discontinuity at $x = 3$; removable discontinuity at $x = -3$.
3.6 (a) $A = 1$; (b) $A = 1$ and $B = 0$ (all n), or for $n \geq 2$ (all B).
3.7 (a) $6x^2 + 4$; (b) $-2/x^3$; (c) $-15\sin 3x$.
3.8 (a) $x^3 e^x + 3x^2 e^x$; (b) $\dfrac{x\cosh x - \sinh x}{x^2}$; (c) $-\dfrac{1}{\sqrt{1-x^2}}$; (d) $\dfrac{1}{\sqrt{1+x^2}}$; (e) $6x^2 e^{2x^3}$; (f) $\dfrac{6x}{1+x^2}$.
3.9 (a) $\dfrac{\cos(\ln x)}{x}$; (b) $\dfrac{(1+x^2)\cos x - x(1-x^2)\sin x}{(1-x^2)^2}$; (c) $\dfrac{2x}{1+(1+x^2)^2}$; (d) $\dfrac{1}{x\ln x}$.
3.10 (a) $x^x(\ln x + 1)$; (b) $x^{\cos x}\left(\dfrac{\cos x}{x} - \sin x \ln x\right)$; (c) $-\dfrac{2\cot(1/x^2)}{x^3}$; (d) $\tan x$.
3.11 (a) $\dfrac{dy}{dx} = (\ln a)a^x = (\ln 2)2^x = (0.693\ldots)2^x$; (b) $1/x\ln a$.
3.12 $\dfrac{-128xy}{(4+x^2y^2)^2(4+3x^2y^2)}$.
3.13 $\theta = \arctan\left(\dfrac{w-gt}{u}\right)$; $\dfrac{dT}{dt} = -mg(w-gt)$.
3.14 $x = -8$ and $y = 8/9$, respectively.
3.15 $dy/dx = -7/6$, $7y = 6x - 21$.
3.16 Three: $f^{(1)}$, $f^{(2)}$ and $f^{(3)}$.
3.17 (a) $2^{n-1}e^{2x} + (-1)^{n+1}2^{n-1}e^{-2x}$; (b) $\dfrac{(-1)^{n+1}(n-1)!}{x^n}$, for all $n \geq 1$.
3.19 (b) $-2/x^2$.
3.20 (a) Minimum at $x = 0$, maxima at $x = \pm 1$.
3.21 Maxima at $x = \pi/4 + 2\pi n$; minima at $x = 5\pi/4 + 2\pi n$.
3.24 Approximate solutions: $x = -0.8, 1.5$ and 3.4.
3.27 Vertex $x = 3a,\ y = 0$.

Problems 4

4.1 Area = 16.41.
4.2 (b) $1 - \pi/4$.

4.3 (a) $\frac{1}{2}(2+x^2)^{5/2}+c$; (b) $-\dfrac{1}{3+x}+\dfrac{3}{(3+x)^2}-\dfrac{3}{(3+x)^3}+c$; (c) $2\ln\left(1+\sqrt{x-1}\right)+\dfrac{2}{1+\sqrt{x-1}}+c$.

4.4 (a) $\dfrac{1}{6}\arcsin(3x)+\dfrac{x}{2}\sqrt{1-9x^2}+c$; (b) $2(3+x-x^2)^{1/2}+c$; (c) $\ln(x-2)-\dfrac{2}{(x-2)}+c$.

4.5 (a) $\sin x-\frac{2}{3}\sin^3 x+\frac{1}{5}\sin^5 x+c$; (b) $\frac{3}{2}x-\sin 2x+\frac{1}{8}\sin 4x+c$.

4.6 (a) $\frac{1}{7}\sinh^7 x+\frac{1}{9}\sinh^9 x+c$; (b) $\frac{6}{7}(3+\sin x)^7-\frac{1}{8}(3+\sin x)^8-\frac{4}{3}(3+\sin x)^6+c$.

4.7 (a) $2\ln(x-2)+3\ln(x+1)+c$; (b) $-\frac{1}{6}\ln(x-1)+\frac{1}{15}\ln(x+2)-\frac{1}{10}\ln(x-3)+c$.

4.8 (a) $-\frac{1}{7}\ln(x+3)-\frac{6}{7}\ln(x-4)+2/(x-4)+c$; (b) $\frac{1}{4}\ln 2+\frac{1}{8}\pi$.

4.9 (a) $-\dfrac{2}{1+\tan(x/2)}+c$; (b) $\dfrac{1}{\sqrt{3}}\tan^{-1}\left(\dfrac{\tan x}{\sqrt{3}}\right)+c$.

4.10 (a) $\tan x\ln(\tan x)-\tan x+c$; (b) $2\tan(x/2)-x+c$.

4.11 (a) $\dfrac{1}{4}\ln|[\sinh(4x)]|+c$; (b) $\dfrac{1}{\sqrt{5}}\ln\left|\left[\dfrac{\tan(x/2)+\sqrt{5}}{\tan(x/2)-\sqrt{5}}\right]\right|+c$.

4.12 (a) $\dfrac{x}{a^2(a^2-x^2)^{1/2}}+c$; (b) $\dfrac{2}{\sqrt{4ac-b^2}}\arctan\left(\dfrac{2ax+b}{\sqrt{4ac-b^2}}\right)+c$.

4.13 (a) $\dfrac{1}{3}\ln 2$; (b) $\dfrac{\pi}{2}-1$; (c) -1.

4.15 $I_3=\dfrac{x}{4(1+x^2)^2}+\dfrac{3}{8}\dfrac{x}{(1+x^2)}+\dfrac{3}{8}\arctan x$.

4.16 (a) $-\frac{1}{2}e^{-x}(\sin x+\cos x)+c$; (b) $\dfrac{x\ln|x|}{(1-x)}+\ln|(x-1)|+c$; (c) $\frac{1}{2}x\left[\sin(\ln x)-\cos(\ln x)\right]+c$.

4.17 (b) At $x=0$ and all negative integers; $\Gamma(-5/2)=-8\sqrt{\pi}/15$.

4.18 564 m.

4.19 (a) Convergent with value 1/2; (b) convergent with value 7/4; (c) divergent; (d) divergent.

4.21 (a) Converges for all $\alpha<-1$; (b) Converges provided $\beta>-1,\alpha<-1$.

4.22 $17k/3a$.

4.23 $\pi/2$.

4.24 $L=2\int_{1.5}^{2.5}\left(\dfrac{2x^2-1}{x^2-1}\right)^{1/2}dx$; $L=3.0896$ (trapezium rule), $L=3.0847$ (Simpson's rule).

4.25 $\pi=3.14294$ for $n=2$ and 3.14170 for $n=4$. Four intervals are needed.

***4.26** $A=2\pi$, $V=2\pi/3$, $S=3\pi$.

***4.27** $V=\frac{4}{3}\pi ab^2$

***4.28** (a) $ma^2/12$; (b) $ma^2/12$.

***4.29** $Mr^2/2$.

***4.30** $3MR^2/10$.

Problems 5

5.1 Sum = 148875.

5.2 $S_N=\ln\left(\dfrac{r^Z}{N+1}\right)$, where $Z=\frac{1}{2}N(N+1)$. No values of r.

5.3 $S_N=\dfrac{e^{-x/2}\left[1-(e^{-x})^{N+1}\right]}{1-e^{-x}}$, convergent as $N\to\infty$.

5.4 $S_N=\dfrac{a-(a+Nx)y^{N+1}}{(1-y)}+\dfrac{xy(1-y^N)}{(1-y)^2}$.

5.5 (a) convergent; (b) divergent; (c) divergent; (d) convergent.

5.6 (a) $|x|<1$; (b) $1/2<x<3/2$; (c) $x<0$.

5.7 (a) 11/2; (b) $-\pi$; (c) -1; (d) 1/2.

5.8 (a) $1/(2\sqrt{5})$; (b) 1.

5.9 $\dfrac{1}{\sqrt{2}} - \dfrac{(x-\pi/4)}{\sqrt{2}} - \dfrac{(x-\pi/4)^2}{2\sqrt{2}} + \dfrac{(x-\pi/4)^3}{6\sqrt{2}} + \cdots$, valid for all x.

5.10 3 terms give $\sin x = 0.56465$, the calculator value is 0.56465.

5.11 (a) $1 + x^2/2 + 5x^4/4! + \cdots$; (b) $x + x^3/3 + 2x^5/15 + \cdots$

5.14
$$\sqrt{x} = \begin{cases} \sqrt{x} = 1 + \dfrac{1}{2}(x-1) - \dfrac{1}{8}(x-1)^2 + \cdots & \text{about } x = 1, \\ & \text{valid for } |x-1| < 1 \\ \sqrt{x} = \sqrt{2} + \dfrac{(x-2)}{2\sqrt{2}} - \dfrac{(x-2)^2}{16\sqrt{2}} + \cdots & \text{about } x = 2, \\ & \text{valid for } |x-2| < 2 \end{cases}$$

5.15 (a) 1/2; (b) −1/2.

5.16 0.838.

5.17 First minimum is at 4.50 rad = 258° to the nearest 1°.

5.20 $x + x^3/6 + 3x^5/40$; valid for $|x| < 1$.

5.21 $1 + x - x^3/3 - x^4/6$; valid for all x.

5.22 (a) conditionally convergent; (b) absolutely convergent for $\alpha \neq k\pi$, where k is an integer; (c) conditionally convergent for $\alpha = 0$.

5.23 (a) conditionally convergent; (b) not convergent; (c) absolutely convergent; (d) absolutely convergent.

Problems 6

6.1 (a) $-(1+2i)$; (b) $-5(1+2i)$; (c) $6(5+7i)$; (d) $4(3-4i)$; (e) $(7/425) + (74/425)\,i$, (f) $(11/37) - (8/37)\,i$.

6.3 (a) $(86/325) + (77/325)\,i$; (b) $(3/10)\,i$; (c) $(18/25) + (24/25)\,i$.

6.5 (a) $r = 1$, $\arg z = \pi/4$; (b) $r = 1/\sqrt{2}$, $\arg z = 5\pi/12$; (c) $r = 2\sqrt{2}$, $\arg z = \pi/12$.

6.7 (a) circle radius 4 centre $(0,3)$; (b) circle radius 1/2 centre $(0,0)$; (c) converges for all z.

6.8 (a) $|z| = \sqrt{5}$ and $\arg(z) = 1.11$ rad; (b) $|z| = \sqrt{2}$ and $\arg(z) = 1.31$ rad; (c) $|z| = 1/4$ and $\arg(z) = -0.939$ rad.

6.9 (a) $-0.101 - 0.346i$; (b) $0.417 + 0.161i$; (c) $1.272 - 0.786i$ and $-1.272 + 0.786i$.

6.10 (a) $0.0313i$, (b) $1.864 + 0.290i$, $-1.183 + 1.468i$ and $-0.680 - 1.758i$, (c) $-i$.

6.11 (a) $0.951 + 0.309i$, i, $-0.951 + 0.309i$, $-0.588 - 0.809i$ and $0.588 - 0.809i$; (b) $0.080 + 0.440i$; (c) $0.920 + 0.391i$.

6.12 (a) $0.1080 + 0.4643i$; (b) $3i/4$; (c) $-0.805 + 1.007i$.

6.13 (a)$-0.266 + 0.320i$; (b) 4.248; (c) $(7/16) + (9/16)i$.

6.14 (a) $\cos(12\theta) + i\sin(12\theta)$.

6.15 (a) $\cos 7\theta - i\sin 7\theta$.

***6.16** $-3/25$.

***6.17** $e^{2\cos x}\sin(2\sin x)$.

***6.18** $2^n\cos^n(x/2)\sin(nx/2)$.

Problems 7

7.3 $\alpha = -1/3$, $\beta = -2/3$, $\gamma = 14/3$; $x + 2y + 3z = 14$.

7.8 (a); (c) and (d)

7.9 (a) $f(x,y) = \dfrac{xy}{(x+y)} + k$; (b) $f(x,y) = x^2\ln(xy) + k$, k an arbitrary constant.

7.10 (a) $y(4x+3y^3)-xy^2(9x+16y)$; (b) –2, (c) $e^{-x}\left[(y-x)\ln\left(\frac{x}{y}\right)+\left(y+\frac{1}{y}+x+\frac{x}{y^2}\right)\right]$.

7.11 (a) satisfied $k=3$; (b) not satisfied; (c) satisfied $k=0$; (d) satisfied $k=1/2$.

7.16 $e^2[1+(x-2)-2(y-1)+\frac{1}{2}(x-2)^2+4(y-1)^2-3(x-2)(y-1)+\cdots]$.

7.17 $x-\frac{1}{2}x^2-xy+\frac{1}{3}x^3+\frac{1}{2}x^2y+xy^2+\cdots$.

7.18 maximum $3\sqrt{3}/8$, minimum $-3\sqrt{3}/8$.

7.19 maximum $(0,0)$, saddle point $(-1,1)$.

***7.20** $(0,-1)$, $(2,1)$, $(-2,1)$.

***7.21** $8abc/3\sqrt{3}$.

***7.22** $(N/n)^n$.

***7.23** (a) $\mathrm{d}I/\mathrm{d}x=-e^{-x}$, $I(x)=e^{-x}-e^{-1}$; (b) $\frac{2t+1}{\ln(t+t^2)}-\frac{2}{\ln(2t)}$.

***7.24** (a) $\sin(ye^y)\left(1+\frac{1}{y}\right)-\frac{2\sin y^2}{y}$.

***7.25** $\frac{1\cdot 3\cdot 5\cdot\ldots\cdot(2k-1)}{(2\pi)^k}\left(\frac{a}{\pi}\right)^{-(2k+1/2)}$.

Problems 8

8.1 $\mathbf{r}=(a/\sqrt{2})(3\,\mathbf{i}+4\mathbf{j}+5\mathbf{k})$; $|\mathbf{r}|=5a$.

8.3 (0, 4, 0).

8.4 (b) $\theta=2.68$ rad $=153.4°$, direction cosines $(0,0,1)$.

8.7 $\hat{\mathbf{n}}=\pm[\mathbf{i}+4\mathbf{j}+5\mathbf{k}]/\sqrt{42}$, area $=\sqrt{21/2}$.

8.9 75/6.

8.10 (a) $-6\mathbf{i}-3\mathbf{j}$; (b) $\mathbf{r}=\mathbf{b}-\frac{(\mathbf{b}\cdot\mathbf{c})}{(\mathbf{a}\cdot\mathbf{c})}\mathbf{a}$.

8.12 (a) $\boldsymbol{\tau}=-9\mathbf{i}+6\mathbf{j}+5\mathbf{k}$; (b) $\boldsymbol{\tau}=3\mathbf{i}-2\mathbf{j}$; (c) $\tau_z=5$; (d) $\tau_l=11/\sqrt{2}$.

***8.15** (a) $\mathbf{a}'=-\frac{1}{5}(2\mathbf{i}+\mathbf{j}-\mathbf{k})$, $\mathbf{b}'=-\frac{1}{5}(-2\mathbf{i}+\mathbf{j}-6\mathbf{k})$, $\mathbf{c}'=-\frac{1}{5}(\mathbf{i}-3\mathbf{j}+3\mathbf{k})$. (b) $\mathbf{a}'=\frac{1}{a}(\mathbf{i}+\mathbf{j}-\mathbf{k})$, $\mathbf{b}'=\frac{1}{a}(-\mathbf{i}+\mathbf{j}+\mathbf{k})$, $\mathbf{c}'=\frac{1}{a}(\mathbf{i}-\mathbf{j}+\mathbf{k})$.

8.16 $\lambda=\frac{\mathbf{a}\cdot(\mathbf{b}\times\mathbf{d})}{\mathbf{a}\cdot(\mathbf{c}\times\mathbf{d})}$, $\mu=\frac{\mathbf{b}\cdot(\mathbf{a}\times\mathbf{c})}{\mathbf{b}\cdot(\mathbf{d}\times\mathbf{c})}$.

8.17 $\sqrt{14}$, D is $(0,1,0)$.

8.19 $x-2y-3z=7$, $\sqrt{7/2}$.

8.20 $x+4y+6z=5$.

8.21 $\theta=1.52$ rad $=87.2°$; $\mathbf{r}=(2\mathbf{i}-3\mathbf{j})+s(\mathbf{i}-2\mathbf{j}+\mathbf{k})$.

8.22 $(2\mathbf{i}+2\mathbf{j}+\mathbf{k})/3$.

8.23 $\mathbf{I}=\mathbf{a}\times\frac{\mathrm{d}\mathbf{a}}{\mathrm{d}t}+\mathbf{c}$, where c is a constant.

Problems 9

9.1 $\mathbf{a}\times\mathbf{b}=7\mathbf{i}+10\mathbf{j}-9\mathbf{k}$, $\mathbf{b}\cdot\mathbf{a}\times\mathbf{c}=52$.

9.2 (a) $16+5i$; (b) –1.

9.3 –105.

9.4 (a) $x=0,\ 1$ or -2; (b) $\Delta_2=(\alpha+\beta+\gamma)(\alpha-\beta)(\beta-\gamma)(\gamma-\alpha)$.

9.5 $\Delta_n=(n+1)(-1)^n$.

9.6 (a) no non-trivial solution; (b) $x : y : z = -3 : 1 : 1$.
9.7 $\alpha = 3$ and 14. For the latter, $x = 2/5,\ y = -6/5$.
9.13 (a) $(\mathbf{A}+\mathbf{B})^3 = (\mathbf{A}^3 + \mathbf{A}^2\mathbf{B} + \mathbf{ABA} + \mathbf{AB}^2 + \mathbf{BA}^2 + \mathbf{BAB} + \mathbf{B}^2\mathbf{A} + \mathbf{B}^3)$; (b) $\mathbf{AB} = \mathbf{BA}$.
9.15 (b) $\mathbf{A}_S = \frac{1}{2}\begin{pmatrix} 2 & 5 \\ 5 & 4 \end{pmatrix},\ \mathbf{A}_{AS} = \frac{1}{2}\begin{pmatrix} 0 & -1 \\ 1 & 0 \end{pmatrix}$.
9.19 $\mathbf{A}^{-1} = \begin{pmatrix} -3 & 5 & -1 \\ -1 & 2 & 0 \\ -2 & 4 & -1 \end{pmatrix}$ and $\mathbf{X} = \begin{pmatrix} 1 & 3 \\ 2 & 2 \\ 3 & 1 \end{pmatrix}$.
9.20 $x = 3/4,\ y = 1/2, z = 7/4$.
9.21 $x = 1/5,\ y = 1/5,\ z = 1$.
9.22 $m_a = 960$ gm, $m_b = 120$ gm.
9.23 (a) $\alpha \neq 3$, independent of the value of β; (b) $x = 1/2,\ y = -5/2,\ z = 3$; (c) (i) $\beta = 6$, a solution exists, but is not unique; (ii) $\beta = 2$, no solutions exist.

Problems 10

10.1 $\lambda_1 = 3,\ \lambda_2 = 1 - \sqrt{2},\ \lambda_3 = 1 + \sqrt{2}$. Normalised eigenvectors are: $\mathbf{u}^{(1)} = \frac{1}{\sqrt{2}}\begin{pmatrix} 1 \\ 1 \\ 0 \end{pmatrix}$,

$$\mathbf{u}^{(2)} = \frac{1}{\sqrt{6+2\sqrt{2}}}\begin{pmatrix} -(1+\sqrt{2}) \\ 1 \\ -\sqrt{2} \end{pmatrix},\ \mathbf{u}^{(3)} = \frac{1}{\sqrt{6-2\sqrt{2}}}\begin{pmatrix} -(1-\sqrt{2}) \\ 1 \\ \sqrt{2} \end{pmatrix}.$$

Eigenvectors not orthogonal.

10.7 Eigenvalues: $\lambda = 0,\ 1$. Linearly independent eigenvectors: $\frac{1}{\sqrt{2}}\begin{pmatrix} 1 \\ 0 \\ 1 \end{pmatrix}$, $(\lambda = 0)$ and $\begin{pmatrix} 0 \\ 1 \\ 0 \end{pmatrix}$, $(\lambda = 1)$.

The matrix is defective.

10.9 (a) Eigenvalues: $\lambda_{1,2} = 1 \pm \sqrt{2}$. Corresponding normalised eigenvectors:

$$\mathbf{u}^{(1)} = \frac{1}{\sqrt{3}}\begin{pmatrix} 1 \\ -i\sqrt{2} \end{pmatrix},\quad \mathbf{u}^{(2)} = \frac{1}{\sqrt{3}}\begin{pmatrix} 1 \\ i\sqrt{2} \end{pmatrix}.$$

Eigenvectors not orthogonal.

(b) Eigenvalues: $\lambda_{1,2} = 1 \pm \sqrt{2}$. Corresponding normalised eigenvectors:

$$\mathbf{u}^{(1)} = \frac{1}{\sqrt{2}}\begin{pmatrix} (1+i)/\sqrt{2} \\ 1 \end{pmatrix},\quad \mathbf{u}^{(2)} = \frac{1}{\sqrt{2}}\begin{pmatrix} -(1+i)/\sqrt{2} \\ 1 \end{pmatrix}.$$

10.10 $\hat{\mathbf{x}}^{(1)} = \frac{1}{\sqrt{2}}(1,\ 0,\ 1),\ \hat{\mathbf{x}}^{(2)} = \frac{1}{\sqrt{3}}(1,\ 1, -1),\ \hat{\mathbf{x}}^{(3)} = \frac{1}{\sqrt{6}}(-1,\ 2,\ 1)$.

***10.12** $\begin{pmatrix} -3 & 1 & 1 \\ 1 & 0 & 0 \\ -5 & \sqrt{2} & -\sqrt{2} \end{pmatrix}$.

***10.15** $x_1(t) = \frac{5}{12}\cos\omega_1 t - \frac{5}{12}\cos\omega_2 t,\ x_2(t) = -\frac{1}{4}\cos\omega_1 t + \frac{5}{4}\cos\omega_2 t$.
***10.16** Principal axes along: $2x + 2y + z = 0,\ 2x - y - 2z = 0,\ x - 2y + 2z = 0$, shortest distance is 2.
***10.17** (a) oblate spheroid; (b) one-sheet hyperboloid.
***10.19** $\theta = \tan^{-1}(-1/2) = 0.42$ rad $\approx 24°$ to the x-axis. The two branches are closest at $(x,\ y) = (2/\sqrt{5}, -1/\sqrt{5}),\ (-2/\sqrt{5},\ 1/\sqrt{5})$.

Problems 11

11.1 (a) 11; (b) 47/3.
11.2 (a) 7/12; (b) 7/12.
11.3 2π.
11.4 (a) -4π; (b) -16.
11.5 (a) 7/3; (b) 3.
11.6 $1 - \ln 2$.
11.7 11.
11.9 (a) $2\ln 2 - 1$; (b) 1/3.
11.10 27.
11.11 0.
11.12 1/2.
11.13 $-3/2$.
11.14 (a) 11; (b) $-3/2$.
11.15 32/3.
11.16 $2a^4b^2/15$.
11.17 $\sqrt{\pi}[1 - \exp(-4a^4)]/8$.
11.20 $4\,\mathbf{i} + 3\,\mathbf{j} - 24\sqrt{2}\,\mathbf{k}$.
11.21 $\pi a^6(e-2)/32$.
11.22 $\pi(R^2h - \frac{1}{3}h^3)$.
11.23 $-11/36$.

Problems 12

12.1 $|\mathbf{E}| = 2\sqrt{5}$ in direction $-2\mathbf{i} + \mathbf{j}$; direction of most rapid decrease at $(-3, 2)$ is $3\mathbf{i} + 2\mathbf{j}$; $\mathrm{d}\phi/\mathrm{d}s = \sqrt{10}$.
12.2 (a) $2\mathbf{i} - 2\mathbf{j} - \mathbf{k}$; (b) $\mathrm{d}\psi/\mathrm{d}s = 5/\sqrt{6}$; (c) $(1,1,1) + (2,-2,1)t$.
12.3 (a) $y\,\mathbf{i} - (3z^3 - 4xz)\mathbf{j}$; (b) $x^2z(3xz^3 + 2y^2)\mathbf{i} + 3x^2yz^2(2x - 3z^2)\mathbf{j} - 3xz^2(y^2 + x^2z)\mathbf{k}$;
(c) $(-4x + 9z^2)\mathbf{i} + (4z - 1)\mathbf{k}$; (d) $-2yz^2\mathbf{i} + z^2(3z^2 - 2x)\mathbf{j} + 4yz(3z^2 - x)\mathbf{k}$; (e) 0.
12.6 (a) $r^2 \sin 2\theta \sin\phi$;
(b) $r\sin\theta\sin\phi(\sin\theta\cos\phi - \cos\theta)\,\mathbf{e}_r + r\sin\theta\sin\phi\,(\cos\theta\cos\phi + \sin\theta)\,\mathbf{e}_\theta + r\sin\theta\cos^2\phi\,\mathbf{e}_\phi$;
(c) $4r\cos\theta\sin\theta\sin\phi\,\mathbf{e}_r + 2r\sin\phi(\cos^2\theta - \sin^2\theta)\,\mathbf{e}_\theta + 2r\cos\theta\cos\phi\,\mathbf{e}_\phi$;
(d) $(\cos\theta - \sin\theta\cos\phi)\mathbf{e}_r - (\sin\theta + \cos\theta\cos\phi)\,\mathbf{e}_\theta + \sin\phi\,\mathbf{e}_\phi$.
12.7 $-2\pi a^4$.
12.8 2π.
12.9 $-(9\pi + 16)/6$.
12.10 (a) $a^3/3$; (b) a^3.
12.11 -1.
12.12 $-y^2\cosh^2(xz) + c$, c constant
12.13 $a = 4$; $b = 2$; $c = -1$, $-\phi = \frac{1}{2}x^2 + 2yx + 4zx - \frac{3}{2}y^2 - zy + z^2 + d$, d constant.
12.14 114.
12.15 2.
12.16 $\frac{4}{3}\pi\sigma a^4$.
12.17 (a) $\frac{2}{5}Ma^2$; (b) $\frac{7}{5}Ma^2$.
12.18 $Ma^2/4 + Md^2/3$.
12.19 1.
***12.23** (b) $\mathbf{E} = \dfrac{Q}{4\pi\varepsilon_0 R^2}\hat{\mathbf{r}}$, $\phi = \dfrac{Q}{4\pi\varepsilon_0 R}$; (c) $c = Q^2/(4\pi\varepsilon_0)$.

*12.24 (b) $\rho_0(\mathbf{r})e^{-\sigma t/\varepsilon}$.
12.27 (a) 0; (b) 0; (c) no.

Problems 13

13.1 $\dfrac{4\pi}{9} - \dfrac{4}{\pi}\sum_{n=1}^{\infty}\dfrac{\sin^2(n\pi/3)}{n^2}\cos(nx)$.

13.2 $\dfrac{1}{\pi}\sum_{n\ \mathrm{odd}}^{\infty}\dfrac{1}{n}\left[\sin nx - \sin 2nx\right]$, sum $= \dfrac{\pi}{4}$.

13.3 $\dfrac{7}{15}\pi^4 + 48\sum_{n=1}^{\infty}\dfrac{(-1)^n}{n^4}\cos(nx)$.

13.5 (a); (b); (d) do not satisfy Dirichlet conditions; (c) does satisfy Dirichlet conditions. (c) $\sinh x = \dfrac{2\sinh\pi}{\pi}\left[\dfrac{\sin x}{2} - \dfrac{2\sin 2x}{5} + \dfrac{3\sin 3x}{10} - \cdots\right]$.

13.6 $\dfrac{2}{3}\pi^2 + 2\sum_{n=1}^{\infty}\dfrac{(-1)^n\cos nx}{n^2} + \sum_{n=1}^{\infty}\left\{\dfrac{2}{\pi n^3}\left[(-1)^n - 1\right] - \dfrac{\pi}{n}\right\}\sin nx$; $\sum_{n=1}^{\infty}\dfrac{1}{n^2} = \dfrac{\pi^2}{6}$.

13.7 $L - \dfrac{2L}{\pi}\sum_{n=1}^{\infty}\dfrac{1}{n}\sin\left(\dfrac{n\pi x}{L}\right)$.

13.8 $\dfrac{1}{2} + \dfrac{2}{\pi}\sum_{n=1}^{\infty}\dfrac{(-1)^{n+1}}{(2n-1)}\cos[(2n-1)\pi x]$.

13.9 (a) $\dfrac{\pi^4}{5} + 48\sum_{n=1}^{\infty}(-1)^n\left(\dfrac{\pi^2}{6n^2} - \dfrac{1}{n^4}\right)\cos nx$, (b) $x^3 = 12\sum_{n=1}^{\infty}(-1)^{n+1}\left(\dfrac{\pi^2}{6n} - \dfrac{1}{n^3}\right)\sin nx$ is a valid series, but $x^2 = 4\sum_{n=1}^{\infty}(-1)^{n+1}\left(\dfrac{\pi^2}{6} - \dfrac{1}{n^2}\right)\cos nx$ is an invalid series.

13.11 $\dfrac{1}{3} + \dfrac{4}{\pi^2}\sum_{n=1}^{\infty}\dfrac{\cos(n\pi x)}{n^2}$, $\sum_n\dfrac{1}{n^4} = \dfrac{\pi^4}{90}$.

13.13 $\dfrac{\pi^2}{3} + \sum_{\substack{n=-\infty\\ n\neq 0}}^{\infty}(-1)^n\left(\dfrac{2}{n^2} + \dfrac{i}{n}\right)e^{inx}$.

13.14 $\dfrac{2}{i\pi}\sum_{\substack{n=-\infty\\ n\ \mathrm{odd}}}^{\infty}\dfrac{1}{n}e^{inx}$.

13.15 $g(k) = \dfrac{2[ka - \sin(ka)]}{i\pi k^2}$.

13.17 $\dfrac{1}{\pi}\dfrac{k}{k^2+1}$.

13.18 $-\dfrac{2}{\pi k^3}\left[3k\cos(3k) + (4k^2-1)\sin(3k)\right]$, integral $= 0$.

13.19 (a) $\dfrac{e^{-ika}e^{-k^2/4\alpha}}{2\sqrt{\pi\alpha}}$; (b) $\dfrac{-ik}{4(\pi\alpha^3)^{1/2}}e^{-k^2/4\alpha}$; (c) $\dfrac{1}{8\sqrt{\pi}}\left(2\alpha^{-3/2} - k^2\alpha^{-5/2}\right)e^{-k^2/4\alpha}$.

13.20 (a) $\dfrac{1}{4i}\left[\delta(k-5) + \delta(k-1) - \delta(k+1) - \delta(k+5)\right]$; (b) $-\dfrac{i}{2\pi a}\sin ka$.

*13.22 $\sigma_f = \sqrt{2}(\sigma_e^2 - \sigma_r^2)^{1/2}$.
*13.23 (a) $I_1 = I_2 = \pi$; (b) $f(k)/2\pi$.

Problems 14

14.1 (a) $\frac{1}{12}(3x^4 + 4x^3 + 24x - 7)$; (b) $(x-2)/(x-1)$.

14.2 (a) $(x - y - 2) - \ln(x + y + 1) = 0$; (b) $\ln(x - \tan^{-1} x + \pi/4)$.

14.4 $\left(\dfrac{y}{x}\right) + \ln\left[\dfrac{c(y-2x)^2(y-3x)}{x^4}\right] = 0.$

14.5 $\tan^{-1}\left(\dfrac{5y-1}{5x+3}\right) - \ln[(5x+3)^2 + (5y-1)^2] + c = 0.$

14.6 (a) not exact; (b) exact, $y^2 \ln x = c$.

14.7 (a) $2x^3 - 6x^2y + 3y^2 - 6y + c = 0$; (b) $\frac{1}{2}(x^2 - x^2y^2) + 4y^2 + c$.

14.8 (a) $x^2(\sin x - x\cos x + c)$; (b) $2y\sin x - \cos^2 x + c = 0$.

14.9 (a) $-\frac{1}{5}\left[2(\sin x + 2\cos x) + e^{x/2}\right]$, (b) $2(x+1)^{5/2} + (x+1)^2$.

14.10 $\dfrac{\sqrt{2}}{[3(2x^2 - 2x + 1) + c\, e^{-2x}]^{1/2}}.$

14.11 (a) $\left[1 + \left(\dfrac{2-e}{e}\right)x\right]e^x$; (b) $e^{-3x}(3\cos 2x + 8\sin 2x)$.

14.14 (a) $a_1e^x + a_2e^{-2x} - \frac{1}{3}xe^{-2x}$; (b) $a_1e^x + a_2e^{2x} + \frac{7}{4} + \frac{3}{2}x + \frac{1}{2}x^2$.

14.15 (a) $(Ax + B)e^{-2x} + 2x^2e^{-2x}$; (b) $Ax + B + Ce^{3x} - \frac{4}{3}x^2 - \frac{4}{3}x^3 - x^4$.

14.16 (a) $-\dfrac{1}{130}(7\cos 3x + 9\sin 3x)$; (b) $y = \dfrac{1}{16}e^{3x}$.

14.17 $-e^{-2x}\left(x^2/2 + x - 1\right)$

14.18 $(A_1 + A_2x + A_3x^2)e^{-x} + \frac{1}{27}(14\cosh 2x - 13\sinh 2x)$.

14.19 $Ae^{-Cx} + Be^{Cx} - \int\limits_{-\infty}^{\infty} \dfrac{e^{ikx}h(k)}{k^2 + C^2}\mathrm{d}k.$

***14.20** $-5e^x + 3e^{2x} + 3x\,e^{2x}$.

***14.21** $\cos(\omega_0 t) + \dfrac{A}{\omega_0(\omega^2 - \omega_0^2)}\left[\omega\sin(\omega_0 t) - \omega_0\sin(\omega t)\right]$

***14.23** $y(x) = 2e^x - e^{2x} + \int\limits_0^x \left[e^{2u} - e^u\right] h(x-u)\,\mathrm{d}u$

***14.24** $Ax + Bx^{-3} + x\ln x + 1$.

***14.25** $\dfrac{(Ax+B)}{x^2} - \dfrac{1}{130}\left[7\sin(3\ln x) + 9\cos(3\ln x)\right]$

Problems 15

15.2 $a_0\left(1 - \dfrac{x^2}{2} + \dfrac{5}{24}x^4 - \cdots\right) + a_1\left(1 - \dfrac{x^3}{3} + \dfrac{x^5}{6} - \cdots\right).$

15.3 $y(x) = a_0\left[1 + \dfrac{(x-1)^2}{2} + \dfrac{3(x-1)^4}{8} + \cdots\right] + a_1\left[(x-1) + \dfrac{2(x-1)^3}{3} + \dfrac{8(x-1)^5}{15} + \cdots\right].$

15.4 $Ax^{3/2}\left(1 - \dfrac{2}{5}x + \dfrac{2}{35}x^2 - \cdots\right).$

15.5 $(A + B\ln x)/x$.

15.6 $A\left[\dfrac{x}{1-x}\right] + B\left[\dfrac{x\ln x}{1-x} + \dfrac{1}{2}(1+x)\right].$

15.8 $a_0 \sum\limits_{r=0}^{m} \dfrac{(-2)^r m!}{(2r+1)!(m-r)!}x^r.$

15.9 $\lambda_n = 1 + n^2\pi^2$, $y_n(x) = A_n\sin(n\pi x)$.

15.10 $\lambda_n^2 = n^2\pi^2$, $y_n(x) = \sqrt{2}x^{-1/2}\sin(n\pi\ln x)$.

***15.11** $q_2(x) = \frac{3}{2}x,\ q_3(x) = \frac{5}{2}x^2 - \frac{2}{3},\ q_4 = \frac{35}{8}x^3 - \frac{55}{24}x$

15.12 $\sum\limits_{k=0}^{\infty} \dfrac{2c_k^2}{2k+1}$.

***15.14** $\dfrac{2e}{4\pi\varepsilon_0 r}\sum\limits_{n=1}^{\infty}\left(\dfrac{p}{r}\right)^{2n} P_{2n}(\cos\theta) = \dfrac{ep^2}{4\pi\varepsilon_0 r^3}(3\cos^2\theta - 1),\ r \gg p.$

***15.17** $c_n = \dfrac{2^n n! n!}{(2n)!}$.

***15.18** $k = 3.8317,\ 7.0156,\ 10.1735,\ 13.3237.$

15.20 $J_2(2) = 0.35283$ to 5 decimal places, 1 extra term for 7 decimal places.

***15.22** $J_{-1/2}(x) = \left(\dfrac{2}{\pi x}\right)^{1/2}\cos x,\ J_{3/2}(x) = \left(\dfrac{2}{\pi x}\right)^{1/2}\left[\dfrac{\sin x}{x} - \cos x\right].$

Problems 16

16.1 $k^2 = k_x^2 + k_y^2 + k_z^2 = (n_x^2 + n_y^2 + n_z^2)(\pi/L)^2,\ n_x,\ n_y,\ n_z$ positive integers,
$u(x,y,z) = A_{n_x n_y n_z}\sin\left(\dfrac{n_x\pi x}{L}\right)\sin\left(\dfrac{n_y\pi y}{L}\right)\sin\left(\dfrac{n_z\pi z}{L}\right).$

16.2 $E_n = \left(\dfrac{2\pi n}{L}\right)^2\dfrac{\hbar^2}{2m} = \dfrac{n^2 h^2}{2mL^2},\ n > 0.$

16.3 $-\dfrac{4u_0 a^2}{\pi^3}\sum\limits_{n=1}^{\infty}\dfrac{[1+2(-1)^n]}{n^3}\dfrac{\sin(n\pi x/a)\sinh(n\pi y/a)}{\sinh(n\pi b/a)}.$

16.5 $\dfrac{8u_0L^2}{\pi^3}\sum\limits_{n=1}^{\infty}\dfrac{\sin[(2n-1)\pi x/L]\exp[-\kappa(2n-1)^2\pi^2 t/L^2]}{(2n-1)^3}.$

16.6 $u(x,y) = \sum\limits_{n=1}^{\infty} E_n e^{-n\pi y/d}\sin\left(\dfrac{n\pi x}{d}\right)$, where $E_n = \dfrac{2}{d}\displaystyle\int_0^d f(x)\sin\left(\dfrac{n\pi x}{d}\right)\mathrm{d}x.$

16.7 $\dfrac{a_0}{2} + \sum\limits_{n=1}^{\infty}\left(\dfrac{r}{R}\right)^n(a_n\cos n\theta + b_n\sin n\theta),\ a_n,\ b_n$ arbitrary constants.

16.9 $-E_0\left(1 - \dfrac{a^3}{r^3}\right)r\cos\theta.$

16.11 $-\dfrac{\hbar^2}{2m}\dfrac{\mathrm{d}^2R}{\mathrm{d}r^2} + \left[V(r) + \dfrac{l(l+1)\hbar^2}{2mr^2}\right]R = ER.$

16.13 $\dfrac{1}{2}(T_U + T_L) + \dfrac{3}{4}\left(\dfrac{r}{R}\right)(T_U - T_L)P_1(\cos\theta) - \dfrac{7}{16}\left(\dfrac{r}{R}\right)^3(T_U - T_L)P_3(\cos) + \cdots$

16.14 $J_m(p\rho)e^{-az}(A\cos m\phi + B\sin m\phi),\ a > 0$ and $p = \sqrt{k^2 + a^2}$.

16.16 4.878.

***16.17** (a) $u(x,t) = \frac{1}{2}\exp[-(x-vt)^2] + \frac{1}{2}\exp[-(x+vt)^2]$; (b) $-\dfrac{1}{4v}\left[e^{-(x+vt)^2} - e^{-(x-vt)^2}\right].$

***16.18** (a) $f(x - \frac{1}{2}y) + g(x + 2y)$; (b) $f(x+3y) + xg(x+3y)$;
(c) $f[x + (4+3i)y/5] + g[x + (4-3i)y/5]$; (d) $f(x - y/2) + g(y)$.

***16.19** $y(y-x) + \dfrac{x\sin(x-y)}{x-y}.$

Index

Mathematics for Physicists, First Edition. B.R. Martin and G. Shaw.

Companion website: www.wiley.com/go/martin/mathsforphysicists

Printed and bound by CPI Group (UK) Ltd, Croydon, CR0 4YY
27/10/2024
14580143-0003